Springer Handbook of Auditory Research

Volume 59

More information about this series at http://www.springer.com/series/2506

The ASA Press

The ASA Press imprint represents a collaboration between the Acoustical Society of America and Springer dedicated to encouraging the publication of important new books in acoustics. Published titles are intended to reflect the full range of research in acoustics. ASA Press books can include all types of books published by Springer and may appear in any appropriate Springer book series.

ASA Press

Jennifer A. Clack · Richard R. Fay
Arthur N. Popper
Editors

Evolution of the Vertebrate Ear

Evidence from the Fossil Record

With 116 Illustrations

ASA Press

Springer

Editors
Jennifer A. Clack
University Museum of Zoology
University of Cambridge
Cambridge
UK

Arthur N. Popper
Department of Biology
University of Maryland
College Park, MD
USA

Richard R. Fay
Marine Biological Laboratory
Woods Hole, MA
USA

ISSN 0947-2657 ISSN 2197-1897 (electronic)
Springer Handbook of Auditory Research
ISBN 978-3-319-46659-0 ISBN 978-3-319-46661-3 (eBook)
DOI 10.1007/978-3-319-46661-3

Library of Congress Control Number: 2016953325

Printed on acid-free paper

This Springer imprint is published by Springer Nature
The registered company is Springer International Publishing AG
The registered company address is: Gewerbestrasse 11, 6330 Cham, Switzerland

Acoustical Society of America

The purpose of the **Acoustical Society of America** (www.acousticalsociety.org) is to generate, disseminate, and promote the knowledge of acoustics. The Acoustical Society of America (ASA) is recognized as the world's premier international scientific society in acoustics, and counts among its more than 7,000 members, professionals in the fields of bioacoustics, engineering, architecture, speech, music, oceanography, signal processing, sound and vibration, and noise control.

Since its first meeting in 1929, the ASA has enjoyed a healthy growth in membership and in stature. The present membership includes leaders in acoustics in the United States of America and around the world. The ASA has attracted members from various fields related to sound including: engineering; physics, oceanography' life sciences' noise and noise control' architectural acoustics; psychological and physiological acoustics; applied acoustics; music and musical instruments; speech communication; ultrasonics, radiation, and scattering; mechanical vibrations and shock; underwater sound; aeroacoustics; macrosonics; acoustical signal processing; bioacoustics; and many more topics.

To assure adequate attention to these separate fields and to new ones that may develop, the Society establishes technical committees and technical groups charged with keeping abreast of developments and needs of the membership in their specialized fields. This diversity and the opportunity it provides for interchange of knowledge and points of view has become one of the strengths of the Society.

The ASA's publishing program has historically included the *The Journal of the Acoustical Society of America, JASA-Express Letters, Proceedings of Meetings on Acoustics*, the magazine *Acoustics Today*, and various books authored by its members across the many topical areas of acoustics. In addition, ASA members are involved in the development of acoustical standards concerned with terminology, measurement procedures, and criteria for determining the effects of noise and vibration.

Series Preface

The following preface is the one that we published in volume 1 of the Springer Handbook of Auditory Research back in 1992. As anyone reading the original preface, or the many users of the series, will note, we have far exceeded our original expectation of eight volumes. Indeed, with books published to date and those in the pipeline, we are now set for over 60 volumes in SHAR, and we are still open to new and exciting ideas for additional books.

We are very proud that there seems to be consensus, at least among our friends and colleagues, that SHAR has become an important and influential part of the auditory literature. While we have worked hard to develop and maintain the quality and value of SHAR, the real value of the books is very much because of the numerous authors who have given their time to write outstanding chapters and to our many co-editors who have provided the intellectual leadership to the individual volumes. We have worked with a remarkable and wonderful group of people, many of whom have become great personal friends of both of us. We also continue to work with a spectacular group of editors at Springer. Indeed, several of our past editors have moved on in the publishing world to become senior executives. To our delight, this includes the current president of Springer US, Dr. William Curtis.

But the truth is that the series would and could not be possible without the support of our families, and we want to take this opportunity to dedicate all of the SHAR books, past and future, to them. Our wives, Catherine Fay and Helen Popper, and our children, Michelle Popper Levit, Melissa Popper Levinsohn, Christian Fay, and Amanda Fay Seirra, have been immensely patient as we developed and worked on this series. We thank them and state, without doubt, that this series could not have happened without them. We also dedicate the future of SHAR to our next generation of (potential) auditory researchers—our grandchildren—Ethan and Sophie Levinsohn, Emma Levit, and Nathaniel, Evan, and Stella Fay.

Preface 1992

The Springer Handbook of Auditory Research presents a series of comprehensive and synthetic reviews of the fundamental topics in modern auditory research. The volumes are aimed at all individuals with interests in hearing research including advanced graduate students, post-doctoral researchers, and clinical investigators. The volumes are intended to introduce new investigators to important aspects of hearing science and to help established investigators to better understand the fundamental theories and data in fields of hearing that they may not normally follow closely.

Each volume presents a particular topic comprehensively, and each serves as a synthetic overview and guide to the literature. As such, the chapters present neither exhaustive data reviews nor original research that has not yet appeared in peer-reviewed journals. The volumes focus on topics that have developed a solid data and conceptual foundation rather than on those for which a literature is only beginning to develop. New research areas will be covered on a timely basis in the series as they begin to mature.

Each volume in the series consists of a few substantial chapters on a particular topic. In some cases, the topics will be ones of traditional interest for which there is a substantial body of data and theory, such as auditory neuroanatomy (Vol. 1) and neurophysiology (Vol. 2). Other volumes in the series deal with topics that have begun to mature more recently, such as development, plasticity, and computational models of neural processing. In many cases, the series editors are joined by a co-editor having special expertise in the topic of the volume.

Arthur N. Popper, College Park, MD, USA
Richard R. Fay, Woods Hole, MA, USA

SHAR logo by Mark B. Weinberg, Bethesda, Maryland, used with permission.

Volume Preface

The evolution of the vertebrate inner ear is a topic that holds great interest for all students of auditory neuroscience. When and in which vertebrate groups did the ear arise and how has it changed over the tens of millions of years since the first vertebrate? While some insight into this history comes from comparing the ears of extant vertebrates, the best view of the origin and evolution of the vertebrate middle and inner ears comes from the growing fossil record. This volume focuses on that record to provide unique insights into this evolution.

This kind of overview is possible because the middle and inner ears of vertebrates are represented by the best and most readily traceable record for a sensory system among fossils. Each major group of bony vertebrates tells its own story of the evolution of its auditory system and, in recent years, more and more information has become available. A combination of traditional paleontological methods of fossil preparation has been employed to great effect, including serial grinding, mechanical and chemical preparation, and advanced methods of virtual reconstruction. Detailed phylogenetic analyses increasingly have refined knowledge of vertebrate phylogeny from both extinct and extant groups, allowing the evolution of the ear region to be traced. This volume presents the most recent findings about the ear region in each of the major bony vertebrate clades.

The clear patterns that emerge are those of convergent evolution and variations on a theme. Based on the same original building blocks, each vertebrate group has refined and utilized them for the differing needs and habits of the animals. Constrained by the physical facts of transmission of sound through air, water, or body tissues, it is no surprise that similar solutions to the problem have evolved several times. Ideas about these processes have been similarly refined as a result of advances in understanding as our knowledge has grown. The similarities observed can be and have been deceptive: a single evolutionary origin for hearing in air has been inferred in the past, whereas deeper knowledge has revealed subtle anatomical differences among the ears of vertebrate groups that are a testament to their divergent history.

In Chap. 1, Clack provides an exciting and important overview of evolution of the vertebrate ear: a guide to the vertebrates in general and to the ear in particular.

She provides an important discussion of the terminology used by anatomists and paleontologists with respect to the latest understanding of the vertebrate lineage. This chapter also describes some of the latest discoveries from fossils that often are preserved poorly and usually do not provide much data on the ear region. These include fossils of the jawless vertebrates, placoderms, and chondrichthyans. The rest of the chapters cover the bony vertebrates or osteichthyans.

In Chap. 2, Friedman and Giles explain how the fossil record of ray-finned fishes shows the main modifications to the otic region and surrounding parts of the braincase. Clack and Ahlberg (Chap. 3) then cover the sarcopterygians, apart from the limbed forms that are known as tetrapods. With the rise of limbed stem tetrapods from the Devonian, the earliest complement of a stapes, fenestra vestibuli, and its supporting structures developed, as described by Clack and Anderson in Chap. 4.

Three chapters consider the evolution of the unique construction of the mammalian ear, detailing the origins of its three-ossicle series of malleus, incus, and stapes, and the coiled cochlea. In the first of these, Kemp (Chap. 5) describes the most basal sector of the mammalian lineage, the synapsids, starting with the late Carboniferous and Early Permian forms sometimes informally called pelycosaurs. In Chap. 6, Luo, Schultz, and Ekdale follow on from the descriptions of basal synapsids in Chap. 5 and show the increasingly mammal-like conditions found in later members of the synapsid clade. The gradual acquisition of mammalian features and the convergent development of some of the mammalian ear's components are key to this chapter. The story of mammalian ears finishes with Chap. 7 in which Ekdale discusses the evidence from fossil representatives of mammals belonging to modern groups. Ekdale shows how the mammalian ear was refined and adapted for different environments.

The next three chapters deal with reptiles (in the phylogenetic sense, which includes birds). In contrast to mammals, reptiles have only the single stapes as a middle ear ossicle. Chapter 8 by Sobral, Reisz, Neenan, Müller, and Scheyer treats the early representatives of the clade. Some of these have no living representatives, such as pareiasaurs and procolophonids, as well as the extinct marine reptiles, such as ichthyosaurs and plesiosaurs. The chapter describes the earliest members of the diapsid clade, which provided the foundation for most living reptiles. These fall into two groups, the lepidosaurs and the archosaurs. Turtles may belong to one of these two groups, and current thinking suggests the archosaurs. Chapter 9 by Evans provides a review of the lepidosaur clade, which includes lizards and snakes, as well as the rhynchocephalian clade that culminates in its only living representative, the extant *Sphenodon*. The archosaurs are covered in Chap. 10 by Sobral and Müller. The group is a large one, containing modern representatives of crocodiles and birds and the extinct members of each. The extinct members of the avian lineage include all the dinosaurs. One other constituent member is entirely extinct: the flying pterosaurs.

The final chapter by Schoch and Anderson (Chap. 11) follows the evolution of the Amphibia. It is treated last to distinguish the group from those animals described in Chap. 4 (Clack and Anderson on early tetrapods). Fossil members of the Lissamphibia arrived quite late on the scene, around the time of the first dinosaurs, and they are similar in many respects to extant lissamphibians. Although the amphibian ear region is superficially similar to that of lizards, it arose quite independently, first among the Paleozoic tetrapods known as temnospondyls, which are considered to be the stem from which modern forms arose.

This volume of the Springer Handbook of Auditory Research (SHAR) is unique in that it is the only book ever to consider the evolution of the vertebrate ear from the perspective of the fossil record. At the same time, evolution of the ear and hearing has been a frequent topic in SHAR through individual chapters and through a series of books on comparative hearing for insects, fishes, amphibians, reptiles and birds, and mammals. Moreover, *Evolution of the Vertebrate Auditory System* (vol. 22, 2004, edited by Manley, Popper, and Fay) focused on principles of the evolution of hearing. This volume focuses on the associated bony anatomy.

Jennifer A. Clack, Cambridge, UK
Richard R. Fay, Woods Hole, MA, USA
Arthur N. Popper, College Park, MD, USA

Contents

Contributors

Per Erik Ahlberg Department of Organismal Biology, Uppsala University, Uppsala, Sweden

Jason S. Anderson Department of Comparative Biology and Experimental Medicine, University of Calgary, Calgary, AB, Canada

Jennifer A. Clack University Museum of Zoology, Cambridge, Cambridge, UK

Eric G. Ekdale Department of Biology, San Diego State University, San Diego, CA, USA; Department of Paleontology, San Diego Natural History Museum, San Diego, CA, USA

Susan E. Evans Department of Cell and Developmental Biology (Anatomy Building), University College London, London, UK

Matt Friedman Department of Earth Sciences, University of Oxford, Oxford, UK; Museum of Paleontology and Department of Earth and Environmental Sciences, University of Michigan, Ann Arbor, MI, USA

Sam Giles Department of Earth Sciences, University of Oxford, Oxford, UK

Tom S. Kemp St John's College and Oxford University Museum of Natural History, Oxford, UK

Zhe-Xi Luo Department of Organismal Biology and Anatomy, The University of Chicago, Chicago, IL, USA

Johannes Müller Museum für Naturkunde Berlin, Berlin, Germany

James M. Neenan Oxford University Museum of Natural History, Oxford, UK

Robert Reisz University of Toronto at Mississauga, Mississauga, ON, Canada

Torsten M. Scheyer Paläontologisches Institut und Museum der Universität Zürich, Zurich, Switzerland

Rainer R. Schoch Staatliches Museum für Naturkunde, Stuttgart, Germany

Julia A. Schultz Department of Organismal Biology and Anatomy, The University of Chicago, Chicago, IL, USA

Gabriela Sobral Departamento de Ecologia e Zoologia, Universidade Federal de Santa Catarina, Florianópolis, Brazil; Museum für Naturkunde Berlin, Berlin, Germany

Chapter 1
Vertebrate Diversity in a Sensory System: The Fossil Record of Otic Evolution

Jennifer A. Clack

Abstract This chapter introduces some major concepts from the world of paleontology, starting with some of the techniques used to investigate fossils. A current consenus phylogeny of the main vertebrate groups and some of the key terms underlying current systematic practice are given. A current geological timescale to be used in conjunction with other chapters is presented. In order to understand the evolutionary history of the otic region, its component parts, and the anatomical context in which those are placed, a basic description of the embryonic origin of skull and braincase anatomy is outlined. To an extent, the fossil record complements and underscores the relationship between ontogeny and phylogeny. Some key examples of early and basal vertebrates are described in brief because, although their otic regions are not generally well preserved, some recently described specimens and taxa are providing unprecedented information. They demonstrate some of the starting conditions for the vertebrate ear region, although there is still insufficient information for them to warrant their own chapter. The last section summarizes the main foci of the remaining 10 chapters. Together they cover the majority of fossil vertebrates—those with an internal skeleton of bone—the bony vertebrates.

Keywords Bony vertebrate groups · Chondrichthyans · Embryonic components · Geological timescale · Hyomandibula · Inner ear structure · Middle ear · Non-osteichthyans · Otic capsule · Stapes · Stem gnathostomes

1.1 General Introduction

The ear of vertebrates —the middle and inner ears—is the sensory system that is represented by the best and most readily traceable record among fossils. Each major group of bony vertebrates has its own story of the evolution of its auditory system, and in recent years, more and more information has become available.

J.A. Clack (✉)
University Museum of Zoology, Cambridge, Downing Street, Cambridge CB2 3EJ, UK
e-mail: j.a.clack@zoo.cam.ac.uk

J.A. Clack et al. (eds.), *Evolution of the Vertebrate Ear—Evidence from the Fossil Record*, Springer Handbook of Auditory Research 59,
DOI 10.1007/978-3-319-46661-3_1

Combinations of traditional paleontological methods of fossil preparation (serial grinding, mechanical and chemical preparation) and advanced methods of virtual reconstruction using high resolution micro-computed tomography (μCT or HRXCT, synchrotron-CT [SRμCT], and neutron-CT scanning) followed by three-dimensional modeling have been employed to great effect. Examples of the use of these methods are seen in Clack et al. (2003), examining the ear region of the early tetrapod *Ichthyostega*; Porro et al. (2015) for the skull of the early tetrapod *Acanthostega* (see Clack and Anderson, Chap. 4); and Laaß (2014) for the ear region of dicynodonts (see Kemp, Chap. 5). Detailed phylogenetic analyses have increasingly refined knowledge of vertebrate phylogeny from both extinct and extant groups, allowing the evolution of the ear region to be traced. This volume presents the most recent findings about the ear region in each of the major bony vertebrate clades.

The clear patterns that emerge are those of convergent evolution and variations on a theme. Each vertebrate group refines and utilizes the same original building blocks in its own way for the differing needs and habits of the animals. Constrained by the physical facts of transmission of sound through air, water, or body tissues, it is no surprise that similar solutions to the problems of sound reception have evolved several times. Ideas about these processes have been similarly refined as a result of advances in understanding as our knowledge has grown. The similarities observed can be, and have been, deceptive. A single answer for facilitating sound reception has been inferred in the past; however, deeper knowledge has revealed subtle differences that are a testament to a divergent history.

There is now a reasonable consensus regarding the phylogenetic relationships of the living groups of vertebrates to one another (Fig. 1.1), although controversies remain over the positions of some of the living groups with respect to the fossil groups. These controversies are explored in the following chapters, and the evidence from the fossil forms is considered with respect to how their auditory and vestibular regions contribute to the debate. Without a sound phylogenetic underpinning, hypotheses of convergent evolution versus close relationship cannot be tested.

Several important terms require explanation: *crown group*, *stem group*, and *total group*. The crown group of a lineage includes all the extant members of that lineage and all fossil relatives that lie within the node encompassing that group. For example, the crown group of tetrapods consists of the two major groups: Amphibia (anurans, urodeles, and caecilians) and Amniota (reptiles including birds, and mammals). The stem group consists of all the fossil forms that are more closely related to that crown group than to any other crown group (Fig. 1.1). For example, *Acanthostega* and *Ichthyostega* (see Clack and Anderson, Chap. 3) cannot be placed among any of the crown group members. Instead, they fall below the split between amphibians and amniotes. Thus, they are stem members of the Tetrapodomorpha, the total group that includes all stem members (in this particular case including some animals with fins and scales that we would classify as "fish") as well as all crown members.

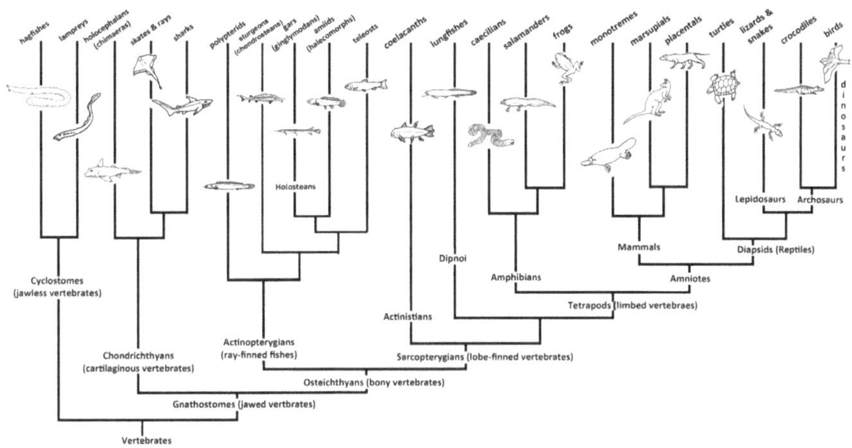

Fig. 1.1 A basic consensus phylogeny of vertebrates. The only node of real controversy is the position of turtles. Alternative hypotheses have placed turtles as a separate lineage of amniotes known as parareptiles, as basal diapsids, as lepidosaurs, or as archosaurs

A geological timescale is also necessary for understanding the relative geological ages of taxa described in this book (Fig. 1.2).

1.2 General Structure of Inner Ear, Middle Ear, Otic Capsule, Hyomandibula, and Stapes: Terminology

There are basic features of the otic region that are common to all vertebrates and a summary of them is given here. A brief introduction to the embryonic origins and development of some skeletal features, such as the otic capsule and the stapes, also are provided. This is necessary for understanding their evolutionary history and is key to understanding the modifications that each feature has undergone among disparate vertebrate groups.

A vestibular region consisting of semicircular canals and their associated ampullae is found on each side of the rear of the braincase and in jawed vertebrates is housed within an otic capsule. During the embryogenesis of jawed vertebrates, the capsules become linked or fused to the braincase to form the otoccipital region. Initially the otoccipital region is separated from the base of the braincase by the lateral otic fissure and separated by the ventral cranial fissure from the anterior component that houses the eyes, olfactory nerves, and the anterior parts of the brain (Goodrich 1930). The anterior component is called the ethmosphenoid (primarily in fishes) or sphenethmoid (primarily in tetrapods). In most adult bony vertebrates, these components further fuse to give a unitary braincase (Fig. 1.3).

The vestibular apparatus of the inner ear of jawed vertebrates consists of three semicircular canals set approximately at right angles to one another: anterior

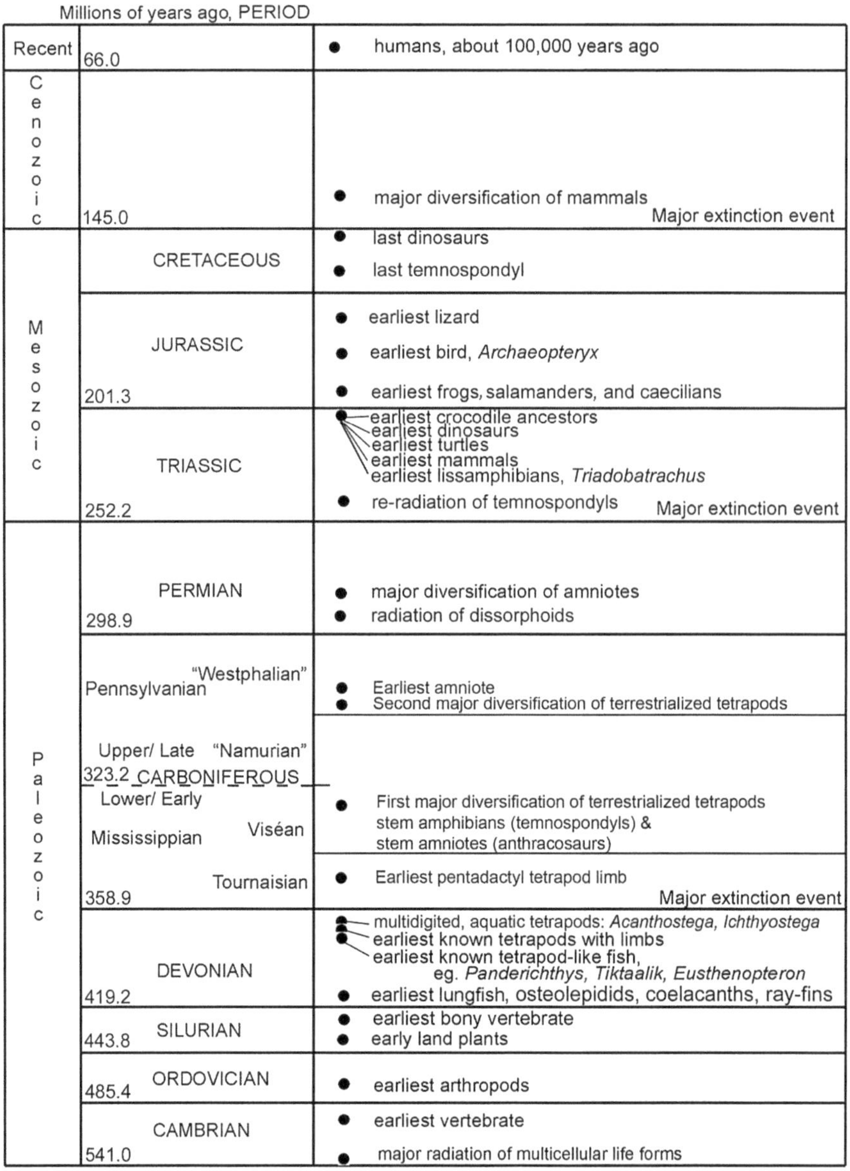

Fig. 1.2 A geological timescale showing major events and origins or extinctions of major groups. (Dating from Gradstein et al. 2012)

vertical, lateral horizontal, and posterior vertical. Each of these contains fluid, and each has at its base an ampulla or region containing sensory hair cells that detect the movement of a gelatinous cupula within the ampulla. The vestibular region also includes the utricle, which is an epithelium (or macula) with sensory hair cells

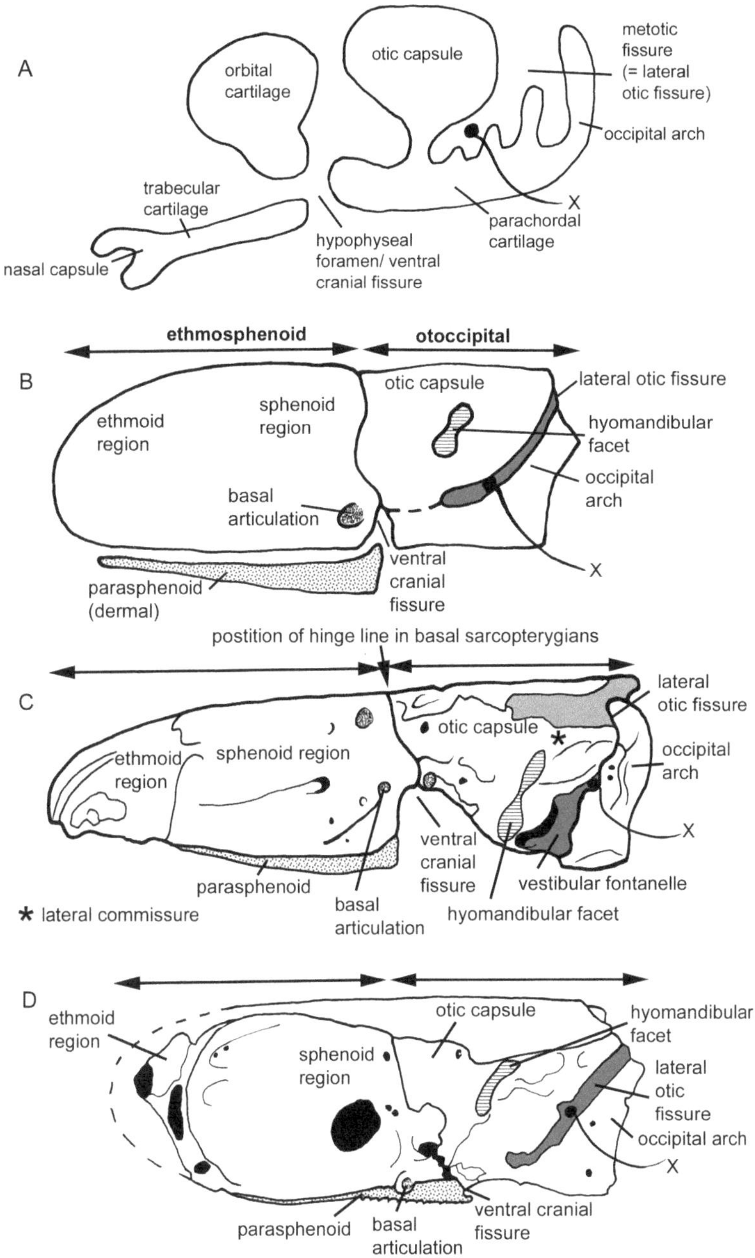

Fig. 1.3 Basic construction of the braincase (neurocranium). **(A)** Embryonic elements. **(B)** Elements as they begin to co-ossify. The *anterior ethmosphenoid* and *posterior otoccipital* are indicated by *double-ended arrows* also in **C** and **D**. The *parasphenoid* is a dermal element underplating the *ethmosphenoid* (although it also underplates the otoccipital in most tetrapods, not shown). **(C)** The braincase of the Devonian sarcopterygian *Eusthenopteron* shows essentially the primitive condition for sarcopterygians. **(D)** The braincase of the Devonian actinoptergyian *Mimipiscus* shows essentially the primitive condition for actinopterygians. (X shows the opening for the vagus, cranial nerve X)

overlain by calcium carbonate crystals (called otoconia) or an otolith. Below or posterior to the vestibular region are the maculae of the saccule and lagena, which are similar structurally to the utricle but may be enclosed in separate pouches. The inner ear components may or may not be reflected in the bone structure of the otic capsule. In origin, the otic region is sensitive to orientation, posture, and movement, and probably low-frequency sound. Only later, during the evolution of different animal groups, does it also become used to detect higher frequencies (Fig. 1.4). Further details of the anatomy of the inner ear apparatus can be found in a vertebrate biology text such as Goodrich (1930), Hildebrand and Goslow (2001), and Kardong (2014).

In fishes, the lateral exterior surface of the otic capsule provides the articulation point for the hyomandibula (Fig. 1.3), a branchial element that controls the movements of the gill apparatus and the opercular flap. It also links to the lower jaw, either directly or via a series of small intermediary bones, and thus the hyomandibula is associated with the jaw joint (see Clack and Ahlberg, Chap. 3). Because of its articulation with the otic capsule, the head of the hyomandibula is inevitably situated close to the enclosed sensory tissue in the capsule, and this proximity was a factor that had profound consequences for the development of hearing in vertebrates.

Embryological and developmental studies have shown that the hyomandibula is homologous, at least in part, to the tetrapod stapes. Whereas the hyomandibula abuts against the wall of the otic capsule, the stapedial footplate penetrates the wall via a hole called the fenestra vestibuli (FV). The condition in tetrapodomorph fishes, such as *Eusthenopteron* (see Clack and Ahlberg, Chap. 3) and early tetrapods (see Clack and Anderson, Chap. 4), suggests that this hole is a remnant of the lateral otic fissure that remained unossified during embryogenesis.

The otic capsule itself is formed from two or more ossifications in the braincase of osteichthyans (Goodrich 1930; Clack 2001). Anteriorly, the prootic houses the

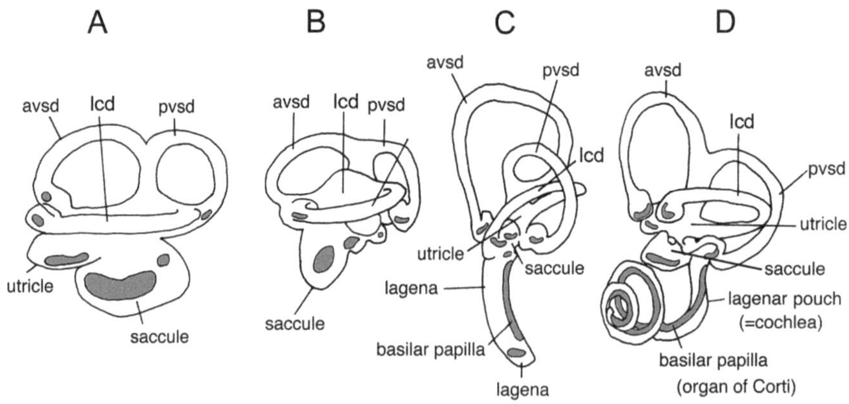

Fig. 1.4 The vestibular regions of representative vertebrates: (**A**) lungfish, (**B**) frog, (**C**) bird, and (**D**) mammal. *Abbreviations*: *avsd*, anterior vertical semicircular duct; *lcd*, lateral canal duct; *pvsd*, posterior semicircular canal duct (modified from Clack 2012)

anterior vertical canal and part of the lateral canal. Posteriorly, the opisthotic houses the posterior vertical canal and part of the lateral canal. In most adult osteichthyans, the suture between the prootic and opisthotic becomes fused. The opisthotic in many taxa extends laterally to form the paroccipital process, which may contact the skull roofing bones. Both the prootic and opisthotic bones contribute to housing the semicircular canals and the saccule and lagena in some forms. There may or may not be an internal wall from these bones separating the otic capsules from the brain space. In some taxa, the braincase is roofed by a supraoccipital, which may also contribute to the otic capsule roof. In amphibians, the exoccipitals (part of the embryonic occipital arch rather than the braincase proper) often take the place of the opisthotics to provide the posterior wall of the capsule and form the paroccipital process, as viewed externally. The basioccipital is the bone forming the base of the braincase and the ventral margin of the lateral otic fissure. That bone may be incorporated into the margin of the FV or it may be lined by the underlying dermal bone of the parasphenoid, such that the latter, rather than the basioccipital, forms the FV ventral margin (see early diapsids, Sobral, Reisz, Neenan, Müller, and Scheyer, Chap. 8).

Other skull bones are involved in the evolution of the ear region, and these include those forming the jaw joint—the quadrate and articular. These bones are derived from the primary jaw elements that are maintained in chondrichthyans in their original condition as the palatoquadrate and Meckel's cartilage, respectively. In osteichthyans they are ossified, but in modern tetrapods they are often the only remnants of those two embryonic structures that are evident. The evolution of the mammalian middle ear involves the incorporation of those bones into the hearing mechanism (see Kemp, Chap. 5; Luo, Schultz, and Ekdale, Chap. 6; Ekdale, Chap. 7). By contrast, in amphibians and in non-mammalian amniotes the quadrate and the more dorsally placed squamosal form parts of the supporting structure for the tympanic membrane, as well as forming the outer wall of the middle ear cavity. The palate itself, an ossified version of the palatoquadrate, contributes to the walls of the cavity in some cases.

Skull roofing bones may form attachment points for the otic capsule to a greater or lesser extent. In early tetrapods, the squamosal and tabular bones are often deeply notched posteriorly, forming what has been called an "otic notch", although in the earliest tetrapods it may have housed an opening of the spiracle (see Clack and Anderson, Chap. 4). The spiracle, a modified gill cleft, is open dorsally in some chondrichthyans and some fishes such as *Polypterus* (bichirs). The structure is considered to be homologous, in part, to the middle ear cavity in tetrapods. The ventral opening of the spiracular cleft into the palate is homologous to the eustachian tube in tetrapods.

Within the literature concerning the otic apparatus in different groups, alternative terminology is often used. For example, the FV is often referred to as the fenestra ovalis (FO) to distinguish it from the fenestra cochleae or fenestra rotunda in mammals and the fenestra pseudorotunda in reptiles. The latter two are pressure relief windows that have evolved convergently in many tetrapods. The FV (mistakenly) is sometimes referred to as the foramen ovalis or ovale, but the latter is a mammalian feature that allows the passage of cranial nerve V. The stapedial shaft is known as a columella or columella auris in a number of groups, including amphibians, lepidosaurs, and birds.

1.3 Non-Osteichthyans: Stem Gnathostomes and Chondrichthyans

The bony vertebrates, or Osteichthyes, comprise the largest group of vertebrates. This group includes not only most of the fishes in the waters of the modern world (the ray-finned fishes, or Actinopterygii) but also those that share their heritage with tetrapods, the lobe-finned vertebrates (coelacanths, lungfishes, and tetrapods, or Sarcopterygii).

Although the literal translation of "Osteichthyes" is "bony fishes," in fact, phylogenetic relationships are now used to define groups. All the osteichthyans are more closely related to each other than they are to chondrichthyans, and they form a natural "monophyletic" group. A monophyletic group includes all the descendants of a common ancestor. A more useful translation would therefore be "bony vertebrates." This form of "natural" classification has been in place since the late 1970s when cladistic analysis came into regular use. It came as a shock to some older members of the vertebrate biology community at that time to realize that a lungfish is more closely related to a cow than it is to a salmon but such is the case.

Non-osteichthyans include jawless vertebrates, of which the most basal are living hagfishes and lampreys. Crownward of them lie the stem gnathostomes and the chondrichthyans (sharks, rays, and holocephalans). Chondrichthyans belong to the jawed vertebrates or crown gnathostomes. Some of the fossil jawless vertebrates show a range of features that foreshadow gnathostome structure and they are termed stem gnathostomes. There are also a number of early bony-jawed vertebrates that have no living relatives. These are generally placed into either "placoderms" or "acanthodians," although recent studies have shown that neither of these groups may be monophyletic (e.g., Davis et al. 2012; Zhu et al. 2013). Non-osteichthyans are treated only briefly in this volume because the fossil record of their ear region is usually poor and data are sparse in general. However, there are some exceptions in which the type of preservation has allowed details of the anatomy to be seen, and references to recent work are provided below.

Some of the earliest fossil "fishes" from the Paleozoic era show some of the basic features of the otic region and its relation to the hyobranchial (or gill) skeleton, which are associated with the braincase and pharynx, respectively. Among these, the anatomy of the inner ear systems of fossil jawless fishes has recently been elucidated using powerful computed tomography techniques, as have those of some of the earliest known jawed vertebrates described in the next few paragraphs.

Some of the best-known fossils of jawless fishes, among them the stem gnathostomes known as the osteostracans and galeaspids, have been revealed by μCT scanning and modeling. From these methods it can be seen that osteostracans, considered in many analyses as the closest jawless vertebrate group to gnathostomes, have at least two semicircular canals per side (e.g., Stensiö 1932, 1963a), which has been considered to represent a condition similar to that of lampreys. However, Stensiö's work was done many decades ago and perhaps deserves to be reviewed in the light of recent work on lampreys. The use of detailed dissection and μCT scanning has shown that lampreys' semicircular canals are considerably more complex than previously appreciated (Maklad et al. 2014).

The work of Gai et al. (2011) on the Chinese galeaspid taxon *Shuyu zhejiangensis* also suggests a close relationship between galeaspids and gnathostomes based on μCT images of the braincase. Anterior and posterior semicircular canals are shown, each with an enlarged ampulla ventrally. However, the main focus of this group's work was on the anterior parts of the braincase; there may be more detail in the scan of the vestibular system than is apparent from the published reconstruction.

Some members of the groups formerly known as "placoderms" and "acanthodians" now are seen as stem members of the osteichthyans and chondrichthyans, but the situation is volatile as new discoveries and analyses are made. This is in contrast to former assessments that considered them each to be monophyletic with acanthodians closer to osteichthyans and placoderms closer to chondrichthyans (e.g., Davis et al. 2012; Zhu et al. 2013). Little is known about the placoderm inner ear region, although the external braincase morphologies of certain arthrodire taxa (a group of placoderms) are known in some detail and confirm the presence of three semicircular canals (e.g., Stensiö, 1963b; Goujet, 2001). In acanthodians, the three semicircular canals are occasionally preserved, as are the otoliths in some taxa. Three solid otoliths comparable to those of osteichthyans are present (Schultze 1990).

The chondrichthyans by their nature do not usually fossilize well since their skeletons are composed of calcified cartilage rather than bone. The skeletons generally fall apart on death or, if they are preserved, are frequently overlooked by fossil collectors who do not recognize them. Braincases are even less frequently preserved than jaws and palates. One exception is the Pennsylvanian-age iniopterygian stem holocephalan *Sibyrhynchus denisoni*, in which the three gnathostome semicircular canals, the bases of endolymphatic ducts, and even details of the brain and cranial nerves can be seen from the reconstructed SRμCT scan series (Pradel et al. 2009). The relationship between the hyomandibular, gill system, and otic capsule of elasmobranchs (as distinct from holocephalans) from the earliest times to the present is essentially similar to that of the osteichthyans, except that the latter have the otic capsule encased in bone.

1.4 The Bony Vertebrate Groups: Review of Chapters

The osteichthyans, or bony vertebrates, comprise the majority of vertebrates. Due to the nature of their skeletal composition, there is a better fossil record of the otic components of the skull for the osteichthyans, allowing some knowledge of the components in the actinopterygians and the sarcopterygians–tetrapodomorphs (the tetrapod lineage), dipnomorphs (lungfishes and their extinct relatives), and actinistians (coelacanths).

Actinopterygians constitute the vast majority of vertebrate diversity with more than 30,000 species. Their otic regions show considerable morphological variation, and some groups hear particularly well. Friedman and Giles in Chap. 2 explain how the fossil record of ray-finned fishes shows the main modifications to the otic region and surrounding parts of the braincase. Some of the earliest bony vertebrates are known from three-dimensional preserved material. Friedman and Giles describe the ear regions in some of the most primitive fishes, moving up the phylogenetic tree and the geological column to show the main changes that occurred among them. They also examine the fossil record of the ear region in extant non-teleost groups such as halecomorphs (bowfins) and ginglymodans (gars). They show how the components of the Weberian apparatus found in the ostariophysans are gradually developed through the phylogenetic changes that occur among precursor taxa.

Clack and Ahlberg (Chap. 3) cover the sarcopterygians apart from the limbed forms that are known as tetrapods. Among the most basal groups of sarcopterygians are the coelacanths or actinistians. The former have an extensive fossil record that goes back to the Early Devonian and are still represented today by the genus *Latimeria*. As a group, coelacanths show a basic sarcopterygian feature: a hinge between the anterior and posterior components of the braincase, which is reflected also in the overlying skull bones. This feature is present in several fossil sarcopterygian groups but is lost in lungfishes and tetrapods. Fossil coelacanths from the Devonian to the present are essentially conservative in their braincase and otic morphology, such that early forms are easily interpreted with the aide of the light provided by the braincase of *Latimeria*.

One major clade among sarcopterygians is the Dipnomorpha, a group that includes lungfishes and their relatives, the porolepiforms. The latter retain the intracranial hinge in a similar way to coelacanths. They became extinct at the end of the Devonian period.

The lungfishes, or dipnoans, lost the intracranial hinge early in their evolutionary history and have consolidated their skulls into a single unit. The lungfish became specialized for buccal pumping and suction feeding using a cranial rib combined with elements of the shoulder girdle. This mechanism helps to enlarge the buccal cavity. In association with this, the hyomandibula became reduced or absent. That bone is a key element of the development of the ear region in tetrapods.

The Tetrapodomorpha is the other major clade among sarcopteryians and includes stem tetrapods. Some current terminology refers to all of the tetrapodomorphs as stem tetrapods even though their paired appendages are fins and not limbs.

Tetrapodomorph fishes, as distinct from limbed forms, show the starting conditions for the evolution of the ear in tetrapods (Clack and Ahlberg, Chap. 3). Several examples of tetrapodomorph fishes provide excellent material from which the braincase and otic region have been described. The best known is still that of *Eusthenopteron*, a genus whose anatomy has been described in detail by Jarvik (1980). For many decades this taxon was considered a good model for the kind of fish that gave rise to limbed tetrapods. However, this has now been superceded by newly discovered fossils of more tetrapod-like forms. Chief among the modifications that they show is the enlargement of the spiracular notch at the back of the skull (see Sect. 1.2), which is associated with enlargement of the spiracular chamber. Part of this apparatus eventually forms the middle ear cavity in tetrapods. Some of the most tetrapod-like of the tetrapodomorph fishes show further developments of modification to the ear region, such as the loss of the intracranial hinge and reduction of the hyomandibular. Strictly speaking then, the tetrapodomorphs make up the majority of animals included in this volume, with the limbed forms contributing by far the most intricate and complex story.

With the rise of limbed stem tetrapods from the Devonian, the earliest complement of a stapes, FV (later the FO), and its supporting structures were developed, as described by Clack and Anderson in Chap. 4. The conversion of the hyomandibula to a true stapes and the development of the FV (into which the latter is inserted) is still a poorly understood process, which the fossil record does not at present represent. However, it would appear that at first the stapes was a bulky bone whose function in sound perception must have been rather limited. Even in the early stages of limbed tetrapod evolution there seems to have been specialization, for example, the bizarre morphology of the Devonian form *Ichthyostega* (Clack et al. 2003). Its ear region is thought to show modifications for underwater sound perception. Once tetrapods had become terrestrially capable in the Carboniferous and Permian, they show middle ear morphology that was highly diverse. The spiracular notch became converted to a support for a tympanic membrane, although the groups in which this had occurred is debatable. Several tetrapod groups that were not closely related shared the primitive, bulky form of stapes. Others, however, such as the small forms known as microsaurs, showed stapes that had very large footplates but very short stubby shafts and lacked a tympanic notch. Other groups such as diadectomorphs had bony plates filling the tympanic notch, suggestive of secondary aquatic habits.

During the late Paleozoic, probably in the early Carboniferous or Mississippian periods, the lineages of Amphibia and Amniota diverged from each other. Stem members of both groups are known from the Carboniferous and Permian periods. The construction of the ear region became more refined during the later parts of the era as stem members of both groups evolved (Clack and Anderson, Chap. 4). In this volume, the Amniota are treated first (in Chaps. 5–10) with Amphibia being dealt with last in Chap. 11 by Schoch and Anderson. This is to emphasize the distinctiveness of Amphibia and to show that they are no more primitive than the early amniotes. Many amniote groups developed tympanic hearing independently; whereas, although amphibians developed it separately from amniotes, they lost it in two of the three modern groups, probably independently. This book describes this diversity and attempts to understand its capabilities.

Diversification among amniotes led to the differentiation of diapsid and synapsid groups, each of which evolved their own middle ear mechanisms, and their inner ears became refined accordingly. The synapsid lineage, which is probably the earliest amniote group to diverge from the stem, ultimately gave rise to mammals, and recent finds and techniques have led to exciting new discoveries in the story that might have been considered well known and understood. The picture becomes more complex the more that is found. The evolution of the ear in synapsids is divided into three chapters, each taking a sector of the phylogenetic tree whose boundaries are defined by key taxa or morphological innovations.

In Chap. 5, Kemp describes the most basal sector, starting with the late Carboniferous and Early Permian forms informally called pelycosaurs (or "mammal-like reptiles"). The phylogenetic tree then covers groups that show increasingly more mammal-like features of the skull and postcranial skeleton. Among these were the reduction of the quadrate, articular, and angular bones of the basic amniote lower jaw and its articulation to become ultimately the incus, malleus, and ectotympanic of the mammalian ear. Established on embryonic evidence, the fossil record shows the progressive reduction of the jaw bones through the phylogenetic tree in a very convincing series. Originally associated with feeding, those jaw bones were reduced as jaw closing muscles moved their insertion points more anteriorly, taking the pressure off the jaw joint. However those early forms had not quite reached the stage of having replaced the quadrate-articular joint with the defining mammalian one of a dentary-squamosal joint.

The dentary-squamosal joint was achieved in the mammaliaforms described in Chap. 6 by Luo, Schultz, and Ekdale. Continuing from the descriptions of basal synapsids in Chap. 5, they show the increasingly mammal-like conditions found in later members of the synapsid clade. They also show how many features were independently derived within different sub-clades leading to true mammals—members of the crown group. They take the middle ear components and show the transition from a form in which the components remained, to varying degrees, attached to elements of the lower jaw, in particular to the primary lower jaw or Meckel's cartilage (the main lower jaw element in chondrichthyans and the embryonic precursor in bony vertebrates). Attachment is gradually lost by the more posterior bones of the amniote lower jaw, including the articular and its corresponding part of the joint in the upper jaw, the quadrate, as they become smaller and start to form the elements of the mammalian ear system.

The ectotympanic of mammals is formed from an expansion of the angular bone in the amniote lower jaw known as the reflected lamina of the angular. This is a thin flange of bone growing back and slightly outwards from the back of the jaw, creating a space between itself and the jaw joint that was probably air filled and that might have had some auditory function. They also show the development of the malleus and its manubrium from the articular and its retroarticular process.

Mammal-like ear regions change incrementally from conditions much like those of primitive reptiles (dealt with in Sobral, Reisz, Neenan, Müller, and Scheyer, Chap. 8; Evans, Chap. 9; Sobral and Müller, Chap. 10) to those of true mammals, but along the way, there is much variability and many convergences with a definitive mammalian type ear appearing several times in the most crownward non-mammalian taxa.

Luo, Schultz, and Ekdale (Chap. 6) then tackle the inner ear, showing how ossifications around the inner ear become increasingly fused and house the internal structures more and more closely. This means that the development of key features, such as the coiling of the cochlea, can be followed among the more derived fossil taxa, especially now that μCT can reveal the internal bony structure of the otic capsule. Drawing all this information together, they can make informed judgments about the hearing capabilities of some of these fossil animals.

The most basal reptilian amniote groups were highly varied in their otic structure, as seen in early tetrapods. The positions of several groups (and even individual taxa) are uncertain in the phylogenetic tree, such as the parareptiles, chelonians (turtles), and the Mesozoic marine reptiles (ichthyosaurs, plesiosaurs, and placodonts). These and basal diapsids are treated in Chap. 8 by Sobral, Reisz, Neenan, Müller, and Scheyer. Some parareptiles developed a tympanic notch and tympanic membrane separately from other amniotes, and their innovation was the earliest attempt at tympanic hearing within the whole amniote group. The earliest stem diapsids, such as the Permian *Captorhinus*, lacked a tympanic notch altogether, and their stapes were often long and stout and, in some cases, contacted or even articulated with the quadrate. It was thought to stabilize the jaw joint in those forms rather than having much to do with hearing. The early diapsid *Youngina* is one of the best known, although opinions vary as to how much of a tympanic notch it actually had. It had an elongate, although slender, stapes set into a large FO whose margins were not well ossified. *Youngina* was once considered to be a stem lepidosaur, but more recently, it has been placed further down the phylogenetic tree below the split between lepidosauromorphs and archosauromorphs. It does show some basic diapsid features that help to understand the starting conditions from which each of those two lineages sprang. Among those two lineages many aspects of their ear structure were later developed independently.

Lepidosaurs comprise lizards, snakes, and their allies, the amphisbaenians and the rhynchocephalians. The latter are represented today only by the single genus *Sphenodon*. Perceptions have been shifted as to what constitutes a "primitive" lizard-type ear, and Chap. 9 by Evans explores current thinking. At one time, the rhynchocephalian *Sphenodon* was regarded as a primitive form because it appeared to show the classic diapsid type of skull in its unmodified form. We now know from the fossil record that its condition is secondary, as is likely also for its ear construction and function. Its features are described in Chap. 9 with those of its fossil relatives, which in the Triassic were the dominant forms among lepidosaurs. Squamates are the main group of modern lepidosaurs and are ecologically diverse, which their ear regions reflect. Many have well-developed tympanic ears, whereas others, such as burrowers, have lost those features. Snakes have their own unique modifications, some of which suggest a burrowing ancestry. Among fossil lepidosaurs, the mosasaurs represent another modification to the ear region, some having developed cartilaginous or bony plates in the tympanic notch.

Archosaurs, considered in Chap. 10 by Sobral and Müller, are probably the most diverse group of amniotes, including crocodiles, pterosaurs, dinosaurs and birds. The earliest archosauromorph, *Prolacerta*, provides the baseline from which other members of the group can be interpreted. It shows the FO to be a relatively large opening to which not only the bones of the otic capsule but also those of the basicranium contribute part of the margin. Unfortunately, the stapes is unknown, which is true for many early members of the group. Among the earliest groups to branch off from the early forms were the crocodylians, whose descendants are around today. They developed a highly complex otic region by isolating the internal elements of the inner ear in a bony capsule in tandem with developing a system of air-filled spaces and canals connected with the pneumatization of the skeleton. The amount and complexity of the bones made their ear regions difficult to study in three dimensions until μCT techniques become common.

Pterosaurs, dinosaurs, and birds constitute the Avemetatarsalia. Pterosaurs, the flying Mesozoic reptiles, developed sophisticated flying capabilities, but the later ones were also quadrupedal walkers. Both these aspects are reflected in the structure of their inner ears.

Among dinosaurs, as shown by Sobral and Müller in Chap. 10, even within a single lineage of dinosaurs, some features of the middle ear have evolved independently, for example, a separate fenestra pseudorotunda. Its presence implies the separate acquisition of impedance matching that would have improved hearing. For example, a separate fenestra pseudorotunda was developed in early ornithischians, although it was apparently lost again in some later members of the group. The fenestra appears to have been absent in sauropods and in some of the theropods (such as the carnosaurs), but it was present in the avialians including early birds.

The Amphibia are covered by Schoch and Anderson in Chap. 11. They are treated last in this volume in order to emphasize the fact that the Amphibia are not closely related to the basal fossil tetrapods that were treated in Chap. 4 by Clack and Anderson. Those basal tetrapods should not be referred to as "amphibians" unless their relationship to stem amphibians in the fossil record can be shown. The identity of the stem group is, on the other hand, somewhat controversial. An alternative view to that given by Schoch and Anderson is found in Marjanović and Laurin (2013), and this should be borne in mind. However for the authors of this volume, the temnospondyl hypothesis, which is accepted in Chap. 11, is the more convincing, and its support comes in part from the structure of the ear region.

Superficially similar middle ear structures in amphibians (e.g., frogs) and reptiles (e.g., lizards) turn out to be convergent, which has been documented by features of the braincase and inner ear in the group known as temnospondyls that have only recently been accessible to study. Although temnospondyls are known from as far back as the early Carboniferous, the fossil record of modern amphibians only really begins in the Jurassic period. The huge time difference between these groups in part explains the existence of such a controversy regarding amphibian origins. By the

Jurassic, specimens are sufficiently similar to modern forms that descriptions of those modern forms are treated in Chap. 11. Most of the fossils from the Mesozoic era are preserved as flattened skeletons that do not yield much information on internal braincase anatomy.

1.5 Summary

This chapter acts first as an introduction to some major concepts from the world of palaeontology: the techniques for investigating fossil material; the crown group, stem group, and total group distinctions in phylogenies; and the geological timescale. Although fossil vertebrate embryos are rarely found, understanding how their braincases form during embryogenesis is important for interpreting fossil material and for understanding the sometimes complex formation of the otic capsule and other related structures of the skull. Section 1.2 explains the embryonic components contributing to the skull and how they gradually evolve into the forms that are seen in diverse vertebrate groups (also see Friedman and Giles, Chap. 2 for more details).

Some early vertebrate groups have very little in the way of a fossil record either because they are usually represented by dissociated or isolated elements or because they have little or no bone to be fossilized. These include various jawless vertebrates, chondrichthyans, and some early bony vertebrates. However, some recent discoveries made by use of new techniques such as μCT scanning and reconstruction have allowed unprecedented access to inner ear structures in some of these animals. Stem gnathostomes (gnathostomes are jawed vertebrates; stem gnathostomes have some features shared with jawed vertebrates but no jaws) include a range of animals with a type of bone that often forms complex head shields. These structures sometimes provide information via μCT techniques of internal structures, including those of the inner ear and brain. This is also the case in some exceptionally well preserved shark-like chondrichthyans. Although they do not have a separate chapter devoted to them, some of the major recent discoveries are introduced in this chapter (Sect. 1.3).

This is followed by a summary of each of the following chapters, setting out some of the main areas of discussion and picking out highlights of the evolutionary story of each major vertebrate group. One of the strongest impressions overall is of the amount of convergent evolution that has gone on both within and between groups. It is an expression of the capacity for evolution to work with the original components derived from the common features of the vertebrate embryo and their ability to respond to the needs of a sensory system used in different ways, in differing environments, by diversely modified taxa.

Compliance with Ethics Requirements J. A. Clack has declared that she has no conflicts of interest.

References

Clack, J. A. (2001). The otoccipital region—origin, ontogeny and the fish-tetrapod transition. In P. E. Ahlberg (Ed.), *Major events in early vertebrate evolution* (pp. 400–413). London: Systematics Association Symposium Volume.

Clack, J. A. (2012). *Gaining ground: The origin and evolution of tetrapods, 2nd ed.* Bloomington, Indiana, Indiana University Press.

Clack, J. A., Ahlberg P. E., Finney, S. M, Dominguez Alonso P. et al. (2003). A uniquely specialised ear in a very early tetrapod. *Nature*. 425, 65–69.

Davis, S. P., Finarelli, J. A., & Coates, M. (2012). *Acanthodes* and shark-like conditions in the last common ancestor of modern gnathostomes. *Nature*, 486, 247–250.

Gai, Z., Donoghue, P. C. J., Janvier, P., & Stampanoni, M. (2011). Fossil jawless fish from China foreshadows early jawed vertebrate anatomy. *Nature*, 476, 324–327.

Goodrich, E. S. (1930). *Studies on the structure and development of vertebrates*. London: Macmillan.

Goujet, D. (2001). Placoderms and basal gnathostome apomorphies. In P. E. Ahlberg (Ed.), *Major events in early vertebrate evolution* (pp. 172–186). London: Systematics Association Symposium Volume.

Gradstein, F. M., Ogg, J. G., Schmitz, M. D., & Gabi, M. (2012) The geologic time scale 2012. Oxford: Elsevier.

Hildebrand, M., & Goslow, G. (2001). *Analysis of vertebrate structure, 5th ed.* New York: John Wiley & Sons, Inc.

Jarvik, E. (1980). *Basic structure and evolution of vertebrates, vol. 1.* London: Academic Press.

Kardong, K. V. (2014). *Vertebrates—comparative anatomy, function and evolution, 7th ed.* New York: McGrawHill Higher Education.

Laaß, M. (2014). Bone conduction hearing and seismic sensitivity of the Late Permian anomodont *Kawingasaurus fossilis*. *Journal of Morphology*. 276, 121–143. doi:10.1002/jmor.20325

Maklad, A., Reed, C., Johnson, N. S., & Fritzsch, B. (2014). Anatomy of the lamprey ear: Morphological evidence for occurrence of horizontal semicircular ducts in the labyrinth of *Petromyzon marinus*. *Journal of Anatomy*, 224, 432–446.

Marjanović, D., & Laurin, M. (2013). The origin(s) of extant amphibians: A review with emphasis on the "lepospondyl hypothesis". *Geodiversitas*, 35(1), 207–272.

Porro, L. B., Rayfield, E. J., & Clack, J. A. (2015). Descriptive anatomy and three-dimensional reconstruction of the skull of the early tetrapod *Acanthostega gunnari* Jarvik, 1952. *PLoS ONE*, e0118882.

Pradel, A., Langer, M., Maisey, J. G., Geffard-Kuriyama, D., et al. (2009). Skull and brain of a 300 million year old chimaeroid fish revealed by synchrotron holotomography. *Proceedings of the National Academy of Sciences of the USA*, 106 (13), 5224–5228.

Schultze, H.P. (1990). A new acanthodian from the Pennsylvanian of Utah, U.S.A., and the distribution of otoliths in gnathostomes. *Journal of Vertebrate Paleontology*, 10, 49–58.

Stensiö, E. (1932). *The cephalaspids of Great Britain*. London: British Museum of Natural History.

Stensiö, E. (1963a). Anatomical studies of the arthrodiran head. *Kungla Svenska Vetenskapsacademiens Handlingar*, 9(2), 1–419 (462 plates).

Stensiö, E. (1963b). *The brain and the cranial nerves in fossil, lower craniate vertebrates*. Oslo: Universitetsforlaget.

Zhu, M., Yu, X., Ahlberg, P. E., Choo, B., et al. (2013). A Silurian placoderm with osteichthyan-like marginal jaw bones. *Nature* 502, 188–193.

Chapter 2
Actinopterygians: The Ray-Finned Fishes —An Explosion of Diversity

Matt Friedman and Sam Giles

Abstract Living ray-finned fishes number approximately 30,000 species, roughly equal to modern lobe-finned fishes plus tetrapods combined. The fossil record of ray-finned fishes extends to the Early Devonian (ca. 415 Ma), although the oldest taxa for which the morphology of the otic region is known in any great detail are 35 million years younger. Early actinopterygians have consolidated saccular otoliths, open vestibular fontanelles, and a persistent otico-occipital fissure. These general osteichthyan conditions are joined by specializations like the lateral cranial canal, a diverticulum that extends through the loop of the posterior semicircular canal and is thought to be unrelated to hearing. This anatomy persists into early members of the dominant modern radiation of ray-finned fishes, Neopterygii. Neopterygian subclades, including teleosts, show parallel losses of persistent fissures, fragmentation of the endocranium into separate ossification centers, and reduced bony enclosure of the inner ear. Teleosts show remarkable specializations for hearing based on connections between the gas-filled swim bladder and the otic capsule. Cretaceous fossils of clupeomorphs (herrings) and otophysans (catfishes, minnows, characins, and Neotropical knifefishes) document the assembly of the specialized anatomical features associated with hearing in both groups.

Keywords Actinopterygii · Cladistia · Chondrostei · Holostei · Inner ear · Neopterygii · Otolith · Skeletal labyrinth · Teleostei · Weberian apparatus

M. Friedman (✉) · S. Giles
Department of Earth Sciences, University of Oxford,
South Parks Road, Oxford OX1 3AN, UK
e-mail: matt.friedman@earth.ox.ac.uk

S. Giles
e-mail: sam.giles@earth.ox.ac.uk

M. Friedman
Museum of Paleontology and Department of Earth and Environmental Sciences,
University of Michigan, 1109 Geddes Avenue, Ann Arbor MI 48109-1079, USA

J.A. Clack et al. (eds.), *Evolution of the Vertebrate Ear—Evidence from the Fossil Record*, Springer Handbook of Auditory Research 59,
DOI 10.1007/978-3-319-46661-3_2

Abbreviations in Figures

ampa	ampulla of the anterior semicircular canal
ampe	ampulla of external semicircular canal
ampp	ampulla of posterior semicircular canal
asc	anterior semicircular canal
cnX	vagus nerve
df	dorsal fontanelle
esc	external semicircular canal
fb	fossa bridgei
hym	hyomandibula
hymf	hyomandibular facet
lcc	lateral cranial canal
oa	occipital arch
oof	otico-occipital fissure
ot	otolith
pqc	palatoquadrate complex
psc	posterior semicircular canal
sac	saccular chamber
spc	spiracular canal
vf	vestibular fontanelle
ssu	sinus superior
utr	utricular recess
vf	vestibular fontanelle

2.1 Introduction

Osteichthyans, the bony fishes, comprise two divisions: actinopterygians, or the ray-finned fishes, and sarcopterygians, including the lobe-finned fishes and tetrapods, which represent the other half of this major vertebrate radiation and form the focus of all subsequent chapters. Actinopterygians and sarcopterygians contain roughly equal numbers of living species, so the account given here of ray-finned fishes is necessarily skeletal in its detail relative to that provided for their lobe-finned cousins. The following sections establish the basic context for the overwhelming diversity of actinopterygians, providing broad outlines of skull anatomy in osteichthyans (Sect. 2.1.1) and an overview of paleontological evidence bearing on the very earliest bony fishes. Figure 2.1 shows a basic family tree of actinopterygian groups.

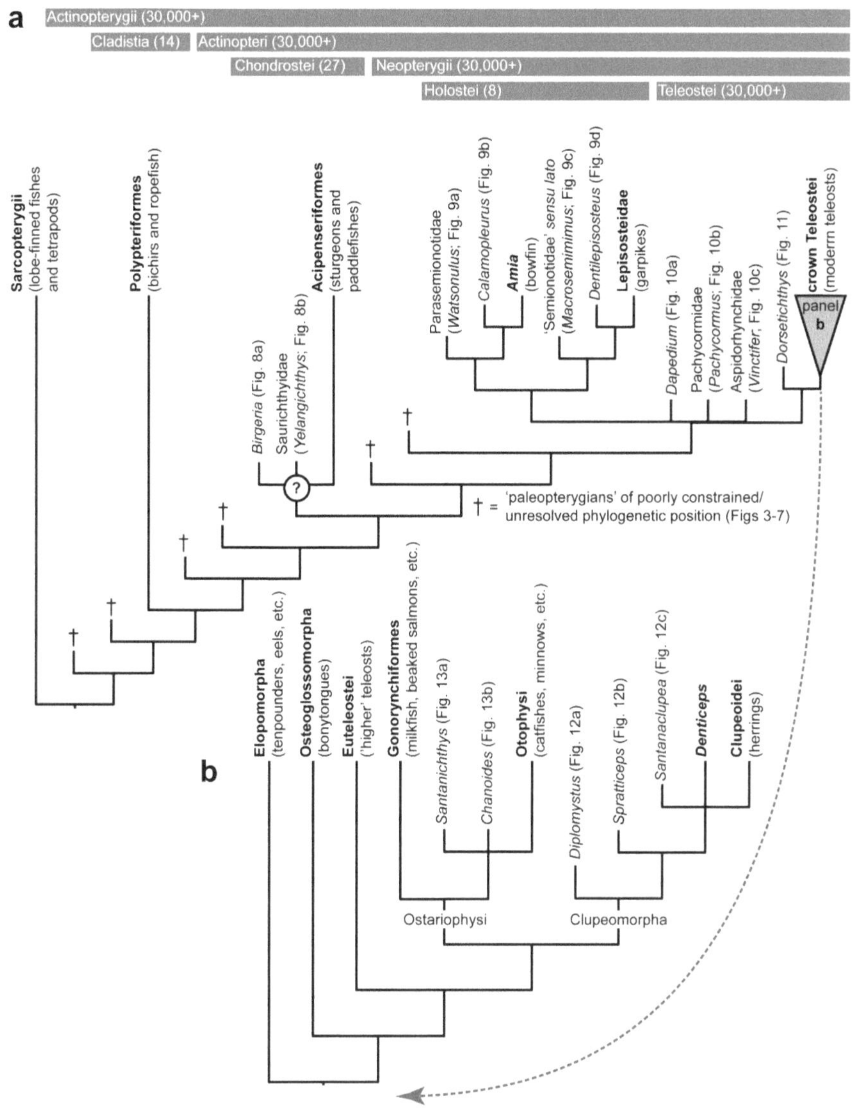

Fig. 2.1 Phylogenetic framework for actinopterygian fishes. (**a**) Relationships among major groups discussed in this chapter. Extant lineages are shown in boldface with fossils in standard text at the tips of truncated branches. Daggers ('†') represent possible positions of anatomically primitive fossil actinopterygians of debated phylogenetic affinities. *Gray bars*: numbers in parentheses after major clade names refer to counts of living species belonging to each of those groups. (**b**) Relationships among modern (crown group) teleosts. Extant lineages are shown in boldface with fossils in standard text at the tips of truncated branches. Topologies represent informal consensus derived from multiple sources

2.1.1 Bony Fish Skull Anatomy: An Overview

Cranial anatomy in early actinopterygians corresponds broadly to that for jawed vertebrates generally (Fig. 2.2) (also see Clack, Chap. 1). The braincase represents the core of the skull with other components of the head suspended from or applied

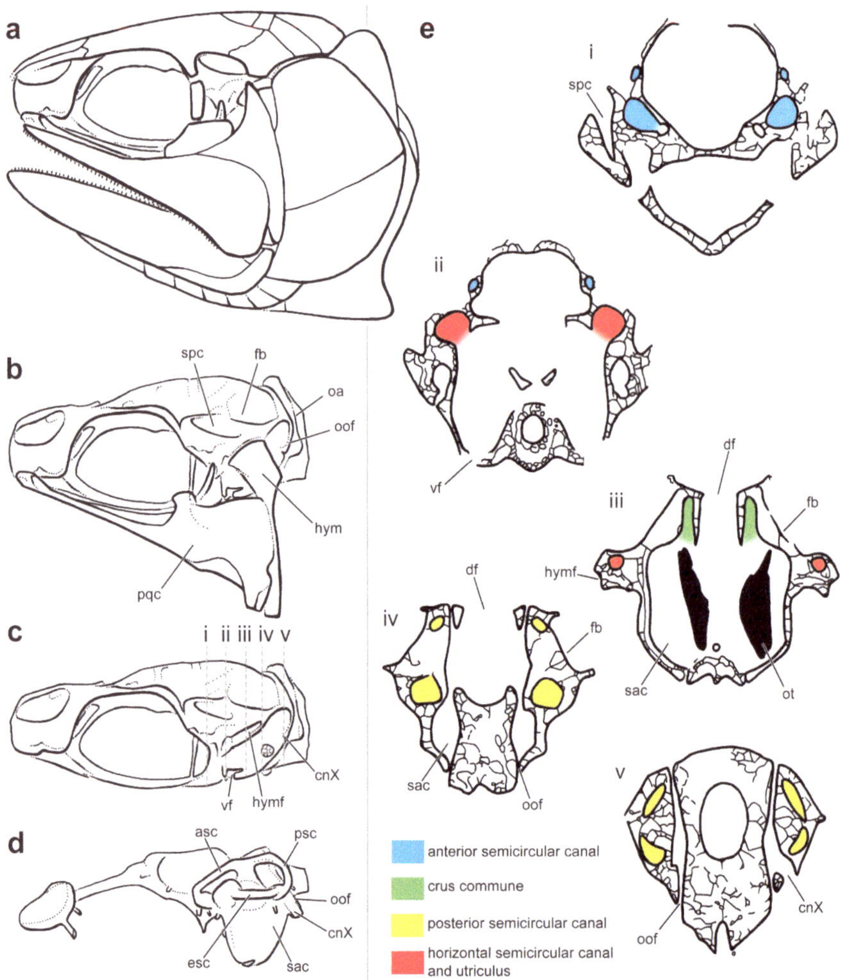

Fig. 2.2 Cranial anatomy in bony fishes, illustrated by the Triassic actinopterygian *Australosomus*. **(a)** Skull showing complete covering of external bones overlying the endoskeleton. **(b)** Skull with external dermal bones, gill arches, and ventral halves of the mandibular and hyoid arches removed. **(c)** Endoskeletal braincase in isolation. **(d)** Cranial endocast (for enlarged version highlighting inner ear, see Fig. 2.7). **(e)** Cross sections through the otic region, with numerals keyed to sections indicated in **c**. Void spaces corresponding to the skeletal labyrinth are color-coded. (Based on Nielsen 1949)

to it. A series of jointed arches hang beneath the braincase. From anterior to posterior, these are: the mandibular arch, the hyoid arch, and the branchial arches. There is a single pair of mandibular arches, one on each side of the skull, and they represent the primary upper and lower jaws. The upper jaw, which is known as the palatoquadrate, shows many patterns of articulation with the braincase, but the primitive pattern in bony fishes includes a basal articulation beneath the eye and an ethmoid articulation beneath the nasal capsule. The hyoid arch lies behind the mandibular arch and is dominated by two principal segments: the dorsally located hyomandibula, which articulates with the otic region of the braincase, serves an important role in supporting the primary upper jaw, and represents the evolutionary precursor of the tetrapod stapes; and the ventrally located ceratohyal, which is a critical component of suction feeding in many fishes.

Primitively, five gill arches trail the hyoid arch, although this number is reduced in many bony fish lineages. The oral surfaces of the mandibular and gill arches are associated with plate-like dermal bones that often bear teeth. This complex series of arches and the central braincase are then covered to varying degrees by external bones of the skull. It is increasingly clear that this extensive exterior cladding represents the primitive style of jawed-vertebrate skull construction, with the reduction and loss of the outer bony plates typical of cartilaginous fishes representing a specialized arrangement (Zhu et al. 2013).

The braincase arises developmentally from a series of embryonic cartilages that are replaced by bone during ontogeny (see Clack, Chap. 1). Traces of these embryonic precursors persist in the braincases of some extinct ray-finned fishes in the form of persistent fissures between ossifications. Most relevant here are fissures in the posterior half of the braincase, which is divided between the otic and occipital regions. The embryonic otico-occipital fissure between these regions can be completely obliterated, can persist as a cartilaginous band between bones, or can be present as an unfilled gap between ossifications. In many early bony fishes, the persistent fissure intersects a broad, unossified area on the ventrolateral face of the chamber housing the sacculus of the inner ear: the vestibular fontanelle. Anterodorsal to this fontanelle, the otic capsule of bony fishes primitively bears a longitudinal canal for the jugular vein, as well as areas of articulation for the hyoid and gill arches. Some ray-finned and lobe-finned fishes show further modifications to the exterior of the otic region, including partial endoskeletal enclosure of the spiracle (a putatively vestigial gill slit associated with the hyoid arch) and various shallow depressions, or fossae, that lie between the roof of the braincase and the overlying dermal skull roof. The homologies of these fossae are uncertain, leading to the proposal of complex nomenclatural schemes to describe patterns of variation across groups (Bjerring 1984).

When lined with bone, the cranial cavity within the otic region can provide broad outlines of the structure of the inner ear. Positive models of this void space—called an endocast—indicate that the inner ear was almost entirely encased by bone in early osteichthyans. Enclosure of individual components of the inner ear is reduced in living ray-finned fishes in comparison to their earliest relatives, meaning that many major features of inner ear geometry cannot be discerned from cranial endocasts.

2.1.2 *The Earliest Bony Fishes*

Definitive actinopterygians and sarcopterygians are joined by a series of Late Silurian–Early Devonian bony fishes of less secure phylogenetic placement. These "basal osteichthyans" include two Early Devonian (ca. 415–400 Ma; see Clack, Chap. 1 for a geological timescale) forms for which some details of the ear region have been described: *Psarolepis* and "*Ligulalepis*" (enclosed in quotes because the definitive material of the genus is scales with no clear links to relevant cranial material). Current understanding of neurocranial anatomy in these important early taxa is limited, despite their antiquity and clear significance.

Psarolepis (Yu 1998) is known from the earliest Early Devonian of China, and its ear structure is known only from flattened otic capsules. The dorsal surface of the cranial endocavity is visible owing to taphonomic loss of the parachordal region due to the presence of a persistent otico-occipital fissure. Despite the additional exposure of internal anatomy this provides, little can be said of ear structure in *Psarolepis* apart from the presence of three large semicircular canals. "*Ligulalepis*" is potentially more informative with respect to ear morphology in very primitive bony fishes, as it is represented by a single incomplete but uncrushed braincase from Australia that has been subjected to acid preparation (Basden and Young 2001). Like *Psarolepis*, "*Ligulalepis*" bears a complete otico-occipital fissure. Here, too, the parachordal plate is not preserved, revealing the detailed internal surface of the brain cavity and apertures to the semicircular canals. However, no virtual endocast is yet available for the bony labyrinth of this important early osteichthyan, so no detailed comparisons can be made with the arrangement in younger forms.

Joining these early osteichthyans of uncertain placement is *Janusiscus*. Originally interpreted as a species of *Dialipina* (and a ray-finned fish), renewed study using X-ray computed tomography (μCT) reveals a braincase lacking a basal fissure and vestibular fontanelles. *Janusiscus* is placed as a bony-fish-like member of the gnathostome stem lineage (Giles et al. 2015c). That is, it is more closely related to living jawed vertebrates (the "crown" group) than the next closest living lineage (in this case jawless vertebrates), but does not belong to either bony fishes or cartilaginous fishes, the two surviving lineages of jawed vertebrates. Rather, it branches from the jawed vertebrate tree before the evolutionary divergence between these two familiar groups. Traces of the bony labyrinth are apparent, but not sufficiently well-preserved to make meaningful comparisons with other early gnathostomes.

2.2 The Fossil Record of Actinopterygian Hearing

Knowledge about ear structure in fossil actinopterygians derives principally from two lines of evidence: endocasts of the bony labyrinth and saccular chamber, and saccular otoliths (also known as the sagitta or saccolith). The latter are composed of either of two forms of calcium carbonite, vaterite or aragonite, and are one of the three pairs of large otoliths or "ear stones" present in most modern ray-finned fishes (a single pair of otoliths characterizes early actinopterygians and sarcopterygians and is likely to be primitive; see Sect. 2.3). These two records are largely non-overlapping in terms of time and phylogeny; a majority of actinopterygian endocasts belong to a modest set of Paleozoic or early Mesozoic non-teleosts that are known in exhaustive detail, while the study of fossil otoliths is almost exclusively limited to latest Mesozoic and Cenozoic teleost remains that are usually found isolated from body fossils. These key lines of anatomical evidence are joined by hard-tissue specializations related to hearing in specific groups, such as the elaborate Weberian system of otophysan teleosts. This overview of actinopterygians will focus principally on aspects of the bony skeleton. The vast otolith record is beyond the scope of this general review, but key aspects are noted below where relevant. More detailed summaries of fossil otoliths can be found in Nolf (2013).

2.3 Devonian Actinopterygians

Despite their dominance in aquatic settings today, ray-finned fishes were neither abundant nor diverse during the Devonian "Age of Fishes" (Friedman 2015). Fragmentary fossils of latest Silurian and Early Devonian age have been aligned with ray-finned fishes in the past (Schultze 2015; see also Sect. 2.1.2), although there is doubt surrounding the veracity of these proposed affinities (Friedman and Brazeau 2010). The earliest taxon reliably identified as an actinopterygian is *Meemannia* (Lu et al. 2016), which is from the same Early Devonian Chinese deposits as *Psarolepis* and was described initially as a lobe-finned fish (Zhu et al. 2006). *Meemannia* is represented by highly incomplete material with partial braincases that permit reconstruction of the dorsal parts of the inner ear. Evidence for the actinopterygian affinity of *Meemannia* includes the presence of a lateral cranial canal, which is a diverticulum that extends from the endocranial cavity through the loop of the posterior semicircular canal (discussed in greater detail for *Mimipiscis* later in this section).

Prior to reinterpretation of *Meemannia*, *Cheirolepis* represented the earliest unequivocal actinopterygian. First appearing in the Middle Devonian (ca. 390 Ma), *Cheirolepis* is represented by hundreds of articulated specimens but details of internal cranial anatomy remain obscure. A badly crushed braincase of the genus reveals a persistent otico-occipital fissure and large vestibular fontanelles like those of some early sarcopterygians (see Clack and Ahlberg, Chap. 3) but no further details of the inner ear (Giles et al. 2015a). The hyomandibula is incompletely known, but it is fused with a splint-shaped dermal ossification known as the dermohyal. This fusion is a common feature among early ray-finned fishes that was lost in later forms. The braincase bears a single articular facet located near the anterior of the otic region, indicating that the hyomandibula showed the single-headed arrangement primitive for jawed vertebrates. The facet appears restricted to the region dorsal of the jugular foramen.

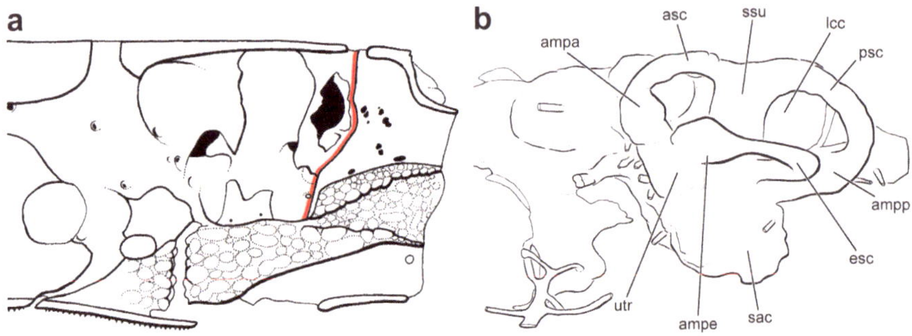

Fig. 2.3 Generalized aspects of otic morphology in actinopterygians, as exemplified by the Late Devonian (Frasnian) *Mimipiscis toombsi*. (**a**) Neurocranium in sagittal section, with anterior to left. *Red* indicates extent of otico-occipital fissure. (**b**) Endocast highlighting morphology of the inner ear, shown in left-lateral view. Lateral cranial canal sectioned in this image so as to not obscure labyrinth (**a** based on Gardiner 1984; **b** based on Giles and Friedman 2014)

The early Late Devonian (ca. 380 Ma) *Mimipiscis* yields the most detailed information available on ear structure for Devonian ray-finned fishes, and an overview of external structure of the otic capsule and inner details of the bony labyrinth and saccular chamber highlights general conditions of these structures in bony fishes (Fig. 2.3). Exceptionally preserved three-dimensional material of this genus from the Gogo Formation of Western Australia has been extracted from surrounding rock by acid preparation and studied both visually (Gardiner 1984) and through μCT (Giles and Friedman 2014).

The external surface of the otic capsule bears a small vestibular fontanelle that is intersected by the otico-occipital fissure. Other early actinopterygians and many early sarcopterygians have large vestibular fontanelles, so the condition in *Mimipiscis* might be specialized (Giles et al. 2015a). Apart from the persistent fissure, there are no indications of separate ossification centers. The hyomandibula makes a single articulation with the otic capsule. The articular facet is located near the anterior of the otic region on a transverse wall of the neurocranium that is joined with the postorbital process. The strap-shaped facet is located dorsal to the level of the jugular canal. The hyoid articulation of *Mimipiscis* and other early osteichthyans contrasts strongly with that found in chondrichthyans, where the posteriorly placed articulation lies ventral to the jugular vein. The top of the braincase is tightly applied to the overlying braincase, with no intervening fossae.

As in other early jawed vertebrates (Jarvik 1980), the region occupied by the bony labyrinth is large, representing nearly one-third of skull length. The osseous labyrinth and saccular chamber communicate directly with the rest of the cranial endocavity. This arrangement also characterizes lobe-finned fishes (Jarvik 1980) as well as chondrichthyans (sharks, rays, and ratfishes) (Maisey 2005) and their fossil relatives known as acanthodians (Davis et al. 2012). By contrast, the inner ear is separated from the brain cavity by a cartilaginous or bony wall in placoderms and armoured jawless fishes (Janvier 1996). Absence of this partition seems to be a specialization of modern jawed vertebrates (Brazeau and Friedman 2014). All three bony semicircular canals bear enlarged ampullae near their anterior (anterior and horizontal canals) or posterior (posterior canal) junctions with the cranial endocavity. Of the three, the anterior canal is the longest. The posterior and anterior bony canals are united at the crus commune and continue into the cranial endocavity as the sinus superior. The horizontal and anterior canals join the endocranial chamber very close to one another in the swollen utricular region. By contrast, the posterior junctions between the horizontal and posterior canals and the endocavity are more broadly separated. As in other bony fishes, the horizontal canal joins the endocranial chamber at the level of the ampulla of the posterior canal (Giles and Friedman 2014). The otic capsules extend far laterally relative to the parachordal plate. The course of the jugular vein extends below—rather than lateral to—the horizontal semicircular canal.

In internal view, the lateral wall of the otic capsule bears a large opening immediately anterior to the otico-occipital fissure. This feature, known as the lateral cranial canal, is unrelated to the bony labyrinth and is interpreted as bearing a hemopoietic organ based on comparisons with modern sturgeons, paddlefish, gars and bowfin (Jarvik 1980). Some have argued for the presence of a lateral cranial canal in basal osteichthyans (Zhu et al. 2010), but these interpretations are problematic (Giles and Friedman 2014) with putative openings to the canal representing the crus commune (Lu et al. 2016).

The saccular chamber extends far below the ventral surface of the chamber for the brain. It is incompletely mineralized posteroventrally, opening to the exterior through the small vestibular fontanelle. The vagus cranial nerve (X) exits from the lateral face of the saccular chamber at roughly the level of the posterior junctions between the horizontal and lateral semicircular canals with the cranial endocavity. A single compact saccular otolith is present but is only known from sectioned skulls; apart from appearing longer than deep, nothing can be said of its morphology and its mineralogy is unknown (Gardiner 1984). The primitive arrangement and composition of actinopterygian—and osteichthyan—otoliths has been the subject of considerable debate (summarized in Coates 1998). However, available evidence from modern and fossil sarcopterygians and early ray-finned fishes suggests that the presence of a single pair of large otoliths is the primitive condition for bony fishes. Modern ray-finned fishes bear three pairs of large otoliths, which is likely a secondary arrangement (Coates 1993).

Outside of *Mimipiscis*, clues about internal ear structure are scant in Devonian ray-finned fishes. Acid-prepared material of the co-occurring *Moythomasia* preserves similar detail of the otic capsules but has not been subject to the same degree of detailed description (Gardiner 1984). Models of the braincase and endocast for *Raynerius* (Giles et al. 2015b) agree with *Mimipiscis* in broad details of the otic region, although there are numerous specific differences elsewhere in the neurocranium. Otoliths in situ are reported for the Middle Devonian *Howqualepis* in association with badly crushed otic capsules (Long 1988) and appear to consist of one large otolith per side. Early–Middle Devonian deposits in the Baltic states, Belarus, and Arctic islands yield isolated otoliths attributed to ray-finned fishes

(Talimaa in Nolf 1985), but these differ in both mineralogy and structure from in situ specimens, leading Coates (1993) to reject placement with actinopterygians.

Cladistia, comprising the modern polypterids *Polypterus* and *Erpetoichthys* (bichirs and reedfish), is traditionally regarded as diverging from other actinopterygians in the Devonian (Patterson 1982). Endocranial structure in the living species differs considerably from that of the earliest ray-finned fishes. Unfortunately, the polypterid fossil record is notoriously thin, with nothing known of the endocranium in extinct members of the group (Friedman 2015).

2.4 "Paleopterygians"

Although the relationships among the principal extant lineages of ray-finned fishes are well-supported by both morphology and molecular data, the placement of Paleozoic and some Mesozoic fossils relative to modern groups remains unclear, as reviewed by Sallan (2014) and Friedman (2015). This assortment includes the "palaeoniscoids" of classical usage: a paraphyletic assemblage of ray-finned fishes recognized on the basis of shared primitive conditions of jaw structure, squamation, and overall body geometry. McCune and Schaeffer (1986) defined the term "Paleopterygii" to encompass the range of fossils that could not be definitively aligned with modern groups, explicitly acknowledging that this assemblage does not constitute a natural group. As this term carries no implication of monophyly, it is applied here as a shorthand reference to early actinopterygians in preference to older names like "palaeoniscoid" that come with considerable historical baggage. Because of uncertainty surrounding relationships, fossil taxa are reviewed in stratigraphic sequence, noting possible affinities with modern taxa where possible.

The oldest post-Devonian actinopterygian for which the otic region is known in any detail is the early Carboniferous (Mississippian) *Kentuckia* (Fig. 2.4a).

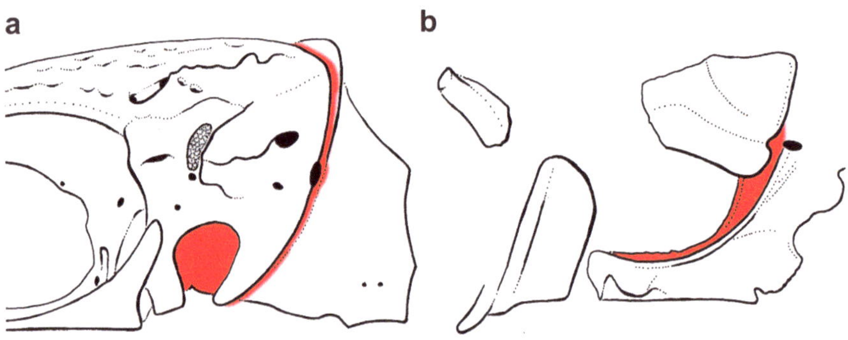

Fig. 2.4 Otic morphology in early Carboniferous (Mississippian) ray-finned fishes. *Red* indicates extent of otico-occipital fissure and vestibular fontanelles and *gray* shading defines the region of the hyomandibular facet. **(a)** Neurocranium of *Kentuckia deani* (Tournaisian-Viséan) shown in left-lateral view. **(b)** Neurocranium of *Cosmoptychius striatus* (Viséan) shown in left-lateral view (**a** based on Rayner 1951; **b** based on Schaeffer 1971)

Represented by three-dimensionally preserved heads from Kentucky, USA, *Kentuckia* (Rayner 1951) served as a key model for primitive ray-finned fish cranial conditions until the discovery and description of Gogo forms like *Mimipiscis* (Schaeffer 1968). Many aspects of the otic region—including the presence of a vestibular fontanelle and a persistent otico-occipital fissure—correspond closely to those found in Devonian taxa. The dorsal surface of the otic region bears a series of irregular depressions thought to be precursors of the paired anterior and posterior fossa bridgei. These depressions create void spaces between the dermal skull roof and the underlying neurocranium. This is unlike the condition in *Mimipiscis*, where the endoskeletal braincase is directly applied to the overlying dermal skull. Dorsal fossae are a prominent feature of many post-Devonian ray-finned fishes.

There are some differences in the endocast of the skeletal labyrinth of *Kentuckia* relative to *Mimipiscis*, the most significant of which is the placement of the crus commune of the anterior and posterior canals ventral to the roof of the endocranial chamber (Giles and Friedman 2014). This is derived relative to the earliest ray-finned fishes and is common to many post-Devonian forms. Information on the otic region is more limited for other early Carboniferous actinopterygians. A juvenile specimen of *Cosmoptychius* bears large vestibular fontanelles and an otico-occipital fissure, but it is unusual among early actinopterygians in having separate paired ossifications of the braincase including paired otic ossifications (Fig. 2.4b) (Schaeffer 1971). Such separate bones are common in stratigraphically younger taxa, especially in smaller individuals before distinct ossification centers become insensibly fused (see Sect. 2.6 and Patterson 1975 for details). No details of the labyrinth are known. A flattened neurocranium of *Woodichthys* from Scotland reveals a persistent otico-occipital fissure and well-developed vestibular fontanelles. This specimen also includes two in situ otoliths, with one each present in the right and left saccular regions. Otoliths are also known in several other specimens, although not in place (Coates 1998). The otoliths are pear shaped, with a flat medial surface and convex lateral surface. Concentric growth rings are present, but there is no central or marginal sulcus. Additional early Carboniferous taxa preserving otoliths include *Melanecta* (Fig. 2.5) (Coates 1998), *Mesopoma* (Coates 1993), *Rhadinichthys* (Moy-Thomas and Bradley Dyne 1938) and possibly *Mansfieldiscus* (Long 1988). Structure of the otoliths is clearest in *Mesopoma* and *Melanecta*. The single pair of large otoliths in *Mesopoma* broadly resembles those of *Woodichthys* in outline, although they bear a clear sulcus. *Melanecta* preserves a pair of large otoliths that differ considerably from these other examples: they are not as elongated, and they bear a prominent ventral rim composed of coalesced, rounded tubercles. More significantly, the two large otoliths of *Melanecta* are joined by three smaller ones in one specimen, suggesting a total of three pairs in the genus (Coates 1998). This complement matches that of extant ray-finned fishes, but is unlike the single pair of other early actinopterygians.

External anatomy of the otic region is known for several late Carboniferous (Pennsylvanian) actinopterygians, including *Amphicentrum* (Bradley Dyne 1939; Fig. 2.6c), *Coccocephalichthys* (Poplin and Véran 1996), and a natural mould of the skeletal labyrinth is preserved in *Mesopoma* (Coates 1999). From what is

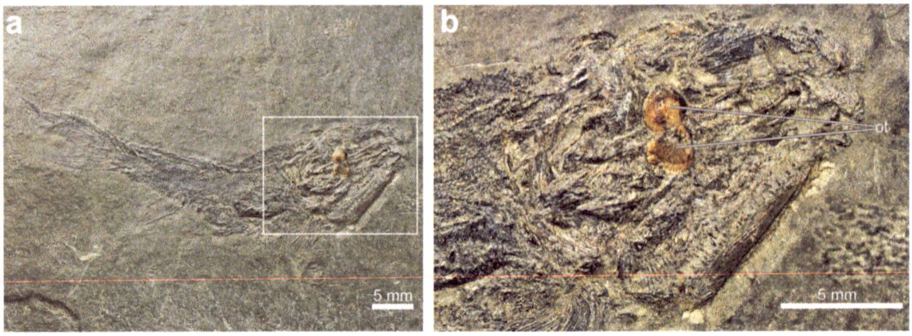

Fig. 2.5 Otolith morphology in early ray-finned fishes, illustrated by in situ examples in the Carboniferous (Serpukhovian) *Melanecta annae* (National Museum of Scotland NMS 1983.33.4). **(a)** Complete specimen. **(b)** Close-up of skull, showing large, paired saccular otoliths. (For an interpretive drawing of this same specimen, see Coates 1998)

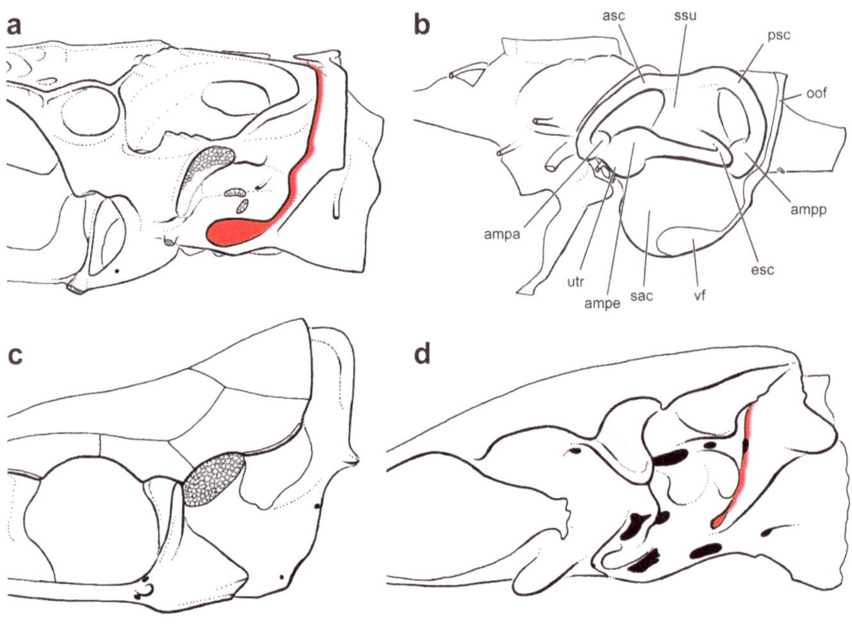

Fig. 2.6 Otic morphology in late Carboniferous (Pennsylvanian) and Permian ray-finned fishes. *Lawrenciella schaefferi* (Kasimovian), **(a)** neurocranium and **(b)** cranial endocast shown in left-lateral view. **(c)** Neurocranium of *Amphicentrum granulosum* (Moscovian) shown in left-lateral view. **(d)** Neurocranium of *Luederia kempi* (Artinskian) shown in left-lateral view. *Red* indicates extent of otico-occipital fissure and vestibular fontanelles, and *gray* shading defines region of the hyomandibular facet. (**a**, **b** based on Hamel and Poplin 2008; **c** based on Bradley Dyne 1939; **d** based on Schaeffer and Dalquest 1978)

visible, these taxa have otic capsules broadly similar to those of other early ray-finned fishes, with the exception of the apparent absence of vestibular fontanelles and an otico-occipital fissure in *Amphicentrum*. While anatomical details for these genera are superficial, the otic region of several late Carboniferous actinopterygians are described in exquisite detail. Best known are *Kansasiella* (Poplin 1974) and *Lawrenciella* [Hamel and Poplin (2008); referred to as "palaeoniscid A" and "palaeoniscid B" in Watson (1925) and other older literature], which are represented by three-dimensional crania preserved within concretions collected in Kansas, US (Fig. 2.6a, b). Material of a third taxon (Moodie 1915; "palaeoniscid C" of Rayner 1951) remains incompletely studied. In terms of external anatomy, the otic region is broadly similar in *Kansasiella* and *Lawrenciella*. There is a modest (*Lawrenciella*) or small (*Kansasiella*) vestibular fontanelle that joins a persistent otico-occipital fissure. There are no indications of separate ossification centers. The hyomandibular facet is smaller than in *Mimipiscis* and is located at the level of the jugular canal. Both taxa have well-formed fossa bridgei comprising anterior and posterior depressions. The skeletal labyrinth and its endocast are broadly similar in the two Kansas forms. Well-developed ampullary recesses are associated with each of the semicircular canals, with those of the anterior and horizontal canals in close contact. The posterior end of the horizontal canal joins the endocavity dorsal to the ventral opening for the posterior canal. The dorsal surface of the labyrinth either lies above (*Lawrenciella*) or below (*Kansasiella*) the level of the roof of the endocranial cavity. Despite the excellent preservation of relevant material, no otoliths are known. However, Gottfried (1994) describes *in situ* examples in an unnamed actinopterygian from a different deposit from the late Carboniferous of Kansas.

In contrast to the relative abundance of material from the Carboniferous, Permian deposits yield few informative specimens. This is symptomatic of a poor Permian record of actinopterygians more generally (Friedman and Sallan 2012). *Luederia* (Fig. 2.6d) and *Progyrolepis* are represented by single specimens from the early Permian of Texas, USA. Only a portion of the otic capsule is exposed in *Progyrolepis* (Dunkle 1946), while *Luederia* comprises a complete, three-dimensionally preserved braincase lacking separate ossification centers (Schaeffer and Dalquest 1978). *Luederia* bears an otico-occipital fissure and vestibular fontanelles, although the latter are very small. No details of the inner walls of the otic capsule are known, so anatomy of the skeletal labyrinth remains uncertain. These early Permian remains are joined by late Permian material of *Palaeoniscum*, which radiographs reveal to have a pair of large otoliths (Haubold and Schaumberg 1985).

Rocks of Triassic age yield a variety of actinopterygians with well-preserved crania that have been subjected to very detailed study. The best known examples derive from Early Triassic deposits in Greenland, Spitzbergen, and Madagascar, which provide broadly similar faunas, including a range of "paleopterygians" as well as taxa branching close to or within the modern neopterygian radiation. This latter group comprises the vast majority of post-Paleozoic actinopterygian fossils and is dealt with in the following sections. Because of the great number of Mesozoic taxa for which detailed information on the otic region is available, accounts are necessarily abbreviated with references provided for more complete descriptions.

The otic morphology of some Triassic actinopterygians is broadly similar to that of Paleozoic forms, bearing well-developed vestibular fontanelles in conjunction with an otico-occipital fissure but lacking separate ossification centers (Figs. 2.2, 2.7).

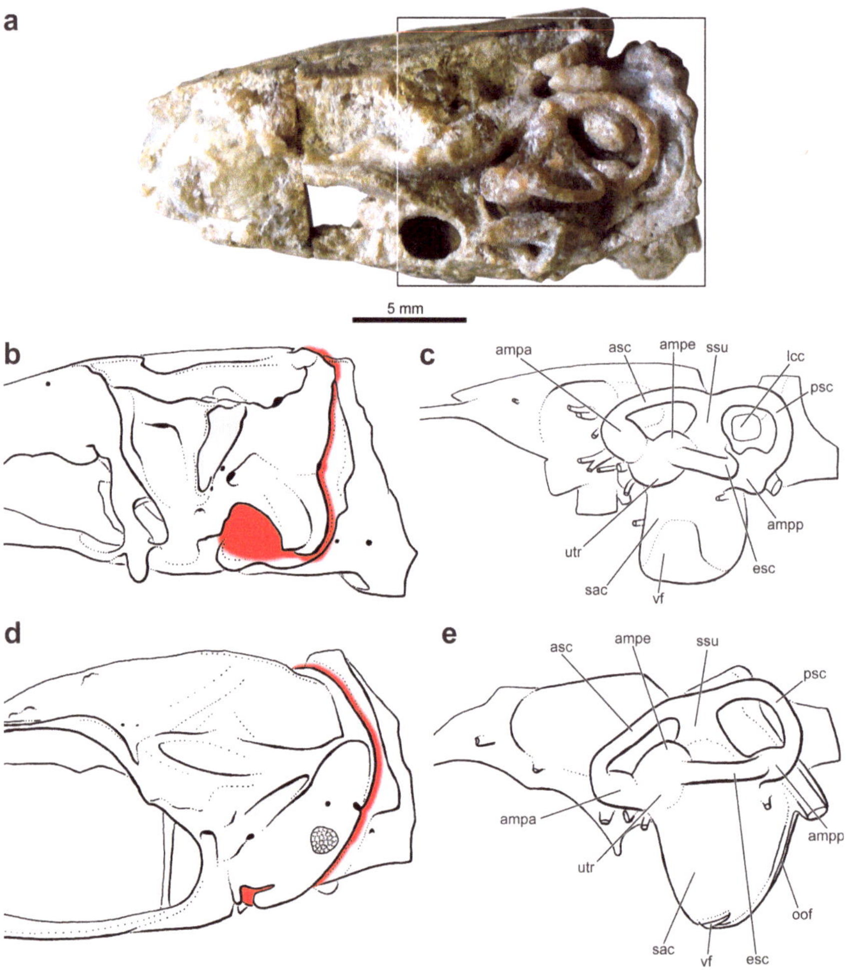

Fig. 2.7 Otic morphology in Triassic ray-finned fishes. *Red* indicates extent of otico-occipital fissure and vestibular fontanelles and *gray* shading defines region of the hyomandibular facet. (**a–c**) *Pteronisculus stensioei* (Induan). (**a**) Mechanically prepared fossil specimen showing natural endocast of skeletal labyrinth, reversed to match orientation of other figures (anterior semicircular canal missing; University Museum of Zoology, Cambridge UMZC GN.851). (**b**) Neurocranium. (**c**) Cranial endocast shown in left-lateral view. (**d, e**) *Australosomus kochi* (Induan). (**d**) Neurocranium. (**e**) Cranial endocast shown in left-lateral view. (**a**, **b** based on Nielsen 1942; **d**, **e** based on Nielsen 1949)

Examples include *Pteronisculus* (Nielsen 1942; referred to as *Glaucolepis* in older accounts) and *Australosomus* (Nielsen 1949), where the otico-occipital fissure sometimes extends anterior to the level of the vestibular fontanelle and joins the ventral fissure, thus completely separating the parachordal region from the otic capsules. Internal details of the otic region are known in great detail from both serial sectioning (e.g., Nielsen 1942) and natural endocasts (e.g., Coates 1999).

2.5 Chondrosteans

Many Triassic actinopterygians deviate from the conservative pattern seen in statigraphically older taxa (Fig. 2.8). The braincase of *Birgeria* comprises several ossification centers with two sets of paired otic ossifications contributing to the postorbital process and the lateral wall of the otic capsule (Fig. 2.8a) (Nielsen 1949). The latter bones carry the posterior and anterior semicircular canals but lie dorsal to the horizontal canal. No detailed model of the labyrinth or associated structures is available for *Birgeria*. *Saurichthys* and other saurichthyids have monolithic endocrania lacking not only separate ossification centers but also vestibular fontanelles or a persistent otico-occipital fissure (Fig. 2.8b) (Stensiö 1925). *Perleidus* shows a similar degree of consolidation of the endocranium (Stensiö 1932), although separate ossification centers can be distinguished in small individuals (Patterson 1975). Despite these striking differences in the degree of neurocranial consolidation, *Saurichthys* and *Birgeria* are sometimes considered closely related to one another as adjacent branches on the stem of Chondrostei (paddlefishes and sturgeons; but see Coates 1999). Unambiguous members of this group lack extensive neurocranial ossification, so nothing can be said of otic capsule morphology in fossil members, some of which are otherwise exceptionally preserved (Grande and Hilton 2006).

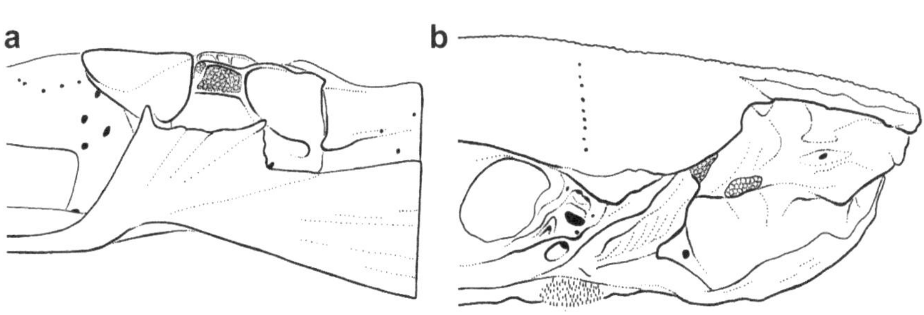

Fig. 2.8 Otic morphology in anatomically specialized Triassic ray-finned fishes often associated with chondrosteans. *Gray* shading defines region of the hyomandibular facet. **(a)** Neurocranium of *Birgeria groenlandica* (Induan), shown in left-lateral view. **(b)** Neurocranium of *Yelangichthys macrocephalus* (Anisian), a close relative of *Saurichthys*, shown in left-lateral view. (**a** based on Nielsen 1949; **b** based on Wu et al. 2013)

2.6 Holosteans

Modern holosteans comprise roughly half a dozen species of Ginglymodi (garpikes; Grande 2010) and a single species of Halecomorphi (bowfin; Grande and Bemis 1998). Together they are the living sister lineage of teleosts and along with them comprise crown group Neopterygii. The limited richness of modern holosteans is complemented by restricted environmental and geographic distribution. However, fossil examples show substantially higher levels of taxonomic diversity and morphological disparity, particularly in the Mesozoic.

2.6.1 Parasemionotids

The earliest unequivocal members of Holostei are also the earliest definitive neopterygians, and they belong to an assemblage known as parasemionotids. The anatomy of the otic region is known in varying detail for this Early Triassic group, with the best information available for *Watsonulus* (Fig. 2.9a) (Olsen 1984) and

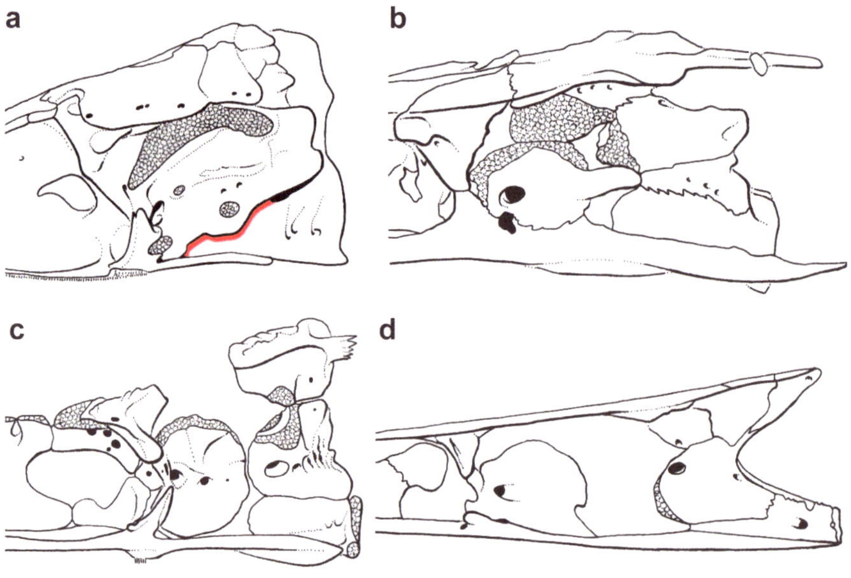

Fig. 2.9 Otic morphology in Mesozoic holosteans. *Red* indicates extent of otico-occipital fissure, which is sutural in *Watsonulus*. **(a)** Neurocranium of *Watsonulus eugnathoides* (Early Triassic: Induan) shown in left-lateral view. **(b)** Neurocranium of *Calamopleurus cylindricus* (Early Cretaceous: Aptian-Albian) shown in left-lateral view. **(c)** Neurocranium of *Macrosemimimus toombsi* (Late Jurassic: Kimmeridgian-Tithonian) shown in left-lateral view. **(d)** Neurocranium of *Dentilepisosteus laevis* (Early Cretaceous: Aptian-Albian) shown in left-lateral view (**a** based on Olsen 1984; **b** based on Maisey 1999; **c** based on Patterson 1975 where the specimen is identified as *Lepidotes*; **d** based on Wenz and Brito 1996 where the specimen is identified as *Obaichthys*)

Ospia (Stensiö 1932). Separate ossification centers of the otic region are visible to a varying extent among parasemionotids with some specimens showing completely co-ossified endocrania and others bearing distinct bones. However, most—if not all—parasemionotids bear vestibular fontanelles and persistent otico-occipital fissures, although the degree of development of both is variable. Rudimentary reconstructions of the skeletal labyrinth are given for *Watsonulus* and *Devillersia* by Beltan (1968), for *Ospia* by Stensiö (1932), and for an unnamed form by Giles et al. (2016), but they show no particularly striking features relative to derived Paleozoic actinopterygians like *Kansasiella* and *Lawrenciella*. The relationships of parasemionotids to other neopterygians have been controversial (Olsen 1984), and their monophyly has not been corroborated. However, there is broad consensus that *Watsonulus* is an early diverging member of the halecomorph stem, and thus a relative of the extant bowfin *Amia* and a crown-group holostean (Grande and Bemis 1998).

2.6.2 Halecomorphs

Ossified endocrania are known for a great variety of halecomorphs, and the most detailed examples are known from the Jurassic and Cretaceous. These include several examples referred to *Caturus* (reviewed in Patterson 1975), *Osteorachis* (Gardiner 1960), and *Ionoscopus* (Maisey 1999) from the Jurassic, and *Macrepistius* (Schaeffer 1971) and *Calamopleurus* (Fig. 2.9b) (Maisey 1991, 1999) from the Cretaceous. These taxa all lack a persistent otico-occipital fissure and vestibular fontanelles, and separate ossification centers of the otic region are clearly delimited, with a few exceptions (Gardiner 1960; Rayner 1948). In *Calamopleurus*, the ossifications of the otic region are separated by broad, cartilage-filled gaps, anticipating the arrangement in the modern *Amia* (Grande and Bemis 1998). Anatomy of the skeletal labyrinth and associated structures has been reconstructed for *Caturus* (Aldinger 1932; Rayner 1948) and to a lesser degree for *Ionoscopus* and *Calamopleurus* (Maisey 1999). The braincase is more completely mineralized in *Caturus*, with the semicircular canals lying completely within bone. By contrast, portions of some or all of the canals were cartilage-enclosed in the two Cretaceous forms. The structure of the saccular chamber is unknown in *Caturus* (its extent is inferred by Maisey 1999), but in *Ionoscopus* and *Calamopleurus* this feature is relatively shallow relative to the height of the skeletal labyrinth. These latter two taxa also show a clearly delimited utricular recess. The lateral cranial canal is complete rather than blind ended in both *Caturus* and *Ionoscopus*, joining the endocavity anterior and posterior to the sinus superior. The canal is absent in *Calamopleurus*, as in *Amia*.

2.6.3 Ginglymodians

The other major division of Holostei is Ginglymodi, the only living representatives of which are two genera of garpike. Fossil members of this clade include two other major assemblages in addition to gars: macrosemiids, which are a clade, and "semionotids" *sensu lato*, which are a grade. Both appear limited to the Mesozoic, although the Permian *Acentrophorus* is often interpreted as a "semionotid." Features of the otic region are known for only a handful of forms in each group. The inner surface of the neurocranium in a flattened specimen of *Macrosemius* (Bartram 1977) provides the most detailed information for macrosemiids. These are slender-bodied fishes with a long-based, high dorsal fin; to draw a modern parallel, macrosemiids can be crudely described as goby-like in their overall body proportions. The otic capsule comprises a series of separate ossifications that are tightly sutured with one another. A vestibular fontanelle is absent. Although the lateral surface of the saccular chamber is preserved, openings to the skeletal labyrinth are not apparent (Bartram 1977). The "semionotids" are broadly carp-like in overall form, although obviously not closely related to that group of modern teleosts. Among "semionotids," the otic region is best shown in material historically deposited in *Lepidotes*, a problematic wastebin taxon (Lopez-Arbarello 2012). Some relevant specimens seem correctly assigned to that genus (e.g., *L. semiserratus* in Rayner 1948), but others have been distributed to the newer taxa *Macrosemimimus* (*L. toombsi* in Patterson 1975) and *Scheenstia* (*L. mantelli* in Gardiner 1960). Otic structure is most exhaustively described in *Macrosemimimus* (Patterson 1975), but this seems largely representative of the other forms (Fig. 2.9c). There are separate ossification centers, many of which are broadly separated by cartilage. Vestibular fontanelles and an otico-occipital fissure are absent. Reduced ossification prohibits any detailed reconstruction of the inner ear. Other "semionotids" preserving portions of the otic region include *Semionotus* (Olsen and McCune 1991) and *Araripelepidotes* (Wenz and Brito 1996), although morphology is less clear in these examples due to crushing or other distortion.

External anatomy of the otic region is known for fossil representatives of both living gar genera, as well as several extinct genera (Fig. 2.9d) (Grande 2010). As in "semionotids," there are separate ossification centers separated by substantial cartilage-filled gaps, but vestibular fontanelles and otico-occipital fissures are absent.

2.7 Teleosts

Numbering almost half of all living vertebrate species, the taxonomic richness and morphological diversity of living teleosts is staggering. The modern variety of teleosts is complemented by a rich fossil record. An exhaustive treatment of otic morphology in fossil teleosts is beyond the scope of this contribution, which instead focuses on a few key examples: early members of the extended teleost branch of the tree of life, paleontological evidence of special innovations related to hearing in some teleost groups, and a broad overview of the otolith-based fossil record.

2.7.1 Stem Teleosts

A number of fossil fish groups have been aligned with teleosts but clearly fall outside of the modern radiation. Several Mesozoic lineages are universally accepted to branch from the teleost stem, but the placement of several others is strongly debated: Pycnodontiformes, Pachycormiformes, Aspidorhynchiformes, and Dapediidae (Fig. 2.10). These controversial groups will be reviewed first, with the acknowledgement that future study might export them to other branches of neopterygian phylogeny or outside the modern radiation entirely (e.g., Poyato-Ariza 2015).

The Jurassic dapediid *Dapedium* provides the most detailed information on the otic region in this group (Fig. 2.10a). The braincase is fully mineralized with

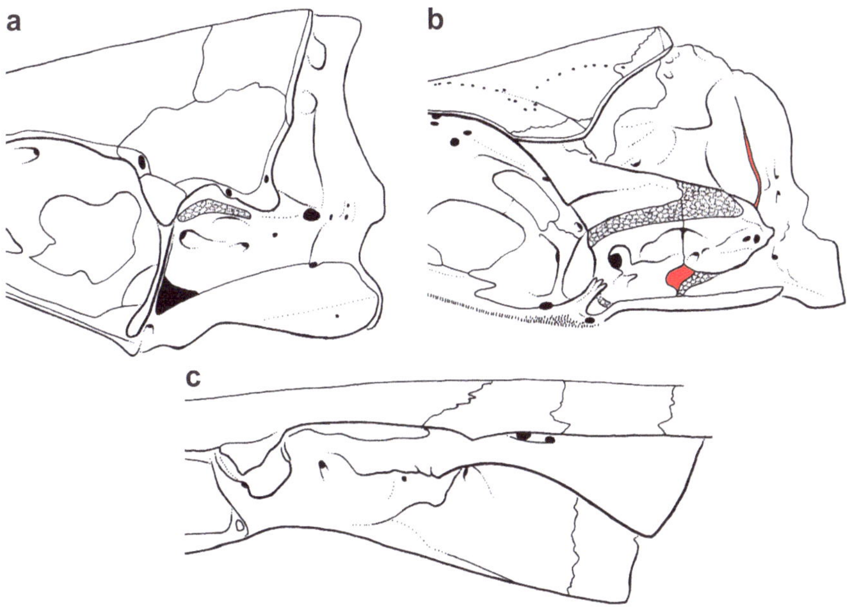

Fig. 2.10 Otic morphology in Mesozoic neopterygian groups associated with the teleost stem. *Red* indicates extent of otico-occipital fissure and vestibular fontanelles and *gray* shading defines the region of the hyomandibular facet. **(a)** Neurocranium of *Pachycormus* sp. (Early Jurassic: Toarcian), shown in left-lateral view. **(b)** Neurocranium of *Dapedium* sp. (Early Jurassic: Sinemurian), shown in left-lateral view. **(c)** Neurocranium of *Vinctifer comptoni* (Early Cretaceous: Aptian-Albian) shown in left-lateral view. (**a**, **b** based on Patterson 1975; **c** based on Maisey 1991)

separate ossification centers lacking (Patterson 1975). Both the otico-occipital fissure and vestibular fontanelles are absent. The morphology of the inner ear has recently been described based on μCT (Giles et al. 2016), but was previously known only from apertures to the skeletal labyrinth and saccular chamber visible on the inner wall of an incomplete braincase (Rayner 1948).

Only external morphology of the otic capsules is known for pycnodontiforms and aspidorhynchiforms. *Mesturus* preserves the most informative neurocranial material among pycnodontiforms (Nursall 1999). Sutures are apparent between different ossification centers but are incomplete, and the full margins between bones cannot be traced. A short otico-occipital fissure is reported, and there is a large, apparently unossified region of the ventrolateral wall of the otic capsule above the dorsal margin of the parasphenoid. Positionally, this corresponds to the vestibular fontanelle but has not been interpreted as such. The only other pycnodontiform in which the otic region is known in any detail is the Cretaceous *Neoproscinetes*, where ossification is greatly reduced relative to *Mesturus* (Nursall and Maisey 1991).

Aspidorhynchiforms are superficially needlefish-like or gar-like fishes restricted to the Jurassic and Cretaceous. The endocranium is best represented in acid-prepared specimens of the Early Cretaceous *Vinctifer* but only from external details (Fig. 2.10c) (Maisey 1991). The braincase is also known from acid-prepared material of the closely related *Richmondichthys*, and it agrees with that of *Vinctifer* in structure (Bartholomai 2004). Separate ossification centers are present and are tightly sutured with one another. There is no otico-occipital fissure or vestibular fontanelle. The braincase attributed to "*Aspidorhynchus*" in the literature (e.g., Rayner 1948) is not, in fact, an aspidorynchiform but rather a halecomorph holostean (Patterson 1975).

Pachycormiforms are a Jurassic–Mesozoic group that includes a series of tuna-like and swordfish-like forms, as well as large-bodied suspension feeders ecologically analogous to baleen whales. Otic capsules are known in several genera, including *Bonnerichthys* (Friedman et al. 2010), but mechanically and acid-prepared material of the early representative *Pachycormus* reveals the most detailed information (Patterson 1975; Fig. 2.10a). Separate ossification centers are present, and occur in conjunction with small, comma-shaped vestibular fontanelles and persistent otico-occipital fissures that become increasingly closed in larger specimens. Some features of the skeletal labyrinth are noted by Patterson (1975) and Mainwaring (1978), but the overall geometry of the inner ear is unknown. In contrast to three-dimensionally preserved material available for *Pachycormus*, known endocrania of other pachycormiforms are heavily compressed and are not described in detail.

A series of lineages of unambiguous teleostean affinity join these putative members of the teleost total group (Fig. 2.11). These include grades like "pholidophoriforms" and "leptolepids" *sensu lato*, as well as well-characterized monophyletic groups like Ichthyodectiformes. The Early Jurassic *Dorsetichthys bechei* (called *Pholidophorus bechei* in older accounts) represents the archetypal "pholidophorid," and its otic region is known in considerable detail on the basis of both

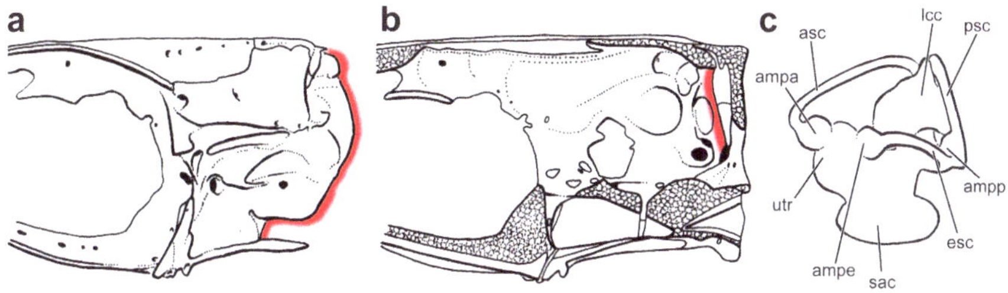

Fig. 2.11 Otic morphology of early teleosts as exemplified by the Early Jurassic (Sinemurian) *Dorsetichthys bechei*. **(a)** Neurocranum shown in left-lateral view, occipital arch removed. **(b)** Neurocranium in sagittal section, with anterior to left. **(c)** Endocast of inner ear and lateral cranial canal shown in left-lateral view. (Figures based on Patterson 1975)

mechanically (Rayner 1948) and acid-prepared (Patterson 1975) material, which has recently been examined via μCT scanning (Giles et al. 2016). A vestibular fontanelle and otico-occipital fissure are present. The latter is lined with perichondral bone dorsal to the level of the vestibular fontanelle. Separate ossification centers are present in small individuals, but individual bones are fused in larger specimens. The semicircular canals and their ampullae are largely encased within bone with the exception of the sinus superior, which is unenclosed medially. The posterior canal is almost transverse in its orientation, and both it and the horizontal canal are substantially shorter than the anterior canal. The posterior end of the horizontal canal joins the ampulla of the posterior canal, and the ampullae of the anterior and horizontal semicircular canals are broadly separated (Fig. 2.11c). As in holosteans, a lateral cranial canal extends through the loops of the anterior and posterior canals, flanking the sinus superior laterally. The utricular recess is particularly well developed and is separated from the endocranial cavity by a constriction. Similarly, the bean-shaped saccular chamber is set off from the remainder of the bony inner ear by a constriction. A Middle Jurassic "*Pholidophorus*" described by Patterson (1975) is broadly similar to *Dorsetichthys*, but the labyrinth cavity is less fully enclosed by bone. The anterior canal lay within the cranial cavity, the posterior and horizontal canals were partially enclosed by cartilage, and the utricular recess is less clearly demarcated. Giles et al. (2016) present a virtual endocast of this specimen.

"Leptolepids" like *Leptolepis coryphaenoides* and "*Leptolepis*" *dubia* lack both otico-occipital fissures and vestibular fontanelles (Patterson 1975). Sutures between separate bones may be present, or the otic region can be represented by a single, continuous ossification. Internally, there appears to be a reduced correspondence between endocavities and the soft-tissue features they would have contained. This stands in contrast to less crownward members of the teleost stem, early holosteans, and Paleozoic ray-fins, but anticipates the morphology of crown-group teleosts. The anterior canal and sinus superior lay completely within the endocranial cavity. Additionally, the bar separating the anterior and posterior opening of the lateral cranial canal is lost, resulting in a simple diverticulum, and the utricular recess is not differentiated from the remainder of the cranial endocavity. However, the posterior and horizontal canals are completely enclosed by bone, matching the condition in living teleosts and attributed by Patterson (1975) to changes in ossification pattern associated with loss of the otico-occipital fissure. Little has been described of the endocranial chamber in taxa crownward of “leptolepids.” Dunkle (1940) provided a diagrammatic reconstruction of the semicircular canals in the pachyrhizodontid *Notelops*, based on serial sectioning of a three-dimensionally preserved specimen.

2.7.2 Crown Teleosts I: Innovations in Hearing

The fossil record of crown teleosts is exceptionally rich in articulated body fossils. In spite of this—or perhaps because of it—there are few accounts presenting detailed descriptions of either internal or external anatomy of the otic region of the braincase, especially when considering the amount of available material. In the light of this descriptive bias and the overwhelming variety of fossil forms, focus will be placed here on paleontological evidence with a bearing on remarkable innovations for hearing found within teleosts. There are a variety of anatomical strategies for enhanced auditory capability in fishes (Braun and Grande 2008) but most are tied to direct or indirect linkages between the swim bladder and otic region (Popper and Coombs 1982). At least some of these modifications are apparent in hard tissues and thus have the potential to be detected in fossil material. The discussion here will review the most concrete examples of these specializations in fossils, noting other groups bearing similar features but for which clear evidence in fossil members is not available.

Modifications for hearing in teleosts are most readily apparent in a clade variously termed Ostarioclupeomorpha or Otocephala. This is a remarkably diverse teleost group, including species as morphologically and ecologically divergent as minnows, catfishes, electric eels, and anchovies. Alepocephaliforms (slickheads and tubeshoulders), an unusual radiation of deep-sea species, represent an unexpected addition to Otocephala arising from molecular studies (Ishiguro et al. 2003). Specifically, many clupeomorphs and ostariophysans show elaborate linkages between the swim bladder and ear, either through direct connections or an elaborate relay of highly modified bones. Otocephalans share a series of modifications that might pre-adapt them to such otophysic connections, most notably, posterior extension of the labyrinth and placement of the saccular and lagenar otoliths in a posteriomedian position (Braun and Grande 2008). Like many deep-sea species, alepocephaliforms lack a swim bladder, but existing accounts of their anatomy do not indicate whether they share other modifications of the otic region common to clupeomorphs and ostariophysans (e.g., Gosline 1969).

Clupeomorpha, or herrings, is the sister lineage of all remaining otocephalans. Extant clupeomorphs show a variety of remarkable specializations of the otic region generally, and acousticolateralis system specifically, that provide key evidence for the monophyly of the group as a whole or its constituent clades. These include the

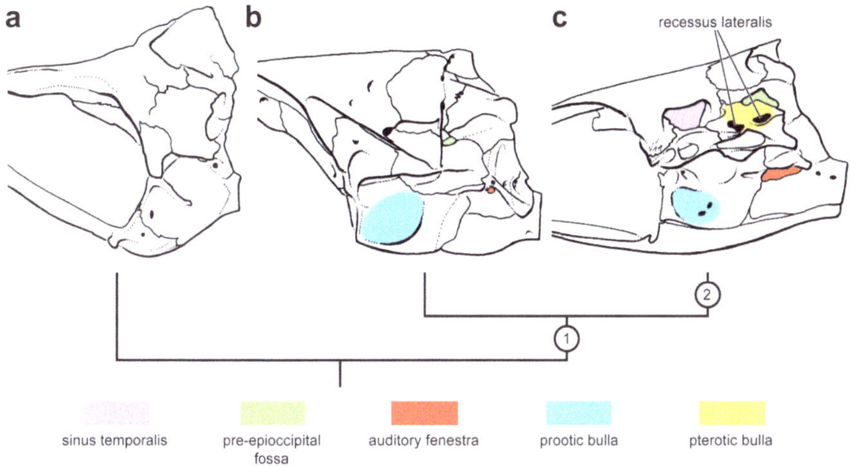

Fig. 2.12 Modifications of cluepomorph skull anatomy associated with elaborations of the acousticolateralis system. *Gray* shading indicates position of hyomandibular facet(s). **(a)** Neurocranium of *Diplomystus* sp. (Late Cretaceous: Cenomanian) shown in left-lateral view. **(b)** Neurocranium of *Spratticeps gaultinus* (Early Cretaceous: Albian) shown in left-lateral view. **(c)** Neurocranium of *Santanaclupea silvasantosi* (Early Cretaceous: Aptian-Albian) shown in left-lateral view. Transformations at *node 1*: development of rudimentary sinus temporalis (remains covered by skull roof), appearance of prootic bulla and auditory fenestra that indicate invasion of otic region by diverticulum of swim bladder, origin of pre-epioccipital fossa. Transformations at *node 2*: exposure of sinus temporalis and origin of pterotic bulla. (**a** based on Forey 2004; **b** based on Patterson 1970; **c** based on Maisey 1993)

sinus temporalis, a sac-like enlargement of the supraorbital sensory canal housed within a fenestration in the temporal region of the skull; the recessus lateralis, a bone-enclosed chamber within the temporal region communicating with the multiple branches of the cranial lateral-line system; the pre-epioccipital fossa (called a pre-epiotic fossa in older literature), a canal of uncertain function that opens through a foramen between the parietal, pterotic and epioccipital; prootic and pterotic bullae, swellings of those bones marking diverticulae of the swim bladder that invade the braincase via the auditory foramen (for more details, see de Dario 2004). Clupeomorphs also have a rich and well-studied fossil record (e.g., Grande 1985) that helps to document the origin of many of these innovations (Fig. 2.12).

Ellimmichthyiformes, represented by the single family Paraclupeidae, are a Cretaceous–Paleogene clade of primitive clupeomorphs known from a range of marine and freshwater settings. Braincase anatomy is most clearly illustrated by a three-dimensionally preserved, acid-prepared example of *Diplomystus* from early Late Cretaceous marine deposits of the UK (Fig. 2.12a) (Forey 2004). This genus shows none of the auditory specializations associated with modern clupeomorphs (Clupeiformes). Any kind of connections between the ear and swim bladder, if present, must have been simple and external.

A series of well-preserved late Early Cretaceous fossils provide the important evidence for the stepwise acquisition of sensory modifications characteristic of modern herrings: *Spratticeps* from the UK (Patterson 1970), and the Brazilian *Pseudoellimma* (de Figueiredo 2009) and *Santanaclupea* (Maisey 1993). *Spratticeps* is known only from isolated braincases, but these show well-developed prootic bullae that indicate the development of an internal swim-bladder inner-ear connection, a rudimentary pre-epioccipital fossa, and an incipient sinus temporalis that remains covered by the roof of the posttemporal fossa (Fig. 2.12b). *Pseudoellimma* seems more closely related to modern clupeiforms than *Spratticeps* and bears evidence of a recessus lateralis. With the additional presence of a pterotic bulla and fully exposed temporal fenestra, *Santanaclupea* shows an essentially complete suite of modern clupeiform auditory specializations (Fig. 2.12c).

Ostariophysans include two divisions: Anotophysi, which includes the milkfish, beaked salmons, and shellears; and Otophysi, which includes minnows and allies, catfishes, characins and allies, and Neotropical knifefishes. Anotophysans show many of the features interpreted as potential preadaptions foreshadowing the more elaborate auditory systems of otophysans (Rosen and Greenwood 1970), but many of these pertain to large scale (e.g., swim bladder structure; Grande and de Pinna 2004) or small scale (e.g., cellular orientation in the lagenar epithelium; Popper and Platt 1983) aspects of soft tissue anatomy that cannot be detected in fossils.

Otophysans famously bear the so-called Weberian apparatus: a series of specialized bones, derived from various components of the anterior axial column, some of which transmit vibrations from the swim bladder to the skull (Weber 1820). These ossicles include, from anterior to posterior (with possible precursors, as identified by comparative and developmental studies, indicated in parentheses): the claustrum, a small ossicle that bears no connection to other Weberian ossicles and thus is not part of the vibration-transmitting chain (homologized with the accessory neural arch of other clupeocephalans); the scaphium, which lies immediately dorsal to the first vertebral centrum (arises developmentally from the first basidorsal in many taxa, but apparently a composite structure in others); the intercalarium, which lies above the second vertebral centrum (derived from the second neural arch) and bears an anterolateral process called the manubrium; and the tripus, a variously triradiate bone that extends ventrolaterally from the third vertebral centrum (formed by the pleural rib and parapophysis of the third vertebra). Joining these is the os suspensiorium, the modified pleural rib of the fourth vertebra. This structure is considered a component of the Weberian apparatus by some and shows varying degrees of modification across modern otophysans (Grande and de Pinna 2004). The presence of Weberian ossicles is associated with further modifications of the anterior vertebral column, including the foreshortening of centra that is readily apparent in fossil material.

As a complicated morphological system, the Weberian apparatus has attracted considerable attention from comparative anatomists attempting to document its evolutionary assembly (e.g., Chardon and Vanderwalle 1997). Paleontological evidence bearing on the origin of this complex morphological system has been the subject of some controversy. Because of their position at the anterior of the

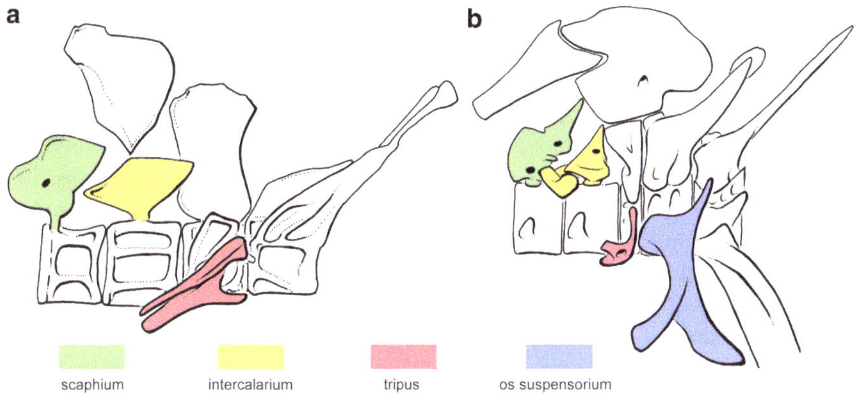

Fig. 2.13 Examples of the Weberian apparatus in stem otophysans help to document the stepwise assembly of this complex auditory system. **(a)** Weberian apparatus of *Santanichthys diasii* (Late Cretaceous: Aptian-Albian) shown in left-lateral view. **(b)** Weberian apparatus of *Chanoides macropoma* (Eocene: Ypresian) shown in left-lateral view. (**a** based on Filleul and Maisey 2004; **b** based on Patterson 1984)

vertebral column, the Weberian ossicles are often overlapped or otherwise disrupted by bones of the skull, opercular series, and shoulder girdle in fossil material. Specimens that are clearly aligned with individual otophysan orders show uncontroversial Weberian ossicles or other associated vertebral modifications closely comparable with those of modern species. Clear examples are present in Characiformes (e.g., Weiss et al. 2014), Siluriformes (e.g., Grande and Lundberg 1988), and Cypriniformes (e.g., Grande et al. 1982). Fossil Gymnotiformes are rare and existing material is too poorly preserved to show the ossicles (e.g., Albert and Fink 2007).

A variety of fossil fishes have been nominated as particularly relevant to understanding the origin of the Weberian apparatus: the Early Cretaceous *Santanichthys* (Fig. 2.13a) (Filleul and Maisey 2004), the Late Cretaceous *Lusitanichthys* (Gayet 1981) and *Clupavus* (Taverne 1995), and the Eocene *Chanoides* (Fig. 2.13b) (Patterson 1984). Cretaceous material previously assigned to *Chanoides* has now been placed in *Nardonoides*, an *incertae sedis* ostariophysan (Mayrinck et al. 2015). Arguments for a Weberian apparatus in the gonorynchiform *Ramallichthys* (Gayet 1986) and the attendant complexities implied for the evolution of this elaborate anatomical system have been comprehensively rejected in detailed rebuttals by Fink et al. (1984) and Grande and Grande (2008). Some of these remaining fossils were initially associated with specific extant otophysan lineages (e.g., *Santanichthys* with characiforms) (Filleul and Maisey 2004), thus suggesting that some features characteristic of the modern Weberian apparatus evolved multiple times. However, more extensive analyses suggest that all fall outside of crown Otophysi and thus do not reject a monophyletic origin of the Weberian apparatus (Grande and de Pinna 2004).

Features of the Weberian apparatus are preserved to varying degrees in these stem otophysans (reviewed in detail by Diogo 2009). Material of *Lusitanichthys* and *Clupavus* is flattened, but both *Santanichthys* and *Chanoides* are known from acid-prepared material that permits detailed description and attempted reconstruction of the Weberian apparatus. In particular, the individual ossicles preserved in these taxa appear to be more primitive than those of crown otophysans. The claustrum is unknown in either. The scaphium and intercalarium of *Chanoides* bear foramina for ventral and dorsal nerve roots, a primitive feature of neural arches not commonly found in these Weberian ossicles among extant otophysans. In addition, the intercalaria of both *Chanoides* and *Santanichthys* lack co-ossified manubria (although a small sesamoid ossification representing this feature is present in *Chanoides*), again, showing a primitive neural-arch-like condition relative to the derived intercalarium morphology of crown otophysans. The claustrum is not preserved in either genus but both genera show unusual morphologies of the tripus. The particularly primitive tripus of *Santanichthys* comprises a separate rib and parapophysis. By contrast, the tripus of *Chanoides* appears to be a single bone, but it is substantially smaller than the equivalent ossification in either *Santanichthys* or modern otophysans. The os suspensorium is unknown in *Santanichthys*; however, in *Chanoides* the os suspensorium is a compound structure formed from the rib and parapophysis of the fourth vertebra. As in modern species, the anatomy of the os suspensorium appears variable among primitive otophysans.

Outside of Otocephala, modifications to the skeleton potentially associated with hearing enhancement are found in a variety of teleost groups and include extracranial bullae in megalopids (tarpons), otic bullae in mormyrids (elephantfishes), holocentrids (squirrelfishes), and cichlids, and suprabranchial chambers in anabantoids (labyrinth fishes) (Braun and Grande 2008). Laterally directed openings associated with bulla-like otic swellings are found in a variety of fossil non-percomorph acanthomorphs, including the polymixiid *Homonotichthys* and the "beryciforms" *Hoplopteryx*, *Caproberyx* (Patterson 1964), and *Holocentrites* (Dunkle and Olsen 1959). However, a similar arrangement is found in modern species with no connection between the otic capsule and swim bladder. Such a connection is present in some living holocentroids, but is marked by very distinctive modifications to the bony structure of the otic capsule not apparent in these fossils (Nelson 1955).

2.7.3 Crown Teleosts II: The Otolith Fossil Record

No morphological account of auditory capsules in fossil actinopterygians would be complete without mention of the rich archive of isolated otoliths. The density of the Cenozoic teleost otolith fossil record has drawn comparisons to the tooth record of mammals (Patterson 1977), but there is an important distinction: teleost otoliths are rarely associated with additional skeletal remains (there are, however, noteworthy exceptions: Nolf 1985). Much of this appears to reflect taphonomic bias against otoliths in depositional settings conducive to the preservation of fish skeletons (Nolf 1985, 2013). As a consequence, fossil otoliths are typically identified through comparison with reference collections of otoliths from modern species. This approach has clear limitations, with taxonomic identifications becoming more difficult in progressively older sediments to the degree that reliable interpretations of Mesozoic otoliths are very difficult. The literature surrounding fossil otoliths is vast, but owing to the isolated nature of these remains, largely separate from that from the body fossil record. However, computed tomography might have the potential to reveal otoliths in a broader range of fish body fossils than are known at present, representing a means of beginning to unify these parallel records of actinopterygian biodiversity (cf. Friedman et al. 2015).

2.8 Summary

1. Excepting specializations of the two groups, internal anatomy of the skull—and the apparatus of hearing—in early ray-finned fishes is broadly comparable between early fossil lobe-finned fishes and ray-finned fishes. Details of otic anatomy are only hazily known in the most ancient osteichthyans. There is substantially more information available for younger fossils that can be clearly aligned with one of the two branches of the bony fish tree.
2. Fossil evidence bearing on the evolution of ear structure in ray-finned fishes stems largely from endocasts of the inner ear and isolated saccular otoliths. These archives are largely non-overlapping both stratigraphically and phylogenetically.
3. Known from exceptionally well-preserved and acid-prepared material, the Late Devonian *Mimipiscis* provides the most detailed model for the primitive geometry of the actinopterygian inner ear.
4. External and internal anatomy of the otic region is known in a variety of Carboniferous and Permian actinopterygians classically identified as "palaeoniscoids" (a taxonomic wastebin of anatomically generalized ray-finned fishes), with the most detailed information available for the Early Triassic based on studies using destructive physical tomography. These taxa generally retain morphologies broadly similar to those of the earliest ray-finned fishes, including the presence of an otico-occipital fissure and vestibular fontanelles and the absence of separate ossification centers.
5. Anatomy of the otic region and inner ear in extinct relatives of chondrosteans (sturgeons and paddlefishes) is limited to highly specialized early Mesozoic forms whose affinities to modern lineages are not well established.
6. Fossil holosteans—members of the group containing garpikes and the bowfin—show a variety of otic morphologies. The most ancient examples lack separate ossification centers but bear a persistent otico-occipital fissure and vestibular fontanelles. Later examples in both the garpike and bowfin lineages show separate ossification centers. In both cases, reduction of the mineralized walls of

the endocranial cavity severely reduces the detail with which the inner ear can be reconstructed in fossils.

7. A series of highly specialized Mesozoic groups have been aligned with teleosts. Details of otic anatomy in these lineages are limited almost exclusively to external features, which vary considerably among the groups.
8. Otic structure in early undisputed members of the teleost clade is broadly similar to that of early holosteans in retention of an otico-occipital fissure and vestibular fontanelles, absence of separate ossification centers, and tight enclosure of the inner ear by bone. These features are lost in more crownward members of the teleost stem lineage.
9. Cretaceous fossils deliver clear evidence for the stepwise origin of hearing specializations in clupeomorphs and ostariophysans.

Compliance with Ethics Requirements Matt Friedman and Sam Giles have declared that they had no conflicts of interest.

References

Albert, J. S., & Fink, W. L. (2007). Phylogenetic relationships of fossil Neotropical electric fishes (Osteichthyes: Gymnotiformes) from the Upper Miocene of Bolivia. *Journal of Vertebrate Paleontology*, 27, 17–25.

Aldinger, H. (1932). Über einen Eugnathiden aus der unteren Wolgastufe von Ostgrönland. *Meddelelser om Grønland*, 86, 1–51.

Bartholomai, A. (2004). The large aspidorhynchid fish, *Richmondichthys sweeti* (Etheridge Jnr and Smith Woodward, 1891) from Albian marine deposits of Queensland, Australia. *Memoirs of the Queensland Museum*, 49, 521–536.

Bartram, A. W. H. (1977). The Macrosemiidae, a Mesozoic family of holostean fishes. *Bulletin of the British Museum (Natural History): Geology*, 29, 137–234.

Basden, A. M., & Young, G. C. (2001). A primitive actinopterygian neurocranium from the Early Devonian of southeastern Australia. *Journal of Vertebrate Paleontology*, 21, 754–766.

Beltan, L. (1968). *La faune ichthyologique de l'Eotrias du N.W. de Madagascar: le neurocrâne*. Paris: Éditions du Centre National de Recherches Scientifique.

Bjerring, H. C. (1984). The term 'fossa bridgei' and five endocranial fossae in teleostome fishes. *Zoologica Scripta*, 13, 231–238.

Bradley Dyne, M. (1939). The skull of *Amphicentrum granulosum*. *Proceedings of the Zoological Society, Series B*, 1939, 195–210.

Braun, C. B., & Grande, T. (2008). Evolution of peripheral mechanisms for the enhancement of sound reception. In J. F. Webb, A. N. Popper, & R. R. Fay (Eds.), *Fish bioacoustics* (pp. 99–144). Berlin: Springer Science+Business Media.

Brazeau, M. D., & Friedman, M. (2014). The characters of Palaeozoic jawed vertebrates. *Zoological Journal of the Linnean Society*, 170, 779–821.

Chardon, M., & Vanderwalle, P. (1997). Evolutionary trends and possible origin of the Weberian apparatus. *Netherlands Journal of Zoology*, 47, 383–403.

Coates, M. I. (1993). New actinopterygian fish from the Namurian Manse Burn formation of Bearsden, Scotland. *Palaeontology*, 36, 123–146.

Coates, M. I. (1998). Actinopterygians from the Namurian of Bearsden, Scotland, with comments on early actinopterygian neurocrania. *Zoological Journal of the Linnean Society*, 122, 27–59.

Coates, M. I. (1999). Endocranial preservation of a Carboniferous actinopterygian from Lancashire, UK, and the interrelationships of primitive actinopterygians. *Philosophical Transactions of the Royal Society of London B: Biological Sciences*, 354, 435–462.
Dario, F. de. (2004). Homology between the *recessus lateralis* & cephalic sensory canals, with the proposition of additional synapomorphies for the Clupeiformes and Clupeoidei. *Zoological Journal of the Linnean Society*, 141, 257–270.
Davis, S. P., Finarelli, J. A., & Coates, M. I. (2012). *Acanthodes* and shark-like conditions in the last common ancestor of modern gnathostomes. *Nature*, 486, 247–250.
Diogo, R. (2009). Origin, evolution and homologies of the Weberian apparatus: A new insight. *International Journal of Morphology*, 27, 333–354.
Dunkle, D. H. (1940). The cranial osteology of *Notelops brama* (Agassiz), an elopid fish from the Cretaceous of Brazil. *Lloydia*, 3, 157–190.
Dunkle, D. H. (1946). A new palaeoniscoid fish from the Lower Permian of Texas. *Journal of the Washington Academy of Sciences*, 36, 402–409.
Dunkle, D. H., & Olsen, S. J. (1959). Description of a beryciform fish from the Oligocene of Florida. *Florida Geological Survey Special Publication*, 2, 1–20.
Figueiredo, F. J. de. (2009). A new clupeiform fish from the Lower Cretaceous (Barremian) of Sergipe-Alagoas Basin, northeastern Brazil. *Journal of Vertebrate Paleontology*, 29, 993–1005.
Filleul, A., & Maisey, J. G. (2004). Redescription of *Santanichthys diasii* (Otophysi, Characiformes) from the Albian of the Santana Formation and comments on its implications for otophysan relationships. *American Museum Novitates*, 3455, 1–21.
Fink, S. V., Greenwood, P. H., & Fink, W. L. (1984). A critique on recent work on fossil ostariophysan fishes. *Copeia*, 1984, 1033–1041.
Forey, P. L. (2004). A three-dimensional skull of a primitive clupeomorph from the Cenomanian English Chalk and implications for the evolution of the clupeomorph acusticolateralis system. In G. Arratia & A. Tintori (Eds.), *Mesozoic fishes, vol. 3. Systematics, paleoenvironments, and biodiversity* (pp. 405–417). Munich: Verlag Dr. Friedrich Pfeil.
Friedman, M. (2015). The early evolution of ray-finned fishes. *Palaeontology*, 58, 213–228.
Friedman, M., & Brazeau, M. D. (2010). A reappraisal of the origin & basal radiation of the Osteichthyes. *Journal of Vertebrate Paleontology*, 30, 36–56.
Friedman, M., & Sallan, L. C. (2012). Five hundred million years of extinction and recovery: a Phanerozoic survey of large-scale diversity patterns in fishes. *Palaeontology*, 55, 707–742.
Friedman, M., Shimada, K., Martin, L. D., Everhart, M. J., et al. (2010). 100-million-year dynasty of giant planktivorous bony fishes in the Mesozoic seas. *Science*, 327, 990–993.
Friedman, M., Beckett, H. T., Close, R. C., & Johanson, Z. (2015). The English chalk and London clay: Two remarkable British bony fish *Lagerstätten*. In Z. Johanson, P. M. Barrett, M. Richter, & M. Smith (Eds.), *Arthur Smith Woodward: His life and influence on modern vertebrate palaeontology. Geological Society of London, Special Publications*, 430, 165–200.
Gardiner, B. G. (1960). A revision of certain actinopterygian and coelacanth fishes, chiefly from the Lower Lias. *Bulletin of the British Museum (Natural History): Geology*, 4, 239–384.
Gardiner, B. G. (1984). The relationships of the palaeoniscid fishes, a review based on new specimens of *Mimia* and *Moythomasia* from the Upper Devonian of Western Australia. *Bulletin of the British Museum (Natural History): Geology*, 37, 173–428.
Gayet, M. (1981). Contribution à l'etude anatomique et systématique de l'ichthyofaune Cénomanien du Portugal, deuxième partie: les ostariophysaires. *Comunicacoes dos Servicos Geologicos de Portugal*, 67, 173–190.
Gayet, M. (1986). *Ramallichthys* GAYET du Cénomanien inférieur marin de Ramallah (Monts de Judée) une introduction aux relations phylogénetiques des Ostariophysi. *Mémoires du Museum naturelle d'Histoire naturelle, Paris, Série C*, 51, 1–81.
Giles, S., & Friedman, M. (2014). Virtual reconstruction of endocast anatomy in early ray-finned fishes (Osteichthyes, Actinopterygii). *Journal of Paleontology*, 88, 636–651.

Giles, S., Coates, M. I., Garwood, R. J., Brazeau, M. D., et al. (2015a). Endoskeletal structure in *Cheirolepis* (Osteichthyes, Actinopterygii), an early ray-finned fish. *Palaeontology*, 58, 849–870.

Giles, S., Darras, L., Clément, G., Blieck, A., & Friedman, M. (2015b). An exceptionally preserved Late Devonian actinopterygian provides a new model for primitive cranial anatomy in ray-finned fishes. *Proceedings of the Royal Society of London B: Biological Sciences*, 282, 20151485.

Giles, S., Friedman, M., & Brazeau, M. D. (2015c). Osteichthyan-like cranial conditions in an Early Devonian stem gnathostome. *Nature*, 520, 82–85.

Giles, S., Rogers, M., & Friedman, M. (2016). Bony labyrinth morphology in early neopterygian fishes (Actinopterygii: Neopterygii). *Journal of Morphology*. doi:10.1002/jmor.20551.

Gosline, W. A. (1969). The morphology and systematic position of the alepocephaloid fishes. *Bulletin of the British Museum (Natural History): Zoology*, 18, 183–218.

Gottfried, M. D. (1994). A subadult palaeoniscoid (Actinopterygii, Osteichthyes) from the upper Pennsylvanian of Kansas, U.S.A. *Pollichia*, 29, 103–112.

Grande, L. (1985). Recent and fossil clupeomorph fishes with materials for revision of the subgroups of clupeoids. *Bulletin of the American Museum of Natural History*, 181, 231–372.

Grande, L. (2010). An empirical synthetic pattern study of gars (Lepisosteiformes) and closely related species, based mostly on skeletal anatomy. The resurrection of Holostei. *American Society of Ichthyologists and Herpetologists Special Publication*, 6, 1–871.

Grande, L., & Bemis, W. E. (1998). A comprehensive phylogenetic study of amiid fishes (Amiidae) based on comparative skeletal anatomy. An empirical search for interconnected patterns of natural history. *Society of Vertebrate Paleontology Memoir*, 4, 1–690.

Grande, L., & Hilton, E. J. (2006). An exquisitely preserved skeleton representing a primitive sturgeon from the Upper Cretaceous Judith River Formation of Montana (Acipenseriformes: Acipenseridae: n. gen., & sp.). *Journal of Paleontology, Supplement*, 65, 1–39.

Grande, L., & Lundberg, J. G. (1988). Revision and description of the genus *Astephus* (Siluriformes: Ictaluridae) with a discussion of its phylogenetic relationships. *Journal of Vertebrate Paleontology*, 8, 139–171.

Grande, L., Eastman, J. T., & Cavender, T. M. (1982). *Amyzon gosiutensis*, a new catostomid fish from the Green River Formation. *Copeia*, 1982, 523–532.

Grande, T., & Pinna, M. de (2004). The evolution of the Weberian apparatus: A phylogenetic perspective. In G. Arratia & A. Tintori (Eds.), *Mesozoic fishes, vol. 3. Systematics, paleoenvironments & biodiversity* (pp. 429–448). Munich: Verlag Dr. Friedrich Pfeil.

Grande, T., & Grande, L. (2008). Reevaluation of the gonorynchiform genera †*Ramallichthys*, †*Judeichthys* and †*Notogoneus*, with comments on the families †Charitosomidae & Gonorynchidae. In: G. Arratia, H.-P. Schultze, & M. V. H. Wilson (Eds.), *Mesozoic fishes, vol. 4. Homology & phylogeny* (pp. 295–310). Munich: Verlag Dr. Friedrich Pfeil

Hamel, M.-H., & Poplin, C. (2008). The braincase anatomy of *Lawrenciella schaefferi*, actinopterygian from the Upper Carboniferous of Kansas (USA). *Journal of Vertebrate Paleontology*, 28, 989–1006.

Haubold, H., & Schaumberg, G. (1985). *Die Fossilien des Kupferschiefers*. Wittenberg: A. Zeimsen Verlag.

Ishiguro, N. B., Miya, M., & Nishida, M. (2003). Basal eutelostean relationships: A mitogenomic perspective on the phylogenetic reality of the 'Protacanthopterygii'. *Molecular Phylogenetics & Evolution*, 27, 476–488.

Janvier, P. (1996). *Early vertebrates*. Oxford: Clarendon Press.

Jarvik, E. (1980). *Basic structure and evolution of vertebrates*. London: Academic Press.

Long, J. A. (1988). New palaeoniscid fishes from the Late Devonian and Early Carboniferous of Victoria. *Memoir of the Association of Australasian Palaeontologists*, 7, 1–63.

Lopez-Arbarello, A. (2012). Phylogenetic interrelationships of ginglymodian fishes (Actinopterygii: Neopterygii). *PLoS ONE*, 7, e39370.

Lu, J., Giles, S., Friedman, M., den Blaauwen, J. L., & Zhu, M. 2016. The oldest actinopterygian highlights the cryptic early history of the hyperdiverse ray-finned fishes. *Current Biology*, 16, 1–7.

Mainwaring, A. J. (1978). Anatomical and systematic review of the Pachycormidae, a family of Mesozoic fossil fishes. Dissertation. Westfield College, London.

Maisey, J. G. (1991). *Calamopleurus* Agassiz, 1841; *Oshunia* Wenz & Kellner, 1986; *Vinctifer* Jordan, 1919. In J. G. Maisey (Ed.), *Santana fossils: An illustrated atlas* (pp. 139–189). Neptune City (NJ): T. F. H. Publications.

Maisey, J. G. (1993). A new clupeomorph fish from the Santana Formation (Albian) of NE Brazil. *American Museum Novitates*, 3076, 1–15.

Maisey, J. G. (1999). The supraotic bone in neopterygian fishes (Osteichthyes, Actinopterygii). *American Museum Novitates*, 3267, 1–52.

Maisey, J. G. (2005). Braincase of the Upper Devonian shark *Cladodoides wildungensis* (Chondrichthyes, Elasmobranchii), with observations on the braincase in early chondrichthyans. *Bulletin of the American Museum of Natural History*, 288, 1–103.

Mayrinck, D., Brito, P. M., & Otero, O. 2015. Review of the osteology of the fossil fish formerly attributed to the genus †*Chanoides* and implications for the definition of otophysan bony characters. *Journal of Systematic Paleontology*, 13, 397–420.

McCune, A. R., & Schaeffer, B. (1986). Triassic and Jurassic fishes: Patterns of diversity. In K. Padian (Ed.), *The beginning of the age of dinosaurs* (pp. 171–181). Cambridge: Cambridge University Press.

Moodie, R. L. (1915). A new fish brain from the Coal Measures of Kansas, with a review of other fossil brains. *Journal of Comparative Neurology*, 32, 135–181.

Moy-Thomas, J. A., & Bradley Dyne, M. (1938). The actinopterygian fishes from the Lower Carboniferous of Glencartholm, Eksdale, Dumfriesshire. *Transactions of the Royal Society of Edinburgh*, 59, 437–480.

Nelson, E. M. (1955). The morphology of the swim bladder and the auditory bulla in the Holocentridae. *Fieldiana: Zoology*, 37, 121–130.

Nielsen, E. (1942). Studies on Triassic fishes from East Greenland. I. *Glaucolepis & Boreosomus*. *Meddelelser om Grønland*, 138, 1–403.

Nielsen, E. (1949). Studies on Triassic fishes from East Greenland. II. *Australosomus* and *Birgeria*. *Meddelelser om Grønland*, 146, 1–309.

Nolf, D. (1985). Otolithi piscium. In H.-P. Schultze (Ed.), *Handbook of paleoichthyology, vol. 10* (pp. 1–145). Stuttgart: Gustav Fischer Verlag.

Nolf, D. (2013). *The diversity of fish otoliths, past and present*. Brussels: Royal Belgian Institute of Natural Sciences.

Nursall, J. R. (1999). The family †Mesturidae and the skull of pycnodont fishes. In G. Arratia & H.-P. Schultze (Eds.), *Mesozoic Fishes, vol. 2. Systematics and fossil record* (pp. 153–188). Munich: Verlag Dr. Friedrich Pfeil.

Nursall, J. R., & Maisey, J. G. (1991). *Neoproscinetes* Figueiredo and Silva Santos, 1987. In J. G. Maisey (Ed.), *Santana fossils: An illustrated atlas* (pp. 125–137). Neptune City (NJ): T. F. H. Publications.

Olsen, P. E. (1984). The skull and pectoral girdle of the parasemionotid fish *Watsonulus eugnathoides* from the Early Triassic Sakamena Group of Madagascar, with comments on the relationships of the holostean fishes. *Journal of Vertebrate Paleontology*, 4, 481–499.

Olsen, P. E., & McCune, A. R. (1991). Morphology of the *Semionotus elegans* species group from the Early Jurassic part of the Newark Supergroup of eastern North America with comments on the family Semionotidae (Neopterygii). *Journal of Vertebrate Paleontology*, 11, 269–292.

Patterson, C. (1964). A review of Mesozoic acanthopterygian fishes, with special reference to those of the English Chalk. *Philosophical Transactions of the Royal Society of London B: Biological Sciences*, 247, 213–482.

Patterson, C. (1970). A clupeomorph fish from the Gault (Lower Cretaceous). *Journal of the Linnean Society of London*, 49, 161–182.

Patterson, C. (1975). The braincase of pholidophorid and leptolepid fishes, with a review of the actinopterygian braincase. *Philosophical Transactions of the Royal Society of London B: Biological Sciences*, 269, 275–579.
Patterson, C. (1977). The contribution of paleontology to teleostean phylogeny. In: M. K. Hecht, P. C. Goody, & B. M. Hecht (Eds.), *Major patterns in vertebrate evolution* (pp. 579–643). New York: Plenum Press.
Patterson, C. (1982). Morphology and interrelationships of primitive actinopterygian fishes. *American Zoologist*, 22, 241–259.
Patterson, C. (1984). *Chanoides*, a marine Eocene otophysan fish (Teleostei: Ostariophysi). *Journal of Vertebrate Paleontology*, 4, 430–456.
Poplin, C. (1974). *Étude de quelques paléoniscidés pennsylvaniens du Kansas*. Paris: Éditions du Centre National de Recherches Scientifique.
Poplin, C. M., & Véran, C. (1996). A revision of the actinopterygian fish *Coccocephalus wildi* from the Upper Carboniferous of Lancashire. In A. R. Milner (Ed.), Studies on Carboniferous and Permian vertebrates. *Special Papers in Palaeontology*, 52, 7–29.
Popper, A. N., & Coombs, S. (1982). The morphology and evolution of the ear in actinopterygian fishes. *American Zoologist*, 22, 311–328.
Popper, A. N., & Platt, C. (1983). Sensory surface of the saccule and lagena in the ears of ostariophysan fishes. *Journal of Morphology*, 176, 121–129.
Poyato-Ariza, F. J. (2015). Studies on pycnodont fishes (I): Evaluation of their phylogenetic position among actinopterygians. *Rivista Italiana di Paleontologia e Stratigrafia*, 121, 329–343.
Rayner, D. M. (1948). The structure of certain Jurassic holostean fishes, with special reference to their neurocrania. *Philosophical Transactions of the Royal Society of London B: Biological Sciences*, 233, 287–345.
Rayner, D. M. (1951). On the cranial structure of an early palaeoniscoid, *Kentuckia* gen. nov. *Transactions of the Royal Society of Edinburgh*, 62, 58–83.
Rosen, D. E., & Greenwood, P. H. (1970). Origin of the Weberian apparatus and the relationships of the ostariophysan and gonorynchiform fishes. *American Museum Novitates*, 2428, 1–25.
Sallan, L. C. (2014). Major issues in the origin of ray-finned fish (Actinopterygii) biodiversity. *Biological Reviews*, 89, 950–971.
Schaeffer, B. (1968). The origin and basal radiation of the Osteichthyes. In T. Ørvig (Ed.), Current problems in lower vertebrate phylogeny. *Nobel Symposium, Stockholm*, 4, 207–222.
Schaeffer, B. (1971). The braincase of the holostean fish *Macrepistius*, with comments on neurocranial ossification in the Actinopterygii. *American Museum Novitates*, 2459, 1–34.
Schaeffer, B., & Dalquest, W. W. (1978). A palaeonisciform braincase from the Permian of Texas, with comments on the cranial fissure and posterior myodome. *American Museum Novitates*, 2658, 1–15.
Schultze, H.-P. (2015). Scales, enamel, cosmine, ganoine, and early osteichthyans. *Comptes Rendus Palevol*, 15, 83–102.
Stensiö, E. (1925). Triassic fishes from Spitzbergen. *Kungliga svenska Vetenskapakademiens Handlingar*, 2, 1–261.
Stensiö, E. (1932). Triassic fishes from East Greenland collected by the Danish Greenland Expeditions in 1929 and 1930. *Meddelelser om Grønland*, 83, 1–305.
Taverne, L. (1995). Description de l'appareil de Weber du téléosteen crétacé marin *Clupavus maroccanus* et ses implications phylogénétiques. *Belgian Journal of Zoology*, 125, 267–282.
Watson, D. M. S. (1925). The structure of certain palaeoniscids and the relationship of that group with other bony fish. *Proceedings of the Royal Society of London B: Biological Sciences*, 1925, 815–870.
Weber E. H. (1820). *De aute et auditu hominis et animalium. Pars I. De aure animalium aquatilium*. Leipzig: Gerhard Fleischer.
Weiss, F. E., Malabarba, M. C., & Malabarba, L. R. (2014). A new stem fossil characid (Teleostei: Ostariophysi) from the Eocene-Oligocene of southeastern Brazil. *Neotropical Ichthyology*, 12, 439–450.

Wenz, S., & Brito, P. (1996). New data about the lepisosteids and semionotids from the Early Cretaceous of Chapada do Araripe (NE Brazil): Phylogenetic implications. In G. Arratia & G. Vihol (Eds.), *Mesozoic fishes—Systematics & paleoecology* (pp. 153–165). Munich: Verlag Dr. Friedrich Pfeil.

Wu, F., Chang, M.-M., Sun, Y., & Xu, G. (2013). A new saurichthyiform (Actinopterygii) with a crushing feeding mechanism from the Middle Triassic of Guizhou (China). *PLoS ONE*, 8, e81010.

Yu, X. (1998). A new porolepiform-like fish, *Psarolepis romeri*, gen. et sp. nov. (Sarcopterygii, Osteichthyes) from the Lower Devonian of Yunnan, China. *Journal of Vertebrate Paleontology*, 18, 261–274.

Zhu, M., Yu, X., Wang, W., Zhao, W., & Jia, L. (2006). A primitive fish provides key characters bearing on deep osteichthyan phylogeny. *Nature*, 441, 77–80.

Zhu, M., Wang., W., & Yu, X. (2010). *Meemannia eos*, a basal sarcopterygian fish from the Lower Devonian of China—expanded description and significance. In D. K. Elliott, J. G. Maisey, X. Yu, & D. Miao (Eds.), *Morphology, phylogeny & paleobiogeography of fossil fishes* (pp. 199–214). Munich: Verlag Dr. Friedrich Pfeil.

Zhu, M., Yu, X., Ahlberg, P., Choo, B., et al. (2013) A Silurian placoderm with osteichthyan-like marginal jaw bones. *Nature*, 502, 188–193.

Chapter 3
Sarcopterygians: From Lobe-Finned Fishes to the Tetrapod Stem Group

Jennifer A. Clack and Per Erik Ahlberg

Abstract The sarcopterygians or lobe-finned fishes is the group that gave rise to tetrapods, and they were the dominant bony fishes of the Devonian period. Their otic regions were constructed similarly to those of both the actinopterygians and chondrichthyans, their structure being the common inheritance of all jawed vertebrates. One distinguishing feature of the primitive sarcopterygian braincase was that the division between the anterior ethmosphenoid and posterior otoccipital sections of the braincase was marked by a flexible hinge joint, which is seen today in the modern coelacanth, *Latimeria*. The hyomandibular was long and projected ventrally with an opercular process that contacted the opercular bone and with the distal end associated indirectly with the jaw joint. It was a key component of the buccal pumping mechanism for breathing and feeding. The braincases of dipnoans (lungfishes) were the most highly modified of sarcopterygian braincases with consolidated fore and aft portions and reduction or loss of the hyomandibula. The utricle was enlarged in several fossil dipnoans, although the reason for this is not clear. The braincases of tetrapodomorph sarcopterygians differed little from the primitive condition in the group. The main modifications were to the more crownward and tetrapod-like forms from the Late Devonian. In these fishes, the hyomandibula was reduced in length, its contact with the opercular bone lost and, ultimately, the opercular bone itself disappeared. The spiracular notch and associated cleft increased in width and volume respectively, possibly resulting in increased air-breathing capacity and reduced use of the gill system.

J.A. Clack (✉)
University Museum of Zoology, Cambridge, Downing Street, Cambridge CB2 3EJ, UK
e-mail: j.a.clack@zoo.cam.ac.uk

P.E. Ahlberg
Department of Organismal Biology, Uppsala University, Norbyvägen 18A, Uppsala 752 36, Sweden
e-mail: Per.Ahlberg@ebc.uu.se

J.A. Clack et al. (eds.), *Evolution of the Vertebrate Ear—Evidence from the Fossil Record*, Springer Handbook of Auditory Research 59,
DOI 10.1007/978-3-319-46661-3_3

Keywords Actinistians · Actinopterygians · Coelacanths · Dipnoans · Dipnomorphs · *Eusthenopteron* · *Latimeria* · Lungfishes · Megalichthyids · *Panderichthys* · Porolepiforms · Tetrapodomorphs · Tristichopterids · *Tiktaalik* · *Youngolepis*

3.1 Introduction to the Sarcopterygians

The sarcopterygians or lobe-fins include the fish groups that dominated the early to mid-Paleozoic, and which ultimately gave rise to the group that dominated the land—the tetrapods. Therefore, because they are descendants of a common ancestor with the lobe-finned fishes, tetrapods are also lobe-fins (whose appendages happen to be modified into limbs with digits) and belong within the Sarcopterygii. A cow and a lungfish are more closely related to each other than either is to a salmon (an actinopterygian, see Friedman and Giles, Chap. 2). Most of the radiation of lobe-finned fishes took place in the Devonian, but many taxa became extinct at or before the end of that period. Only a few survivors, apart from tetrapods, persisted into later periods (see Fig. 1.2 in Chap. 1 by Clack for a geological timescale). Included in this chapter are members of the extant groups, the lungfishes or dipnoans, and the coelacanths or actinistians, as well as extinct lobe-finned fishes.

Crown group sarcopterygians comprises these two groups plus tetrapods. A crown group is defined as the clade containing all extant members of a group, plus all extinct forms that fall among them. The extinct forms that fall outside this crown group, but are more closely related to it than to any other crown group, constitute the associated stem group. Together, the crown group and its associated stem group form a total group. For example, the crown group Tetrapoda includes the extant Amphibia and Amniota, their last common ancestor, and all extinct forms that fall within this clade (and are thus descendants of that same common ancestor). The closest living relatives of the Tetrapoda are the lungfishes, Dipnoi. Any fossils that fall outside crown group Tetrapoda but are closer to them than to the Dipnoi belong to the tetrapod stem group. Together the tetrapod stem and crown group comprise the tetrapod total group, which are called the Tetrapodomorpha. The tetrapodomorphs thus include both lobe-finned fishes with paired fins and limbed vertebrates with digits the latter are treated in Chaps. 4–11 of this volume. Fig. 3.1 shows the relationships between the major sarcopterygian groups. The sister group of Tetrapodomorpha, the Dipnomorpha, consists of modern, crown group dipnoans, plus their extinct fossil relatives.

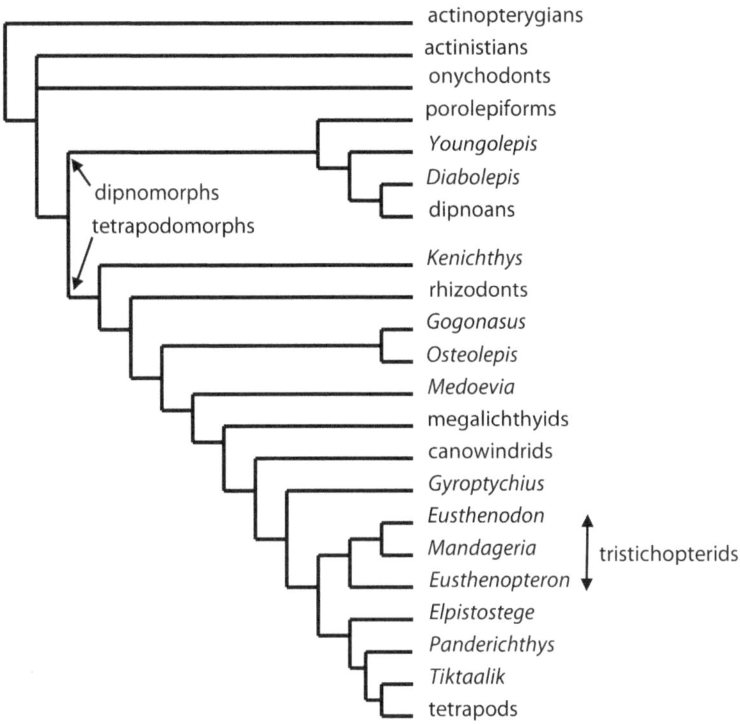

Fig. 3.1 Consensus cladogram showing the relationships of the major tetrapodomorph taxa

3.2 Actinistians and Onychodonts

In cladistic analyses, these two groups usually form the most basal clades among the sarcopterygians, either as sister-taxa or two separate groups. Their relationships to Tetrapodomorpha and Dipnomorpha are therefore uncertain.

3.2.1 Actinistians

In the cladogram in Fig. 3.1, actinistians (more commonly known as coelacanths) are regarded as among the most basal groups. Actinistians first appear in the fossil record during the Early Devonian, and the earliest known actinistian from good fossil material is *Euporosteus* from China (Zhu et al. 2012). Fossil coelacanths are found throughout the rest of the Paleozoic and Mesozoic, but the record ends with the close of the Cretaceous period. As a consequence, they were considered extinct from that time until a living example was found in 1938 (Smith 1956). Initially considered to be closely related to tetrapods, analyses of both the fossil record

(Ahlberg 1991; Cloutier and Ahlberg 1996) and molecular phylogeny (Liang et al. 2013) suggest instead that dipnomorphs are closer to tetrapods than are actinistians and that modern lungfishes are the closest living relatives of tetrapods, contra Fritzsch (1987) and Zhu and Schultze (2001). Coelacanths are sometimes considered to be related to a poorly known group of fossil sarcopterygians, the onychodonts, which will be considered briefly at the end of this section.

Coelacanths reached their peak of diversity during the Paleozoic, rivaling actinopterygians in their adaptations to different ecological niches. Later, for most of the Mesozoic, their body forms became essentially conservative. During the Carboniferous, they are thought to have been euryhaline because they were found in both marine and continental deposits. A review of fossil coelacanths was published by Forey (1998).

The otic anatomy of the modern coelacanth *Latimeria* is known in some detail, and although much of the otic capsule remains cartilaginous, it can be compared with those of fossil forms in which it is ossified and preserved. The posterior portion of the the braincase, the otoccipital, is a separate unit from the anterior portion, the ethmosphenoid, and the two are hinged together in a mobile intracranial joint (note that for all the animals in this volume, anterior = rostral, posterior = caudal, these terms are synonymous in all animals except humans and other bipeds). This feature is characteristic of primitive sarcopterygians, lost in dipnoans and tetrapods, but retained in coelacanths. In most fossil coelacanths, only the external features of the braincase have been described, but they largely resemble their equivalents in *Latimeria*, the living species. The neurocranium of the Late Devonian *Diplocercides* (aka *Nesides*) is one of the best known. In later forms, the otic capsule is rarely preserved, suggesting that, as in *Latimeria*, ossification of the capsule was reduced. The hyomandibula is double-headed as in most fossil sarcopterygians (Forey 1998).

The saccule of *Latimeria* includes a large calcitic otolith (Carlström 1963) and, likewise, fossil coelacanths from the Carboniferous to the Jurassic have been discovered that preserve single, large, presumably saccular otoliths (Fig. 3.2). In the Jurassic genus *Undina*, they can even be found in the two intrauterine embryos that lie within the body cavity of the fossil female fish (Clack 1996).

3.2.2 Onychodonts

The sarcoptergyian group Onychodontiformes is confined to the Devonian. Although a few good specimens exist, they are otherwise poorly known and their relationship to other sarcopterygian groups is uncertain. Braincases in onychodonts were formed from two separate parts in which the posterior otoccipital was hinged to the anterior ethmospenoid, allowing relative movement between the two in a similar way to the retained plesiomorphic condition in modern coelacanths.

The earliest known genus to preserve a braincase is *Qingmenodus* from the Early Devonian of China, in which the external morphology of the otoccipital shows a

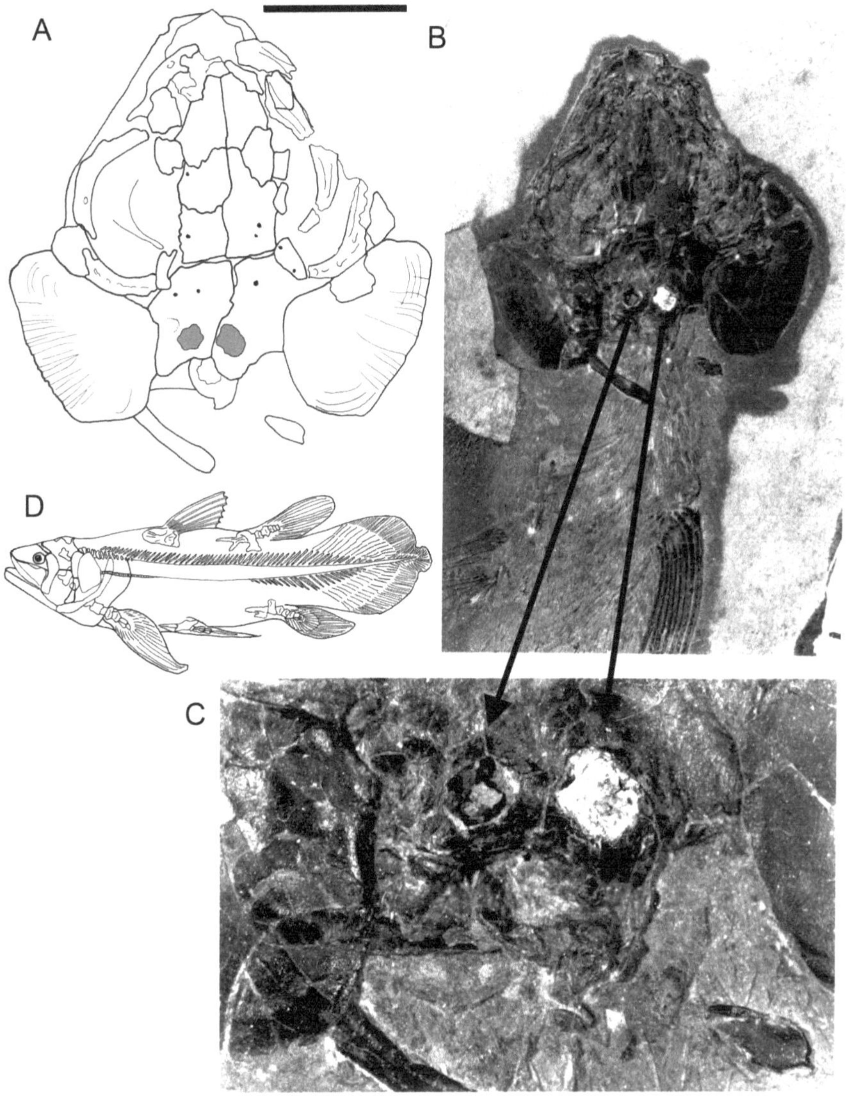

Fig. 3.2 Coelacanths. (**A–C**) The Triassic coelacanth *Diplurus*. (**A**) Line drawing of skull in dorsal view, showing position of otoliths. (**B**) Photograph of the same skull. (**C**) Enlargement of posterior region of same skull; *arrows* mark otoliths. The left otolith has mainly disintegrated, but there are remains in its natural mold. The right otolith is present as a calcitic mass. (**D**) Skeleton of the extant coelacanth *Latimeria*. (Original photographs and drawings by JAC)

saccular bulge in its sidewall (Lu and Zhu 2010). The best preserved onychodont is *Onychodus* from the Late Devonian of Australia, although the braincase of this genus is poorly ossified and was probably mainly cartilaginous (Andrews et al. 2006). A pair of partly preserved otic capsules is known in which the semicircular canal tracts and ampullae can be traced because they are lined with perichondral bone. There is a large vestibular fontanelle positioned below the lateral (horizontal) canal.

3.3 Dipnomorphs: Dipnoans and Porolepiforms

The Dipnomorpha is the dipnoan total group, comprising a crown group (the clade delimited by the extant lungfishes *Neoceratodus*, *Protopterus*, and *Lepidosiren*) and a stem group. One branch of the stem group consists of the Porolepiformes, a group of entirely extinct fossil taxa that did not survive beyond the end of the Devonian. The other branch consists of taxa that sit below the crown group of dipnoans but have most of the characteristics of the crown group.

3.3.1 Stem Dipnoans and Porolepiforms

Modern lungfishes have an extensive suite of derived characters, including autostyly and a palatal bite. Many of these characters can be traced down to Devonian members of the stem group, such as *Dipterus*, *Holodipterus,* and *Dipnorhynchus* (see Clack et al. 2011 for descriptions; Clement and Ahlberg 2014; Challands 2015 for more recent work), which have long been recognized as "early lungfishes". Below these taxa in the stem group lies a single transitional form from the earliest Devonian of China, *Diabolepis* (Ahlberg 1991; Chang 1995), which has a characteristic dipnoan dentition and palatal bite but not the autostylic jaw suspension of later forms. The three most basal taxa in the stem group below *Diabolepis* are the two Early Devonian genera *Youngolepis* from China and *Powichthys* from Canada and the small Devonian clade Porolepiformes (Jarvik 1972; Jessen 1980; Chang 1982). These taxa have few morphological characteristics in common with "early lungfishes," but they are rather generalized sarcopterygians with marginal dentitions of conical teeth. However, they almost certainly represent the type of sarcopteryian from which the lungfishes evolved, and their inner ear morphology is considered here.

Porolepiforms have an intracranial joint similar to that of coelacanths (see Sect. 3.2.1 and Clack, Chap. 1), separating the otic capsules from the more anterior part of the braincase, but in *Powichthys* this joint is not mobile and it is absent in *Youngolepis*. Porolepiforms tend to have weakly ossified otoccipital regions that usually collapse during preservation. This problem meant that Jarvik (1972) could not reconstruct the otoccipital of the Middle Devonian *Glyptolepis groenlandica* from his grinding series. Jarvik (1972, plate 10) figured an otoccipital of *Porolepis*, but the specimen had been subjected to rather heavy-handed mechanical preparation. Therefore it provided no detailed information about the inner ear. In general, it resembles the otoccipital of a tetrapodomorph such as *Eusthenopteron* (see Sect. 3.4.4). The only published braincase of *Powichthys* shows an otoccipital with a pair of poorly preserved, but probably large, vestibular fontanelles (Jessen 1980).

Of the three basal taxa in the stem group leading to dipnoans, only *Youngolepis* has inner ear morphology that has been described. In that animal, in contrast to porolepiforms, the otic region is well-ossified and is known from a number of well-preserved specimens. One of these was investigated by Chang (1982). She was able to build a wax model of the internal spaces and create a detailed reconstruction of the cranial cavity (Chang 1982, Figs. 17, 18, 19). The semicircular canal tracts are rather poorly preserved and are shown with dashed outlines, but their positions and shapes are known. The inner ear (Fig. 3.3) has a rather generalized osteichthyan morphology without any of the specializations that distinguish "early

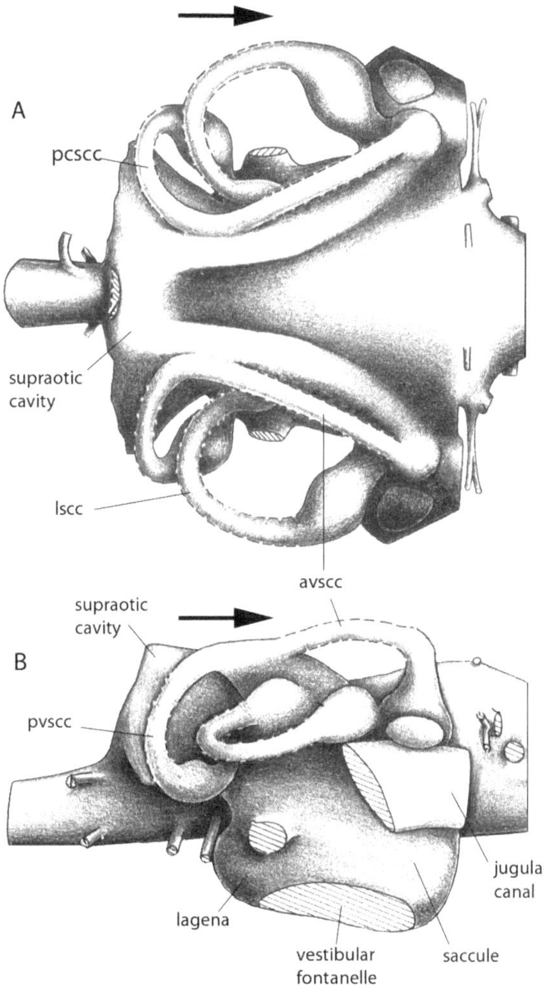

Fig. 3.3 Labyrinth cavities and associated spaces of *Youngolepis* (modified from Chang 1980) in dorsal view (**A**) and lateral view (**B**). *Arrows* point anteriorly. *avscc*, anterior vertical semicircular canal; *lscc*, lateral semicircular canal; *pvscc*, posterior vertical semicircular canal

Fig. 3.4 Extant and extinct lungfish. (**A**) Head skeleton of *Protopterus*, an extant lungfish, ▶ showing position of the otic capsule, and the highly modified skull, gill apparatus, cranial rib, and pectoral girdle. (**B**) Skull and hyoid arch (*gray*) of the Devonian lungfish *Griphognathus*. Note that the hyomandibula does not reach the region of the jaw joint. (**C, D**) Cranial cavity of the Devonian lungfish *Rhinodipterus* with the labyrinth cavities and associated structures highlighted, in dorsal view (**C**) and lateral view (**D**). *Arrows* point anteriorly. *Abbreviations* as in Fig. 3.3. (**B** compound image modified from Miles 1977; **C**, **D** modified from Clement et al. 2016)

lungfishes" and crown-group dipnoans (see Sect. 3.3.2). Although the utricular recess has not been fully figured, it appears to be small. The sacculo-lagenar cavity is bag-shaped and undivided with a small lagenar bulge posteriorly. No otoliths are known but, considering that they are found in all extant vertebrates, they are likely to have been present in *Youngolepis*. The vestibular fontanelle occupies most of the ventral face of the sacculo-lagenar cavity, which means that this part of the inner ear was separated from the pharynx only by a wall of soft tissue. However, the double-headed hyomandibula articulated with a lateral commissure attached to the side wall of the braincase and had no direct contact with the inner ear.

3.3.2 Dipnoans

Dipnoans, or lungfishes, are defined here as comprising the crown group and most of the stem group of the Dipnomorpha (see Sect. 3.3.1). They have highly modified skulls and dentitions, which led to the fossil forms not having been recognized as related to the living forms until 1871. Unlike the braincases of other early sarcopterygians, those of lungfishes are unitary. The intracranial joint is completely eliminated, concomitant with the consolidation and unique modification of the skull roofing bones into an alternating pattern of two bones followed by a single midline one—a "hopscotch" pattern. Again, in association with these modifications, the cranial rib and shoulder girdle were incorporated into a mechanism to operate a buccal force pump that filled the lungs. The jaw joint moved far forward in the skull and lost contact with the hyomandibula with the result that, in later forms, the hyomandibula was reduced and ultimately lost (Fig. 3.4).

As with coelacanths, the peak diversity of lungfishes lay in the mid-Paleozoic, especially in the Middle and Late Devonian. As with coelacanths and some tetrapodomorphs, ossification of their braincases became reduced during the later Paleozoic and subsequent periods. Thus, the most known about their ear regions comes from Devonian forms.

Among the most completely described of Devonian lungfish braincases are those from the Lower Devonian Taemas Limestone Formation and the Upper Devonian Gogo Formation of Australia, including *Dipnorhynchus* (Campbell and Barwick 1982, 2000), *Griphognathus*, "*Chirodipterus*," *Holodipterus*, and *Rhinodipterus* (see Clack et al. 2011 for descriptions) (see Clement and Ahlberg 2014 for more recent work). These fossils are preserved undistorted and in three dimensions as

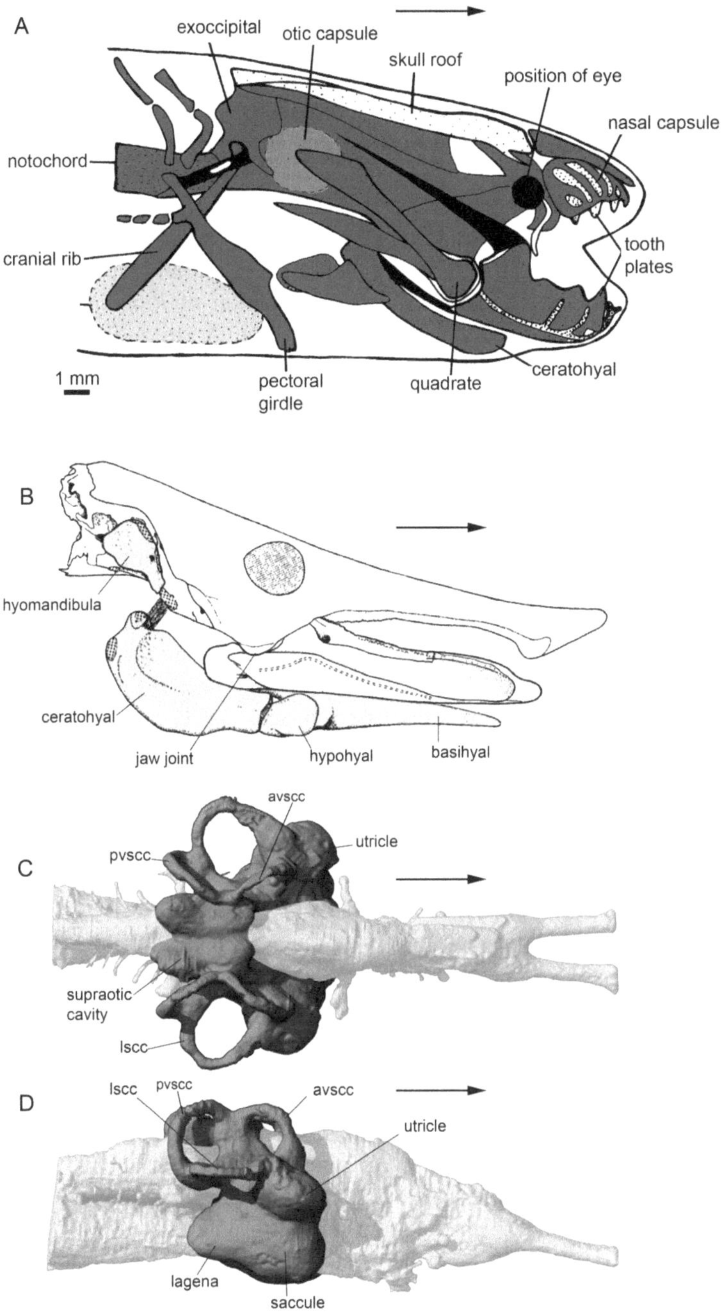
A
exoccipital
otic capsule
skull roof
position of eye
nasal capsule
notochord
cranial rib
tooth plates
1 mm
pectoral girdle
quadrate
ceratohyal
B
hyomandibula
ceratohyal
jaw joint
hypohyal
basihyal
C
avscc
utricle
pvscc
supraotic cavity
lscc
D
lscc
pvscc
avscc
utricle
lagena
saccule

revealed by acid digestion of the limestone matrix in which they are preserved. Much of the work on Devonian lungfish braincases was performed prior to the use of high-resolution computed microtomography (CT) so, in general, only the external morphology was accessible. In a few cases, access to the internal spaces was reconstructed from sections or from incomplete specimens. In recent years μCT has been used to reveal the internal anatomy of fossils, including lungfish braincases (Clement and Ahlberg 2014).

Reconstructions of the vestibular organs have been attempted for several genera and the otic capsules of each described in detail. More recently, endocasts produced from μCT imaging have shown the vestibular region of *Rhinodipterus* in great detail (Fig. 3.4C, D) (Clement and Ahlberg 2014). A striking feature of many lungfish vestibular regions is the enlarged utricular pouch. Among the extant genera, this enlargement is extreme in *Protopterus* and *Lepidosiren*, wherein the utricle is similar in size to the sacculo-lagenar pouch, but is slightly less pronounced in *Neoceratodus* (Clement and Ahlberg 2014). The Devonian fossils reveal that this feature began to develop very early in lungfish evolution. In *Youngolepis* and apparently *Dipnorhynchus*, the utricle is not enlarged, but in slightly later forms, such as *Dipterus* (Challands 2015) and *Rhinodipterus* (Clement and Ahlberg 2014), it is enlarged to varying degrees, although always less than in *Neoceratodus* (Fig. 3.4). It is not yet possible to be sure whether the enlargement occurred as a long-term directional trend, with incremental changes along the lungfish stem lineage, or whether the pattern was more complex.

Unlike the basal dipnomorphs discussed in Sect. 3.3.1, extant and fossil lungfishes never have open vestibular fontanelles. The saccular pouch of *Rhinodipterus* has a shallow notch in its ventral margin that may represent the differentiation of a lagenar pouch (Fig. 3.4). Recent μCT study of *Dipterus* has revealed a similar feature (Challands 2015); however, this is not seen in later genera (Clement and Ahlberg 2014).

Gauldie et al. (1986) described otoliths from *Neoceratodus* in the saccule and utricle, but unlike those of coelacanths or actinopterygians, they are composed of a mass of small crystals of calcite, aragonite, and another form of calcium carbonate bound together in a proteinaceous matrix. These usually disaggregate upon the death of the animal and, therefore, have been thought unlikely to be preserved in fossil lungfishes. The recent μCT study of *Dipterus* by Challands (2015) shows evidence for the existence of pasty otoconial masses in both the saccule and the utricle. He also described the presence of a lining of small pyritic nodules that appear to mimic the inferred position of the saccular macula.

Recent work on the hearing capabilities of lungfishes has shown that they can pick up waterborne and substrate-transmitted sound by use of their lungs (Christensen et al. 2015). It might be reasonable to suppose that early lungfishes could also do the same, assuming they had acquired the capability for air-breathing.

Osteological correlates of air breathing were first noted in living forms (Bishop and Foxon 1968) and have been studied in fossils by Long (1993) and Clement et al. (2016). Long showed that some of the osteological correlates were evident in fossil forms: the development of the cranial rib and a mobile pectoral girdle, a posteriorly elongate parasphenoid reflecting an enlarged buccal cavity, and a space between the tooth plates to accommodate a tongue pad that acts as a stop valve when air is being forced into the lungs. The earliest development of these features is seen in Middle Devonian forms such as *Dipterus*. In the Late Devonian they are even seen in some marine forms such as *Rhinodipterus* from Gogo (Clement and Long 2010; Clement et al., 2016). We can thus infer that the ability to process sound by lungfishes dates back at least to the Late Devonian.

3.4 Tetrapodomorph Fishes

The clade Tetrapodomorpha (Ahlberg 1991) is equivalent to the tetrapod total group. It encompasses the tetrapod crown group (amphibians and amniotes) and a stem group that includes both "early tetrapods" such as *Acanthostega* and a range of sarcopterygian fishes. The internal relationships of the included groups are in need of thorough review, although some parts of the tree are relatively stable. One of the most recent analyses is that of Snitting (2008), which is followed here with some additions.

3.4.1 Introduction to Tetrapodomorphs

The otic region of the braincase has not been described for the most basal tetrapodomorph taxa, the Early Devonian forms from China such as *Kenichthys* (Chang and Zhu 1993; Zhu and Ahlberg 2004). Moving crownward, they are followed by the clade Rhizodontida, a distinctive group that gave rise to some of the largest freshwater predators that have ever existed but that died out before the end of the Carboniferous. Their braincases are unknown because in almost all examples their skulls fall apart on death, and the endoskeleton is not preserved. The next segment of the stem is occupied by a paraphyletic and poorly resolved array of "osteolepiforms" (Fig. 3.1). The megalichthyids were the longest lived of the tetrapodomorph fish groups, persisting into the Early Permian (Romer 1937). The most crownward "osteolepiform" group is the clade Tristichopteridae, which is placed as the sister group to *Panderichthys* and all more crownward tetrapodomorphs. The Tristichopteridae includes the best known tetrapodomorph fish, *Eusthenopteron*, represented by beautifully preserved material from the Late Devonian of Canada that was described by Jarvik, largely on the basis of a grinding series of a single skull (Jarvik 1980). *Panderichthys* and the somewhat similar but more crownward *Elpistostege* and *Tiktaalik* together form a paraphyletic array, informally known as elpistostegids. They are the most crownward tetrapodomorphs that retain paired fins rather than limbs with digits (Ahlberg and Johanson 1998; Daeschler et al. 2006). Limbed tetrapods are herein referred to as "tetrapods," irrespective of whether they belong to the stem group or crown group: the early forms are treated in Chap. 4 (Clack and Anderson).

3.4.2 Osteolepis, Medoevia, *and* Gogonasus

Osteolepis is one of the best known "osteolepiforms" and has been collected from Scottish Middle Devonian localities for nearly 200 years. Unfortunately, perhaps because of its ubiquity, its anatomy and taxonomy have been neglected for most of that time. However, as well as being one of the earliest of the "osteolepiforms," it appears to be one of the most phylogenetically basal after *Kenichthys*. In addition, because most of the specimens are flattened and the endoskeletons often poorly preserved, the braincase is not well known despite the abundance of the fish in the fossil record.

One of the few studies of the ear region is by Watson (1925), who described an exceptionally preserved specimen from Tynet Burn in Scotland. In this specimen of *O. macrolepidotus*, the anterior part of the animal was preserved in three dimensions. With careful preparation, the specimen revealed endocasts of the three semicircular canals. He noted the "perfectly normal nature of its structure." A photograph of the specimen is given in Fig. 3.5.

The skull and only specimen of *Medoevia lata* was found as an isolated nodule probably from the Upper Devonian of Khazakstan (O. A. Lebedev, personal communication). The material has been acid-prepared to reveal its three dimensional morphology, which is almost uncrushed. The ethmosphenoid and otoccipital units are separate and there is a large vestibular fontanelle. The general morphology of the units and of the hyomandibula are typical of basal tetrapodomorphs (Lebedev, 1995). Both *Osteolepis* and *Medoevia* have been reconstructed with narrow, slot-like openings for the spiracle with a contact between the opercular bone and the extratemporal bones.

Gogonasus, like *Medoevia*, is known from exquisitely preserved skull material revealed by acid preparation, but in this case there are several specimens and the locality is known in detail. It is from the early Late Devonian locality of Gogo in Australia. In most respects is it a typical tetrapodomorph in terms of otic construction, but in contrast to many other taxa, such as *Eusthenopteron*, the hyomandibula is described as being short and horizontally oriented in the skull (Long et al. 1997). One other difference from most "osteolepiforms" lies in the spiracular notch which, in this case, is broadly open with a rounded anterior margin (Fig. 3.6A) (Long et al. 2006). An externally open spiracle has been associated with air breathing (Clack 2007) and with the lowered oxygen concentration in the atmosphere in the Middle and early Late Devonian. Although Gogo is notable as representing a marine environment, in which anoxic water would be unusual, air breathing has also been suggested for some of the lungfishes from that locality (see Sect. 3.3.2).

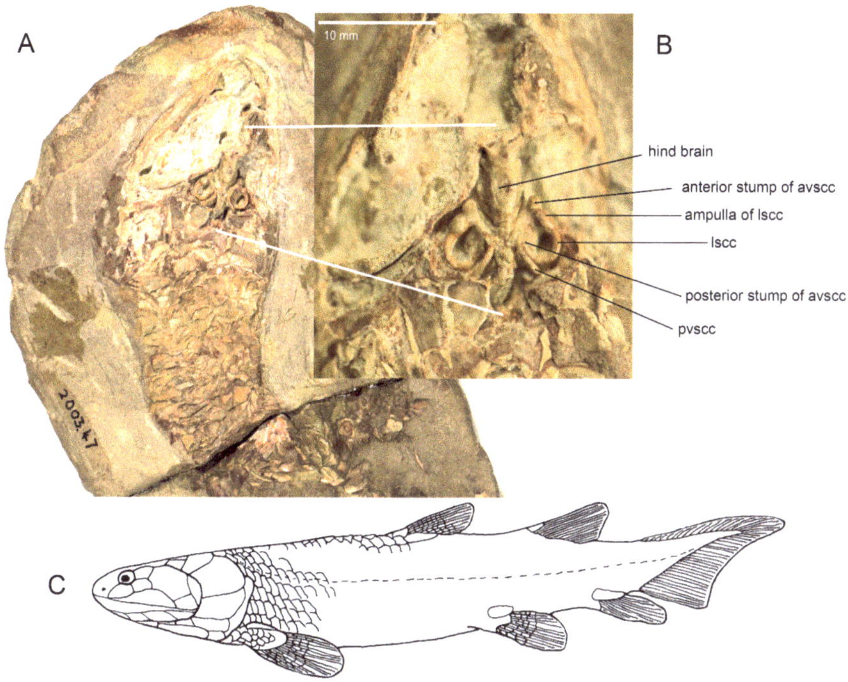

Fig. 3.5 *Osteolepis macrolepidotus*. (**A–B**) University Museum of Zoology Cambridge specimen UMZC 2003.47. (**A**) Anterior half of specimen. (**B**) Close-up of semicircular canal region. *Abbreviations* as for Fig. 3.3. *Scale bar*, 10 mm. (**C**) Reconstruction of *Osteolepis*. (Photographs in **A**, **B** from R. Stebbings, UMZC with permission; **C** original by JAC)

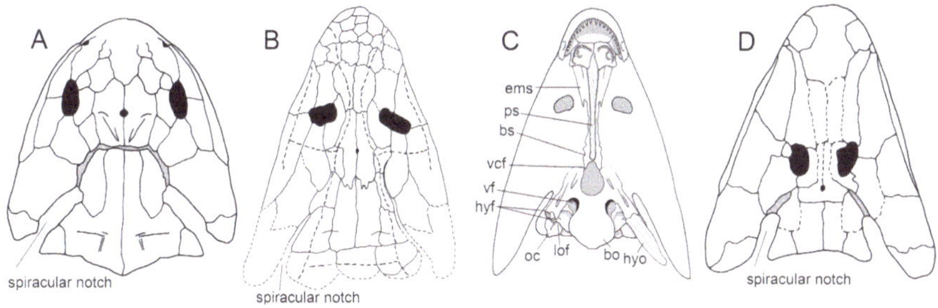

Fig. 3.6 Skulls of tetrapodomorphs: (**A**) *Gogonasus*. (**B**) *Panderichthys*. (**C**) *Panderichthys*. (**D**) *Tiktaalik*. (**A**, **B**, **D**) in dorsal view to show spiracular notch; (**C**) in ventral view to show braincase and hyomandibula. *bo*, basioccipital; *bs*, basisphenoid; *ems*, ethmosphenoid; *hyf*, hyomandibular facets; *hyo*, hyomandibula; *lof*, lateral otic fissure; *oc*, otic capsule; *ps*, parasphenoid; *vcf*, ventral cranial fissure; *vf*, vestibular fontanelle. (**A** original drawing based on Long et al. 2006; **B–D** modified from Clack 2012)

3.4.3 Megalichthyids

The braincase and ear region of the megalichthyid *Megalichthys nitidus* was described in some detail by Romer (1937). This species comes from the Permo-Carboniferous of the southern United States. His description was based on several partial braincase specimens and one more complete one that was sectioned vertically at half-millimeter intervals. The sections were then etched, acetate peels were made from them, and the peels were mounted between glass sides. In essence, the braincase construction is very similar to that of a tristichopterid such as *Eusthenopteron* (Clack, Chap. 1, Fig. 1.3B, and Sect. 3.4.4), though slightly different in proportion. The otic capsule was well ossified and from sections, broken specimens, and casts from the cavities, it was possible for Romer to reconstruct the vestibular region (Fig. 3.7). One important observation was the presence of a large "otolith" in the saccular cavity. New photographs of the specimen show this more likely to have been the remains of a mass of otoconia in a pasty matrix that has collapsed into the saccular cavity (Fig. 3.7). Pasty otoconia are characteristic of tetrapods, and it could be that in fact, all tetrapodomorphs and dipnoans bore pasty otoconia rather than the solid ones apparently found in coelacanths. Romer (1937) also noted the presence of a utricular otolith but that cannot be confirmed from the available evidence.

3.4.4 Tristichopterids

The otoccipital of *Eusthenopteron* was well ossified, allowing detailed study of the otic capsule and the inner ear cavity (Fig. 3.8). These were described by Jarvik (1980) from a grinding series of a three-dimensional skull. The vestibular region is in most respects similar to that of *Youngolepis* (see Sect. 3.3.1) with a small utricular recess and a bag-shaped sacculo-lagenar cavity that ends ventrally in an open vestibular fontanelle. As in *Youngolepis*, the hyomandibula is double-headed. It is an elongate bone that curves ventrally, following the line of the posterior margin of the palatoquadrate, and is a major component of the hyoid arch. Dorsally, it articulates with a lateral commissure that attaches to the lateral wall of the otic capsule and straddles the jugular vein. The ventral hyomandibular articulation is positioned close to the anterior margin of the vestibular fontanelle but does not enter the fontanelle. There is a dorsally placed opercular process on the hyomandibula for contact with the opercular bone. The lateral otic fissure is still patent and passes into the vestibular fontanelle.

The only tristichopterid otic region known to differ significantly from that of *Eusthenopteron* belongs to *Mandageria* from the late Devonian of Canowindra, Australia (Johanson and Ahlberg 1997; Johanson et al. 2003). *Mandageria* is a derived tristichopterid, and it has a derived long-snouted cranial morphology that in certain respects parallels elpistostegids and tetrapods (Ahlberg and Johanson 1998;

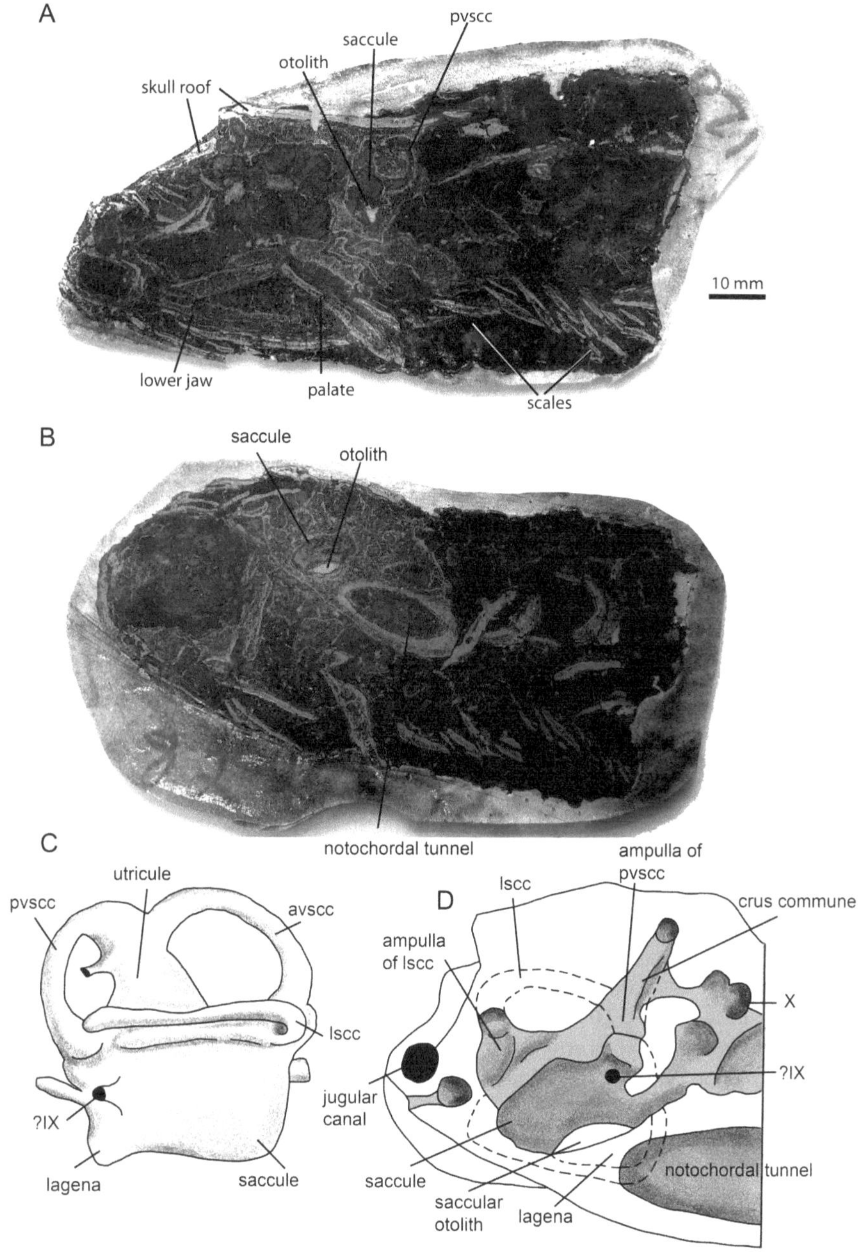

Fig. 3.7 *Megalichthys nitidus*. (**A–B**) Thin sections through a specimen from Museum of Comparative Zoology, Harvard MCZ 6494. (**C**) Reconstruction of inner ear. (**D**) Thick reconstructed section of inner ear region. *Abbreviations* as in Fig. 3.3, plus cranial nerve X and possibly IX. (Photographs in **A**, **B** by JAC; **C**, **D** redrawn and modified from Romer 1937)

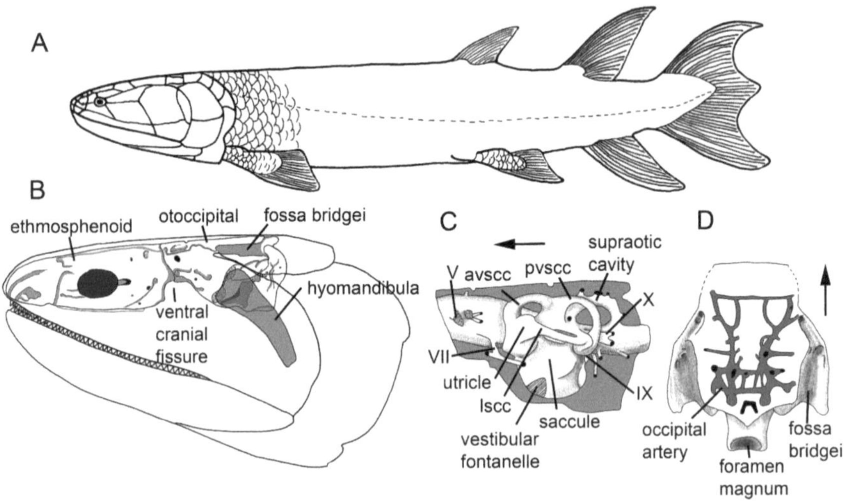

Fig. 3.8 *Eusthenopteron foordi.* **(A)** Reconstruction of fish. **(B)** Skull with roofing bones drawn as transparent to show braincase and hyomandibula. **(C)** Lateral view of otoccipital region of braincase with endoskeletal elements drawn as transparent to show inner ear region. *Abbreviations* as in Fig. 3.3, plus cranial nerves V, VII, IX, and X. **(D)** Dorsal view of otoccipital region to show occipital arteries and fossa bridgei. (All modified from Jarvik 1980)

Johanson et al. 2003). The otic region of the braincase was rather lightly ossified and is poorly preserved. Although it is represented only by natural molds, it is nevertheless informative. The vestibular fontanelle is very large. The ventral articulation of the hyomandibula has shifted posteriorly so that it lies approximately midway along the lateral margin of the vestibular fontanelle, but it is clearly separated from the fontanelle and faces dorsally. Anterior to the level of the fontanelle, two structures in the walls of the spiracular cleft—a curving ridge on the lateral wall of the otic capsule and a bulge with a dorsally facing pocket on the palatoquadrate—may indicate the presence of some type of specialized spiracular diverticulum (Johanson et al. 2003). This is interesting given that otic gas bladders are known to be used in hearing by mormyrid fishes (McCormick and Popper 1984; Yan and Curtsinger 2000) and a spiracular diverticulum has been hypothesized to have had an auditory role in the early tetrapod *Ichthyostega* (Clack et al. 2003).

3.4.5 Elpistostege, Panderichthys, *and* Tiktaalik

The otic braincase of *Elpistostege* remains completely unknown. In both *Panderichthys* and *Tiktaalik* the otic region has been described in ventral view from reasonably well-preserved specimens (Brazeau and Ahlberg 2006; Downs et al. 2008). Interestingly, the two genera share certain distinctive features that are not present in other tetrapodomorph fishes, given their phylogenetic position as immediate successive sister groups to tetrapods, it is probable that these features are aspects of the ancestral condition for the tetrapod inner ear region. *Panderichthys* and *Tiktaalik* differ from a representative tetrapodomorph fish such as *Eusthenopteron* in a number of ways. In *Panderichthys*, the spiracular notch is open posteriorly, although it is apparently rather narrow (compression may have contributed to the narrowing) (Fig. 3.6B). By contrast, in *Tikaalik*, the spiracular notch is broadly open throughout its extent and resembles very closely the condition in the tetrapod *Ventastega* (Clack and Anderson, Chap. 4). Both *Panderichthys* and *Tiktaalik* also have broader basioccipitals than more typical tetrapodomorphs, with a vaguely violin-shaped outline, a smaller and more anteriorly positioned vestibular fontanelle, and a posteriorly positioned lateral commissure (Fig. 3.6B, C, D). The effect of the latter two characteristics is to increase the distance between the ventral hyomandibular articulation and the vestibular fontanelle compared to *Eusthenopteron*. The hyomandibula is also strikingly different. In *Panderichthys*, the bone is straight and ends at the opercular facet (Brazeau and Ahlberg 2006), suggesting that the distal half of the hyomandibula has been reduced in size. *Tiktaalik*, which lacks an opercular bone, has an even shorter hyomandibula (Downs et al. 2008). Thus, the early stages of the transformation from fish hyomandibula to tetrapod stapes may have involved a progressive shortening of the distal end of the bone that was associated with the loss of opercular gill ventilation. The shape of the spiracular tract also began to approach that of the earliest tetrapods, with an elongate spiracle and a short, wide, and straight passage between palatoquadrate and braincase (Brazeau and Ahlberg 2006). This may reflect increased reliance on spiracular air breathing (Brazeau and Ahlberg 2006).

3.5 Summary

From the descriptions in previous sections, it can be seen that the otic region of sarcopterygians is generally very conservative and that there is little structural difference between them and other osteichthyans. Major modifications, however, occur in dipnoans with the consolidation of the braincase, a condition acquired independently from that which occurred in the more crownward tetrapodomorphs. Ultimately in dipnoans, this consolidation resulted in the great reduction and eventual loss of the hyomandibula and the associated modifications to the skull and shoulder girdle consequent upon their unique method of air breathing and suction feeding (see Clack et al. 2011). The main difference between the braincases of most sarcopterygians and those of actinopterygians is that the ventral cranial fissure, which divides the anterior from the posterior sections of the braincase, is elaborated into an intracranial joint in the former but remains a fissure in the latter. The division between the two parts represents a remnant of an embryonic condition (see Clack, Chap. 1, Fig. 1.4). A process of paedomorphosis could be invoked to explain the persistence of this feature.

The foundations for the structure of the otic region in bony vertebrates were established very early on in their evolution, at the inception of bony vertebrates, although the main features were already present in all jawed vertebrates (see Clack, Chap. 1). The main developments to otic structure in tetrapodomorphs occur only toward the crown of the tetrapodomorph tree, involving changes to the proportions of the components. The ethmosphenoid lengthened as the otoccipital shortened. The ventral cranial fissure and its corresponding dermal hinge line ultimately became sealed, the hyomandibula reduced in length and eventually lost its connection with the operculum. Ultimately, the stapes developed a footplate that inserted into the fenestra vestibuli, and the shaft was perforated by a stapedial foramen (Clack and Anderson, Chap. 4). The spiracular notch widened, and the cavity increased in volume to become, at least in part, the air-filled cavity of the middle ear in tetrapods. All these features could have developed in concert with an increased capability for air breathing, which in turn could have led to an increased likelihood of airborne-sound perception.

Studies in the developmental biology of the otic region in living taxa combined with further discoveries of transitional fossil taxa would help to interpret more precisely how the lateral commissure was lost, how the hyomandibula became transformed into a stapes, and how the vestibular fontanelle contributed to a fenestra vestibuli. None of these transformations, crucial though they are for tetrapod hearing, is currently clearly understood.

Compliance with Ethics Requirements J A. Clack and P. E. Ahlberg have declared they had no conflicts of interest.

References

Ahlberg, P. E. (1991). A re-examination of sarcopterygian interrelationships, with special reference to the Porolepiformes. *Zoological Journal of the Linnean Society,* 103, 241–287.

Ahlberg, P. E., & Johanson, Z. (1998). Osteolepiforms and the ancestry of tetrapods. *Nature*, 395, 792–794.

Andrews, S. M., Long, J., Ahlberg, P., Barwick, R., & Campbell, K. (2006). The structure of the sarcopterygian *Onychodus jandemarrai* n. sp. from Gogo, Western Australia: With a functional interpretation of the skeleton. *Transactions of the Royal Society of Edinburgh: Earth Sciences*, 96, 197–307.

Bishop, I. R., & Foxon, G. E. H. (1968). The mechanism of breathing in the South American lungfish, *Lepidosiren paradoxa*: A radiological study. *Journal of Zoology*, 154, 263–271.

Brazeau, M. D., & Ahlberg, P. E. (2006). Tetrapod-like middle ear architecture in a Devonian fish. *Nature*, 439, 318–321.

Campbell, K. S. W., & Barwick, R. E. (1982). The neurocranium of the primitive dipnoan *Dipnorhynchus sussmilchi* (Etheridge). *Journal of Vertebrate Paleontology*, 2, 286–337.

Campbell, K. S. W., & Barwick, R. E. (2000). The braincase, mandible and dental structures of the Early Devonian lungfish *Dipnorhynchus kurikae* from Wee Jasper, New South Wales. *Records of the Australian Museum*, 52, 103–128.

Carlström, D. (1963). A crystallographic study of vertebrate otoliths. *Biological Bulletin*, 125, 441–463.

Challands, T. J. (2015). The cranial endocast of the Middle Devonian dipnoan *Dipterus valenciennesi* and a fossilized dipnoan otoconial mass. *Papers in Palaeontology*, 1(3), 289–317.
Chang, M.-M. (1982). The braincase of *Youngolepis*, a Lower Devonian crossopterygian from Yunnan, South-Western China. Dissertation, University of Stockholm.
Chang, M.-M. (1995). *Diabolepis* and its bearing on the relationships between porolepiforms and dipnoans. *Bulletin du Museum national d'Histoire naturelle, Paris*, 17, 235–268.
Chang, M.-M., & Zhu, M. (1993). A new Middle Devonian osteolepidid from Qujing, Yunnan. *Memoirs of the Association of Australasian Palaeontologists*, 15, 183–198.
Christensen, C. B., Christensen-Dalsgaard, J., & Madsen, P. T. (2015). Hearing of the African lungfish (*Protopterus annectens*) suggests underwater pressure detection and rudimentary aerial hearing in early tetrapods. *Journal of Experimental Biology*, 218, 381-387.
Clack, J. A. (1996). Otoliths in fossil coelacanths. *Journal of Vertebrate Paleontology*, 16, 168–171.
Clack, J. A. (2007). Devonian climate change, breathing, and the origin of the tetrapod stem group. *Integrative and Comparative Biology*, 47, 510–523.
Clack, J. A., Ahlberg, P. E., Finney, S. M., Dominguez Alonso, P., Robinson, J., & Ketcham, R. A. (2003). A uniquely specialized ear in a very early tetrapod. *Nature*, 425, 66–69.
Clack, J. A., Sharp, E. L., & Long, J. A. (2011). The fossil record of lungfishes. In J. M. Jørgensen & J. Joss (Eds.), *The biology of lungfishes* (pp. 1–42). Enfield (NH): Science Publishers Inc.
Clement, A. M., & Ahlberg, P. E. (2014). The first virtual cranial endocast of a lungfish (Sarcopterygii: Dipnoi). *PLoS ONE*, 9(11), e113898.
Clement, A. M., Long, J. A., Tafforeau, P., & Ahlberg, P. E. (2016). The dipnoan buccal pump reconstructed in 3D and implications for air breathing in Devonian lungfishes. *Paleobiology*. doi:10.1017/pab.2015.4.
Clement, A. M., & Long, J. A. (2010). Air-breathing adaptation in a marine Devonian lungfish. *Biology Letters*, 6, 509–512.
Cloutier, R., & Ahlberg, P. E. (1996). Morphology, characters and the interrelationships of basal sarcopterygians. In M. L. J.Stiassny, L. R. Parenti, & G. D. Johnson (Eds.), *Interrelationships of fishes* (pp. 445–479). London: Academic Press.
Daeschler, E. B., Shubin, N. H., & Jenkins, F. A., Jr. (2006). A Devonian tetrapod-like fish and the evolution of the tetrapod body plan. *Nature*, 440, 757–763.
Downs, J. P., Daeschler, E. B., Jenkins, F. A., Jr., & Shubin, N. H. (2008). The cranial endoskeleton of *Tiktaalik roseae*. *Nature*, 455, 925–929.
Forey, P. L. (1998). *History of the coelacanth fishes*. London: Springer-Verlag.
Fritzsch, B. (1987). Inner ear of the coelacanth fish *Latimeria* has tetrapod affinities. *Nature*, 327, 153–154.
Gauldie, R. W., Dunlop, D., & Tse, J. (1986). The remarkable lungfish otolith. *New Zealand Journal of Marine and Freshwater Research*, 20, 81–92.
Jarvik, E. (1972). Middle and Upper Devonian Porolepiformes from East Greenland with special reference to *Glyptolepis groenlandica* n. sp. *Meddelelser om Grønland*, 187(2), 1–307.
Jarvik, E. (1980). *Basic structure and evolution of vertebrates, vol. 1*. London: Academic Press.
Jessen, H. L. (1980). Lower Devonian Porolepiformes from the Canadian Arctic with special reference to *Powichthys thorsteinssoni* Jessen. *Palaeontographica (A)*, 167, 180–214.
Johanson, Z., & Ahlberg, P. E. (1997). New tristichopterid (Osteolepiformes: Sarcopterygii) from the Mandagery Sandstone (Famennian) near Canowindra, N.S.W., Australia. *Transactions of the Royal Society of Edinburgh: Earth Sciences*, 88, 39–68.
Johanson, Z., Ahlberg, P. E., & Ritchie, A. (2003). The braincase and palate of the tetrapodomorph sarcopterygian *Mandageria fairfaxi*: Morphological variability near the fish-tetrapod transition. *Palaeontology*, 46, 271–293.
Lebedev, O. A. (1995). Morphology of a new osteolepidid fish from Russia. In H. Lelièvre & P. Janvier (Eds.), *Études sur les vértebres inférieurs* (pp. 287–342). *Bulletin du Muséum national d'Histoire naturelle, Paris*, 17(C), 1–4.

Liang, D., Shen, X. X., & Zhang, P. (2013). One thousand two hundred ninety nuclear genes from a genome-wide survey support lungfishes as the sister group of tetrapods. *Molecular Biology and Evolution*, 30, 1803–1807.

Long, J. A. (1993). Cranial ribs in Devonian lungfishes and the origin of dipnoan air-breathing. *Memoirs of the Association of Australasian Palaeontologists*, 15, 199–209.

Long, J. A., Barwick, R. E., & Campbell, K. S. W. (1997). Osteology and functional morphology of the osteolepiform fish *Gogonasus andrewsae* Long 1985, from the Upper Devonian Gogo Formation, Western Australia. *Records of the Western Australian Museum, Supplement*, 57, 1–89.

Long, J. A., Young, G. C., Holland, T., Senden, T. J., & Fitzgerald, E. M. G. (2006). An exceptional Devonian fish from Australia sheds light on tetrapod origins. *Nature*, 444, 199–202.

Lu, J., & Zhu, M. (2010). An onychodont fish (Osteichthyes, Sarcopterygii) from the Early Devonian of China, and the evolution of the Onychodontiformes. *Proceedings of the Royal Society of London B: Biological Sciences*, 277, 293–299.

McCormick, C. A., & Popper, A. N. (1984). Auditory sensitivity and psychophysical tuning curves in the elephant nose fish, *Gnathonemus petersii*. *Journal of Comparative Physiology A: Neuroethology, Sensory, Neural, and Behavioral Physiology*, 155, 753–761.

Miles, R. S. (1977). Dipnoan (lungfish) skulls and the relationships of the group: A study based on new species from the Devonian of Australia. *Zoological Journal of the Linnean Society*, 61, 1–328.

Romer, A. S. (1937). The braincase of the Carboniferous crossopterygian *Megalichthys nitidus*. *Bulletin of the Museum of Comparative Zoology*, 82, 1–73.

Smith, J. L. B. (1956). Old four legs. The story of the coelacanth. London: Longman, Green & Co.

Snitting, D. (2008). A redescription of the anatomy of the Late Devonian *Spodichthys buetleri* Jarvik, 1985 (Sarcopterygii, Tetrapodomorpha) from East Greenland. *Journal of Vertebrate Paleontology*, 28, 637–655.

Watson, D. M. S. (1925). The internal ear of *Osteolepis*. *Journal of Anatomy*, 59, 385–386.

Yan, H. Y., & Curtsinger, W. S. (2000). The otic gasbladder as an ancillary auditory structure in a mormyrid fish. *Journal of Comparative Physiology A: Neuroethology, Sensory, Neural, and Behavioral Physiology*, 186, 595–602.

Zhu, M., & Schultze, H. P. (2001). Interrelationships of basal osteichthyans. In P. E. Ahlberg (Ed.), *Major events in early vertebrate evolution: Palaeontology, phylogeny, genetics and development* (pp. 289–314). New York: Taylor & Francis.

Zhu, M., & Ahlberg, P. E. (2004). The origin of the internal nostril of tetrapods. *Nature*, 432, 94–97.

Zhu, M., Yu, X., Lu, J., Qiao, T., et al. (2012). Earliest known coelacanth skull extends the range of anatomically modern coelacanths to the Early Devonian. *Nature Communications*, 3, 772. doi:10.1038/ncomms1764

Chapter 4
Early Tetrapods: Experimenting with Form and Function

Jennifer A. Clack and Jason S. Anderson

Abstract This chapter describes the ear regions of tetrapods from the Paleozoic. In the past couple of decades, understanding of their morphology has been increased by discoveries of fossils from the Late Devonian and early Carboniferous (Mississippian) periods. The primitive condition, as found in several only distantly related taxa, consisted of a bulky stapes with a large footplate and a stout wing-shaped distal portion. In some, the stapes seems to have been a strut supporting the braincase. The hearing capabilities of these taxa were probably poor, confined to low-frequency detection primarily in water but, possibly, also in air. *Ichthyostega* had a highly modified version probably specialized for aquatic audition. Taxa from the later Carboniferous (Pennsylvanian) and Permian show a great deal of variation, especially among more terrestrially adapted forms. These include the larger stem amniotes, such as seymouriamorphs and diadectomorphs. All these taxa have some kind of a notch or embayment at the back of the skull, which in early forms was probably part of a spiracular mechanism but, in later ones, might have housed a tympanic membrane that closed off an air-filled middle ear cavity. The small recumbirostran microsaurs and the earliest amniotes had no such notch. Microsaur stapes had large footplates and short, stubby shafts. These and the earliest amniotes probably had no middle ear cavity. The earliest amniote stapes were robust, long, and laterally or downwardly projecting. In some cases they contacted the cheek bones or the jaw joint, likely precluding good aerial sound reception.

Keywords Aïstopod · Baphetid · Diadectid · Dorsal process · Embolomere · Lepospondyl · Nectridean · Opisthotic · Otic notch · Prootic · Recumbirostran · Seymouriamorph · Stapes · Stem tetrapod

---

J.A. Clack (✉)
University Museum of Zoology, Cambridge, Downing Street, Cambridge CB2 3EJ, UK
e-mail: j.a.clack@zoo.cam.ac.uk

J.S. Anderson
Department of Comparative Biology and Experimental Medicine,
University of Calgary, 3330 Hospital Drive, Calgary, AB T2N 4N1, Canada
e-mail: janders@calgary.ca

J.A. Clack et al. (eds.), *Evolution of the Vertebrate Ear—Evidence from the Fossil Record*, Springer Handbook of Auditory Research 59,
DOI 10.1007/978-3-319-46661-3_4

4.1 Introduction

Early tetrapods, or stem tetrapods, encompass a wide range of forms that lived in the middle to late Paleozoic. They include a mix of conventional tetrapod morphologies and some that resemble nothing alive today (for a review see Clack and Milner 2015). These tetrapods sit outside the node separating Amniota from Amphibia and, thus, are not part of the crown. The crown group of a lineage includes all the extant members of that lineage and all fossil relatives that lie within the node encompassing that group. For example, the crown group of tetrapods consists of the two major groups Amphibia (anurans, urodeles, and caecilians) and Amniota (reptiles including birds, and mammals). The stem group consists of all the fossil forms that are more closely related to that crown group than to any other crown group (Fig. 4.1). *Acanthostega* and *Ichthyostega* do not fit into either of the

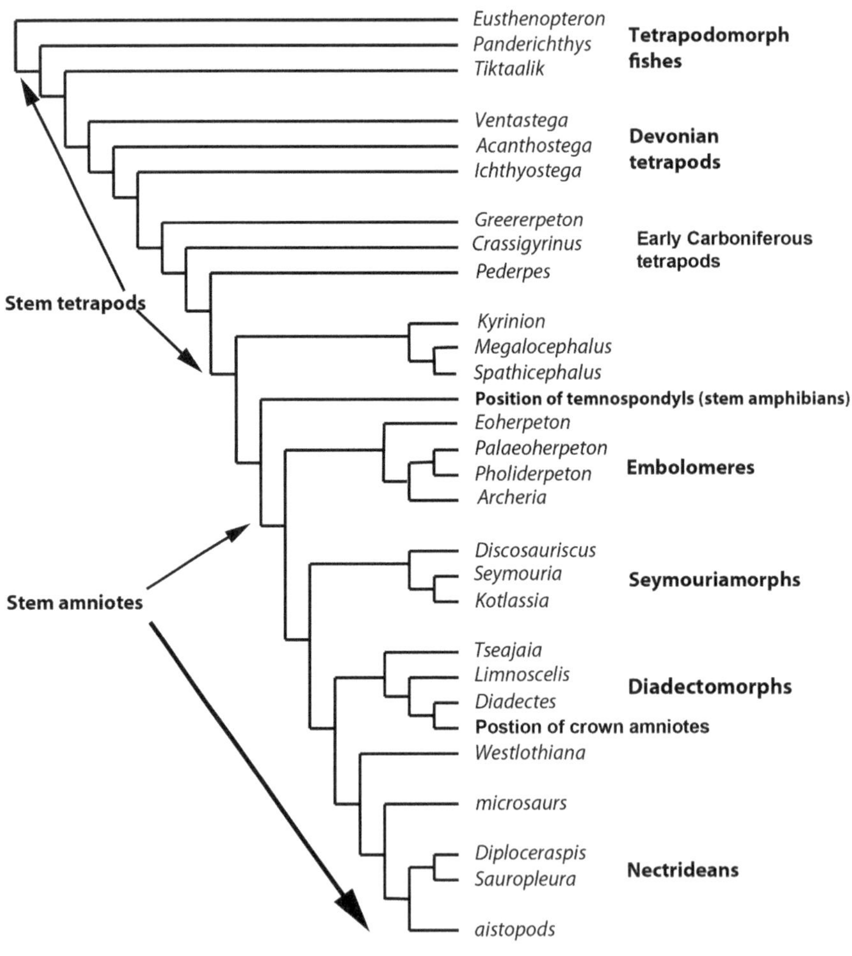

Fig. 4.1 A cladogram of most of the taxa mentioned in this chapter (see text for sources)

crown group members but fall below the split between amphibians and amniotes. Thus they are stem tetrapods. The ear regions found among these early forms show a bewildering variety of forms, few of which conform to the pattern seen among either amniotes or amphibians. Their capacities for audition are thus difficult to assess, but some inferences are discussed where possible in this chapter.

A basic scheme of relationships is shown in the cladogram in Fig. 4.1. However, the phylogeny of early tetrapods is poorly resolved, and their relationships are uncertain and subject to almost constant revision, generating a certain amount of controversy. The cladogram is taken from a number of sources, including Ruta and Coates (2007) and Coates et al. (2008). Figure 4.2 gives skeletal reconstructions of representative examples of the taxa or groups discussed in this chapter. This chapter covers a range of Devonian, Carboniferous, and Permian taxa in which the otic region is at least partially known (see Clack, Chap. 1, Fig. 1.2 for a geological timescale).

4.2 Devonian Stem Tetrapods

The earliest tetrapod (limbed tetrapodomorph) ear regions to be well preserved in the fossil record are those of the Late Devonian taxa *Acanthostega gunnari* and *Ichthyostega* spp. (Figs. 4.2, 4.3, 4.4). They come from East Greenland and were contemporaries and have been found in rocks between 363 and 360 million years (Ma) old. They appear to have inhabited different environments, and they were certainly very different from one another in terms of body morphology, including their otic regions. In most, but not all, phylogenetic analyses, *Acanthostega* is resolved as the more primitive, and its braincase morphology is more similar to that of a tetrapodomorph fish, such as *Eusthenopteron*, than is that of *Ichthyostega* (Clack 2012; Clack and Milner 2015).

4.2.1 Acanthostega

Acanthostega is notable for having paddle-like limbs, each bearing eight digits, and a tail fin bearing long fin rays along both the dorsal and ventral midlines (Fig. 4.2A). It had short straight ribs and its vertebrae were more or less uniform along the column. *Acanthostega* seems to have been highly adapted to an aquatic environment and has been considered primitively so—that is to say, there were no ancestors that were more terrestrial. This has become somewhat controversial in the light of more recent finds (e.g., Niedzwiedzki et al. 2010).

In many respects, the braincase of *Acanthostega* is comparable to that of *Eusthenopteron*. For example, the otic capsules are roofed by the conjoined pairs of opisthotics and prootics, and the capules lay entirely under the elongate postparietals. However, unlike *Eusthenopteron*, but as in all other tetrapods, a lateral

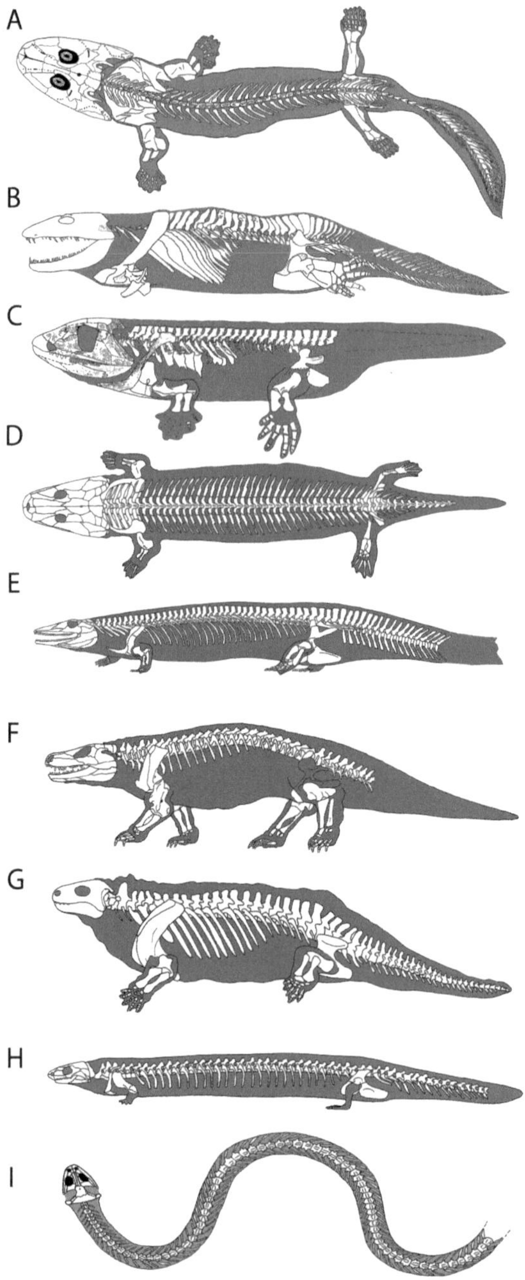
A
B
C
D
E
F
G
H
I

◂**Fig. 4.2** Skeletal reconstructions of some of the animals represented in this chapter. **(A)** *Acanthostega gunnari*; skull length about 120 mm, Sect. 2.1. **(B)** *Ichthyostega stensioi*; skull length up to 250 mm, Sect. 2.2. **(C)** *Pederpes finneyae*; skull length about 200 mm, Sect. 3.1. **(D)** *Greererpeton burkemorani*; skull length about 200 mm, Sect. 3.3. **(E)** *Archeria crassidisca*; skull length about 200 mm, Sect. 3.4. **(F)** *Seymouria baylorensis*; skull length about 150 mm, Sect. 4.1. **(G)** *Diadectes sideropelicus;* skull length about 180 mm, Sect. 4.2. **(H)** *Micraroter erthyrogeios;* skull length about 50 mm, Sect. 4.3.1. **(I)** *Lethiscus stocki*; skull length about 56 mm, Sect. 4.3.2. (**A**, **D**, **E–G**, **I** modified from Clack 2012; **B** modified from Ahlberg et al. 2005; **C** modified from Clack and Finney 2005; **H** modified from Carroll et al. 1998)

commissure (see Clack, Chap. 1; Clack and Ahlberg, Chap. 3) had been eliminated. The otic capsule was poorly ossified internally and only impressions of the anterior and posterior vertical semicircular canals can be seen. There are no ossifications of the floor or internal walls of the capsule. A large unossified space between the ventral margin of the capsule and the underlying basioccipital forms the opening of the fenestra vestibuli (FV). This space is confluent with the junction between the exoccipitals and the otoccipital, in other words, confluent with the lateral otic fissure. This is one factor that suggests that the FV is partially derived from that fissure (Fig. 4.3). The ventral part of the lateral otic fissure is expanded into the vestibular fontanelle in fishes such as *Eusthenopteron* (Clack and Ahlberg, Chap. 3)

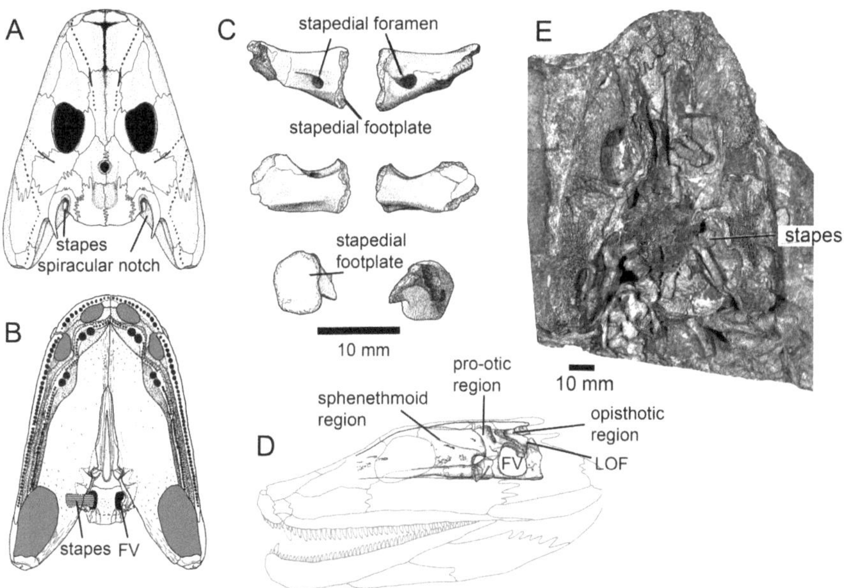

Fig. 4.3 *Acanthostega gunnari*, Natural History Museum, Copenhagen specimen MGUH 29019. **(A)** Skull roof showing the positions of the spiracular notch and stapes; the skull is about 120 mm long. **(B)** Palatal view of skull showing the positions of the stapes and FV. **(C)** Left stapes in six views. *top pair*: dorsal, ventral; *middle pair*: posterior, anterior; *bottom pair*: medial, lateral. **(D)** Braincase of *Acanthostega* shown within an outline of the skull roof, showing the positions of the FV and lateral otic fissure. **(E)** Photograph of the skull of a specimen of *Acanthostega* showing the right stapes in position. The left stapes is also present. *FV*, fenestra vestibuli; *LOF* lateral otic fissure. (**A**, **B** modified from Clack 2012; **C–E** original)

and probably provided part of the FV. It is unclear whether the basiparasphenoid contributed to the ventral margin of the FV or whether only the basioccipital did so. In *Acanthostega*, the parasphenoid is short and is separated from the basioccipital by a suture representing the remnant of the ventral cranial fissure (Clack 1998).

The stapes is a robust bone with a figure-eight-shaped footplate, although the footplate is much smaller than the FV. Whether or not the FV was completed in cartilage cannot be established with certainty but in several specimens the stapes remains in situ, suggesting that it was held in place by firm soft tissue (Clack 1994, 1997). The more distal part of the stapes is a wing-shaped plate with a distal margin not lined with periosteal or cortical bone (i.e., it is “unfinished”). The stapes is penetrated by a large stapedial foramen toward the proximal end (Fig. 4.3). It cannot be assumed that the foramen formed a passage for the stapedial artery, although that is what the stapedial foramen houses in modern tetrapods (see Sect. 4.6). The footplate would have been in close apposition to the perilymphatic space via the FV, and *Acanthostega* was the earliest tetrapod (though not the most primitive, see Sect. 4.2.3 *Ventastega*) to have had this configuration.

Rather than a single notch at the back of the skull, *Acanthostega* has two, separated by a prong-like extension to the lateral part of the tabular. The more lateral of these is most likely to be the spiracular notch, lying at the junction between the squamosal and tabular.

There is some uncertainty as to the spatial relationships between the stapes and other parts of the skull. In dorsoventrally compressed specimens, the stapes lies against the mediodorsal face of the palatoquadrate ossification. If that relationship were retained in life, it would effectively eliminate a recess for a middle ear cavity. However, this apparent close association may be the result of compression. The stapes may have been more free-standing in life, which may mean it was freer to vibrate and transmit sound to the inner ear. Nevertheless, the stapes was robust and relatively massive compared with stapes adapted for higher frequency hearing. It may have been capable of transmitting low-frequency sound from the water or from the substrate (Christensen et al. 2015), and the stapes also may have had a role in the operation of a spiracular mechanism during air breathing but almost certainly did not have a tympanic membrane.

4.2.2 Ichthyostega

Ichthyostega has quite a different postcranial anatomy from that of *Acanthostega* and, in some ways, this animal seems more terrestrially adapted (Figs. 4.2B, 4.4). It has robust forelimbs with apparently weight-bearing lower arm bones and elbow joint. Its vertebral column was also very well ossified and differentiated into regions comparable to that of mammals. The ribs were thick, flanged, and overlapping. By contrast, the hind limbs of *Ichthyostega* were paddle-like and bore seven digits in an unusual pattern, with three small, closely spaced digits at the leading edge and four more robust ones posteriorly. The third digit was minute. The limb appears not to

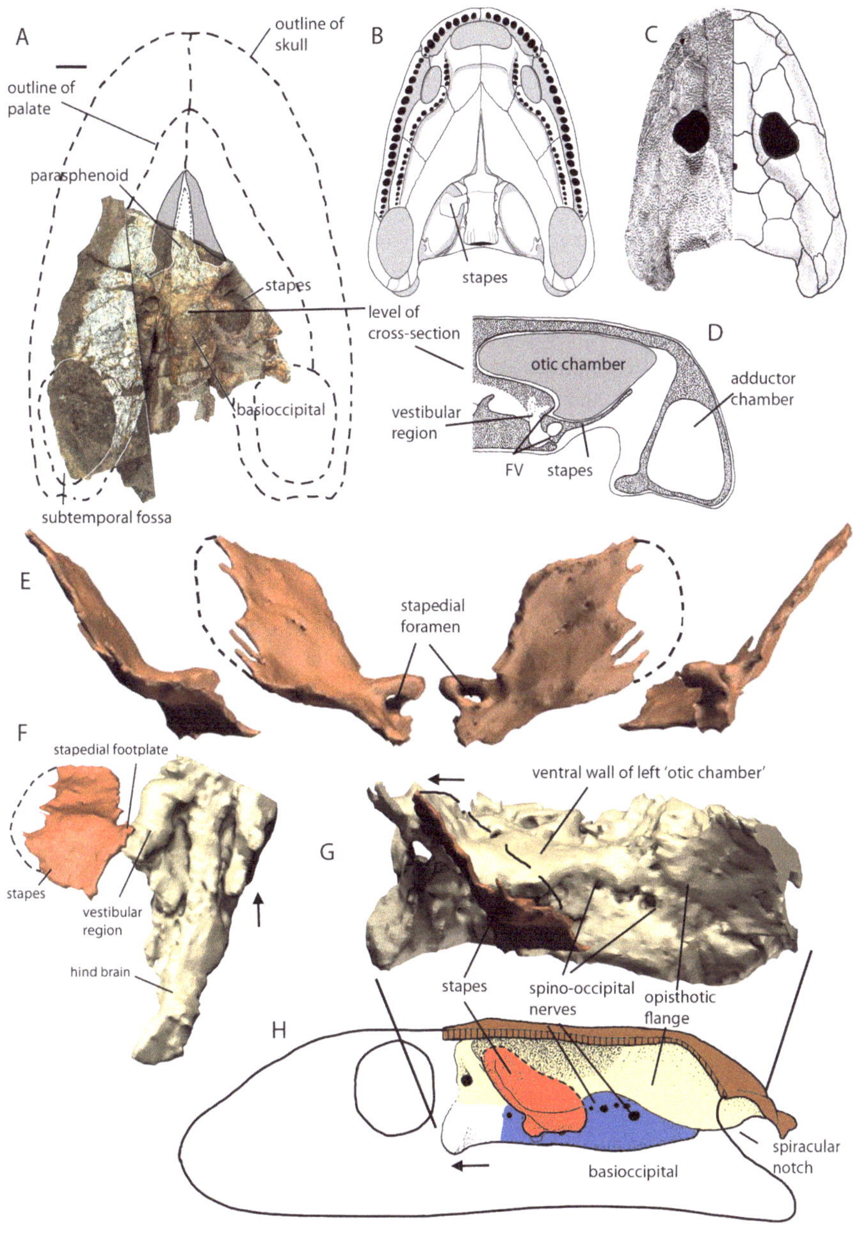

Fig. 4.4 *Ichthyostega stensioi*, Natural History Museum, Copenhagen specimen MGUH VP 6018. **(A)** Palatal view. Outlines of the missing parts of the skull are in *dotted lines*. **(B)** Palatal view of skull. **(C)** Dorsal view of skull. **(D)** Reconstructed cross section of skull as shown in A, seen in posterior view. **(E)** Left stapes in four views. *Left to right*: lateral, posterior, anterior, and medial. **(F)** Rendering of the brain cavity and associated stapes in ventral view from μCT scans of MGUH f.n. 180. **(G)** Rendering of otoccipital in left-lateral view with the stapes in place from μCT scans. **(H)** Interpretive drawing of otoccipital and stapes, with outline of skull indicated. *FV*, fenestra vestibuli; *arrows* point anteriorly; *scale bar* in A, 10 mm. (**A**, **D** modified from Clack et al. 2003; **B**, **C**, **E–H** original)

have had weight-bearing joints at the knee or ankle. Its tail, like that of *Acanthostega*, had fin rays but many fewer and much shorter. Studies have shown that *Ichthyostega* could not walk in the usual manner, and the conflicting range of adaptations make its lifestyle difficult to interpret (Pierce et al. 2012). Nonetheless, the structure of its humerus, more primitive in some respects than that of *Acanthostega*, raises questions about the polarity of other aspects of its anatomy (Callier et al. 2009).

The anatomy of the otic region of *Ichthyostega* remained a mystery until recent investigations used micro-CT (μCT) scanning to model it in three dimensions (Clack et al. 2003). The otic region resembles that of no other early tetrapod, tetrapodomorph fish, or even any modern taxon (Fig. 4.4). Because of its bizarre morphology, earlier descriptions were forced to use anatomically neutral terms such as "longitudinal ridge" and "striated process" to describe its features. An almost circular structure flanking the elongated basioccipital was interpreted as the base of the saccular vesicle (Jarvik 1996). Scanning and three-dimensional modeling was able to solve some of the problems, although not all. The circular structure was resolved as the stapes, a thin-walled spoon-like element, with its convex surface facing ventrally (Fig. 4.4A–E). The proximal end close to the footplate was penetrated by a large stapedial foramen. The footplate articulated with the basoccipital via a ventral component, and a larger, more dorsal component inserted into a foramen in the region of the lateral otic fissure. Other foramina were situated along the fissure or in the wall of the basioccipital, presumably for some of the cranial nerves (Fig. 4.4F–H). The stapes was much more anteriorly placed than in *Acanthostega* or any other tetrapod.

The position and size of the vestibular apparatus could not be resolved in detail, but it appears to have been proportionately very small. The stapes appears to have formed part of the ventral wall of a chamber whose lateral, anterior, dorsal, and medial walls were formed by the palate, skull roof, and otic capsule. The back wall was formed by a posterior flange of bone from the tabular. The chamber was interpreted to have been filled with air, a derivative of the spiracular chamber, and to be part of an aquatically adapted ear. The air-filled chamber is thought to be analogous to the ostariophysan swim bladder. Sound vibrations from the water passed through the air-filled chamber and caused the delicate stapes to pivot on its ventral process, allowing the dorsal process to transmit the vibrations to the inner ear. A unique scroll-like structure known as "Säve-Söderbergh's process" (Jarvik 1996) descends into the chamber from the skull roof and is thought to have been part of some kind of valve mechanism controlling air flow. At the back of the skull, between the squamosal and tabular bones, *Ichthyostega* has a conspicuous notch that would have communicated with the chamber. As far as has been determined, the otic region lay entirely beneath the large, single postparietal bone and the tabulars (Clack et al. 2003).

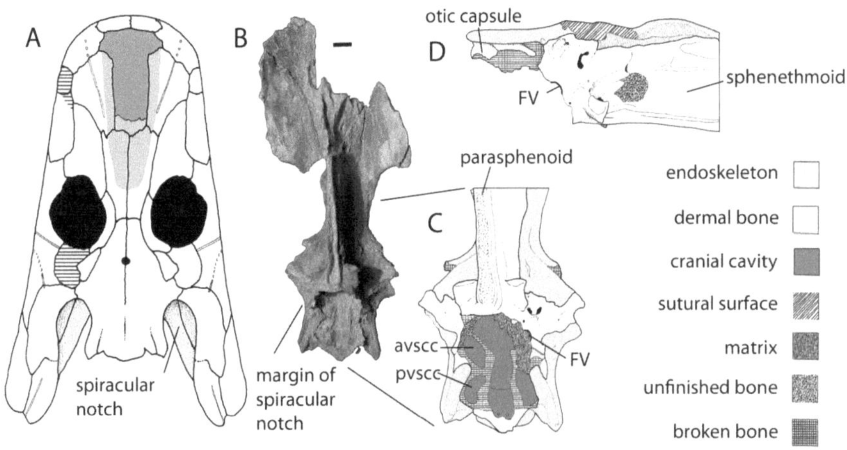

Fig. 4.5 *Ventastega curonica*. **(A)** Reconstruction of skull roof showing spiracular notch. **(B)** Photograph of partial skull roof and preserved braincase in ventral view Latvian Natural History Museum Specimen LDM G81/775. **(C)** Interpretive drawing of the posterior part of the braincase in dorsal view showing internal structure. **(D)** Interpretive drawing, left-lateral view. *avscc*, anterior vertical semicircular canal; *FV*, fenestra vestibuli; *pvscc*, posterior vertical semicircular canal. *Scale bar* in B, 10 mm. (**A**, **C**, **D** modified from Ahlberg et al. 2008; photo in **B** courtesy of P. E. Ahlberg)

4.2.3 Ventastega

A Late Devonian tetrapod from Latvia is another taxon that shows some aspects of the otic region (Fig. 4.5). A little later in the Devonian than *Acanthostega*, *Ventastega curonica* is in some ways more primitive, though in other respects it is similar. The skull is known from exquisite fossils with bones preserved in three dimensions. Much of the braincase is preserved, including the posterior region of the sphenethmoid, the prootic, and the dorsal parts of the otic capsule. Like *Acanthostega*, *Ventastega* lacks a lateral commissure. The presence of an FV is indicated by a smooth curved margin of the prootic that provided its anterodorsal segment. From these two factors, the presence of a stapes rather than a hyomandibula can be inferred. It is the most primitive tetrapod so far to have an FV. Impressions of the anterior and posterior semicircular canals are present in the ventral view of the prootic and opisthotic (Ahlberg et al. 2008).

4.3 Carboniferous Stem Tetrapods

The otic regions of several early Mississippian (early Carboniferous) stem tetrapods are known in some detail. One of the first to be described was that of *Greererpeton burkemorani* (Sect. 4.3.3) from the mid-Carboniferous of the USA (Carroll 1980;

Smithson 1982). The discovery of several intact skull specimens was influential in changing ideas about the stapes of early tetrapods and in helping to interpret those of other taxa.

4.3.1 Whatcheeriids

Whatcheerids have been identified in localities as widely distributed geographically as Australia, Scotland, and the USA and, geologically, from the Tournaisian to the late Viséan. However, further finds may show them to be a "grade group" exhibiting similarly primitive characters, but having no derived characters in common. Only one, *Pederpes finneyae* from Scotland, preserves any otic material and that is a stapes (Clack and Finney 2005). It is similar in overall shape to that of *Acanthostega* but both absolutely larger and relatively smaller (compared to the skull). This is further evidence that the robust and broad shape is plesiomorphic for tetrapods. The skull is deep, in contrast to the shallower earlier tetrapod skulls such as those of *Acanthostega*. In *Pederpes*, the spiracular notch is wide, with the rear margin of the skull sloping but essentially straight (Figs. 4.2C, 4.6A, B).

4.3.2 Baphetids

In the nineteenth century, baphetids (formerly known as loxommatids) were among the first Carboniferous tetrapods to be discovered (Fig. 4.6C–G). They remain enigmatic, in part because only one specimen has been found that preserves significant amounts of postcranial material. Otherwise, only skulls are known and only two of these include stapes, namely *Kyrinion martilli* (Clack 2003) and *Spathicephalus mirus* (Beaumont and Smithson 1998). The animals were predatory forms in the coal-swamp faunas of the time, between approximately 335–310 million years ago (Mya), contemporary with colosteids such as *Greererpeton* (Sect. 4.3.3). Most of the specimens come from Scotland, and there is only a fragment of a skull from North America.

One of the commonest and best-known baphetids is *Megalocephalus*. The braincase and skull roof material of this species show the attachment points of the otic capsules. The prootic process containing part of the anterior vertical canal extended laterally under the supratemporal and medially for a short distance under the parietals. Together, the capsules provided a solid roof to the braincase space and vestibular regions, and there is no evidence of a supraoccipital (Beaumont 1977).

Kyrinion preserves both stapes in situ, and some of the lateral parts of the otic capsule are visible. The FV was typically large for an early tetrapod, with contributions from the underlying parasphenoid, with which the ventral margin of the

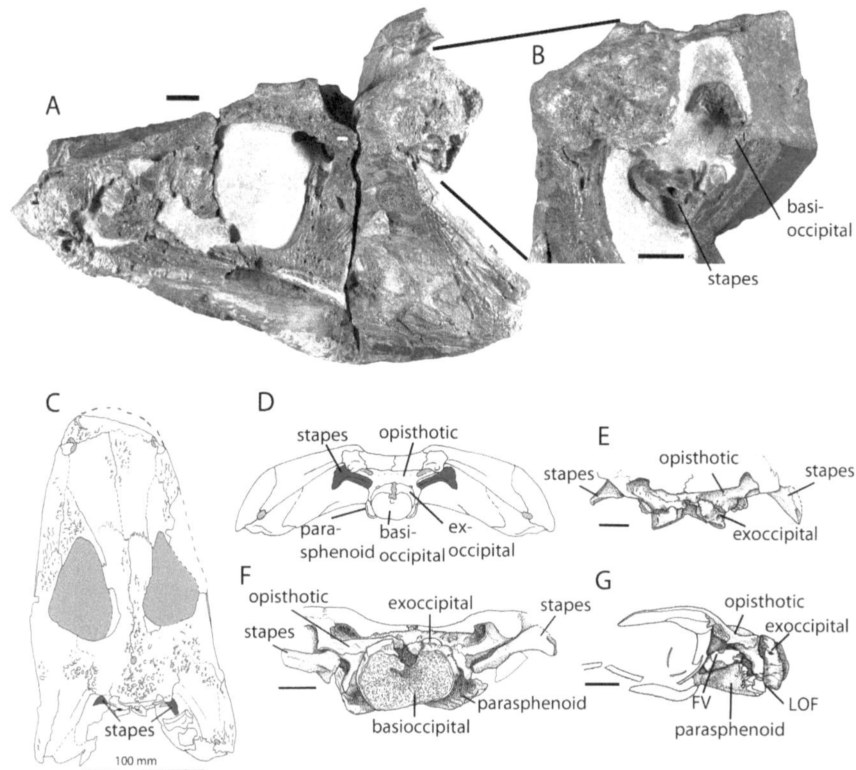

Fig. 4.6 *Pederpes finneyae* and *Kyrinion martilli*. **(A–B)** *Pederpes finneyae* Hunterian Museum Glasgow specimen GLAHM 100815. **(A)** Photograph of skull in left-lateral view. **(B)** Enlargement of skull roof and spiracular notch showing the stapes as preserved. **(C–G)** *Kyrinion martilli*. **(C)** Skull roof of only specimen. **(D)** Reconstruction of occipital view. **(E)** Specimen drawing of dorsal view of occiput. **(F)** Specimen drawing of occipital view of skull. **(G)** Specimen drawing of otoccipital and parasphenoid in left-lateral view. *FV* fenestra vestibuli; *LOF* lateral otic fissure. Scale bars 10 mm unless otherwise indicated. (**A**, **B** modified from Clack and Finney 2005; **C–G** modified from Clack 2003)

stapedial footplate articulated. The lateral otic fissure separated the basioccipitals and exoccipitals from the otic capsule and the parasphenoid and was confluent with the FV. The stapes were similar in form to those of *Acanthostega* and *Greererpeton*, although somewhat more slender and with a distinctly longer waisted region between the footplate and the distal expansion (Clack 2003; Robinson 2005, 2006). There was no stapedial foramen in contrast to most other early tetrapods, although this is the common condition among stem amphibians (temnospondyls; see Schoch and Anderson, Chap. 11). The distal part of the stapes was very poorly ossified, but most of the posterior margin was finished with cortical bone. The footplate was a dorsoventrally elongated oval and was divided into a ventral concave component and a more dorsal convex one. The latter inserted into the FV. Whether or not the stapes was free to vibrate at its junction with the parasphenoid is not clear, but it does appear that there was space for a middle ear cavity. Some low-frequency sound transmission may have been possible as in other early tetrapods, although, like them, it seems unlikely that there was a tympanic membrane. Rather, the notch between the supratemporal and squamosal may have been occupied by a spiracular opening. Despite some compression of the skull, the stapes did not contact the palataoquadrate ossifications and did not form a supporting element for the braincase, which was otherwise firmly sutured to the palate and braincase.

Spathicephalus has an extraordinarily broad and flat skull in which the flattening cannot have been simply the result of compression. Its otic capsules, though well preserved, do not show much anatomical detail except that they are extremely short but very wide. The proportions are reflected in those of the skull roof. The stapes is preserved in place in one specimen and is essentially similar to that of *Kyrinion*, *Acanthostega*, and *Greererpeton*.

4.3.3 Greererpeton

Greererpeton burkemorani is an aquatic tetrapod belonging to the group known as colosteids (Figs. 4.2D, 4.7). They were components of the coal-swamp fauna of the Carboniferous period and have been found in the USA and Scotland. The family is known from rocks 335–310 Ma. The placement of the family within the tetrapod tree is labile. They have been linked with either temnospondyls, which are the stem group of Amphibia (Romer 1947, 1968; Carroll 1987; see Schoch and Anderson, Chap. 11); adelogyrinids, which are a group of early Mississippian, long-bodied, almost limbless forms whose otic capsules and stapes are poorly known (Ruta and Coates 2007); or most commonly, the family is treated as a separate stem plesion, which is a taxon that is on a branch of the cladogram all on its own (e.g., Laurin 1998; Ruta et al. 2003). As a group, the colosteids sometimes appear basal to whatcheeriids and the single taxon of *Crassigyrinus*, and sometimes they appear crownward of them. The cladogram in Fig. 4.1 places *Greererpeton* basal to *Crassigyrinus* and *Pederpes*, respectively. *Greererpeton* is a long-bodied animal with relatively short limbs and an elongate, flattened, and bluntly triangular skull. It had no spiracular or otic notch at the rear of the skull, which could be explained, on the one hand, as the loss of a tympanic ear in an aquatic animal, or by the idea that *Greererpeton* had returned to gill breathing for its main intake of oxygen and thus had no need of a spiracle (Clack 2012).

The otic capsule had no independent ossification in its roof; a gap between the two antimeres was capped entirely by the dermal bones of the skull. The opisthotics were poorly ossified so that the paroccipital processes were incomplete laterally. The FV was large, taking up much of the lateral wall of the capsule (Fig. 4.7). An unfinished region at the posteroventral margin exposed part of the internal

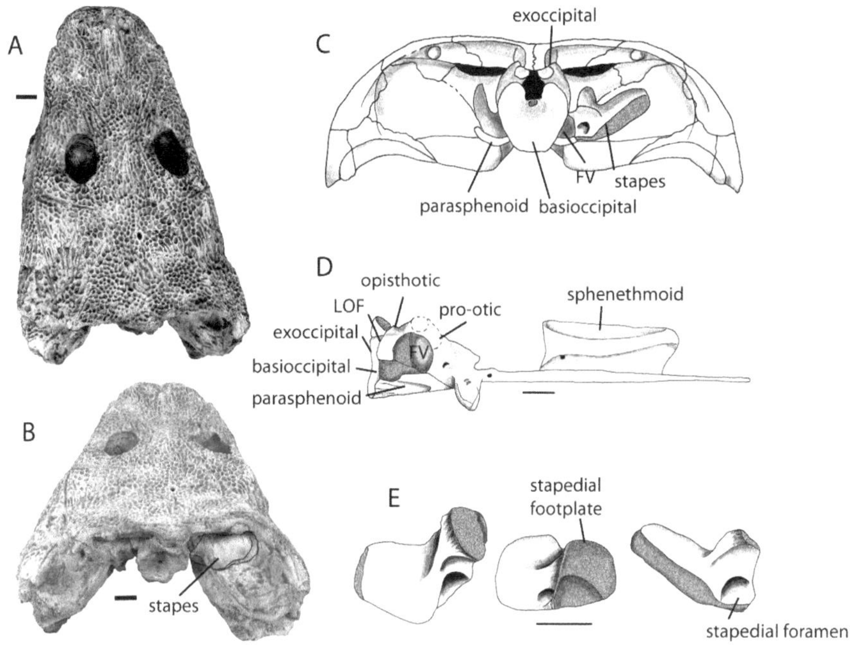

Fig. 4.7 *Greererpeton burkemorani.* **(A)** Cast of a skull; contrast heightened with ammonium choride. **(B)** Oblique view of the skull to show the position of the right stapes. **(C)** Reconstruction of occipital view of skull. **(D)** Braincase in right-lateral view. **(E)** Left stapes in three views, *left to right*: dorsal, medial, posterior. *FV* fenestra vestibuli; *LOV* lateral otic fissure. *Scale bars*, 10 mm. (**A**, **B** original photos by JAC; **C–E** modified from Smithson 1982)

basioccipital, leaving a gap between the opisthotic and the parasphenoid that merged into the lateral otic fissure between the capsule and the exoccipitals. The ventral margin of the FV was formed by the basiparasphenoid, as seen in lateral view. The stapes was a large bone with a footplate that more or less filled the FV. The distal part of the bone was similar in shape to that of *Acanthostega*: a wing-shaped plate, flattened ventrally, slightly convex dorsally, and with a groove running dorsally over the stapes from anterior to posterior close to the position of the stapedial foramen. The distal and posterior margins of the plate were likewise unfinished, suggesting extensions in cartilage (Smithson 1982). Proportionally, it was even more massive than that of *Acanthostega*.

As in *Acanthostega*, the spatial relationship between the stapes and the bones of the palate has been disputed. The majority of *Greererpeton* specimens are flattened, which could either be natural or the result of compression during fossilization. In these, the bulky stapes is lying closely appressed to the bones of the palate. There appears to be no possibility of a middle ear cavity, and the stapes has been interpreted as a support for the braincase within the skull (Carroll 1980). The braincase otherwise has no direct ossified connection with the skull roof. On the other hand, one skull from a different locality is preserved as a deeply three-dimensional structure in which the stapes does not reach the palate but is independently suspended. It would not function as a support for the braincase, and there may have been space for a middle ear cavity (Schultze and Bolt 1996). Nevertheless, the absolute size and bulk of the stapes does seem to rule it out as an efficient means of transmitting sound vibrations, certainly from air.

The condition in *Greererpeton* was one of the first indications that the early tetrapod ear was not primitively adapted for aerial hearing, and it was not equipped with an "otic notch" and a tympanic membrane. Around the same time as its description, a study of the conditions in modern tetrapods had suggested a similar idea (Lombard and Bolt 1979). The subsequent discoveries of the stapes of *Pholiderpeton* and *Palaeoherpeton* (see Sect. 4.3.4), and later *Acanthostega*, were convincing evidence that the early tetrapod ear, if adapted for hearing at all, was capable of detecting only low-frequency sound vibrations in water or through the substrate. The "otic notch" of many early tetrapods has become reinterpreted as a "spiracular notch" associated with breathing (Clack 1992). The suggestion that the stapes played a role in support of the braincase is one that emerges again from other groups, including early amniotes.

4.3.4 Embolomeres

Embolomeres were long-bodied, aquatic, crocodile-like piscivores ranging from the Late Mississippian to the Early Permian of Europe and North America about 330–290 Mya (Figs. 4.2E, 4.8). Embolomeres are much rarer in the fossil record than the almost ubiquitous temnospondyls (stem amphibians; see Schoch and Anderson, Chap. 11). Phylogenetic analysis usually places them above stem amphibians, closer to amniotes than to modern amphibians and related directly or indirectly to more terrestrial forms from the Late Pennsylvanian. They are often placed in a large group known as anthracosaurs. By contrast with most of the taxa described so far, the skulls of these animals were relatively deep dorsoventrally. The otic capsules are known in several taxa, such as the mid-Carboniferous *Eoherpeton* (Smithson 1985) and the Early Permian *Archeria* (Clack and Holmes 1988), but stapes are known in only two: *Pholiderpeton scutigerum* and *Palaeoherpeton decorum* (Clack 1983).

The otic capsules were well ossified dorsally and sat under both the short postparietals and the posterior part of the parietals. The capsules met in the midline but there does not seem to have been a supraoccipital. The opisthotics were visible at the back of the skull and, as in *Acanthostega*, colosteids, and baphetids, the exoccipitals did not contribute to the capsules but were separated from them by the lateral otic fissure. The region of the FV is not well preserved in any specimen and

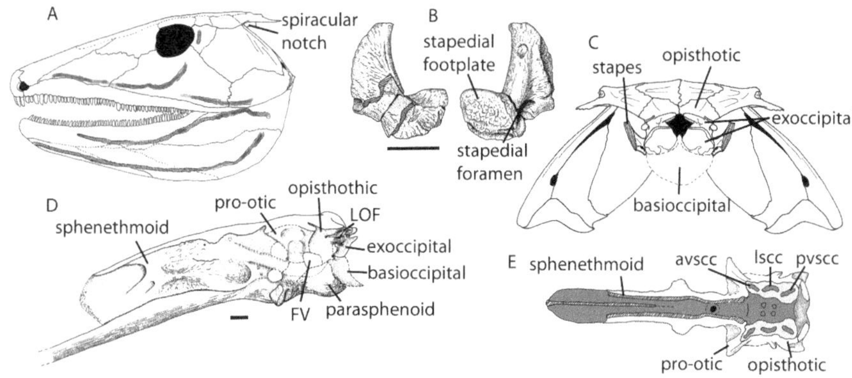

Fig. 4.8 Embolomeres. (**A**) Lateral view of the reconstructed skull of *Pholiderpeton scutigerum*. (**B**) Stapes of *P. scutigerum*: left dorsal view, right ventral view. (**C**) Occipital view of skull of *P. scutigerum*. (**D**) Left side of reconstructed braincase of *P. scutigerum*. (**E**) Ventral view of horizontal section through braincase of *P. scutigerum*. *avscc*, anterior vertical semicircular canal; *FV*, fenestra vestibuli; *LOF*, lateral otic fissure; *lscc*, lateral semicircular canal; *pvscc*, posterior vertical semicircular canal. *Scale bars*, 10 mm. (Figures modified from Agnew 1984)

may, as in *Acanthostega*, colosteids, and baphetids, have been large and confluent with the lateral otic fissure.

The stapes of anthracosaurs, including embolomeres, remained unknown for many decades despite a few fairly complete and articulated specimens. The animals were originally interpreted as having ear regions and stapes that were already adapted for aerial hearing. They were classed as "labyrinthodonts", which was a group that included temnospondyls and other large early tetrapods (see Sect. 4.4.3; also see Schoch and Anderson, Chap. 11). Thus they were assumed to show the same pattern as temnospondyls in terms of their ear structure. An elongate and rod-like stapes was attributed to them, and the notch situated at the back of the skull was assumed to have housed a tympanic membrane (Thomson 1966).

The specimen of *Pholiderpeton* preserves a small bone with a hole through it, whose identity was at first in doubt. However, the discovery and full description of the stapes of *Greererpeton* demonstrated not only that it was a stapes but also how it was oriented. Shortly after that, it became clear that a specimen of *Palaeoherpeton* also preserved a stapes that had hitherto gone unrecognized (Clack 1983). In the early-mid 1980s when these specimens were described, it was becoming clear that anthracosaurs and temnospondyls were not closely related and, in fact, belong to the disparate stem groups of Amniota and Amphibia, respectively. Thus there was no justification for interpreting the ear of the former in the light of the latter.

In shape, the stapes is much like a more slender version of that of *Greererpeton* and *Acanthostega*. It has a relatively large footplate offset dorsally from the main body of the bone, with a large stapedial foramen placed close to the footplate. The distal part was likewise a wing-shaped plate, although narrower in outline than in the latter two taxa. As in the stapes of *Greererpeton* and *Acanthostega*, those of *Pholiderpeton* and *Palaeoherpeton* had unfinished posterior and distal edges. In the reconstructed skull, the stapes contacts the palatoquadrate ossification, although the contact may have been slight or even absent (Clack 1987). It certainly does not appear to have formed a significant support for the braincase, which has a broad contact with the skull roofing bones.

4.4 Later Carboniferous and Permian Tetrapods

During the later Carboniferous (Pennsylvanian), the diversity of tetrapods increased greatly. Many older groups expanded, others arose, and many of the latter came to dominate faunas of the Early Permian. Groups that are recognized as stem amniotes became important components of the fauna and the first herbivores evolved. Some of these groups have no living analogs. Similar faunas existed across the USA and Eurasia. Ear structure among these groups was correspondingly diverse. The first true amniotes have been recognized from Pennsylvanian in localities in Canada.

4.4.1 Seymouriamorphs

Seymouriamorphs are a group of stem amniotes from the latest Pennsylvanian and Early Permian of the USA, Europe, and Russia, from about 300 to 290 Mya (Figs. 4.2F, 4.9A–C) (Berman et al. 2000). Before the advent of cladistic analysis, their relationships to other fossil vertebrates were controversial because they have very amniote-like postcranial skeletons but a skull structure that is clearly not that of an amniote (White 1939; Laurin 2000). For example, they have widely open "otic" notches at the rear of the skull. *Kotlassia* is a terrestrial form from Russia that had interlocking dermal scutes on its back and lived in an arid environment (Bystrow 1944). Ontogenetic series of several taxa from eastern Europe, known as discosauriscids, showed that the larvae were aquatic and had external gills for part of their development, so they were clearly non-amniote (e.g., Klembara 1995, 1997). Seymouriamorphs were often linked with embolomeres as part of the group of anthracosaurs. Cladistic analyses now place them firmly among the more crownward of stem amniotes, not particularly closely allied with anthracosaurs or embolomeres (e.g., Ruta and Clack 2006).

The posterior part of the skull (the skull table) in all seymouriamorphs is short but broad and reflects a suite of unusual features of their otic regions. The ventral surface of the dermal roofing bones shows that the braincase and otic capsules lay beneath the tabulars, postparietals, part of the supratemporal, and the posterior half of the parietals. The otic capsules were laterally broad, spanning the skull table almost from side to side but, like the skull table, they were anteroposteriorly short. Even in those taxa like the discosauriscids in which endoskeletal braincase and otic

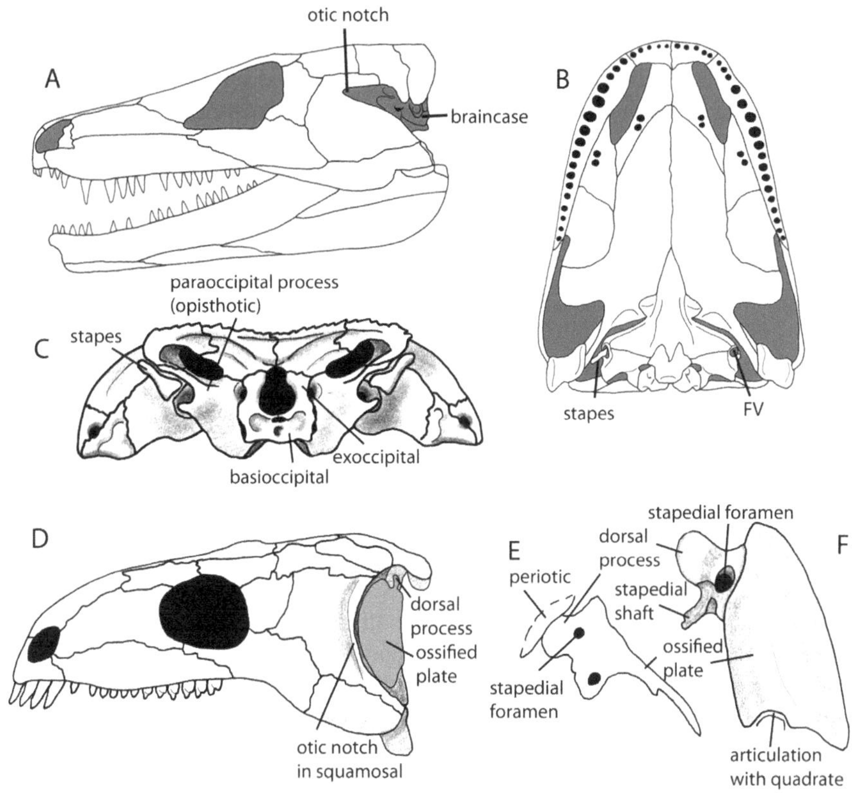

Fig. 4.9 Seymoriamorphs and diadectomorphs. **(A)** Lateral view of skull of *Seymouria baylorensis* showing otic notch; skull length about 120 mm. **(B)** Palatal view of *S. baylorensis* showing stapes and FV. **(C)** Occipital view of *Kotlassia prima*; skull width about 120 mm. **(D)** Left-lateral view of skull of *Diadectes sideropelicus*; skull length about 200 mm. **(E)** Section through stapes of *D. sideropelicus*. **(F)** Stapes and possible extrastapes or ossified plate of *D. sideropelicus*. (**A**, **B** reproduced with permission from Clack 2012; **C** modified from Bystrow 1944; **D**, **F** modified from Olsen 1966; **E** modified from Olsen 1966 and Berman et al. 1998)

capsule are not preserved, some of these unique features can be inferred from the lateral extensions of the dermal parasphenoid that underplated them.

The best-known seymouriamorph taxon, and one with well-preserved otic regions and braincases, is *Seymouria baylorensis* from the Early Permian of Texas and Oklahoma, although they are also visible in *Kotlassia*. The otic notch in *Seymouria* was greatly extended anteriorly and bordered by a deeply excavated but relatively narrow squamosal embayment for much of its margin. Dorsally, the supratemporal and tabulars formed a downwardly projecting flange with a complex internal structure to which the otic capsule attached. Posteriorly, the tabular continued to form much of the back wall of the skull along with the postparietal. There was no separate supraoccipital ossification. The otic capsule was produced laterally into a tube at the end of which lay the FV. Part of this structure was formed by the opisthotic, which also formed the paroccipital process, and part by the prootic under which lay the lateral wings of the parasphenoid. The exoccipitals were far removed from the FV, which thus had no connection with the lateral otic fissure. A small, cylindrical, and poorly ossified bone has been identified as the stapes in *Seymouria* (White 1939; Laurin 2000), but a better-preserved one is described in *Kotlassia* (Bystrow 1944). It is a robust bone with a broad footplate and a curved shaft with a small, proximally placed stapedial foramen (Fig. 4.9D). In the fossil, there was no contact between the stapes and the FV, suggesting that it was loosely held in life. A stapes has also been identified in *Discosauriscus*. It is a crescentic bone that is narrow at the presumed distal end and broader at the presumed proximal end with a poorly developed process in a proximodorsal position. It is poorly ossified and only covered in cortical bone along the ventral surface, with both ends unfinished (Klembara 1997).

In the more terrestrial forms, such as *Seymouria* and *Kotlassia*, the assumption often has been that there was a tympanic membrane present in the otic notch. In the discosauriscids, the crescentic notch occupied the entire posterodorsal margin of the squamosal and was bordered dorsally by the supratemporal and tabular. This is the case even in the larval individuals and is one reason why the discosauriscids were originally considered to be placed among the branchiosaurs (see Schoch and Anderson, Chap. 11). These larval temnospondyls had similarly configured posterior embayments in the back of the skull. Whether the notches in these aquatic animals also housed a tympanic membrane raises interesting and, so far, unanswered questions (see Sect. 4.6).

4.4.2 Diadectomorphs

Diadectomorphs were large barrel-bodied herbivores with crushing dentitions and heavily ossified skulls, and they lived across Euramerica during the Late Pennsylvanian and Early Permian (Figs. 4.2G, 4.9) (Case 1905; Olson 1947). They are among the earliest herbivorous forms to be found in the fossil record. Roughly contemporary with seymouriamorphs, they are sometimes found in the same deposits, such as the Bromacker Quarry in Germany (Berman et al. 2004). There are three families: the diadectids, the tseajaiids, and the limnoscelids. The latter two are monotypic genera represented by only a few specimens.

The diadectids are known from a number of genera of which the best known is *Diadectes*. They are found in the southern USA and Germany in a similar distribution to that of seymouriamorphs, part of a widespread community of terrestrial forms. Although the skull of *Diadectes* has received numerous different interpretations of its suture pattern over the years, new specimens have improved our understanding. At the back of the skull, the dermal bones were excavated into a dorsal embayment, revealing the posteriorly exposed postparietals and a combined supraoccipital and opisthotic. There was a broadly crescentic otic notch formed by the excavated margin of the squamosal posteriorly and the supratemporal dorsally. The occipital plate contacted the large supratemporal, but the tabular was small and superficial. Their otic regions are known from specimens that preserve excellent detail, including the dorsal otic capsule. The semicircular canals were encased in bone, enabling their courses and orientations to be studied. Serial sections permitted a wax model to be constructed, and further information could be gained by μCT scanning, although this has not yet been done. Based on these data, it is clear that the opisthotic formed a robust paroccipital process that contacted the tabulars laterally and carried the opening of the FV into a lateral position via a canal, although shorter than in seymouriamorphs. The FV was oriented with the opening facing ventrolaterally. As in seymouriamorphs, the FV was quite separate from the remnant of the lateral otic fissure, which was represented only by the gap between the exoccipitals and the basioccipital. Anterior and dorsal to the FV and its associated otic tube was another canal of uncertain function (Olson 1947, 1966).

Diadectids have probably the most peculiar stapedial and tympanic structure of any tetrapod, and their homologies have been interpreted in different ways over the decades. The region has been described from serial sections of a *Diadectes* skull that clarified some aspects of their anatomy, although not all (Olson 1966). The stapes had a massive knob-like dorsal process that abutted against the wall of the otic capsule and through which a stapedial foramen passed. Below this a narrow, curved stapedial process is thought to have carried the footplate, though the specimen is incomplete and the process did not appear to reach the FV. Below this, a second rounded process was pierced by a second canal of uncertain function, and between this and the stapedial shaft, there was a groove thought to have housed the chorda tympani nerve VII. Distally, the body of the stapes expanded into, or was sutured to, an extensive thin plate of bone that filled the otic notch and has been interpreted as an ossified tympanic membrane (Olson 1966; Berman et al. 1998) (Fig. 4.9D, E). Despite its peculiarity, this structure is seen in other genera, including *Orobates* from Germany (Berman et al. 2004). The functional implications of this arrangement are not understood. One suggestion might be that it is analogous to the plate-like extrastapes of aquatic turtles (Sobral and Müller, Chap. 9), or the calcified cartilage disc-like extrastapes in some mosasaurs (aquatic lepidosaurs; see Evans, Chap 8), although there seems no other evidence that diadectids were aquatic.

Tseajaia campi has a similarly constructed, though shallower, otic notch and an otic capsule that is also similar to that of *Diadectes* as far as can be established, but there is no evidence of the peculiar stapes and ossified plate seen in that animal (Moss 1972; Berman et al. 1992). *Limnoscelis paludis* appears to lack a notch entirely (Romer 1946; Fracasso 1987). The squamosal has a convexly curved dorsal border that fits under the concave ventral margin of the overlying postorbital and supratemporal bones. The combined posterior margin of these bones is convex, not concave. However, the only skull of this species is compressed dorsoventrally, and there is a distinct groove where the overlying skull table bones lap over the squamosal. There is also a posterior wall to the back of the skull that has been interpreted as formed by the palatoquadrate ossification (JAC, personal observation). These regions would be worth re-investigating using modern scanning methods to establish their identity securely.

4.4.3 "Lepospondyls"

At one time, Carboniferous tetrapods were placed into two groups: labyrinthodonts were large animals, with the eponymous folded enamel in their teeth, compound vertebrae, and an "otic notch"; lepospondyls were small animals without folded enamel in their teeth, had unitary vertebrae, and no otic notch. Subsequent cladistic analysis has demolished the Labyrinthodontia and, more recently, the "Lepospondyli" are increasingly emerging as non-monophyletic (e.g., Anderson 2007). Nonetheless, the groups that once comprised the lepospondyls are usually placed among the stem amniotes (e.g, Ruta et al. 2003), although some analyses place some of them as stem amphibians (Laurin 1998, 2004). Most of the forms preserve very little of their otic apparatus and so will be introduced only briefly here. A few specimens of two groups, the recumbirostrans and the aïstopods, by contrast, show very well-preserved braincases and otic regions that have been revealed by recently collected material, further preparation, and μCT scanning (Fig. 4.10).

Nectrideans are a diverse group of Pennsylvanian and Early Permian tetrapods of uncertain affinities, and they may not be a monophyletic group. Some taxa had enormously extended tabulars forming delta-shaped or boomerang-shaped skulls. One of the most extreme was *Diplocerapsis* from the Late Pennsylvanian and Early Permian of the USA. A few specimens, such as *Sauropleura* and *Diplocerapsis*, include preserved opisthotic bones, and some include basioccipitals, but in none is a stapes known (Carroll et al. 1998).

Lysorophids were once considered a separate group of lepospondyls and, like nectrideans, were also from the Late Pennsylvanian and Early Permian of the USA. They had elongate bodies, tiny limbs, and highly modified skulls. Their phylogenetic relationships are uncertain.

4.4.3.1 Recumbirostran "Microsaurs"

Recumbirostra is a monophyletic subgroup of what had previously been called tuditanomorph microsaurs. These animals share massively ossified skulls and an overturned snout. Some groups also have elongate bodies, and together this suggests varying degrees of head-first burrowing behavior. Most members of the group are less than about 300 mm and, in some cases, only about 70 mm in body length. The majority of taxa are found from the Early Pennsylvanian to the Early Permian of the USA and Europe.

The difficulty in studying "lepospondyl" ears is that a large proportion of them are preserved as flattened impressions in coal. Since the bone of these fossils is

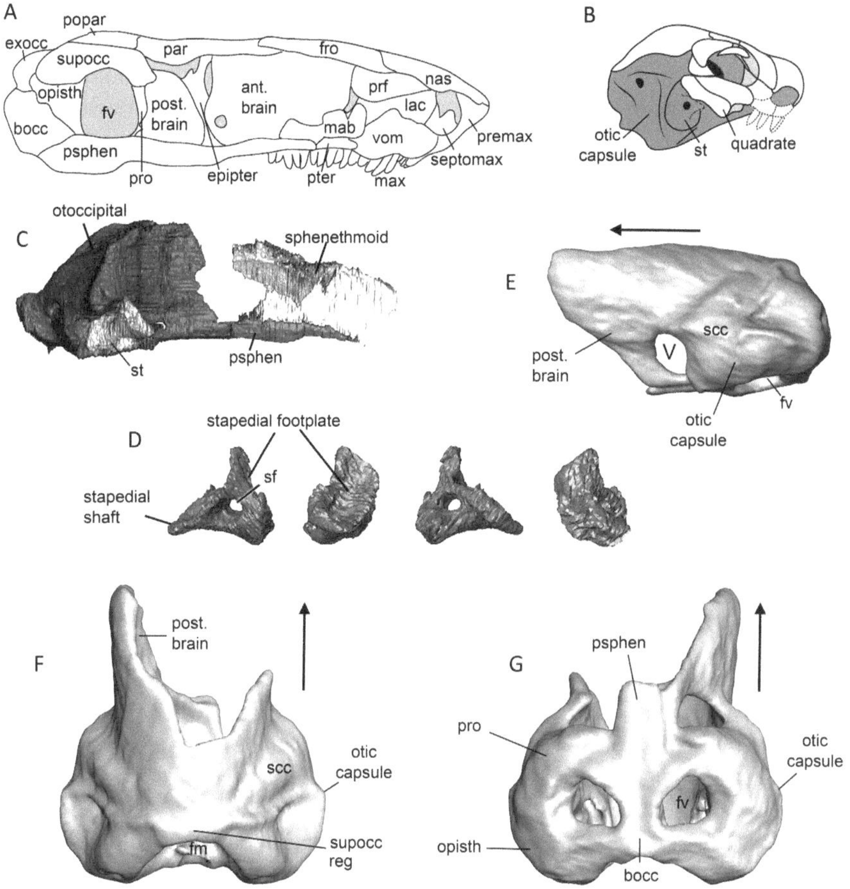

Fig. 4.10 Representative "lepospondyls." (**A**) Lateral view reconstruction of the braincase of recumbirostran *Rhynchonkos* with external dermal skull elements removed from lateral view of braincase; skull length about 14 mm. (**B**) Lateral skull reconstruction of recumbirostran *Quasicaecilia* note the stapes bracing the quadrate; skull length about 15 mm. (**C**) Lateral view of the braincase of recmbirostran *Carrolla*, with stapes in place in the FV; skull length about 16 mm. (**D**) Stapes of *Carrolla* in anterior, medial, posterior, and lateral views. (**E–G**) The phlegethontid aïstopod *Sillerpeton*; skull length 2.2 mm. (**E**) Left-lateral view, (**F**) dorsal view, and (**G**) ventral view. *Arrows* point anteriorly. *ant. brain*, anterior portion of braincase; *bocc*, basioccipital; *epipter*, epipterygoid; *exocc*, exoccipital; *fro,* frontal; *fm*, foramen magnum; *fv,* fenestra vestibuli; *lac*, lacrimal; *mab*, median anterior braincase bone; *max*, maxilla; *nas*, nasal; *opisth*, opisthotic; *par*, parietal; *post. brain*, posterior portion of braincase; *popar*, postparietal; *premax*, premaxilla; *prf*, prefrontal; *pro*, prootic; *psphen*, parasphenoid; *pter*, pterygoid; *reg*, region; *scc*, traces of semicircular canals; *septomax*, septomaxilla; *sphen*, sphenethmoid; *sf*, stapedial foramen; *st*, stapes; *supocc*, supraoccipital; *V*, foramen for cranial nerve V (trigeminal); *vom*, vomer. (**A** from Szostakiwskyj et al. 2015, commons license; **B** from Pardo et al. 2015, commons license; **C**, **D** modified from Maddin et al. 2011)

often poorly preserved, these specimens are often studied by removing the bone with acids. Rubber peels are then produced from the resulting natural molds (Baird 1955). Fortunately, recumbirostrans are generally preserved in three dimensions in flood plain deposits or other more terrestrial environments, and thus are well-suited to study using μCT.

Understanding the patterns of braincase morphology of recumbirostrans in a phylogenetic context is in its early stages, but a few trends can be seen. One is a trend toward consolidation of the braincase ossifications into a single, massive structure. Coupled with this is another trend of reduction of dermal ossifications in the skull roof, presumably as the consequently enlarged braincase assumes more of a support role. Some recumbirostrans show a doming of the posterior skull roof (Fig. 4.10A), creating a large occipital surface for epiaxial muscle insertion, while others seem to accentuate the hypaxial series (Pardo et al. 2015).

Several recumbirostrans had their ears described in the years before access to μCT scanning was available. Carroll and Gaskill (1978) described the ears of several taxa, although in the case of some (e.g., *Rhynchonkos*) the descriptions were based on specimens belonging to several different taxa. Two of the better described taxa are the broad-headed ostodolepids *Pelodosotis* and *Micraroter* (Fig. 4.2H). The footplate of the stapes of *Pelodosotis* was described as "palmate." Instead of being circular, its margin has a series of thickened areas that radiate from the base of the stapedial shaft (known as the columella in these taxa), out toward its margins. There is also a distinct, short, dorsal process from the base of the imperforate stapes that itself projects laterally to contact the quadrate ramus of the pterygoid. A dorsal process of the stapes is also seen in amniotes and, given the common placement of recumbirostrans on the amniote stem, this might suggest that amniote-styled hearing was in the process of being acquired at this evolutionary stage. The FV is not well shown in these taxa but its construction appears consistent with other taxa. The opisthotics are formed of thin bone, the prootics meet at the dorsal midline, and the exoccipital wraps around the posterior margin laterally. The parasphenoid floors the FV but it is incomplete, with a rounded medial emargination ventral to the stapes. A fragment of bone previously attributed to an "accessory ossicle" (Carroll and Gaskill 1978) overlies the columella posteriorly but may be a misidentified part of the hyoid or shoulder girdle.

Similarly, *Micraroter* seems to have had an emargination of the parasphenoid ventral to the FV, although in this case the emargination is partially comprised of the basisphenoid as well, the two bones suturing at a level approximately even with the midpoint of the FV (Carroll and Gaskill 1978). *Micraroter* has a more standard stapes than *Pelodosotis*, with a round footplate that completely fills the FV and a foramen at the base of the columella. The columella projects laterally and probably articulated with the quadrate. The great distance between the lateral wall of the braincase and the quadrate means that the columella is longer than in most recumbirostrans, and it has a more gracile appearance than is typical for them.

Romer (1969) provided a partial description of the ear region of *Pantylus*, a large-headed form with crushing dentition, a little less than 300 mm in body length. The FV is circumscribed by the opisthotic and prootic, which extend laterally, and by the parasphenoid ventrally. However, much of the prootic and the ventral part of the opisthotic is overlapped by the parasphenoid, so that in lateral view it appears to form much of the anterior margin of the FV. The medial wall of the otic capsule was unossified. Although no stapes were found in place, an isolated element identified by Romer has a broad footplate and a prominent dorsal process that would have projected toward the dorsal surface of the otic capsule. The columella would have projected laterally toward the quadrate and pterygoid, but it is unknown whether they made contact.

Micro-CT data have shown that *Rhynchonkos* has a well ossified braincase (Fig. 4.10A) (Szostakiwskyj et al. 2015). Also, as seen in the generalized form *Huskerpeton* (Huttenlocker et al. 2013), the otic capsules are floored by the parasphenoid, which are offset from the occiput by a recess for the basicranial fenestra. Distinct stout opisthotics and prootics frame the FV on the remaining margins, meeting at the midpoint of the dorsal margin at a straight suture. The relatively wide FV is completely filled by a stapes with a broad, plate-like footplate that is concave on its medial surface. A stapedial foramen is present on the short, stout columella, which projects anterolaterally toward the quadrate. The stapes does not articulate with the quadrate because of the intervening quadrate ramus of the pterygoid. No "accessory ossicle", such as that previously identified (Carroll and Gaskill 1978), was found. Traces of the horizontal semicircular canal are seen on both the prootic and opisthotic ventral to the dorsal shelf, and a lagenar crest is present between the FV and horizontal canal.

A few brachystelechid recumbirostrans have also been studied using μCT because some analyses have suggested a sister-group relationship with caecilians (see Schoch and Anderson, Chap. 11). Brachystelechids are diminutive recumbirostrans, with highly consolidated braincases and greatly reduced dermal skull roofs (postparietals are absent). The jaw articulations are placed anterior to the otic capsules, which represent the widest point of the skull. *Carrolla* (Fig. 4.10C) is the best known (Maddin et al. 2011) and, whereas its morphology is highly derived with respect to other recumbirostrans, it is still recognizably "microsaurian." The braincase has been compared with the os basale of caecilians, representing a coossification of the basioccipital, exoccipitals, parasphenoid, prootics, and opisthotics (also possibly a supraoccipital). The medial wall of the otic capsules is mostly absent, but a partial foramen is present on its dorsal margin that probably represents the perilymphatic foramen. This opening leads to a channel on the inside of the dorsal braincase that may be the path of the endolymphatic sac. The FV is situated ventrolaterally and the footplate of the robust stapes completely fills it. The columella is robust, has a stapedial foramen at its base, and is directed anterolaterally toward the quadrate, which possesses a roughened facet for the articulation (Fig. 4.10D).

Because the braincase is so well ossified it was possible to produce a digital rendering of the endocast of the inner ear. The auditory and vestibular regions are well represented. All three semicircular canals and their ampullae are well shown. They are highly arched, and the horizontal canal appears nearly circular in dorsal view. The anterior canal is slightly taller than the posterior, but the posterior canal passes below the level of the horizontal canal. Nestled between all three canals is a dorsal projection from the common crus that is interpreted as a dorsal sinus. The ventral auditory region is narrower than the vestibular region, and is slightly taller than wide, terminating at the FV and the footplate of the stapes. No impressions for possible sensory epithelia or foramina for branches of the auditory nerve can be seen because the medial wall is mostly unossified. Despite the high resolution of the endocast, Maddin et al. (2011) were unable to determine whether the auditory region corresponded more closely with the amniote (enlarged basilar papilla) or amphibian (saccule with multiple papillae) patterns. What was noted was that the enlarged ventral region of the inner ear (relative to frogs and salamanders) was consistent with a burrowing lifestyle.

Another brachystelechid was recently described via μCT. *Quasicaecilia* is even more extreme in its cranial specializations than *Carrolla* (Fig. 4.10B) (Pardo et al. 2015). It has a similarly consolidated otoccipital of the braincase, but it is uniquely expanded laterally, and the highly reduced complement of dermal skull roof is greatly foreshortened, giving the skull a brachycephalic morphology. Although reconstructed by Carroll (1990) with a massive stapes with a large dorsal process, *Quasicaecilia* is not as well preserved as *Carrolla* and the stapes is present only as fragments.

4.4.3.2 Aïstopods

Aïstopods are elongate, limbless nonamniotes with over 200 total vertebrae in some species. *Lethiscus* is the oldest (Wellstead 1982), from the Early Mississippian of the United Kingdom (Fig. 4.2I). In many ways it is the most mysterious aïstopod and has been placed at the base clade, but sister-group relationships remain difficult to understand because of aïstopods' highly derived morphology (Anderson et al. 2003).

Most aïstopods are known from cannel coals. As a result, braincase and stapedial morphology are usually obscured by compression. One specimen, *Sillerpeton*, however is exceptionally well preserved and its otoccipital region is very well described (McGinnis 1967; Lund 1978). This little specimen presents an entire otoccipital region (Fig. 4.10E–G). It is fully co-ossified, comprising exoccipitals, opisthotics, prootics, possibly a supraoccipital, basisphenoid, and parasphenoid. It has two large FVs located mostly ventrally and slightly laterally. Other foramina present include the prootic fenestra, and foramina for the jugular vein, carotid artery, and numerous cranial nerves up to and possibly including the posterior margins for cranial nerve II. Being miniaturized, the traces of the semicircular canals are visible on the exterior bone surface. Seen from a digital endocast, the vestibular portion is very well-developed with ovoid arcs to the canals, but the acoustic region is nearly absent (JSA, personal observation) and is dominated by the impression of the FV.

This little braincase is consistent with portions of the braincase seen in a later-occurring aïstopod, *Phlegethontia*, which is otherwise only known from impressions in either flat coals or nodules (Anderson 2002). Aspects of stapes morphology can nevertheless be determined from them. The FV is completely infilled by the stapedial footplate. The stapes lacks a columella; instead, there is a ridge that runs more or less transversely across the ventral surface of the footplate. Previous workers have suggested this might represent a scar for the insertion of an opercularis-type muscle arising from the remnants of the pectoral girdle (Anderson 2002).

4.5 Earliest Amniotes

Westlothiana lizziae has been suggested as a stem amniote placed above diadectomorphs in phylogeny. It shows some of the skull characters seen in true amniotes, including the reduction of some of the posterior temporal series and, in keeping with this, it lacks any kind of posterior temporal or spiracular notch (Smithson et al., 1994). However, the known specimens are generally poorly preserved, and *Westlothiana* lacks some of the palatal characters shared by true amniotes. It has also been resolved close to the "microsaurs", although since at least some of these appear to be stem amniotes, there is no great conflict implied by this position. Its age—Late Mississippian (mid-Carboniferous)—makes it one of the earliest members of either group and much older than the uncontroversial stem amniotes of the Pennsylvanian (late Carboniferous).

Late Carboniferous stem amniotes date from the Langsettian stage of the Early Pennsylvanian, the best known being *Hylonomus lyelli* from the fossil tree stump locality of Joggins, Nova Scotia (Carroll 1964). Its cranial elements are disarticulated but are considered to have the typically "notchless" condition of early amniotes. This condition is also described in slightly younger examples including *Paleothyris* from Florence, Nova Scotia (Carroll 1969) and *Coelostegus* from Nýřany in the Czech Republic (Carroll and Baird 1972). Stapes and partial otic capsules are described from both of the latter and feature a unitary supraoccipital and a downwardly directed, robust stapes with a dorsal process. All of these specimens are poorly preserved and more or less dissociated. They require recent study with more advanced techniques. They appear to demonstrate the primitive condition of the amniote otic region, but it remains a problem as to how this evolved from something more like the primitive broad stapes and spiracular notch of Devonian and early Carboniferous tetrapods.

4.6 Early Developments in the Tetrapod Hearing System

There are a number of developments that can be traced during the early evolution of the tetrapod ear from those of the tetrapodomorph fishes that deserve both comment here and further study in the future. During the evolution of these early forms, the external wall of the otic capsule was reorganized, including the disappearance of the crista parotica and lateral commissure (Clack, Chap. 1, Fig. 1.3; also see *Eusthenopteron*, Clack and Ahlberg, Chap. 3) of tetrapodomorph fishes through which the jugular vein was routed. The stapedial foramen appears at a comparable level to the old jugular canal and so may have contained the jugular vein rather than the stapedial artery. The loss of the lateral commissure forms part of a reconfiguration of the braincase during the evolution of tetrapods that is not well understood in terms of its embryonic development. Few modern fishes have been studied in this respect, and the most likely to yield relevant information, coelacanths, are not amenable to developmental studies. Furthermore, the homology between the stapes and the hyomandibula also is understood poorly. Recent studies regarding the contribution of the neural crest to both the otic capsule and the stapes suggest that neural crest from the second arch as well as mesodermal tissue contribute to both the stapes and to the otic capsule. It is as if part of the old hyomandibula was incorporated into the capsule: the FV does not form properly if the stapes is absent (O'Gorman 2005; Thompson et al. 2012).

Early studies showed that the otic region of the skull roof reduces in length across the fish–tetrapod transition and, furthermore, the reduction continues into the origin of amniotes (Westoll 1938; Romer 1941). Those studies, however, used *Ichthyostega* and an embolomere to represent the transitional stage between fish and amniotes. It is now known that *Ichthyostega* had a highly specialized ear that is probably unrepresentative of other tetrapods (Clack et al. 2003) and that embolomeres are probably stem amniotes rather than stem tetrapods (Ruta et al. 2003). Thus the observed gradual reduction in otic size is probably oversimplified. Nonetheless, additional data appear to corroborate the early observations (Clack 2012). Reduction in size of the underlying otic capsules is implicated, though the nature of the relationship and the timing of the changes remain unclear. In tetrapodomorph fishes, the size of the otic capsules scales closely to the length of the overlying postparietal bones of the skull roof (Fig. 4.11).

Among tetrapods, this relationship breaks down. In early tetrapods and on into amniotes, the postparietal bones reduce in size until they disappear; whereas, in some temnospondyl lineages, the postparietals become enormously elongated. In other words, development of the dermal skull roof becomes decoupled from that of the underlying neurocranium. This can be seen in the comparison between *Eusthenopteron* (Clack and Ahlberg, Chap. 3) and an embolomere such as *Pholiderpeton*. At what point this decoupling occurred could be studied by close investigation of fossil taxa, but also by developmental and genetic studies of modern amphibians and amniotes. The longstanding but currently unresolved question of skull bone homologies between the otic regions of fish and tetrapods likely stems directly from this developmental decoupling.

The reduction in size of the otic capsules during the origin of tetrapods may reflect a reduction in size of the semicircular canals. An early study noticed that tetrapods have smaller semicircular canals relative to body mass than do fishes (Jones and Spells 1963). This hypothesis was later tested for fossil taxa including "osteolepiforms" (the closest then-known fossil fish relatives of tetrapods) and two

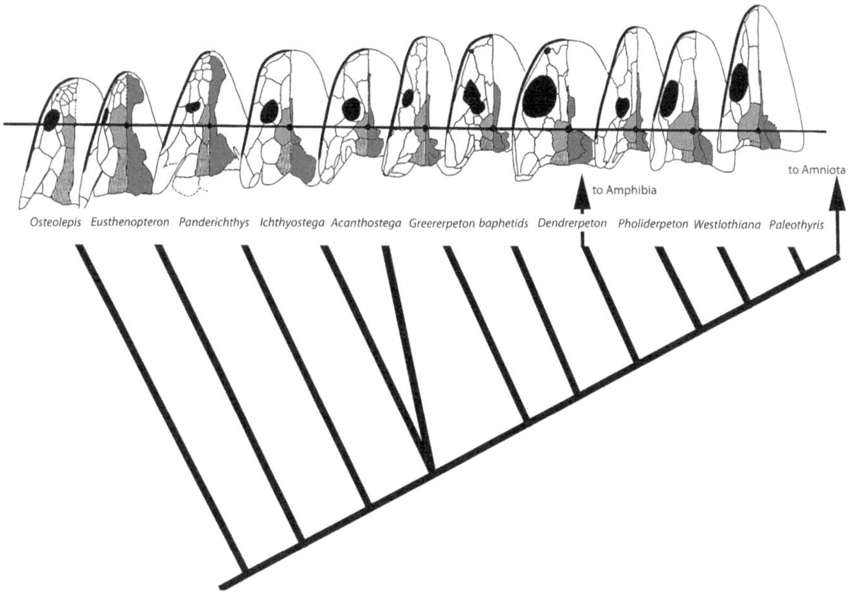

Fig. 4.11 Cladogram of early tetrapods with the position of Amphibia and Amniota indicated. The skulls are aligned on the parietal foramen and to the same snout-quadrate length to show changing proportions of the braincase and otic regions across the fish–tetrapod transition. Right-hand side, *dark gray* shows the braincase, left-hand side shows the postparietal (*vertical hatch*) and parietal (*light gray*) bones. (Modified from Clack 2012)

fossil amphibians, *Edops* and *Eryops*, using an alligator as a model for body mass. This study found that while all the fishes fell approximately on a single regression line, the two tetrapods fell on a clearly different and lower one (Bernacsek and Carroll 1981). However, only two fossil tetrapod data points were available for this study, and the statistical significance of the difference discovered was not tested. Furthermore, these two fossil tetrapod taxa were both Early Permian temnospondyls (stem amphibians), not particularly primitive nor representative of tetrapods in general. The study should now be repeated and enlarged to investigate its validity and wider relevance.

Several hypotheses were put forward to explain the reduction in canal size among tetrapods, including the suggestion that the separation of the skull from the shoulder girdle, the subsequent evolution of a neck, and operation in the lower viscosity of air compared with water increased the angular acceleration that the head would undergo, and thus the sensitivity of the canals needed to be reduced (Bernacsek and Carroll 1981). Furthermore, it is known that among modern tetrapods, agile species have canals with larger arcs and, in consequence, greater sensitivity than more sluggish species (Spoor 2003). Fish, dealing with life in three dimensions, may be more agile and as a result may require greater sensitivity and larger canals than early tetrapods that in effect dealt with only two dimensions. The fact that *Ichthyostega* seems to have had a very small vestibular region might suggest that it had begun the process of reduction, although the middle ear, if adapted for underwater hearing, seems to contradict this idea.

Consideration of another major habitat and environmental transition—from land to water by cetaceans—brings up another enigma. During that transition, rather than echoing the reduction in vestibular dimensions across the water-land transition, in cetaceans the vestibular region becomes proportionately smaller, the exact opposite (Spoor et al. 2002; Kandel and Hullar 2010).

The presence or absence of a tympanic membrane versus a spiracular opening in the notch at the rear of nearly all of these early forms is a long debated and continuing problem. In the earliest forms such as *Acanthostega* and baphetids, the relative size of the stapes suggests its absence. In *Greererpeton*, it is obvious from the lack of a notch that there was neither a tympanic membrane nor the alternative of a spiracular opening. The lack of a notch in microsaurs would similarly suggest that there was no tympanic membrane. However, there is an argument from the Extant Phylogenetic Bracket (EPB) (Witmer 1995) that where a notch was present, it was closed by a membrane or skin of some kind rather than open as in the spiracle of *Polypterus*. In both members of the EPB, namely dipnoans and crown group tetrapods, there was no opening to the spiracular cleft, so to argue that an opening was present in early tetrapods would run counter to the principle of the EPB. Furthermore, as in the larval forms with notches, it becomes hard to say at what point, either in the development of the organism or in the course of cladogenesis, an open spiracular notch would become a closed tympanic ear.

One fact seems to be well established, and that is that the temporal or spiracular notch was lost early in amniote evolution with no trace of it in the posterior part of the skull. The dermal bones forming the temporal series in early tetrapods, including some of those that bounded the dorsal margin of the notch, became greatly reduced in size or were lost altogether. The posterior margin of the skull in these early amniotes was smoothly rounded. In the few Pennsylvanian early amniotes for which the stapes is known, they were robust and ventrally oriented bones that approached or articulated with the quadrate. Their role in hearing might have been confined to detecting substrate-borne vibrations. The contrast between the condition in these earliest amniotes and even the most basal members of the crown group is stark, and it is open to speculation as to how the gulf might be bridged.

One common feature among the early tetrapods covered in this chapter is the apparent large size of the FV. Its boundaries were defined dorsally by the otic capsule bones, the prootic and the opisthotic, but the ventral boundary was variously contributed to by the basisphenoid, basioccipital, and/or parasphenoid. Although the margins are often poorly defined and incompletely ossified, the dimensions of the stapedial footplate often indicate a large opening. Throughout the evolution of the FV in more derived tetrapods, this opening became restricted to within the otic capsule bones.

The condition in recumbirostran microsaurs in some ways resembles that of salamanders except that in microsaurs, the stapedial footplate fills the FV; whereas, in salamanders there is an opercular bone that partially fills that space in addition to the footplate. That condition appears to relate to partitioning between high and low-frequency sound systems (Schoch and Anderson, Chap. 11). Perhaps a better analogue might be that of burrowing lepidosaurs (Evans, Chap. 8), which are adapted for picking up low-frequency or ground-borne vibrations, in that they have large footplates inserted into an FV that intrudes into the more ventral braincase bones and a short stapedial shaft or columella. In some cases, the microsaur condition is indeed associated with taxa that show other adaptations for burrowing (see Sect. 4.4.3.1), although this not may be true in all cases.

None of the early tetrapods described in this chapter had stapes quite like those of any modern taxon. Instead we see great disparity in form among the different groups. The evolutionary history of several anatomical systems, in both vertebrates and invertebrates, often shows early variability followed by later stabilization: compare numbers of digits on the limbs, patterns of lateral line canals, vertebral counts, tooth and bone histology, appendage morphology, and segment counts in arthropod groups. It may not be too speculative to suggest that early tetrapods were "experimenting" with ear structure before stabilization into those seen in modern forms.

4.7 Summary

The ear regions of early tetrapods were not well-adapted for hearing in air, though they may have picked up low-frequency ground-borne or water-borne sound. The primitive form of the tetrapod stapes was a bulky bone and large relative to the skull. It had a large stapedial footplate, relative to the rest of the braincase, and the "shaft" was robust and flared distally. Although it was pierced by a stapedial foramen, it is not clear what passed through that.

This form of stapes is found in several early tetrapods belonging to a range of taxa that are not closely related phylogenetically. These include the Devonian *Acanthostega* and the early and mid-Carboniferous *Pederpes*, *Greererpeton*, *Kyrinion*, and the embolomeres. Although there are no stapes known for the Devonian *Ventastega*, it shows evidence for the presence of a fenestra vestibuli and is the most primitive tetrapod to do so. *Ventastega* had a particularly widely open spiracular notch, resembling that of some tetrapodomorph fish, such as *Gognasus* and *Tiktaalik* (Clack and Ahlberg, Chap. 3).

The spiracular notch was present in most of these early tetrapods, and it was probably open to the external surface, allowing intake either of water or air, and was associated with breathing. Exceptions include *Greererpeton* and its colosteid relatives in which the notch was absent and was assumed to have been lost. In *Greererpeton*, the stapes was massive and probably served more to support the braincase than it served to conduct sound. The stapes as a structural support in the skull may have been an early role for the bone and is a suggestion made also for that bone in early amniotes. Another early variation on the general condition in early tetrapods is found in *Ichthyostega*, in which the stapes and otic region are highly modified into what appears to have been a specialized organ for underwater audition. Baphetids also show a modified version of the general condition in that their skulls were solidly integrated, and they did not require any structural support for the braincase but nevertheless retained a massive stapes with a large footplate.

Generally speaking, with the exception of *Ichthyostega*, the FV of early tetrapods was a large hole, surrounded by four (or sometimes five) separate braincase bones, and occuring between the dorsal otoccipital and ventral basioccipital regions. Throughout the evolution of both amniote and amphibian lineages, the size of the stapedial footplate and the FV reduced in size, becoming restricted to within the otic capsule bones (see Sobral, Reisz, Neenan, Müller and Scheyer, Chap. 8).

Among more terrestrialized forms, the otic region takes a wide range of differing morphologies. The stem amniotes show the greatest diversity, most of which do not resemble that of any modern taxon. The seymouriamorphs, with widely spaced otic capsules, the FV positioned at the distal end of a tube, and a slot-like "otic" notch, provide one extreme. In this group, the juveniles appear to have been fully aquatic with notches that might have been spiracular rather than otic in function, but presumably these morphed into hearing organs as they grew and became terrestrial adults. The diadectomorphs had a bony sheet that lay within their extensive notches, which has been described as an extrastapes or part of the stapes itself. Although the arrangement in some ways resembles the adaptations seen in some aquatic turtles and in the extinct aquatic mosasaurs, other aspects of diadectomorph anatomy do not obviously suggest that they were aquatic. Their barrel-shaped bodies and crushing dentitions shows them to have been herbivores and, furthermore, the other taxa in the deposits in which diadectomorphs occur are predominantly terrestrial.

One group whose otic anatomy is being more accurately understood as a result of μCT scanning are the recumbirostran microsaurs. These terrestrial forms show variations on the theme of a stapes with a very large footplate but a very short stubby shaft. The stapes itself resembles that of salamanders, although there is no operculum as there is in those modern forms. Its absence suggests that the microsaurs did not have the adaptations that salamanders have for distinguishing low-frequency from high-frequency sounds. No microsaur has a posterior notch at the back of the skull, suggesting that neither a tympanic membrane nor an open spiracle was present. The condition in these microsaurs could have resulted from the burrowing habit evident in some or in the miniaturization they underwent that might have allowed them to become burrowers.

The earliest amniotes had yet another variation of the otic region, and it is still not understood how it arose from the general and primitive condition. In early amniotes, in stem forms that do not fall into the crown group, and in basal diapsids and synapsids, the stapes is robust but elongate and ventrally or laterally directed. There is usually a dorsal process that contacts the braincase above the FV, and the distal end may contact, or approach, the quadrate portion of the jaw joint or the cheek bones laterally. The construction suggests that the stapes was more likely to have been a structural support for the jaw joint in the dorsoventrally deep skulls of these animals. As with microsaurs, no early amniote had a posterior notch at the back of the skull, again suggesting the lack of a tympanic membrane or spiracle.

The contrast between the condition in early tetrapods and that in early amniotes is not bridged by taxa showing obvious transitional morphologies. Since the earliest amniotes are Early Pennsylvanian in age, it appears that such transitional forms must have occurred earlier, perhaps much earlier, during the early Mississippian. Unfortunately, fossils of tetrapods from that period are still very rare, and they usually do not have much of the otic region preserved. Such transitional forms, therefore, are still to be discovered.

The advent of μCT technologies will ultimately allow access to much more information about the otic regions of forms like seymouriamorphs and diadectomorphs, when those studies are eventually undertaken. Likewise, those techniques also may be used to track the decoupling of the otic region of the braincase from the dermal skull roofing bones across the fish–tetrapod transition and to track the changes in proportion to the otic region itself and the semicircular canals that it contains. Such studies may ultimately lead to better understanding of the capabilities of the ear region for aerial hearing in these early forms.

Compliance with Ethics Requirements J. A. Clack and J. S. Anderson have declared that there are no conflicts of interest.

References

Agnew, J. A. (1984). *Pholiderpeton scutigerum*, an amphibian from the Yorkshire Coal Measures. Dissertation, University of Newcastle upon Tyne, UK.

Ahlberg, P. E., Clack, J. A., Lukševičs, E., Blom, H., & Zupins, I. (2008). *Ventastega curonica* and the origin of tetrapod morphology. *Nature*, 453, 1199–1204.

Anderson, J. S. (2002). Revision of the aïstopod genus *Phlegethontia* (Tetrapoda, Lepospondyli). *Journal of Paleontology*, 76, 1029–1046.

Anderson, J. S. (2007). Incorporating ontogeny into the matrix: A phylogenetic evaluation of developmental evidence for the origin of modern amphibians. In J. S. Anderson & H.-D. Sues (Eds.), *Major transitions in vertebrate evolution* (pp. 182–227). Bloomington: Indiana University Press.

Anderson, J. S., Carroll R. L., & Rowe, T. (2003). New information on *Lethiscus stocki* (Tetrapoda, Lepospondyli, Aistopoda) from high-resolution computed tomography and a phylogenetic analysis of Aistopoda. *Canadian Journal of Earth Sciences*, 40, 1071–1083.

Baird, D. (1955). Latex micro-molding and latex-plaster molding mixture. *Science*, 122 (3161), 202.

Beaumont, E. I. (1977). Cranial morphology of the Loxommatidae (Amphibia: Labyrinthodontia). *Philosophical Transactions of the Royal Society of London B: Biological Sciences*, 280, 29–101.

Beaumont, E. I., & Smithson, T. R. (1998). The cranial morphology and relationships of the aberrant Carboniferous amphibian *Spathicepalus mirus* Watson. *Zoological Journal of the Linnean Society*, 122, 187–209.

Berman, D. S., Sumida, S. S., & Lombard, R. E. (1992). Reinterpretation of the temporal and occipital regions in *Diadectes* and the relationships of the diadectomorphs. *Journal of Paleontology*, 66 (3), 481–499.

Berman, D. S., Sumida, S. S., & Martens, T. (1998). *Diadectes* (Diadectomorpha: Diadectidae) from the Early Permian of central Germany, with description of a new species. *Annals of the Carnegie Museum*, 67, 53–93.
Berman, D. S., Henrici, A. C., Sumida, S. S., & Martens, T. (2000). Redescription of *Seymouria sanjuanensis* (Seymouriamorpha) from the Lower Permian of Germany based on complete, mature specimens with a discussion of paleoecology of the Bromacker locality. *Journal of Paleontology*, 20, 253–268.
Berman, D. S., Henrici, A. C., Kissel, R. A., Sumida, S. S., & Martens, T. (2004). A new diadectid (Diadectomorpha), *Orobates pabsti*, from the Early Permian of central Germany. *Bulletin of the Carnegie Museum of Natural History*, 35, 1–36.
Bernacsek, G. M., & Carroll, R. L. (1981). Semicircular canal size in fossil fishes and amphibians. *Canadian Journal of Earth Sciences*, 18, 150–156.
Bystrow, A. P. (1944). *Kotlassia prima* Amalitzky. *Bulletin of the Geological Society of America*, 55, 379–416.
Callier, V., Clack, J. A., & Ahlberg, P. E. (2009). Contrasting developmental trajectories in the earliest known tetrapod forelimbs. *Science*, 324, 364–367.
Carroll, R. L. (1964). The earliest reptiles. *Journal of the Linnean Society of London (Zoology)*, 45, 61–83.
Carroll, R. L. (1969). A middle Pennsylvanian captorhinomorph, and the interrelationships of primitive reptiles. *Journal of Paleontology*, 43, 151–170.
Carroll, R. L. (1980). The hyomandibula as a supporting element in the skull of primitive tetrapods. In A. L. Panchen (Ed.), *The terrestrial environment and the origin of land vertebrates* (pp. 293–317). New York: Academic Press.
Carroll, R. L. (1987). *Vertebrate paleontology and evolution*. New York: W. H. Freeman.
Carroll, R. L. (1990). A tiny microsaur from the Lower Permian of Texas size constraints in Paleozoic tetrapods. *Palaeontology*, 33(4), 893–909.
Carroll, R. L., & Baird, D. (1972). Carboniferous stem reptiles of the family Romeriidae. *Bulletin of the Museum of Comparative Zoology*, 143 (5), 321–364.
Carroll, R. L., & Gaskill, P. (1978). The Order Microsauria. *Memoirs of the American Philosophical Society*, 126, 1–211.
Carroll, R. L., Bossy, K. A., Milner, A. C., Andrews, S. M., & Wellstead, C. F. (1998). *Lepospondyli*. In P. Wellnhofer (Ed.), *Handbuch der paläoherpetologie, teil 1*. Munich: Friedrich Pfeil.
Case, E. C. (1905). The osteology of the Diadectidae and their relations to the Chelydosauria. *Journal of Geology*, 13, 126–159.
Christensen, C. B., Christensen-Dalsgaard, J., & Madsen, P. T. (2015). Hearing of the African lungfish (*Protopterus annectens*) suggests underwater pressure detection and rudimentary aerial hearing in early tetrapods. *Journal of Experimental Biology*, 218, 381–387.
Clack, J. A. (1983). The stapes of the Coal Measures embolomere *Pholiderpeton scutigerum* Huxley (Amphibia: Anthracosauria) and otic evolution in early tetrapods. *Zoological Journal of the Linnean Society*, 79, 121–148.
Clack, J. A. (1987). *Pholiderpeton scutigerum* Huxley, an amphibian from the Yorkshire coal measures. *Philosopical Transactions of the Royal Society of London B: Biological Sciences*, 318(1188), 1–107.
Clack, J. A. (1992). The stapes of *Acanthostega gunnari* and the role of the stapes in early tetrapods. In D. B. Webster, R. R. Fay, and A. N. Popper (Eds.), *Evolutionary biology of hearing* (pp. 405–420). New York: Springer-Verlag.
Clack, J. A. (1994). Earliest known tetrapod braincase and the evolution of the stapes and fenestra ovalis. *Nature*, 369, 392–394.
Clack, J. A. (1997). The evolution of tetrapod ears and the fossil record. *Brain, Behavior and Evolution*, 50, 198–212.
Clack, J. A. (1998) The neurocranium of *Acanthostega gunnari* and the evolution of the otic region in tetrapods. In D. B. Norman, A. R. Milner, and A. C. Milner (Eds.), *A study of fossil vertebrates*. *Zoological Journal of the Linnean Society*, 122, 61–97.

Clack, J. A. (2003). A new baphetid (stem tetrapod) from the Upper Carboniferous of Tyne and Wear, and the evolution of the tetrapod occiput. *Canadian Journal of Earth Sciences*, 40, 483–498.

Clack, J. A. (2012). *Gaining ground: The origin and evolution of tetrapods, 2nd ed.* Bloomington, Indiana: Indiana University Press.

Clack, J. A., Ahlberg, P. E., Finney, S. M., Dominguez Alonso, P., Robinson, J., & Ketcham, R. A. (2003). A uniquely specialised ear in a very early tetrapod. *Nature*, 425, 65–69.

Clack, J. A., and Finney, S. M. (2005). *Pederpes finneyae*, an articulated tetrapod from the Tournaisian of western Scotland. *Journal of Systematic Palaeontology*, 2, 311–346.

Clack, J. A., & Holmes, R. (1988). The braincase of the anthracosaur *Archeria crassidisca* with comments on the interrelationships of primitive tetrapods. *Palaeontology*, 31(1), 85–107.

Clack, J. A., & Milner, A. R. (2015). *Basal Tetrapoda*. In H-D Sues (Ed.), *Handbook of paleoherpetology, vol. 3A1*. Munich: Friedrich Pfeil.

Coates, T. J., Richter, L., & Caceres, C. (2008). Behavioural strategies to reduce HIV transmission: how to make them work better. *The Lancet*, 372(9639), 669–684.

Fracasso, M. A. (1987). Braincase of *Limnoscelis paludis* Williston. *Postilla*, 201, 1–22.

Huttenlocker, A. K., Pardo J. D., Small, B. J., & Anderson, J. S. (2013). Cranial morphology of recumbirostrans (Lepospondyli) from the Permian of Kansas and Nebraska, and early morphological evolution inferred by micro-computed tomography. *Journal of Vertebrate Paleontology*, 33 (3), 540–552.

Jarvik, E. (1996). The Devonian tetrapod *Ichthyostega*. *Fossils and Strata*, 40, 1–206.

Jones, G. M., & Spells, K. E. (1963). A theoretical and comparative study of the functional dependence of the semicircular canal upon its physical dimensions. *Proceedings of the Royal Society of London B: Biological Sciences*, 157, 403–419.

Kandel, B. M., & Hullar, T. E. (2010). The relationship of head movements to semicircular canal size in cetaceans. *Journal of Experimental Biology*, 213, 1175–1181.

Klembara, J. (1995). The external gills and ornamentation of skull roof bones of the Lower Permian tetrapod *Discosauriscus* (Kuhn 1933) with remarks to its ontogeny. *Paläontologische Zeitschrift*, 69, 265–281.

Klembara, J. (1997). The cranial anatomy of *Discosauriscus* Kuhn, a seymouriamorph tetrapod from the Lower Permian of the Boskovice Furrow (Czech Republic). *Philosophical Transactions of the Royal Society of London B: Biological Sciences*, 352, 257–302.

Laurin, M. (1998). The importance of global parsimony and historical bias in understanding tetrapod evolution. Part 1. Systematics, middle ear evolution and jaw suspension. *Annales des Sciences Naturelles*, 1, 1–42.

Laurin, M. (2000). Seymouriamorphs. In H. Heatwole & R. L. Carroll (Eds.), *Amphibian biology, vol. 4* (pp. 1064–1080). Chipping Norton, AU: Surrey Beatty.

Laurin, M. (2004). The evolution of body size, Cope's rule and the origin of amniotes. *Systematic Biology*, 53, 594–622.

Lombard, R. E., & Bolt, J. R. (1979). Evolution of the tetrapod ear: An analysis and reinterpretation. *Biological Journal of the Linnean Society*, 11, 19–76.

Lund, R. (1978). Anatomy and relationships of the family Phlegethontiidae (Amphibia, Aïstopoda). *Annals of the Carnegie Museum*, 47, 53–79.

Maddin, H. C., Olori J. C., & Anderson, J. S. (2011). A redescription of *Carrolla craddocki* (Lepospondyli, Brachystelechidae) based on high-resolution CT, and the impacts of miniaturization and fossoriality on morphology. *Journal of Morphology*, 272, 722–743.

McGinnis, H. J. (1967). The osteology of *Phlegethontia*, a Carboniferous and Permian aïstopod amphibian. *University of California Publications in Geological Sciences*, 71, 1–46.

Moss, J. L. (1972). The morphology and phylogenetic relationships of the lower Permian tetrapod *Tseajaia campi* Vaughn (Amphibia: Seymouriamorpha). *University of California Pubications in Geological Sciencies*, 98, 1–63.

Niedzwiedzki, G., Szerk, P., Narkiewicz, K., Narkiewicz, M., & Ahlberg, P. E. (2010). Tetrapod trackways from the early Middle Devonian. *Nature*, 463, 43–48.

O'Gorman, S. (2005). Second branchial arch lineages of the middle ear of wild-type and *Hoxa2* mutant mice. *Developmental Dynamics*, 234, 124–131.
Olson, E. C. (1947). The family Diadectidae and its bearing on the classification of reptiles. *Fieldiana, Geology*, 11, 1–53.
Olson, E. C. (1966). The relationships of *Diadectes*. *Fieldiana, Geology*, 14, 199–227.
Pardo, J. D., Szostakiwskyj, M., & Anderson, J. S. (2015). Cranial morphology of the brachystelechid "Microsaur" *Quasicaecilia texana* Carroll provides new insights into the diversity and evolution of braincase morphology in recumbirostran "microsaurs". *PLoS ONE*, 10 (6), e0130359.
Pierce, S. E., Clack, J. A., & Hutchinson, J. R. (2012). Three-dimensional limb joint mobility in the early tetrapod *Ichthyostega*. *Nature*, 486, 523–526.
Robinson, J. (2005). The early tetrapod *Kyrinion martilli* and the transition to hearing airborne sound. *Journal of Vertebrate Paleontology*, 25, 106A.
Robinson, J., (2006). The evolution of the early tetrapod middle ear and associated structures. Dissertation, University College London, UK.
Romer, A.S. (1941). Notes on the crossopterygian hyomandibular and braincase. *Journal of Morphology*, 69, 141–160.
Romer, A. S. (1946). The primitive reptile *Limnoscelis* restudied. *American Journal of Science*, 244, 149–188.
Romer, A. S. (1947). Review of the Labyrinthodontia. *Bulletin of the Museum of Comparative Zoology*, 99, 3–366.
Romer, A. S. (1969). The cranial anatomy of the Permian amphibian *Pantylus*. *Breviora*, 314, 1–37.
Romer, A. S. (1968). *Notes and comments on vertebrate paleontology*. Chicago: University of Chicago Press.
Ruta, M., & Clack, J. A. (2006). A review of *Silvanerpeton miripedes*, a stem amniote from the Lower Carboniferous of East Kirkton, West Lothian, Scotland. *Transactions of the Royal Society of Edinburgh: Earth Sciences*, 97, 31–63.
Ruta, M., & Coates, M. (2007). Dates, nodes and character conflict: Addressing the lissamphibian origin problem. *Journal of Systematic Palaeontology*, 5, 69–122.
Ruta, M., Coates, M. I., & Quicke, D. L. J. (2003). Early tetrapod relationships revisited. *Biological Reviews*, 78, 251–345.
Schultze, H.-P., & Bolt, J. R. (1996). The lungfish *Tranodis* and the tetrapod fauna from the upper Mississippian of North America. *Special Papers in Palaeontology*, 52, 31–54.
Smithson, T. R. (1982). The cranial morphology of *Greererpeton burkemorani* (Amphibia: Temnospondyli). *Zoological Journal of the Linnean Society*, 76, 29–90.
Smithson, T. R. (1985). The morphology and relationships of the Carboniferous amphibian *Eoherpeton watsoni* Panchen. *Zoological Journal of the Linnean Society*, 85, 317–410.
Smithson, T. R., Carroll, R. L., Panchen, A. L., & Andrews, S. M. (1994). *Westlothiana lizzia*e from the Viséan of East Kirkton, West Lothian, Scotland. *Transactions of the Royal Society of Edinburgh: Earth Sciences*, 84(3 and 4), 417–431.
Spoor, F. (2003). The semicircular canal system and locomotor behaviour, with special reference to hominin evolution. *Courier Forschunginstitut Senckenberg*, 243, 93–104.
Spoor, F., Bajpai, S., Hussain, S. T., Kumar, K., & Thewissen, J. G. M. (2002). Vestibular evidence for the evoluition of aquatic behaviour in early cetaceans. *Nature*, 417, 163–166. doi:10.1038/417163a
Szostakiwskyj, M., Pardo J. D., & Anderson, J. A. (2015). Micro-CT study of *Rhynchonkos stovalli* (Lepospondyli, Recumbirostra), with description of two new genera. *PLoS ONE*, 10 (6), e0127307.
Thompson, J., Ohazama, A., Sharpe, P. T., & Tucker, A. S. (2012). The origin of the stapes and relationship to the otic capsule and oval window. *Developmental Dynamics*, 241, 1396–1404. doi:10.1002/dvdy.23831
Thomson, K. S. (1966). The evolution of the tetrapod middle ear in the rhipidistian-amphibian transition. *American Zoologist*, 6, 379–397.

Wellstead, C. F. (1982). A Lower Carboniferous aïstopod amphibian from Scotland. *Palaeontology*, 25 (1), 193–208.

Westoll, T. S. (1938). Ancestry of the tetrapods. *Nature*, 141, 127–128.

White, T. E. (1939). Osteology of *Seymouria baylorensis* Broili. *Bulletin of the Museum of Comparative Zoology*, 85, 325–409.

Witmer, L. M. (1995). The Extant Phylogenetic Bracket and the importance of reconstructing soft tissues in fossils. In J. Thomason (Ed.), *Functional morphology in vertebrate paleontology* (pp. 19–33). Cambridge (UK): Cambridge University Press.

Chapter 5
Non-Mammalian Synapsids: The Beginning of the Mammal Line

Tom S. Kemp

Abstract The homologies between the malleus, incus, and ectotympanic bones of the mammalian middle ear and the articular, quadrate, and angular bones of the tetrapod palatoquadrate, respectively, are well-established by embryonic evidence. The evolutionary question of how the implied transition occurred is less clear and, historically, there have been two general views. The first view is that the ancestral state was a reptile-like tympanic membrane behind the quadrate that activated the stapes, and that the jaw hinge bones were subsequently incorporated between tympanic membrane and stapes. The second view is that in the ancestor, low-frequency ground-borne sound was received by the lower jaw and transmitted via the hinge bones and stapes to the fenestra ovalis. The anatomy of the middle ear region of the known sequence of fossil stem-group mammals—pelycosaurs, basal therapsids, and several cynodonts—is reviewed in this chapter. As with almost all recent authors, the interpretation offers support for the second view. Within the cynodont grades, decreasing mass of the postdentary bones relative to the dentary is part of a complex of changes in the feeding mechanism but also implies increasing sensitivity to airborne sound. As long as the jaw hinge continued to perform a mechanical role in mandibular function, the bones were too massive to be receptive to higher frequency sound, and therefore an air-filled tympanic cavity and tympanic membrane were unlikely to have evolved. This latter stage awaited the origin of the new mammalian jaw hinge between the dentary and squamosal bones in mammaliaforms.

Keywords Cynodonts · Dicynodonts · Evolution of hearing · Origin of mammals · Pelycosaurs · Synapsids · Synapsid ear · Therapsids · Therocephalians

---

T.S. Kemp (✉)
St John's College and Oxford University Museum of Natural History, Oxford OX1 3JP, UK
e-mail: tom.kemp@sjc.ox.ac.uk

J.A. Clack et al. (eds.), *Evolution of the Vertebrate Ear—Evidence from the Fossil Record*, Springer Handbook of Auditory Research 59,
DOI 10.1007/978-3-319-46661-3_5

5.1 Introduction

The presence of a chain of three middle ear ossicles, the malleus, incus, and stapes in the mammals compared to the single columella auris in living amphibians, reptiles, and birds, offers a classic example of the power of comparative embryology to reveal obscure homologies, as iconic in its context as, say, *Archaeopteryx* is in the context of the power of paleontology. The pre-Darwinian anatomist Karl Reichert (1837) is credited with the discovery that the malleus and incus of the mammals develop from the same parts of the first visceral arch that become the jaw hinge bones, the articular and quadrate respectively, in the reptiles (Fig. 5.1A). The mammalian stapes and the reptilian columella auris are formed within the second visceral arch, so they too were inferred to be homologous. As reviewed by a number of authors (e.g., Anthwal et al. 2013; Maier 1990; Takechi and Kuratani 2010), further comparative anatomical and embryological evidence subsequently accrued that continued to support this conclusion, such as the pattern of innervation and the musculature. Most notably, Gaupp (1913) demonstrated the additional homologies of the angular bone of the reptilian jaw with the mammalian ectotympanic bone, and the reptilian prearticular bone with the mammalian goniale that is attached to the malleus (Fig. 5.1B).

The homologies of the parts of the middle ear originally were established at a time when the cause and meaning of homology was not clear. The evolutionary question only subsequently arose as to how the transformation from the basal amniote condition of jaw hinge bones to the mammalian accessory ear ossicles occurred. This is a functional as well as an anatomical question and embraces the extent to which, if at all, the tympanic ear of the mammal is homologous as a sound-receiving structure with those of other tetrapods. Two alternative broad scenarios emerged during the 20th century. One (e.g., Goodrich 1930; Parrington 1946; Hopson 1966) presumed that the mammalian tympanic membrane is a postquadrate tympanum homologous with that of modern reptiles and birds. In the common ancestor, the tympanic membrane would have been connected directly by the stapes to the fenestra ovalis (FO), and only later in the mammalian lineage did a reduced articular and quadrate became incorporated, which enhanced the impedance matching between the airborne sound waves received by the tympanic membrane and the waves transmitted into the fluid-filled cochlea.

The second evolutionary scenario (e.g., Gaupp 1913; Tumarkin 1955; Allin 1975) is that the amniote-grade common ancestor of reptiles and mammals lacked a tympanic membrane, and that in the mammalian lineage sound waves were originally received by the lower jaw and transmitted via the angular, articular, and quadrate bones to the stapes and FO. In this view, the evolution of the definitive mammalian ear had to await the origin of a new jaw articulation between the dentary and the squamosal, from which point there followed a reduction in the mass of the postdentary bones and quadrate and loss of their connection to the jaws, but no change in their essential topological relationships to one another. The

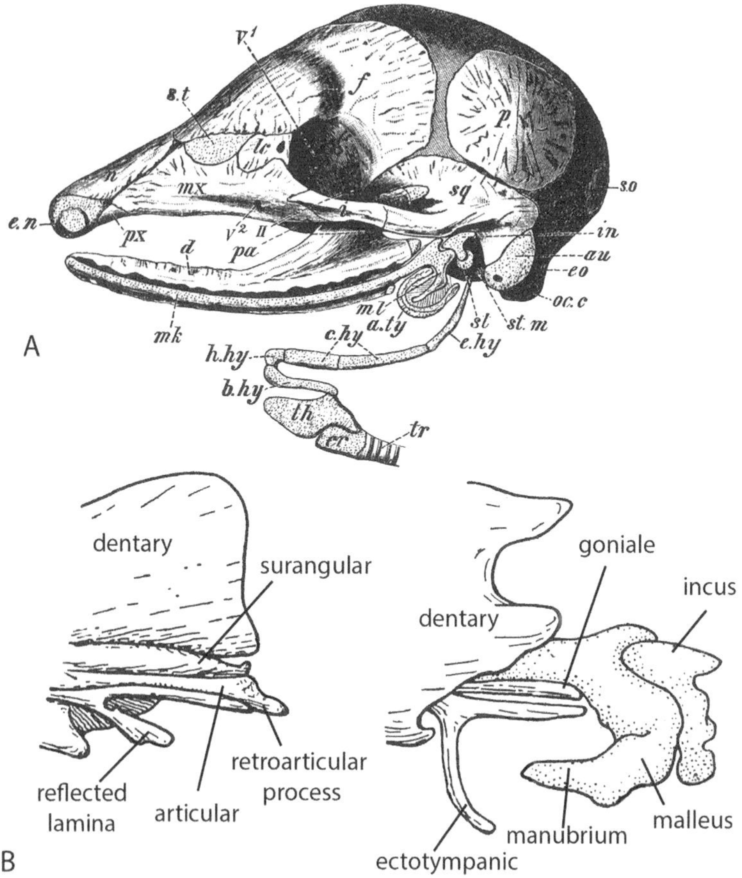

Fig. 5.1 **(A)** Skull of the embryo of an armadillo (*Tatusia hybrida*) showing Meckel's cartilage (*mk*) with manubrium (*ml*), incus (*in*), stapes (*st*), and ectotympanic (*a.ty*). For other abbreviations, see original publication. **(B)** Comparison of the posterior end of a cynodont jaw (left) with the posterior end of the jaw and ear ossicles of a late embryo of the marsupial *Perameles* (right). (**A**, **B** from Goodrich 1930)

mammalian tympanic membrane is seen, therefore, as a neomorph and not homologous with that of any other living tetrapod taxon.

More recent developmental and genetic evidence supports the second view—the mammalian tympanic membrane is not homologous with the tympanic membrane of reptiles and birds. Takechi and Kuratani (2010) showed that certain genes associated with development of the malleus and incus also affect the tympanic membrane, suggesting a developmental relationship between these elements that is presumably absent in non-mammalian amniotes. Going further, Kitazawa et al. (2015), by inactivating certain genes involved in jaw development, found that formation of the tympanic membrane of the mouse was associated with correct lower jaw development but was associated with upper jaw development in the chick. This is what would be expected if the mammalian tympanic membrane was historically attached to the angular bone, while that of the non-mammalian amniotes to the quadrate of the upper jaw.

Thompson and Tucker (2013) described the development of the middle ear cavity in mammals. It involves breakdown of part of the endodermal epithelial lining, invasion by a mass of neural crest mesenchyme cells, and subsequent cavitation of the latter. None of these processes occur in the non-mammalian amniotes, which implies that the cavity of the two respective taxa is not homologous and, therefore, the tympanic membrane bounding the cavity must have had a separate evolutionary origin.

Fossil members of the mammalian lineage, colloquially referred to as "mammal-like reptiles" and more formally as the paraphyletic taxon non-mammalian Synapsida, were first collected from Upper Permian rocks in South Africa in the middle of the 19th century and described by Richard Owen (1845), who initially regarded them only as a new kind of reptile. By the time of the publication of Owen's catalogue of fossil reptiles in the British Museum (Natural History), a considerably greater diversity of what are now referred to as therapsids was known and, furthermore, their mammalian affinity had become recognized (Owen 1876). Meanwhile in North America, members of even more basal synapsids, the peleycosaurs, were being discovered in Upper Carboniferous and Lower Permian rocks and described by Cope (1877) and later by Case (1907). However, it was not until 1910, when the South African paleontologist Robert Broom visited the USA that the opportunity arose to compare the North American and South African fossils. He demonstrated beyond any subsequent serious doubt that the two are related and together illustrate stages in the evolution of the mammalian lineage: the stem mammals (Broom 1910).

Over the intervening century, the non-mammalian synapsid record has expanded into numerous taxa representing some dozen grades (Fig. 5.2) between the basal-most pelycosaurs, which possessed only two or three definitive mammalian characters, and the earliest mammaliaforms. This has had profound consequences for understanding how the structure of the middle ear evolved in the mammalian lineage, to culminate in the ancestral mammaliamorph grade represented by, for example, the Upper Triassic *Morganucodon*.

However, anatomy aside, there have been very few serious attempts at functional analyses of hearing in the fossil synapsids. The majority of opinions expressed have been based on little more than subjective assumptions about whether there is evidence for the attachment of, or indeed enough space to accommodate, a tympanic membrane large enough to have activated the stapes in question. There have also been a number of vague assertions about the possible extent of non-tympanic and seismic sound reception. Exceptions include the studies by Hotton (1959), using a physical model of a pelycosaur middle ear; by Kemp (2007), who attempted to quantify some of the acoustic properties of a eucynodont; and by Laass (2014) who considered a dicynodont skull from the point of view of design for various possible

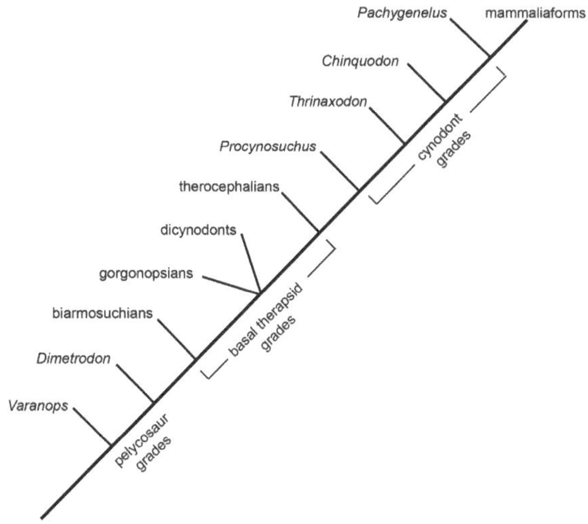

Fig. 5.2 Phylogenetic relationships of the synapsids

modes of sound reception. The issue of non-mammalian synapsid hearing is colored, inevitably perhaps, by considerations of the transition to the mammalian system. It is rarely addressed in its own right.

5.2 The Pelycosaur Grade

The stapes of pelycosaurs (Fig. 5.3) is invariably described as massive in comparison with the relatively slender structure of those found in subsequent synapsids and is presumed to be the ancestral condition for the lineage. In the case of *Dimetrodon* (Romer and Price 1940), a roughly circular footplate fits over the FO from which the shaft extends ventrolaterally toward, but does not reach, the lower part of the quadrate (Fig. 5.3D). This end is tapered and consists of unfinished bone, suggesting continuation by cartilage in life. There is a stout dorsal process near to the footplate that articulates with the overlying paraoccipital process and a stapedial foramen pierces the stapes. In the relatively few cases known, other pelycosaurian stapes have a similar form, although in smaller species the stapes is more horizontally disposed (Fig. 5.3E) (Hotton 1959).

As reviewed comprehensively by Hotton (1959), there has been much speculation in the past about the function of the stapes in pelycosaurs, mainly concerning whether a tympanic membrane existed at all and, if so, where it was located. In their definitive description, Romer and Price (1940) felt that the mass of the stapes excluded it from a sound transmission function altogether, although they are silent on what precisely the function of the bone might have been. Watson (1952) also believed that there was inadequate space for a sufficiently large tympanic membrane and that there is no positive osteological indication of where one might have been

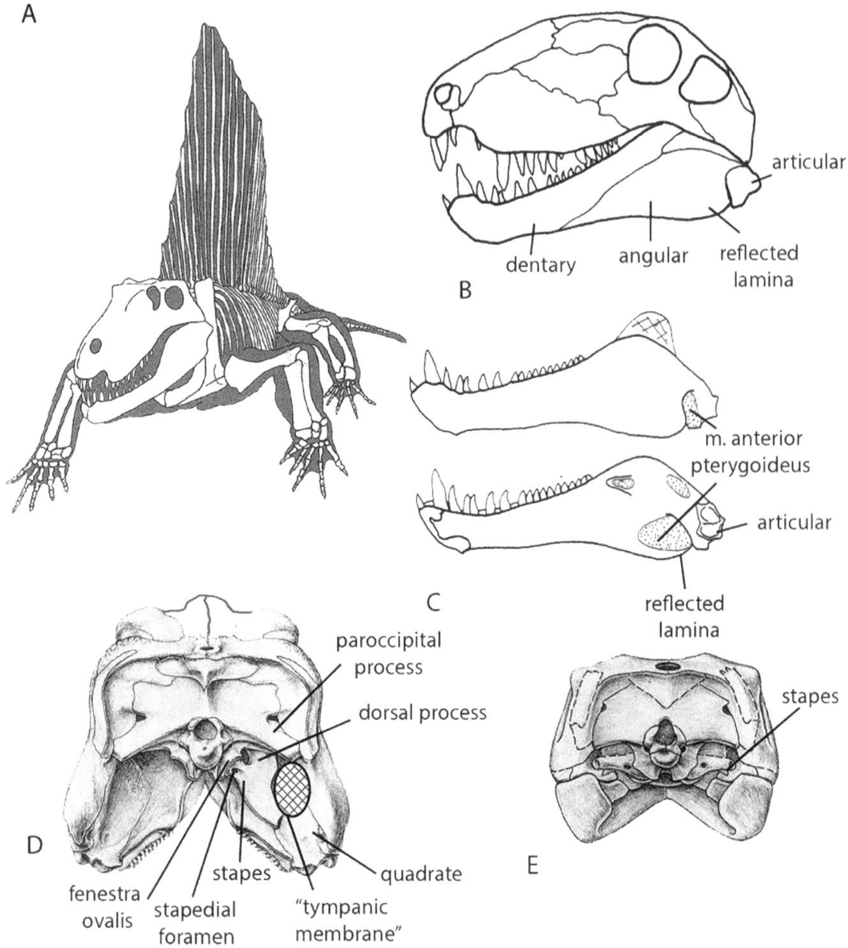

Fig. 5.3 Pelycosaurian grade. *Dimetrodon*. **(A)** Reconstruction of the skeleton. **(B)** Skull in left-lateral view. **(C)** Lateral (*above*) and medial (*below*) views of the lower jaw. **(D)** Posterior view of the skull with Hotton's interpretation of the position of a tympanic membrane added on the right. **(E)** Posterior view of the skull and jaws of *Varanops*. (**A** from Clack 2012, used with permission; **B**, **C** original from Kemp; **D**, **E** from Romer and Price 1940 and Hotton 1959)

attached. On the other hand, Parrington (1955, 1979) proposed that the distal end of the stapes was continued as a cartilaginous extrastapes reaching to a tympanic membrane low down behind the quadrate, a view influenced by his belief that both the non-synapsid predecessors and the therapsid successors of pelycosaurs possessed a postquadrate tympanic membrane.

Hotton (1959) constructed a mechanical model to test whether a stapes of the mass found in pelycosaurs could be activated to a significant extent by a membrane of a size that could be accommodated medial or posterior to the quadrate. His results showed that the model membrane did in fact enhance sound conduction via the stapes, though not to any great degree. However, there are serious difficulties in interpreting the results, for example the extent to which the parts of the model realistically match the physical properties of the living system, and the inability to distinguish between sound transmission due to the vibration of the stapes as a whole and intramolecular sound transmission within it. Actually fitting a similarly sized tympanic membrane perpendicular to the stapes shaft in the actual skull is also difficult, and Hotton's reconstruction has the membrane of *Dimetrodon* rather improbably in a more or less transverse plane, attached along the dorsolateral margin of the stapes shaft rather than terminally (Fig. 5.3D).

The reflected lamina of the angular of the lower jaw first appears in the sphenacodont pelycosaurs such as *Dimetrodon*, as a ventral keel that is set slightly laterally to the major plane of the jaw (Fig. 5.3B). Generally, its function is regarded as an area for the insertion of internal mandibular musculature, particularly the anterior pterygoideus muscle descending posteroventrally from the palate (Barghusen 1968) and, perhaps, intermandibular musculature. The external surface of the angular and its lamina, in common with the external surface of the whole of the postdentary region of the jaw, is finely rugose, indicative of tightly applied skin rather than either muscle or any form of tissue that could be regarded as a mandibular tympanic membrane. The quadrate is firmly sutured to the squamosal and, like the stapes, is of a size to render it impossible that it could have vibrated as a whole bone to transmit sound.

Further work on the possible functioning of the stapes and associated structures in sound conduction in pelycosaurs is long overdue. Currently the most reasonable view is that no tympanic membrane was present, and low-frequency, high-amplitude ground- and airborne sound were received and transmitted via intra-bone conduction, from mandible to quadrate and quadrate to stapes. However, certain anatomical features are not readily explained by this hypothesis, notably the indication of a cartilaginous rather than a bony connection between the quadrate and stapes.

5.3 Basal Therapsids

Therapsida is a more derived clade whose closest relatives are the sphenacodont pelycosaurs (Fig. 5.2) with which they share several synapomorphic characters. Most prominently, these include the reflected lamina of the angular, which had become greatly enlarged in therapsids, and the enlargement of teeth in the canine region of both upper and lower jaws. Indeed, therapsid ancestry probably lies within the Sphenacodontia, rendering the latter technically paraphyletic (Benson 2012; Huttenlocker and Rega 2012).

During the Middle Permian, therapsids radiated into several lineages more or less simultaneously. The interrelationships among these lineages are not completely resolved and the phylogenetic pattern approaches the condition of a soft polytomy (Fig. 5.2), the phenomenon in which there are inadequate shared characters to distinguish with confidence between a finely spaced sequence of dichotomous branchings and a truly simultaneous multiple split (Kemp 2009a, 2012). The implication of this pattern of therapsid radiation is that the taxon rapidly differentiated into a series of different specialist ecotypes, some carnivore and some herbivore. This is indicated by the considerable variation found in the anatomy of the dentition, jaws, reconstructed jaw musculature, and jaw articulation. The structure of these parts of the animal clearly reflects the function in their primary role as the feeding mechanism: collecting different kinds of food, generating bite and triturating forces, manipulating food in the oral cavity, etc.

Given this variation in the postdentary bones and quadrate, if these also had a role in sound reception, then it must have involved a mode of hearing in which the precise design was not particularly critical. Indeed, the stapes too varies to a surprising extent in size and structure both among and within different therapsid taxa. Detection of high-frequency airborne sound does require an adequately designed system, and so it follows that hearing in basal therapsids was probably limited to detection of low-frequency vibrations via intra-bone conduction. This requires a greatly less critical design than does the tympanic-activated, whole bone vibration type of middle ear of modern amniotes including mammals. The lack of a close correspondence between the form of an intra-bone low-frequency sound transmission route and its potential effectiveness, however, does make it difficult to assess the hearing performance of therapsids.

5.3.1 *Biarmosuchia*

Ignoring a few very poorly preserved specimens, the most basal therapsids are the biarmosuchians (Fig. 5.4). From them is inferred the ancestral therapsid structure of the jaw, jaw articulation, and middle ear (Ivakhnenko 1999; Sidor and Rubidge 2006). The stapes is very much reduced compared to the pelycosaur condition, and it extends laterally from the FO to contact the medial side of the quadrate as a fairly delicate rod. The presence of a stapedial foramen appears to be variable (Sidor and Rubidge 2006). The quadrate has not been fully described in any specimen, but appears to be relatively large and firmly fixed to the squamosal behind. The most prominent new character in this region of the skull is the huge expansion of the reflected lamina of the angular. It consists of a broad, but very thin, vertical sheet that extends below the level of the ventral margin of the body of the angular to which it is attached anteriorly. The posterodorsal and posterior margins are free edges. A very narrow cleft exists between the lamina externally and the body of the jaw internally, which is termed the angular recess. The lamina itself is marked by two corrugations radiating from around the center and presumably designed to increase the strength and/or stiffness of the structure. The function of the reflected lamina of therapsids, including its possible role in sound reception, remains a matter

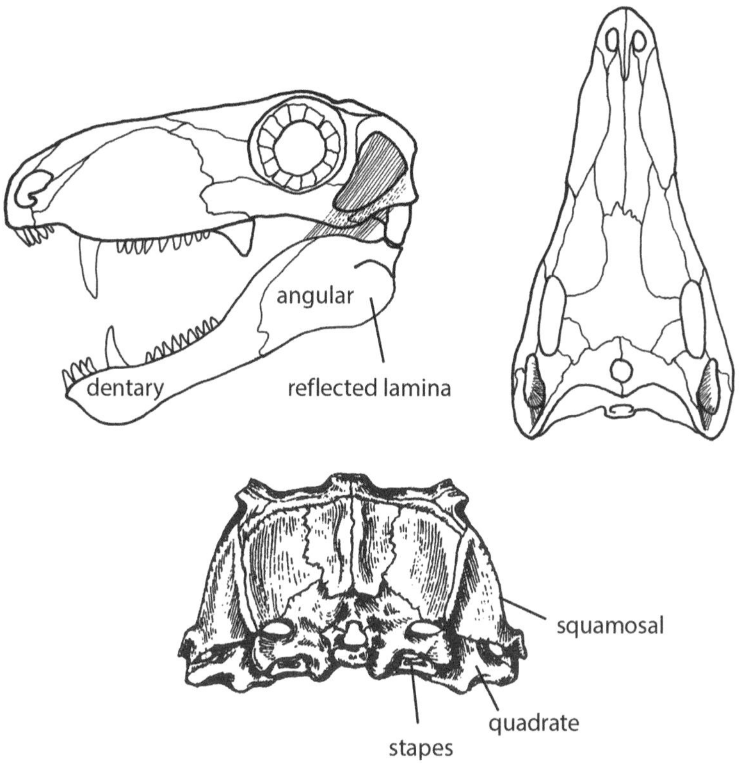

Fig. 5.4 The biarmosuchian *Biarmosuchus tener*. (Left-lateral and dorsal views redrawn from Ivakhnenko 1999; posterior view reproduced from Ivakhnenko 1999)

of unresolved debate. In part, this is because there is no comparable structure on the jaw of any living amniote and, in part, because of its status as the homologue of the otherwise entirely dissimilar, tiny ectotympanic bone of mammals that supports the tympanic membrane.

No functional consideration of hearing specifically in biarmosuchians has yet been presented, although since all the more derived therapsid taxa are assumed to have evolved from a generalized biarmosuchian morphotype, comments about function in the former may well apply also to the latter.

5.3.2 Gorgonopsia

The gorgonopsians (Fig. 5.5) were hypercarnivores of the basal therapsid radiation (Kemp 1969). The dentition was modified by a huge enlargement of the upper and lower canine teeth for prey disablement, along with facultative interdigitation between prominent lower and upper incisors for food processing. A highly

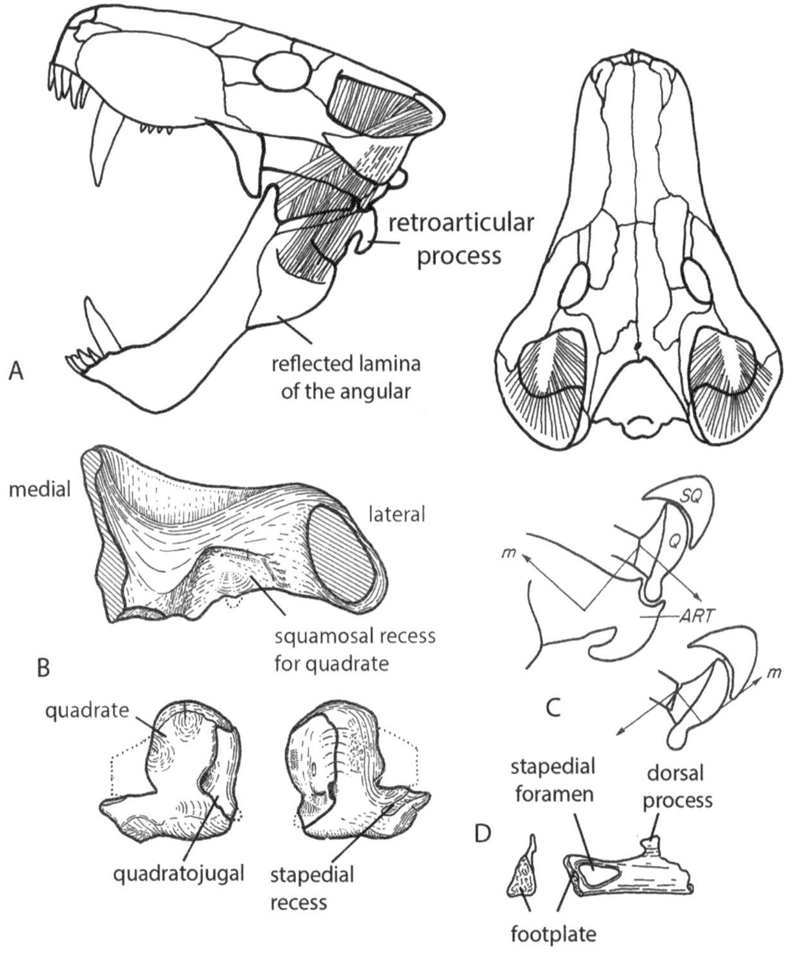

Fig. 5.5 Gorgonopsian grade. **(A)** Reconstruction of the skull in left-lateral and dorsal views showing main jaw adductor musculature. **(B)** Anterior view of the left squamosal showing the recess in which the quadrate complex lay, with anterior (*left*) and posterior (*right*) views of the corresponding quadrate below. **(C)** Diagram of the lateral view of the quadrate rotation in the squamosal recess to achieve a forward shift of the quadrate condyles. **(D)** Medial and anterior views of the left stapes of a gorgonopsian. *ART*, articular; *Q*, quadrate; *SQ*, squamosal; *m*, muscle vectors causing the rotation. (**A** original drawing; **B–D** from Kemp 1969)

specialized jaw articulation met the simultaneous requirements for a gape of at least 90°, which was necessary for the canines to clear one another during food capture and for propalinal (back and forth) movement of the lower jaw so the lower incisors could become engaged with the upper incisors. To achieve the wide gape there is a screw-shaped jaw articulation (Fig. 5.5B) between the quadrate and articular (Parrington 1955). The propaliny resulted from the quadrate being moveably attached to the squamosal (Fig. 5.5C) in a fashion that allowed the jaw hinge to rotate forward and backward (Kemp, 1969). The reflected lamina of the angular possessed a strong vertical ridge on its lateral face to which attached a large new part of the adductor musculature, originating from the squamosal anterolaterally to the jaw articulation. Although comparable in position and function to the masseter of the mammals, this muscle evolved convergently in gorgonopsids.

The stapes (Fig. 5.5D), in the few cases where it has been adequately described (Parrington 1955; Kemp, 1969), is a flat bone lying in a transverse plane and extending laterally to a recess on the medial side of the quadrate. There is a substantial stapedial foramen occupying the medial half of the bone and a prominent dorsal process toward the distal end of the shaft in contact with the overlying paroccipital process.

Parringon (1955) was convinced that there was a small, reptilian tympanic membrane in gorgonopsians located at the terminal end of the stapes, low down and immediately behind the quadrate. This position coincides with the inner, ventral termination of a broad, open sulcus, or channel, where the squamosal sutures with the lateral edge of the tabular bone, which he interpreted as the route of a tubular external auditory meatus. The arguments against the existence of the postquadrate tympanic membrane are, as in other basal synapsids, that the tympanic membrane would have been too small in area to have been functional and that there is no osteological indication of its support. The moveable nature of the quadrate in its socket in the squamosal could perhaps be regarded as a precondition for eventually evolving a whole-bone vibrating element for sound transmission, but at this stage the quadrate was far too massive to have functioned in this manner. On the other hand, the form of the squamosal sulcus remains unexplained if it did not house an external auditory meatus: possibly it was related to the architecture of the occipital and neck musculature.

5.3.3 *Dicynodontia*

The specialized herbivorous taxon Dicynodontia (Fig. 5.6) was by far the most abundant and diverse lineage of therapsids of the Upper Permian. The stapes lacks a stapedial foramen in almost all cases except for the basal genus *Eodicynodon* (Rubidge 1990) and the specialist digging genera *Cistecephalus* and *Kawingosaurus* (Cox 1972; Laass 2014), in which a presumably secondarily derived perforate condition may be associated with subterranean sound detection. The orientation of the stapes is approximately perpendicular to the sagittal plane, and both the footplate and distal end are expanded. The latter abuts firmly up against the medial side of the quadrate. The quadrate, along with the intimately attached quadratojugal on its lateral side, is large, flattish, and set in a shallow fossa on the more or less transverse anterior face of the squamosal. There is no firm suture

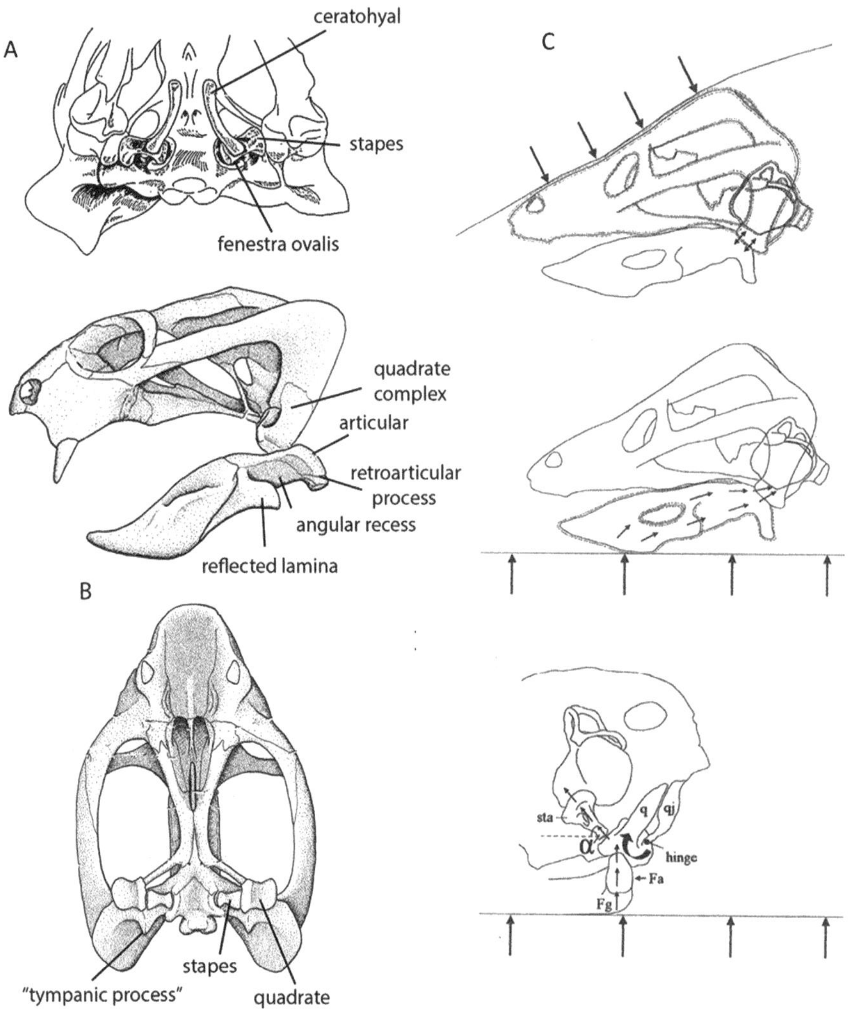

Fig. 5.6 Dicynodont grade. (**A**) Ventral view of the posterior half of the skull of *Lystrosaurus*, showing the preserved ceratohyal. (**B**) left-lateral and ventral views of *Dicynodontoides* (*Kingoria*). (**C**) Possible routes of seismic sound reception by bone conduction to the ear (*shaded*) according to Laass (2014). Top: *arrows* indicate compressional bone conduction from the burrow wall to the skull roof detected due to the inertia of the mandible causing relative movement between the articular and the quadrate (*double arrows*). Center: *arrows* indicate conduction from the ground via the lower jaw, articular, quadrate, and stapes, relying on inertia of the rest of the skull. Bottom: *arrows* indicate sound conduction from the ground or the air via the mandible, jaw articulation, and stapes. F_a, direction of air borne sound pressure; F_g, direction of ground-borne sound pressure; *q*, quadrate; *qj*, quadratojugal; *sta*, stapes. (**A** from Barry 1968; **B** from Cluver and King 1983; **C** from Laass 2014; all used with permission)

between the quadrate complex and squamosal, suggesting that a small degree of movement between the two was possible. The articulation between the quadrate and the articular is relatively elongated and open, allowing some anteroposterior motion of the lower jaw (Crompton and Hotton 1967) in association with the action of the horny tooth-plates that are assumed to have occupied the jaws in lieu of most or all of the dentition. The dicynodont reflected lamina of the angular is large and smooth surfaced, enclosing a substantial angular recess. The lower part tends to curve medially below the ventral margin of the jaw.

In an early contribution to the functioning of the middle ear in dicynodonts, Barry (1968) described what he took to be an ossified ceratohyal attached toward the inner end of the stapes and extending forward into the throat region in the Lower Triassic *Lystrosaurus* (Fig. 5.6A). In comparison with the condition in the modern tuatara *Sphenodon*, Barry (1968) concluded that it was involved in seismic sound reception and, furthermore, that no tympanic membrane was present. However, as no such ossification has been reported in other dicynodonts, and as *Lystrosurus* is a specialized, burrowing form, its middle ear must be assumed to be specialized for that particular mode of life. In contrast, Cox (1959) noted a small, posteriorly projecting process on the occiput of *Dicynodontoides* (*Kingoria*). It lies close to the lateral end of the paroccipital process (Fig. 5.6B), and he interpreted it as partial support for a tympanic membrane. However, in the light of its absence from other dicynodont taxa, and the evidence otherwise against the presence of a postquadrate tympanic membrane in basal therapsids generally, it is more probable that the process represents the point of attachment of a ligament associated with head support.

The most recent and the most detailed functional study of a dicynodont middle ear is that of the digging dicynodont *Kawingasaurus* by Laass (2014), based on his neutron tomography investigation of the structure. He concluded that there were at least two routes by which seismic sound could be detected (Fig. 5.6C). One was by inertial bone conduction from the flat, massive skull roof in contact with the wall of a burrow. In this case, the non-vibrating, fixed reference structure was the mandible, which was acoustically isolated from the cranium by the relatively highly compliant articular-quadrate contact. The second route was a direct one, and it consisted of seismic sound waves being received by the ventral margin of the mandible and transmitted via the quadrate and stapes to the FO. Support for the existence of this route is the ventrolateral orientation of the stapes toward the quadrate attachment, the relatively large size of the stapedial footplate, and the massiveness of the quadrate and stapes.

For these latter reasons, Laass (2014) also concluded that detection of airborne sound was poor and limited at best to low frequencies. As well as this high inertia of the bones, he calculated that there could have been a pressure transformer ratio of no more than 2–3, based on the estimated area of the angular and its reflected lamina as the presumed site of airborne sound reception (25–47 mm^2) and the area of the stapes footplate (13.2 mm^2). This figure is well below that of any mammal and also of eucynodonts, in which airborne sound probably was received and transmitted by the postdentary bones, as discussed in Sect. 5.4.3.

Although no similar study has been made for non-burrowing dicynodonts, their anatomy is sufficiently comparable to *Kawingasaurus* to suggest that detection of seismic sound and only low-frequency airborne sound was the norm. The difference is that in non-burrowing forms the stapes is more or less horizontal and so perpendicular to the plane of the postdentary bones, which might imply that airborne sound was of somewhat greater significance.

5.3.4 Therocephalia

The therocephalians (Fig. 5.7) constitute the next grade in the mammaliaform lineage known from the fossil record and represent the presumed condition of the middle ear region ancestral to that of the cynodonts (Kemp 1972b; Maier and van den Heever 2002). Notably the reflected lamina is particularly large and thin compared to other therapsids, and differs from those of gorgonopsians and dicynodonts in being free of the body of the angular bone along all of its upper margin. It is strengthened by a substantial pattern of corrugations, which led Kemp (1972a) to propose that its function was primarily to provide insertion for part of the mandibular muscle complex. The glenoid fossa of the articular bone is broad, widely open, and faces posteriorly and medially. The quadrate–quadratojugal complex (Fig. 5.7B) appears to have been moveably attached, but in a quite different manner from that of gorgonospsids. The quadratojugal component was held in a slit in the squamosal in such a way that rotation of the complex about a longitudinal axis was possible (Fig. 5.7C). Kemp (1972a) proposed that both the orientation of the jaw articulation surfaces (Fig. 5.7D) and the form of the mobility of the quadrate complex were designed to resist a large bite force that was generated by adductor musculature running dorsomedially from its insertion on the mandible. As it happens, the ability of the quadrate to rotate about a longitudinal axis, whatever its initial mechanical function may have been, is a necessary precursor to later stages in the evolution of cynodont and eventually mammalian hearing as described in Sect. 5.4 (Luo and Crompton 1994). The therocephalian stapes (Fig. 5.7E) is an imperforate, dumbbell-shaped bone that has a broad contact with the medial side of the quadrate (Kemp, 1972a; van den Heever 1994; Maier and van den Heever 2002). There appears to have been no possible room for a postquadrate tympanic membrane near the distal end of the stapes.

Whatever role the angular and its reflected lamina may have had in jaw mechanics does not exclude a secondary role in sound reception (Allin 1975; Maier and van den Heever 2002). The open form of the jaw articulation and the considerable mass of the hinge bones militate against any role in sound transmission by whole bone vibrations from the angular via the articular, quadrate, and stapes to the FO, but intra-bone conduction of low-frequency sound collected by the reflected lamina would have been possible. Maier and van den Heever (2002) calculated the ratio of the area of the reflected lamina to that of the stapes footplate as about 4.0, giving a very poor potential sound pressure level ratio, one that could only have

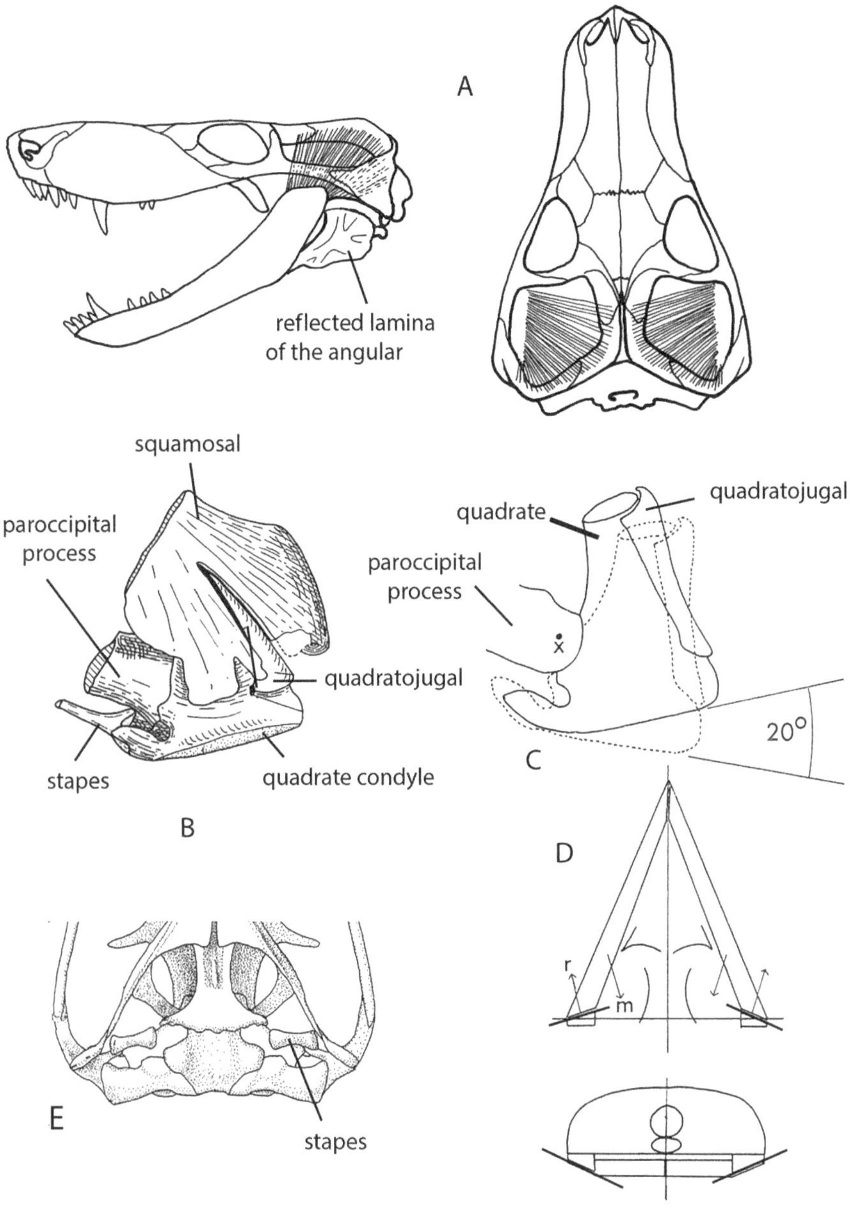

Fig. 5.7 Therocephalian grade: **(A)** Left-lateral and dorsal views of the skull with principal jaw adductor musculature shown. **(B)** Posterior view of the right quadrate complex and its contacts. **(C)** Diagram of the mobility of the quadrate complex about a longitudinal axis at *x*, as viewed from behind of *Theriognathus* (*Whaitsia*). **(D)** Diagram of the dorsal (*above*) and posterior (*below*) views of the skull and jaws to show the orientation of the axis of rotation of the jaw articulation. Because of the oblique nature of the axis, mobility of the quadrate on the squamosal is necessary to permit opening and closing of the jaws. **(E)** Ventral view of the posterior part of the skull of *Tetracynodon*. *m*, muscle force; *r*, hinge reaction force. (**A** original drawing; **B**, **C** and **D** from Kemp 1972a; **E** from Sigurdsen et al. 2012)

been associated with sensitivity to low-frequency, high amplitude sound transmission from postdentary bones to stapes.

5.4 Non-Mammalian Cynodonts

Much attention has been paid to the middle ear of the non-mammalian cynodont therapsids because it is within them that the transition from the basal therapsid anatomy to one recognizably approaching mammalian anatomy can be followed. In the earliest members of the Cynodontia, the quadrate and postdentary bones are not greatly reduced compared to therocephalians, although the reflected lamina of the angular is much smaller and the stapes has a large stapedial foramen. However, within the taxon there was a progressive reduction in these elements as illustrated by several grades of fossils up to the definitive mammaliamorph stage. By this time, the new, mammalian jaw hinge between dentary and squamosal had evolved and the function of the articular and quadrate, now referred to as the malleus and incus, was exclusively in sound reception.

5.4.1 Procynosuchus

Procynosuchus (Figs. 5.8, 5.9) is by far the best known of the basalmost cynodonts (Kemp 1979). The stapes possesses a very large stapedial foramen between quite delicate anterior and posterior arms (Fig. 5.9C). The footplate is a little larger than the FO (Fig. 5.9B) and so made contact with the bony margin of the latter, no doubt by means of a cartilaginous cap. The distal end also lacks a periosteal layer and partially fits into a shallow recess in the lower medial part of the quadrate but with about half of its face extending behind the quadrate. A ridge on the dorsal surface of the stapes lying distal to the foramen represents a dorsal process contacting the overlying paroccipital process. The quadrate-quadratojugal complex lies against the anterior face of a shallow squamosal recess (Fig. 5.9B) without forming a suture and is similar to, but relatively larger than, that of the better known *Thrinaxodon*, described in Sect. 5.4.2. The articulation of the quadrate with the articular bone of the lower jaw is somewhat similar to that of therocephalians. There are two quadrate condyles that face anteroventrally rather than ventrally, and the corresponding posterodorsally facing articular condyle lacks a hind rim. The angular bone is dominated by a broad shallow recess over its lateral face, indicating that adductor musculature had expanded its insertion over the external surface of the jaw. There is a very reduced and simplified reflected lamina of the angular restricted to the anteroventral region of the main angular bone and, therefore, leaving freely exposed most of what had been the angular recess.

Fig. 5.8 **(A)** Preserved skeleton of *Procynosuchus delaharpae* from the Upper Permian of Zambia (TSK collection field number TSK34, Natural History Museum, London). **(B)** Model of *Procynosuchus* in life based on the skeleton, by Richard Hammond

Kemp (1979) argued that the stapes had a dual role. One concerned the mechanics of the jaws and was to give some support to the mobile quadrate. The second, as inferred from its reduced mass and direct contact with the quadrate, was the conduction of sound waves to the FO. Following the theory of Allin (1975), Kemp proposed that this included ground-borne and low-frequency airborne sound, detected by the postdentary bones and transmitted by the small, loosely attached quadrate. More improbably, Kemp also thought it possible that a tympanic membrane existed behind the lower end of the quadrate, even though there is no

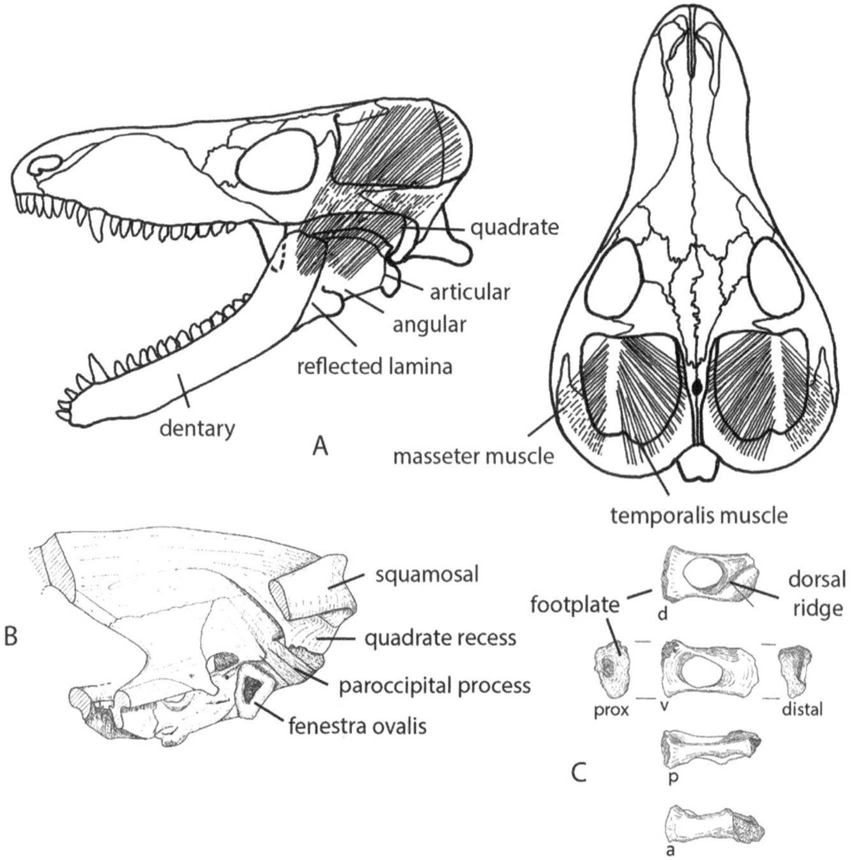

Fig. 5.9 *Procynosuchus*. **(A)** Reconstruction of the lateral and dorsal views of the skull with the principal jaw adductor muscles. **(B)** Ventro-lateral view of the skull to show fenestra ovalis. **(C)** Stapes in dorsal (*d*), ventral (*v*), posterior (*p*), anterior (*a*), proximal (*prox*), and distal views. (**B**, **C** from Kemp 1979)

indication of where exactly it might have attached, on the grounds that this would explain why half the distal end of the stapes lay behind the quadrate and shows signs of completion in cartilage, presumably as a cartilaginous extrastapes.

5.4.2 Thrinaxodon

The posterior jaw and middle ear region of the epicynodont *Thrinaxodon* (Fig. 5.10) (Parrington 1946; Luo and Crompton 1994) are similar to those of *Procynosuchus* except that the postdentary bones and jaw hinge are relatively smaller compared to the dentary (Fig. 5.11A). The dorsoventrally flattened stapes

(Fig. 5.10B), with a large stapedial foramen, abuts against the medial face of the quadrate in the same way. The quadrate lies against the anterior face of the squamosal, but there is a significant notch in the ventral border of the latter bone exposing part of the back of the quadrate (Fig. 5.10B). The manner in which the quadratojugal attached is also further modified. It is a tall, narrow blade wrapped around the lateral edge of the quadrate, and it lies in a near vertical extension of the quadrate notch. As in *Procynosuchus*, the articulating region of the quadrate consists of lateral and medial condyles aligned to face anteroventrally and corresponding to the posterodorsally facing articular condyle. The reflected lamina of the angular is small, thin, and ventrally placed, exposing almost all of the lateral face of the main body of the angular. The postdentary complex is not strongly sutured to the dentary but must have been attached only by soft, relatively compliant tissue, giving it the potential for vibrational movements independently of the rest of the jaw.

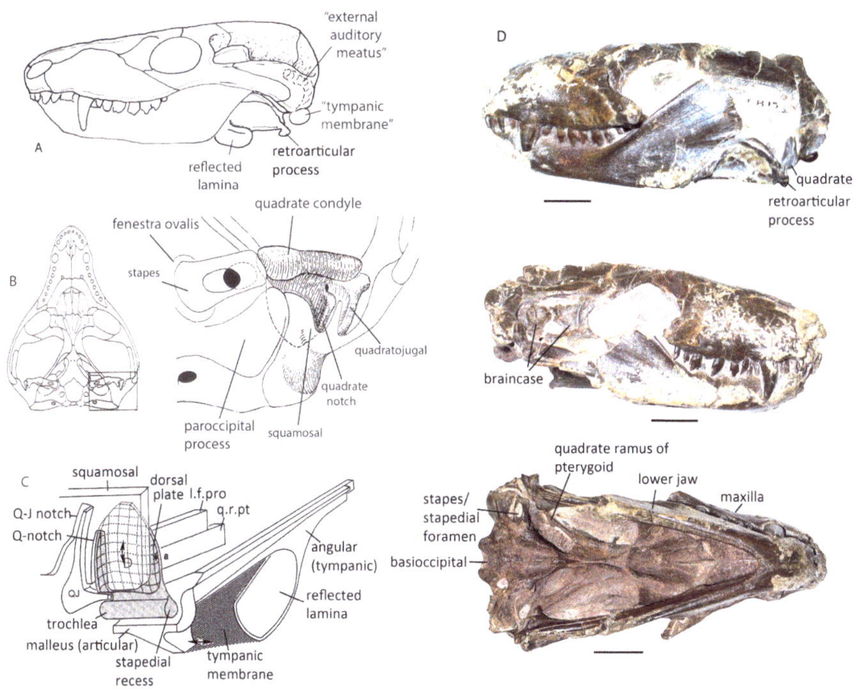

Fig. 5.10 *Thrinaxodon*. **(A)** Lateral view of the skull with Hopson's reconstruction of an external auditory meatus leading to a post-quadrate tympanic membrane. **(B)** Ventral view of the skull with an enlargement of the postero-lateral corner. **(C)** Luo and Crompton's interpretation of the ear region as a model showing the vibrational motion of the quadrate as activated by a tympanic membrane supported by the angular and its reflected lamina; see source for abbreviations. **(D)** Photograph of a skull in left-lateral (*top*), right-lateral (*middle*), and ventral views (*bottom*). (**C** from Luo and Crompton 1994, used with permission; **D** specimen T815, Cambridge University Museum of Zoology, photographed by J.A. Clack)

Parrington (1946), in his classic description of the skull of *Thrinaxodon*, was convinced that, as in all cynodonts, the stapes possessed a cartilaginous extrastapes attached to a tympanic membrane positioned low down behind the quadrate (Fig. 5.11A). Furthermore, he accepted the interpretation of the sulcus on the back of the squamosal as the route of an external auditory meatus leading to the supposed tympanic membrane. In his view, the angular and its reflected lamina were not involved in sound reception. Hopson's (1966) reconstruction of the middle ear of *Thrinaxodon* was essentially the same with a tympanic membrane attached to the squamosal, completely behind the quadrate (Fig. 5.10A).

In contrast to this view, Allin (1975) developed the idea of a postdentary tympanic membrane attached to the upper edge of the angular bone, further supported by the reflected lamina below and the retroarticular process of the articular behind (Fig. 5.11B). The squamosal sulcus was interpreted as the site of origin of a depressor mandibuli muscle, admittedly surprisingly large, from the occiput to the retroarticular process of the articular. He assumed there was an air-filled recessus mandibularis, equivalent to a tympanic cavity, bounded medially by the lateral face of the articular. Allin's interpretation, like the various forerunners of the general idea, is predicated upon the particular homologies of the individual bones with those of mammals, but his version also includes an explicit claim of functional continuity. The angular and its reflected lamina had the same function of supporting the tympanic membrane as the incompletely circular ectotympanic bone of the mammal, while the retroarticular process of the articular had the function of the mammalian manubrium of attaching to the center of the tympanic membrane as a lever arm of the incus. The nature of the quadrate-stapes contact is inferred to have been functionally the same as in mammals. Allin never completely abandoned the possibility of some sort of postquadrate tympanic membrane also being present, a view earlier proposed by Watson (1952). Indeed, in later publications (Allin 1986; Allin and Hopson 1992), he reconstructed a rather awkward-looking arrangement in which both kinds of tympanic membrane were present and suggested, with little evidence, that they may have been sensitive to different respective frequency ranges.

The most recent contribution to the function of the quadrate of *Thrinaxodon* is that of Luo and Crompton (1994). They demonstrated that the quadrate would have been capable of rotating about a longitudinal axis whose fulcrum was the contact with the posteriorly extending parts of the prootic and pterygoid bones (Fig. 5.10C). Rotational vibration of the bone about this axis would be caused by sound waves vibrating the postdentary bones and the tympanic membrane it supported. The quadrate vibrations would then be transmitted directly to the stapes where it abutted onto the quadrate's medial side and thence via the footplate to the FO. As so interpreted, the functional relations of postdentary bones, quadrate, and stapes were fundamentally the same as the ectotympanic, malleus, incus, and stapes of mammals.

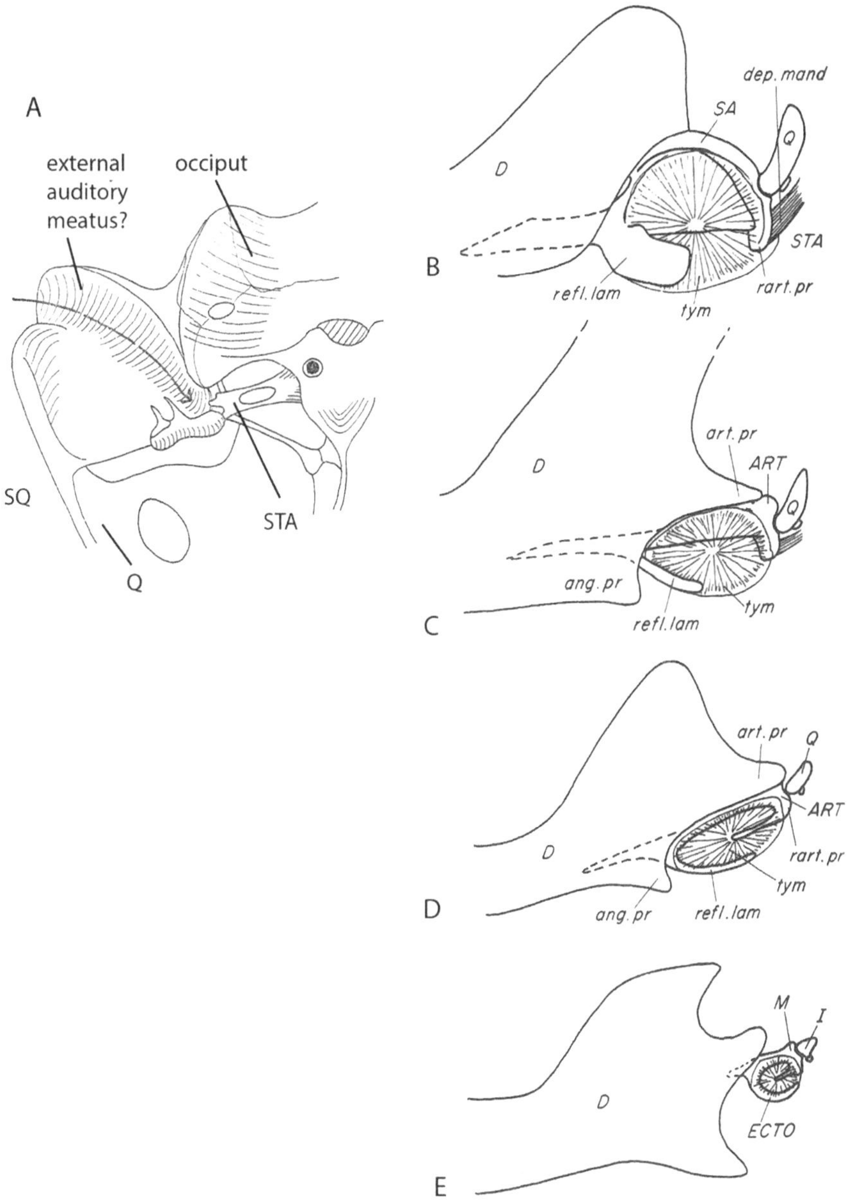

Fig. 5.11 (**A**) Parrington's interpretation of the ear region of a cynodont. (**B–E**) Allin's interpretation of the ear region of cynodonts: (**B**) *Thrinaxodon* grade, (**C**) eucynodont grade, (**D**) basal mammaliamorph grade, and (**E**) modern mammal grade. *ang.pr*, angular process of the dentary; *ART*, articular; *art.pr*, articular process of the dentary; *D*, dentary; *dep.mand*, depressor mandibuli muscle; *ECTO*, ectotypmpanic; *I*, incus; *M*, malleus; *Q*, quadrate; *rart.pr*, retroarticular process; *refl.lam*, reflected lamina of the angular; *SA*, surangular; *SQ*, squamosal; *STA*, stapes; *tym*, tympanic membrane. (**A** from Parrington 1979, used with permission; **B–D** from Kemp 1982, modified from Allin 1975)

5.4.3 Eucynodontia

The clade Eucynodontia (Figs. 5.11C, 5.12) is the next grade in the trend in mandibular and jaw articulation structure seen through therocephalians, *Procynosuchus* and *Thrinaxodon*. The postdentary bones are reduced to a small, narrow, transversely flattened rod preserved loosely in a horizontal trough along the lower medial side of the dentary (Figs. 5.11C, 5.12B). It was held in place in life by soft tissue rather than firm bone-to-bone suturing. The reflected lamina of the angular, on the rare occasions when it is completely preserved, appears as a thin, delicate process extending backwards from the anteroventral corner of the body of the angular adjacent to the dentary. What had once been the angular recess between the main body and the sheet-like reflected lamina of the angular is by this stage completely open laterally. The quadrate bone (Fig. 5.12D) is similarly reduced in relative size, although attached in essentially the same way as in *Thrinaxodon* (Luo and Crompton 1994). The principal qualitative development in this region of the skull (Figs. 5.12C, D) is a new ligamentous contact between a small boss on the surangular of the postdentary rod and a facet on a flange of the squamosal immediately lateral to the quadrate-articular hinge (Crompton 1972; Kemp 2007).

Historically, the range of functional interpretations of the middle ear of eucynodonts is much the same and associated with the same authors as for *Thrinxodon*. One of the contentious anatomical points is whether the sulcus at the back of the squamosal housed an internal auditory meatus leading to a tympanic membrane (Watson 1952; Hopson 1966; Parrington 1979) or, alternatively, was the site of origin of a depressor mandibuli muscle that inserted onto the back of the articular bone (Allin 1975). Neither of these explanations is entirely satisfactory. Within the Eucynodontia, the sulcus in some taxa, such as *Diademodon,* was extraordinarily large, which provokes the argument that an associated depressor mandibuli muscle occupying the sulcus would have been far too large, either for an insertion on the very small retroarticular process of the angular or, indeed, for any feasible functional requirements of the lower jaw. However, the size and the curiously twisted course of the sulcus appears equally inappropriate for an air-filled external auditory meatus. It may be the case, therefore, that the sulcus had a quite different

Fig. 5.12 Eucynodont grade. (**A**) Lateral view of a cast of the skull of the Middle Triassic South African *Trirachodon* (original specimen South African Museum No. K12168). (**B**) Kemp's reconstruction of the postdentary bones and quadrate of *Chiniquodon* as a unit vibrating about an axis. (**C**) Ventral view of the posterolateral part of the skull of *Chiniquodon* showing lower jaw in situ. (**D**) Ventral view of the skull with enlargement of the posterolateral region of *Probainognathus*. *ang.pr.d.*, angular process of the dentary; *bear*, bearing point of postdentary complex on dentary; *BSP*, basisphenoid; *COR*, coronoid; *D*, dentary; *ext.au.me*, possible site of an external auditory meatus; *fen.ov.*, fenestra ovalis; *jug.f.*, jugular foramen; *p.pr.*, paroccipital process; *Q*, quadrate; *QJ*, quadratojugal; *q.ra.pt.*, quadrate ramus of the pterygoid; *rart.pr.*, retroarticular process; *refl.lam.*, reflected lamina of the angular; *sa.boss*, boss on surangular for ligament attachment; *sa.lig.*, surangular ligament; *SPL*, splenial; *SQ*, squamosal; *STA*, stapes; *'tymp.m'.'* functional tympanic membrane. (**B**, **C** from Kemp 2007; **D** from Luo and Crompton 1994, used with permission)

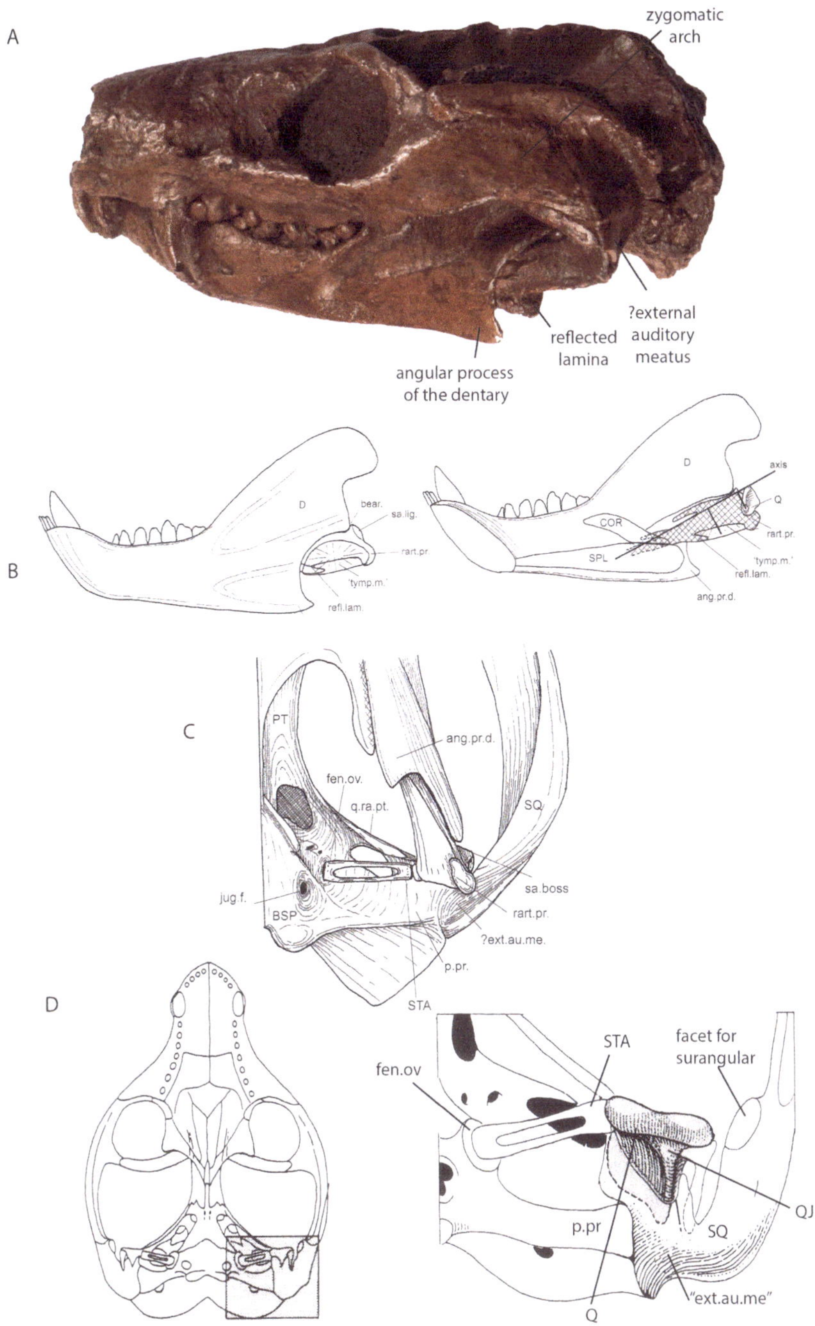
A
zygomatic arch
?external auditory meatus
reflected lamina
angular process of the dentary
B
D
bear.
sa.lig.
rart.pr.
'tymp.m.'
refl.lam.
axis
Q
COR
SPL
ang.pr.d.
C
PT
ang.pr.d.
fen.ov.
q.ra.pt.
SQ
jug.f.
BSP
sa.boss
rart.pr.
?ext.au.me.
p.pr.
STA
D
STA
facet for surangular
fen.ov
QJ
p.pr
SQ
Q
"ext.au.me"

explanation. Its anatomy is highly variable in different eucynodonts, from the extreme prominence in diademodontids that have grinding herbivorous postcanine teeth, to a more moderate form in the carnivorous chiniquodontids, and the virtual absence from the high occiput of the apparently hypercarnivorous cynognathids. The variation in its architecture suggests that the form of this region of the skull reflects differing geometry of the jaw musculature and force patterns within the temporal fenestra and occiput in different kinds of feeding design rather than having anything to do with hearing.

For Allin (1975), an important key to understanding the middle ear of eucynodonts (Fig. 5.11C) is his argument that the reduction of the relative size of what were to become the mammalian middle ear ossicles—ectotympanic, malleus, and incus (Fig. 5.11B–E)—was the result of positive selection for sensitivity to higher frequency airborne sound, a shift permitted by the reduced mechanical requirements of the jaw hinge consequent upon the evolution of greater differentiation of the adductor mandibuli musculature (Crompton 1963; Kemp 1980). The temporalis muscle attached high up on the very prominent coronoid process, and had a posteriorly directed force vector, while the masseter muscle attached to the lateral face of the angular region of the dentary, giving it a dorsally directed force vector. The effect of this reorientation of the adductor musculature was to concentrate all the generated muscle force between occluding upper and lower postcanine teeth, with no reaction at all at the jaw hinge. With little reaction stress to be resisted, the articular and quadrate were free to reduce in size. Luo and Crompton (1994) demonstrated in more detail that the quadrate of a eucynodont such as *Probainognathus* (Fig. 5.12D) was not only reduced in size, but also its contact with the squamosal was modified to make it more mobile and, therefore, more able to be set into vibration about the longitudinal axis.

In a detailed consideration of the possible auditory function of the postdentary bones in the eucynodont *Chiniquodon*, Kemp (2007) showed that the postdentary rod was only loosely attached anteriorly by connective tissue within the dentary trough, while its posterior end was supported by the newly evolved ligament connecting the surangular with the squamosal, adjacent to the articular-quadrate articulation. Therefore, the rod was largely acoustically isolated from the dentary by the soft tissue connection and could have undergone rotational vibration about a longitudinal axis along its dorsal margin (Fig. 5.12B), just as the malleus essentially does in mammals. The quadrate, as noted, is also moveably attached by soft tissue to the squamosal. In principle, therefore, the postdentary rod and quadrate together formed a compliant malleus-incus unit as in mammals. Airborne sound falling on the angular part acting like a tympanic membrane would set the unit as a whole vibrating about its long axis, and this movement would be passed via the quadrate part to the head of the stapes. The stapes in turn would vibrate like a piston in the FO.

Based on this proposed mechanism, the impedance matching function of the middle ear of the specimen can be calculated as a pressure transformer ratio. The ratio between the area of the lateral surface of the angular bone acting as a tympanic membrane and the area of the FO is about 20. There is also a lever effect arising from the force arm from the estimated position of the axis of rotation of the postdentary bones to the center of the tympanic membrane being 1.5 times as long as the lever arm from the axis to the point at which the quadrate contacts the end of the stapes. The effective transformer ratio, as the product of these two values, is therefore around 1:30, which is within the range of modern mammals. On the basis of this calculation, the postdentary rod, quadrate, and stapes are well designed to transform airborne sound waves into fluid waves in the vestibule.

However, the postdentary rod and quadrate still had some residual mechanical functions associated with the jaw mechanics. Actively opening the jaw necessarily generates stress at the hinge, as does biting using the more anterior teeth and, in the case of carnivores, grasping struggling live prey between the jaws. For this reason, the quadrate and articular had to retain far greater mass than would be optimal in their use for sound reception alone. The actual mass of the postdentary rod plus the quadrate, as estimated from the volume of the elements detached from the specimen, is about 0.5 gm, which is around ten times the value found in typical mammals. The implication is that the inertia of the system would have been too high for effective transmission of high-frequency vibrations, although it would have functioned adequately at lower frequencies, perhaps up to about 2 kHz.

A further conclusion from Kemp's (2007) study is that no air-filled tympanic cavity would have been present at this stage. Such a cavity lateral to the angular, as envisaged in Allin's (1975) reconstruction, is problematic because the role of a tympanic cavity is to increase the compliance of the system for increased sensitivity to higher frequency sound. If the high moment of inertia of the rest of the system forbade higher frequency sound reception anyway, then a tympanic cavity would have been functionally redundant. Low-frequency sound waves acting on the surface of the tissue covering the angular bone and the small ventral continuation supported by the reflected lamina would have caused enough vibration of the bones to be detectable.

Tritheledontids (Fig. 5.13) are small, highly derived eucynodonts with a number of new mammalian characters, such as loss of the postorbital bar and the presence of prismatic enamel on the teeth. They are generally regarded as the sister group of mammaliaforms (Bonaparte et al. 2003). Most significantly, the dentary extended a little further posteriorly and made a simple contact with the squamosal, which is an incipient version of the new mammalian jaw hinge. The quadrate is reduced further in size compared to earlier eucynodonts, and it was attached even more loosely to the squamosal (Luo and Crompton 1994). Presumably, the effect was to increase its compliance, which, along with its reduced moment of inertia, would have enhanced sensitivity to high-frequency sound waves. From this point, the new squamosal-dentary contact was free to evolve into a fully mammalian glenoid articulation and so relieve the quadrate and articular from any further masticatory function. It may be speculated that acquisition of this newly acquired sensitivity to higher frequencies was the point at which an air-filled tympanic cavity would be beneficial, increasing the compliance, and therefore sensitivity, of what had become a dedicated chain of auditory ossicles.

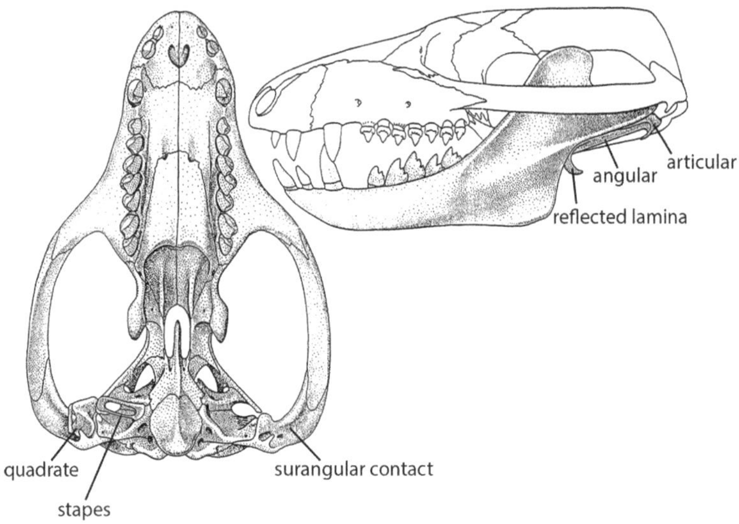

Fig. 5.13 Tritheledontid grade. Ventral and left-lateral views of the skull of *Pachygenelus*. (From Bonaparte et al. 2003, used with permission)

5.5 Summary

There are formidable difficulties in the way of a detailed understanding of hearing in non-mammalian synapsids. First, although the formal homologies of the relevant entities are well established, their actual anatomy is profoundly different from any living tetrapod, amphibian or amniote. For example, no living form has anything resembling the reflected lamina of the angular bone or a stapes that firmly abuts against the quadrate. There are simply no modern functional analogues with which to compare the mammal-like reptiles.

Second, the functional design for intra-bone sound conduction of low-frequency sound waves within skulls is not very critical: to some degree virtually any bony structure can transmit vibrations to anywhere with which it is sufficiently stiffly connected. Therefore, it is difficult to infer transmission routes of such sound from the anatomy. Yet most probably the ancestral synapsid mode of auditory function largely, if not entirely, consisted of just such intra-bone transmission of low-frequency seismic or airborne sound waves.

Third, the structures in question, stapes, articular, and quadrate hinge bones, and postdentary bones, also had mechanical functions in the context of mandibular action in feeding, so any design for hearing must have been extensively compromised by the more stringent optimal requirements for this second biological role. Finally, interpretations have been influenced by prejudgments concerning the transition from the basal tetrapod system to the mammalian condition—different assumptions point to different conclusions.

Most of the earlier hypotheses about how basal synapsids heard were based on little more than speculation about the position and the size of a possible tympanic membrane within the skull and how massive a stapes it would have been capable of actuating. This follows from an early presumption that a tympanic membrane and terrestrial hearing were primitive for tetrapods, dating from the origin of the group in the mid-Paleozoic (see Schoch and Anderson, Chap. 11; Sect. 5.2). In fact, about the only point of universal agreement is that a stapes whose footplate is applied to an open FO leading into the vestibular cavity must have been involved in some way in the mechanoreception of sound waves. Yet whether the stapes acted as the vibrating element relative to the braincase as the fixed element, or acted as the fixed element relative to vibrating cranial elements, may be debated.

Notwithstanding earlier beliefs of several authors, there seems little doubt now that the stapes of pelycosaurs, representing the ancestral synapsid condition, primarily served a mechanical role in the structure of the skull, although exactly what role remains unclear. The high mass of the stapes and consequent high moment of inertia, plus its firm attachment to the surrounding cranial bones and consequent low compliance, indicate that it would take enormous energy to set it into whole-bone vibration. Furthermore, there is no indication of either room for, or bony support of, a tympanic membrane likely to have been large enough to be capable of activating it. All of this points to hearing that was limited to the detection of low-frequency seismic and, perhaps, loud airborne sound waves. Although it is difficult to identify any specific route, the stapes presumably acted as the final element in transmitting such sound waves by intra-bone conduction from the lower jaw and other parts of the skull to the inner ear.

Evolution of the therapsid grade saw the start of major modifications to the jaw mechanics with larger and more complex adductor musculature generating increased and more controlled bite forces. The rapid differentiation of the therapsids into several specialized lineages is manifested by the variation amongst them of the anatomy and inferred functioning of the jaw musculature. The differences in the enlarged reflected lamina probably related to different patterns of intermandibular and internal adductor muscles, and the form of the jaw hinge reflected such aspects of feeding as the directions of the forces of the major adductor muscles and whether there was a requirement for horizontal as well as adductor movement of the mandible. Although the changes in these structures were primarily adaptations for feeding strategies, in the light of their subsequent evolution, it appears there was also a serendipitous effect on their role in sound detection. Indeed, the subsequent evolution of the middle ear and sound detection ability can only be fully appreciated in the context of correlated evolution of dentition, jaw musculature, jaw hinge, and stapes. Two new developments were of particular significance. The enlargement of the reflected lamina of the angular, for whatever primary reason, generated a greater area for receiving airborne sound waves able to set in motion the part of the mandible behind the dentary that could later become partially acoustically isolated from the dentary. At the same time, the tendency to evolve a moveable quadrate as a means of elaborating biting and food manipulation created the initial conditions for eventual acoustic isolation of this bone, along with the attached postdentary bones and stapes, from the rest of the skull. Thanks to these various modifications, significant detection of low-frequency airborne and seismic sound had probably become a significant part of the sensory repertoire of the more basal therapsids, such as gorgonopsians, dicynodonts, and therocephalians.

The trend initiated within therapsids continued within the cynodonts. From the relatively therocephalian-like procynosuchian grade there was continued enlargement of the adductor muscles and their geometrical reorganization into a posteriorly oriented temporalis muscle on the coronoid process and dorsally oriented masseter muscle on the angular region of the dentary. The effect of these changes was to allow a stronger but also more precisely applied bite action in association with larger, more complex occluding teeth, enhancing the function of food assimilation. It also reduced the stresses acting at the jaw articulation during biting and mastication.

Another trend in the cynodonts was increasingly to shift the insertion of the remaining parts of the adductor musculature that inserted on the postdentary bones onto the dentary. This relative enlargement of the dentary and reduction of the postdentary complex is hard to account for solely in terms of jaw mechanics because, as far as the size and geometry of the muscles is concerned, it is their attachment to appropriate mandibular lever arms that matters. It should be irrelevant whether those lever arms are part of the same bone or parts of separate but strongly sutured bones. However, the trend can be explained by the potential of these changes to enhance the sensitivity of airborne sound reception. By reducing the feeding stresses at the jaw hinge, the postdentary bones and the quadrate could reduce in mass and become loosely attached as a unit to the dentary, increasing their compliance relative to the latter. To this end, a secondary articulation between the surangular and the squamosal was first established as a means of further isolating the postdentary rod from the dentary. At the same time, the quadrate reduced in mass and increased in compliance relative to the squamosal and other cranial contacts by loosening its attachment. By the full expression of the eucynodont, the system was able to detect and transmit somewhat higher frequency, lower amplitude sound waves to the FO by means of whole bone vibration of these elements and the stapes.

The next stage in the transition to the mammalian middle ear is represented by tritheledontids. The dentary expanded posteriorly until it came to span what had already become a relatively very small gap to the squamosal. A simple contact between the two bones formed, from which moment this new squamosal-dentary contact was free to evolve into a fully mammalian glenoid articulation and so relieve the quadrate and articular from any remaining masticatory function. Rapid reduction in both the relative and the absolute mass of the hinge bones was now possible, along with the appearance of the definitive mammalian air-filled tympanic cavity. Still higher frequency sound could henceforth be detected, a conclusion supported by the beginning of the enlargement of the cochlear recess in mammaliaforms, and possibly associated, along with the enlarged cerebral hemispheres, with more complex activities such as nocturnal foraging and social behavior (Kemp 2009b).

Compliance with Ethics Requirements T. S. Kemp has declared there were no conflicts of interest.

References

Allin, E. F. (1975). Evolution of the mammalian ear. *Journal of Morphology*, 147, 403–438.

Allin, E. F. (1986). The auditory apparatus of advanced mammal-like reptiles and early mammals. In N. I. Hotton, P. D. MacLean, J. J. Roth, & E. C. Roth (Eds.), *The ecology and biology of mammal-like reptiles* (pp. 283–294). Washington: Smithsonian Institution Press.

Allin, E. F., & Hopson, J. A. (1992). Evolution of the auditory system in Synapsida ("mammal-like reptiles" and primitive mammals) as seen in the fossil record. In D. B. Webster, R. R. Fay, & A. N. Popper (Eds.), *The evolutionary biology of hearing* (pp. 587–614). New York: Springer-Verlag.

Anthwal, N., Joshi, L., & Tucker, A. S. (2013). Evolution of the mammalian middle ear and jaw: Adaptations and novel structures. *Journal of Anatomy*, 222, 147–160.

Barghusen, H. R. (1968). The lower jaw of cynodonts (Reptilia, Therapsida) and the evolutionary origin of mammal-like adductor jaw musculature. *Postilla*, 116, 1–49.

Barry, T. H. (1968). Sound conduction in the fossil anomodont *Lystrosaurus*. *Annals of the South African Museum*, 50, 275–281.

Benson, R. B. J. (2012). Interrelationships of basal synapsids: Cranial and postcranial partitions suggest different toplogies. *Journal of Systematic Palaeontology*, 10(4), 601–624.

Bonaparte, J. F., Martinelli, A. G., Schultz, C. L., & Rubert, R. (2003). The sister group of mammals: Small cynodonts from the Late Triassic of southern Brazil. *Revista Brasiliera de Paleontologia*, 5, 5–27.

Broom, R. (1910). A comparison of the Permian reptiles of North America with those of South Africa. *Bulletin of the American Museum of Natural History*, 28, 197–234.

Case, E. C. (1907). *Revision of the Pelycosauria of North America*. Washington, D.C.: Carnegie Institute.

Cluver, M. A., & King, G. M. (1983). A reassessment of the relationships of Permian Dicynodontia (Reptilia: Theraapsida) and a new classification of dicynodonts. *Annals of the South African Museum*, 91, 195–272.

Cope, E. D. (1877). Descriptions of extinct Vertebrata from the Permian and Triassic formations of the United States. *Proceedings of the American Philosophical Society*, 17, 182–193.

Cox, C. B. (1959). On the anatomy of a new dicynodont genus with evidence of the position of the tympanic membrane. *Proceedings of the Zoological Society of London*, 132, 321–367.

Cox, C. B. (1972). The pectoral region and habits of a new digging dicynodont. In K. A. Joysey & T. S. Kemp (Eds.), *Studies in vertebrate evolution* (pp. 173–189). Edinburgh: Oliver and Boyd.

Crompton, A. W. (1963). On the lower jaw of *Diarthrognathus* and the origin of the mammalian lower jaw. *Proceedings of the Zoological Society of London*, 140, 697–753.

Crompton, A. W. (1972). The evolution of the jaw articulation of cynodonts. In K. A. Joysey & T. S. Kemp (Eds.), *Studies in vertebrate evolution* (pp. 231–251). Edinburgh: Oliver and Boyd.

Crompton, A. W., & Hotton, N. I. (1967). Functional morphology of the masticatory apparatus of two dicynodonts. *Postilla*, 109, 1–51.

Gaupp, E. (1913). Die Reichertsche Theorie. *Archive fur Anatomie und Entwicklungsgeschichte*, 1912(Supplement), 1–416.

Goodrich, E. S. (1930). *Studies on the structure and development of vertebrates*. London: Macmillan.

Hopson, J. A. (1966). The origin of the mammalian middle ear. *Americal Zoologist*, 6, 437–450.

Hotton, N. I. (1959). The pelycosaur tympanic membrane and early evolution of the middle ear. *Evolution*, 13(1), 99–121.

Huttenlocker, A. K., & Rega, E. (2012). The paleobiology and bone microstructure of pelycosaurian-grade synapsids. In A. Chinsamy-Turan (Ed.), *Forerunners of mammals: Radiation, histology, biology* (pp. 91–119). Bloomington: Indiana University Press.
Ivakhnenko, M. F. (1999). Biarmosuches from the Ocher Faunal assemblage of Eastern Europe. *Paleontological Journal*, 33, 289–296.
Kemp, T. S. (1969). On the functional morphology of the gorgonopsid skull. *Philosophical Transactions of the Royal Society of London B: Biological Sciences*, 256, 1–83.
Kemp, T. S. (1972a). The jaw articulation and musculature of the whaitsiid Therocephalia. In K. A. Joysey & T. S. Kemp (Eds.), *Studies in vertebrate evolution* (pp. 213–251). Edinburgh: Oliver and Boyd.
Kemp, T. S. (1972b). Whaitsiid Therocephalia and the origin of cynodonts. *Philosophical Transactions of the Royal Society of London B: Biological Sciences*, 264, 1–54.
Kemp, T. S. (1979). The primitive cynodont *Procynosuchus*: Functional anatomy of the skull and relationships. *Philosophical Transactions of the Royal Society of London B: Biological Sciences*, 285, 73–122.
Kemp, T. S. (1980). Aspects of the structure and functional anatomy of the Middle Triassic cynodont *Luangwa*. *Journal of Zoology*, 191, 193–239.
Kemp, T. S. (1982). *Mammal-like reptiles and the origin of mammals*. London: Academic Press.
Kemp, T. S. (2007). Acoustic transformer function of the postdentary bones and quadrate of a non-mammalian cynodont. *Journal of Vertebrate Paleontology*, 27(2), 431–441.
Kemp, T. S. (2009a). Phylogenetic interrelationships and pattern of evolution of the therapsids: Testing for polytomy. *Palaeontologica Africana*, 44, 1–12.
Kemp, T. S. (2009b). The endocranial cavity of a nonmammalian eucynodont, *Chiniquodon theotenicus*, and its implications for the origin of the mammalian brain. *Journal of Vertebrate Paleontology*, 29(4), 1188–1198.
Kemp, T. S. (2012). The origin and radiation of therapsids. In A. Chinsamy-Turan (Ed.), *Forerunners of mammals: Radiation, histology, biology* (pp. 3–28). Bloomington: Indiana University Press.
Kitazawa, T., Takechi, M., Hirasawa, T., Adachi, N., et al. (2015). Developmental genetic bases behind the independent origin of the tympanic membrane in mammals and diapsids. *Nature Communications*, 6(6853), 1–7.
Laass, M. (2014). Bone-conduction hearing and seismic sensitivity of the Late Permian anomodont *Kawingasaurus fossilis*. *Journal of Morphology*, 276, 121–143.
Luo, Z., & Crompton, A. W. (1994). Transformation of the quadrate (incus) through the transition from non-mammalian cynodonts to mammals. *Journal of Vertebrate Paleontology*, 14, 341–374.
Maier, W. (1990). Phylogeny and ontogeny of mammalian middle ear structures. *Netherlands Journal of Zoology*, 40, 55–74.
Maier, W., & van den Heever, J. (2002). Middle ear structures in the Permian *Glanosuchus sp.* (Therocephalia, Therapsida), based on thin sections. *Mitteilungen aus dem Museum für Naturkunde in Berlin Geowissenschaftliche Reihe*, 5, 309–318.
Owen, R. (1845). Description of certain fossil crania, discovered by A. G. Bain, Esq., in sandstone rocks at the south-eastern extremity of Africa, referable to different species of an extinct genus of Reptilia (*Dicynodon*), and indicative of a new tribe or sub-tribe of Sauria. *Transactions of the Geological Society of London*, 7, 59–84.
Owen, R. (1876). *Descriptive and illustrated catalogue of the fossil Reptilia of South Africa in the collection of the British Museum*. London: The British Museum, Natural History.
Parrington, F. R. (1946). On the cranial anatomy of cynodonts. *Proceedings of the Zoological Society of London*, 116, 181–197.
Parrington, F. R. (1955). On the cranial anatomy of some gorgonopsids and the synapsid middle ear. *Proceedings of the Zoological Society of London*, 125, 1–40.
Parrington, F. R. (1979). The evolution of the mammalian middle and outer ears: A personal view. *Biological Reviews*, 54, 369–387.

Reichert, K. B. (1837). Über die Visceralbogen der Wirbelthiere im allgemeinen und deren Metamorphosen bei den Vögeln und Suäugethieren, 1837. *Archive für Anatomie, Physiologie, und wissenschaftliche Medizin*, 1837, 120–220.

Romer, A. S., & Price, L. W. (1940). Review of the Pelycosauria. *Special Papers of the Geological Society of America*, 28, 1–538.

Rubidge, B. S. (1990). Redescription of the cranial morphology of *Eodicynodon oosthuizeni* (Therapsida: Dicynodontia). *Navorsinge van die Nasionale Museum Bloemfontein*, 7, 1–25.

Sidor, C. A., & Rubidge, B. S. (2006). *Herpetoskylax hopsoni*, a new biarmosuchian (Therapsida: Biarmosuchia) from the Beaufort Group of South Africa. In M. T. Carrano, T. J. Gaudin, R. W. Blob, & J. R. Wible (Eds.), *Amniote paleobiology: Perspectives on the evolution of mammals, birds, and reptiles* (pp. 76–113). Chicago: Chicago University Press.

Sigurdsen, T., Huttenlocker, A. K., Modesto, S. P., Rowe, T. B., & Damiani, R. (2012). Reassessment of the morphology and paleobiology of the therocephalian *Tetracynodon darti* (Therapsida), and the phylogenetic rerlationships of Baurioidea. *Journal of Vertebrate Paleontology*, 32(5), 1113–1134.

Takechi, M., & Kuratani, S. (2010). History of studies on mammalian ear evolution: A comparative morphological and developmental perspective. *Journal of Experimental Zoology Part B: Molecular and Developmental Evolution*, 341B: 417–433.

Thompson, H., & Tucker, A. S. (2013). Dual origin of the epithelium of the mammalian middle ear. *Science*, 339, 1453–1456.

Tumarkin, A. (1955). On the evolution of the auditory conducting apparatus: A new theory based on functional considerations. *Evolution*, 9(3), 221–243.

van den Heever, J. (1994). The cranial anatomy of the early Therocephalia (Amniota: Therapsida). *Annals of the University of Stellenbosch*, 1994(11), 1–59.

Watson, D. M. S. (1952). The evolution of the mammalian ear. *Evolution*, 7, 159–177.

Chapter 6
Evolution of the Middle and Inner Ears of Mammaliaforms: The Approach to Mammals

Zhe-Xi Luo, Julia A. Schultz, and Eric G. Ekdale

Abstract Transformations of ear structures in the evolution of early mammals can be studied with the fossils of mammaliaforms. The middle ear is fully attached to the mandibles in mammaliaforms; however, in Mesozoic eutriconodont and spalacotherioid mammals, it is only connected to the mandible by an ossified Meckel's cartilage, with the ectotympanic and malleus already displaced from the mandible. Recent morphogenetic studies have shown that the developmental potential for ossification of Meckel's element is conserved in extant mammals. New fossils further revealed that this pattern actually evolved in mammaliaform phylogeny and that disconnection of the ear from the mandible occurred independently in monotremes, in therians, and in multituberculate mammals. The inner ear of mammaliaforms is derived in having a single petrosal bone enclosing the entire inner ear and a promontorium for an elongate cochlear canal. Mammaliaforms and most Mesozoic mammals had ancestral features of a simple cochlear canal with a single cochlear nerve foramen but no interior bony laminae nor did they have a bony canal for the cochlear ganglion. The sieve-like foramina for cochlear nerve fibers to enter the cochlear canal evolved independently three times in Mesozoic mammals. Cochlear canal curvature is homoplastic among mammaliaform groups, and a curvature beyond 270° only evolved in cladotherians, accompanied by Rosenthal's canal for the cochlear ganglion. The homoplasies of ear structures in

Z.-X. Luo · J.A. Schultz
Department of Organismal Biology and Anatomy, The University of Chicago, 1027 East 57th Street, Chicago, IL 60637, USA
e-mail: zxluo@uchicago.edu

J.A. Schultz
e-mail: jaschultz@uchicago.edu

E.G. Ekdale (✉)
Department of Biology, San Diego State University, 5500 Campanile Drive, San Diego, CA 92182, USA
e-mail: eekdale@mail.sdsu.edu

E.G. Ekdale
Department of Paleontology, San Diego Natural History Museum, San Diego, CA 92182, USA

J.A. Clack et al. (eds.), *Evolution of the Vertebrate Ear—Evidence from the Fossil Record*, Springer Handbook of Auditory Research 59,
DOI 10.1007/978-3-319-46661-3_6

early mammalian evolution, although seemingly complex, are consistent with the new understanding of a labile morphogenesis of mammalian ears under a complex developmental genetic network.

Keywords Cochlear canal · Cochlear innervation · Development · Evolutionary homoplasies · Inner ear · Meckel's cartilage · Mesozoic mammaliaforms · Middle ear

6.1 Introduction

The evolution of the mammalian middle and inner ear is among the best known and documented morphological transitions in vertebrate history. The phylogenetic transformation of the middle and inner ear structures can be examined by tracing the precursor conditions of the mammalian ear characteristics in Mesozoic mammals and their closest ancestors known collectively as mammaliaforms (Fig. 6.1). Mammaliaformes is a clade of extant mammals plus the proximal relatives to extant Mammalia, modified here from Rowe (1988) to include *Sinoconodon* (see Kemp 2005). The crown group Mammalia (i.e., all extant mammals) is covered by Ekdale in Chap. 7. Mammaliaforms have a fossil record spanning approximately 220 million years with over 450 Mesozoic genera currently recognized (Kielan-Jaworowska et al. 2004; Close et al. 2015), all of which shed important light on mammalian origins (Rowe 1993; Kemp 2005).

Mammaliaforms are distinguished from the pre-mammaliaform cynodonts (hereafter shortened to "cynodonts") (see Kemp, Chap. 5) by many derived features (apomorphies), among which are key characters of the cranio-mandibular joint, the middle ear, and the inner ear (Figs. 6.2, 6.3) (Kielan-Jaworowska et al. 2004; Luo 2007). For example, mammaliaforms have an apomorphic joint between the dentary condyle and the squamosal glenoid, in addition to the plesiomorphic joint of the quadrate (incus) and the articular (malleus), to form a "double jaw joint" (Kermack et al. 1981; Luo 2011). The middle ear is in full contact with the dentary via a Meckel's element, a plesiomorphic feature as in cynodonts (Kermack et al. 1973; Sues 1986). However, the middle ear elements of mammaliaforms are proportionally smaller and more gracile, and the cranial suspension of the middle ear is also more mobile than those of cynodonts (Luo and Crompton 1994). In addition, the mammaliaform inner ear has an elongate and tubular cochlear canal with complete osseous housing by the petrosal bone, which is derived with respect to cynodonts (Rodrigues et al. 2013; Ruf et al. 2013).

Recent discoveries of well-preserved fossils in critical phylogenetic positions have revealed significant homoplasies within the ears of mammals, both evolutionary convergencies and reversals, via the transition of extinct mammaliaforms from cynodonts. Those homoplasies are analogous to evolutionary experiments in form and function during the Mesozoic Era, and they can serve to test the mechanistic hypotheses of the process that underlined the patterns of mammal evolution.

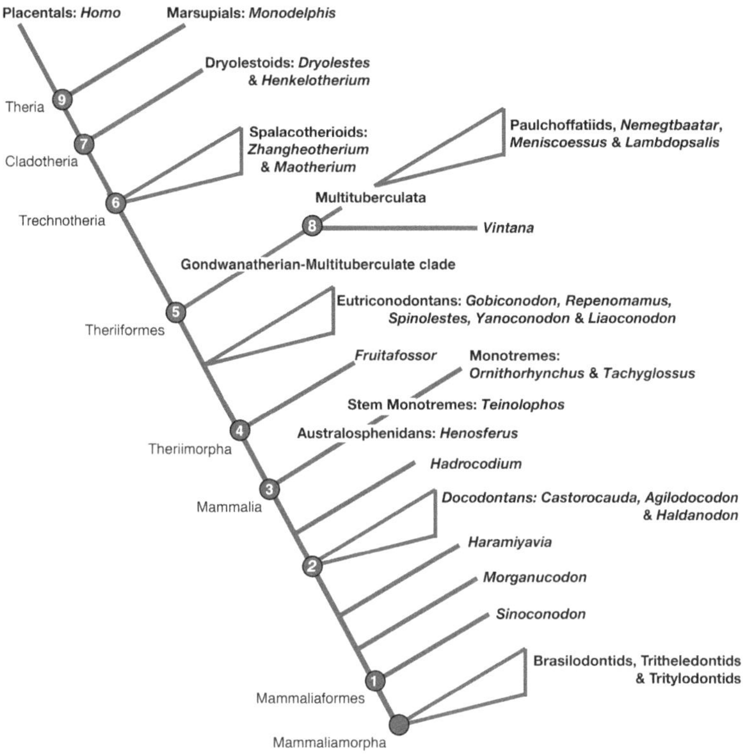

Fig. 6.1 Phylogenetic tree of mammaliaforms and the extinct clades of Mesozoic mammals discussed in the chapter. Advanced cynodont relationships modified from Kemp (2005) with information from Luo (2007) and Liu and Olsen (2010). Relationship of Mesozoic clades follow Luo et al. (2015a), and Krause et al. (2014); node names following Rowe (1988) or Luo et al. (2002) but definition may be modified to accommodate the new taxa discovered in the last 20 years. Phylogenetic trees in subsequent figures all follow the same designation of clades and numbering of their nodes: (*1*) Mammaliaformes, (*2*) Unnamed clade of docodontans+crown mammals (excluding *Morganucodon* and *Sinoconodon*), (*3*) Mammalia (*sensu* Rowe 1988), (*4*) Theriimorpha (modified from Rowe to include *Fruitafossor*; Luo and Wible 2005), (*5*) Theriiformes, (*6*) Trechnotheria, (*7*) Cladotheria, (*9*) Theria (crown therians), (*8*) Clade of Gondwanatheria+Multituberculata (*sensu* Krause et al. 2014)

6.2 Middle Ear Evolution

Transformation of the middle ear from premammalian cynodonts through mammaliaforms to modern mammals has been documented by an extensive fossil record (Allin 1975; Kemp 2005). In the last 20 years, discoveries of many new fossils of Mesozoic mammaliaforms with well-preserved middle ear structures revealed a more detailed picture of this evolutionary process (e.g., Luo et al. 2007; Meng et al. 2011). These new fossils raised new questions for the paleontological understanding of mammaliaform middle ear evolution (Luo 2011), and they provided

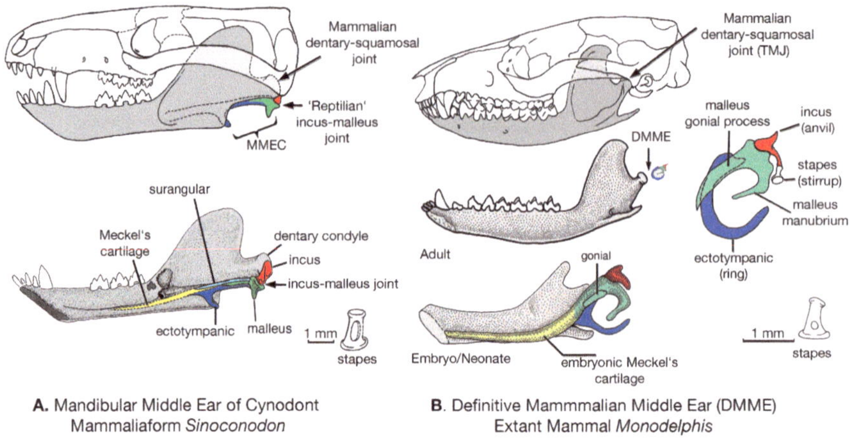

Fig. 6.2 Plesiomorphic versus apomorphic structures of stem mammaliaforms and crown mammals. **(A)** Mandibular middle ear of cynodonts (MMEC) of mammaliaform *Sinoconodon*. **(B)** Definitive mammalian middle ear (DMME) of extant mammal *Monodelphis* (stapes from *Didelphis*). Stapes on different scale. (Original drawings by the authors)

new stimulation for understanding the developmental genetics of mammalian middle ears (Takechi and Kuratani 2010; Anthwal et al. 2013).

6.2.1 Middle Ear Evolution in Mammaliaforms

Co-evolution of the middle ear and the cranio-mandibular joint occurred in an incremental pattern when mapped on large-scale synapsid phylogenies (Luo 1994; Sidor 2001). The mandibular middle ear of cynodonts (MMEC, Fig. 6.2A) (*sensu* Luo 2011) and the definitive mammalian middle ear (DMME, Fig. 6.2B) as exhibited by all extant mammals (*sensu* Allin and Hopson 1992) represent the ends of a morphological continuum, with a partial mammalian middle ear (PMME) of Mesozoic mammal clades as an intermediate condition (Fig. 6.3B, C). In the PMME (also known as "transitional mammalian middle ear" of Meng et al. 2011), the middle ear is mediolaterally displaced from the posterior part of the mandible, yet remains connected to the mandible by the Meckel's cartilage anteriorly (Fig. 6.3), a precursor pattern of the DMME (Fig. 6.2B).

The ancestral MMEC, as seen in *Thrinaxodon* and *Cynognathus*, is characterized by large middle ear bones that are fully attached to the mandible, as well as a large and plate-like angular bone that is interpreted to function as a sound receiver (Allin 1975; Maier and van den Heever 2002; Kemp, Chap. 5). The Meckel's element itself was likely cartilaginous and unossified (Kermack et al. 1973; Kemp 2007), and the Meckel's sulcus is either covered by the splenial bone or exposed in taxa that exhibit splenial reduction (Allin and Hopson, 1992; Bonaparte et al. 2005).

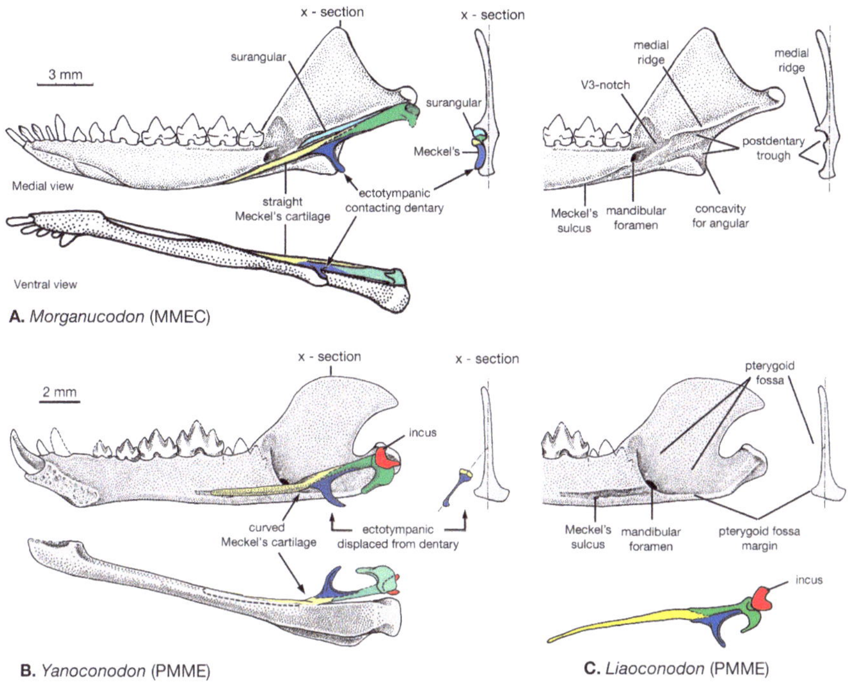

Fig. 6.3 Comparative morphology of the middle ear of stem mammaliaforms and extinct clades of crown mammals. **(A)** *Morganucodon* with mandibular middle ear of cynodonts (MMEC): the ectotympanic (= angular) nestled in the concavity of angular, the surangular and the prearticular (gonial) part of malleus nestled in the postdentary trough under the medial ridge, the Meckel's element extends from the postdentary trough into the Meckel's sulcus (Luo et al. 2007). **(B)** *Yanoconodon* and **(C)** *Liaoconodon* with the partial mammalian middle ear (PMME) (or "transitional mammalian middle ear [TMME]"): the mandible lacks a postdentary trough but has a neomorphic pterygoid fossa; the ectotympanic-malleus complex is mediolaterally displaced from the posterior region of the mandible, made feasible by the curved Meckel's cartilage. The middle ear is still connected anteriorly to the mandible by an ossified Meckel's cartilage fit into the Meckel's sulcus. *x-section* = location of cross section and the cross-sectional view to the right. (Original drawings by authors; *Liaoconodon* is redrawn from Luo's personal observation of the original specimen, modified from Meng et al. 2011)

Stem mammaliaforms show a broad range of morphological variations in the structures that accommodated the middle ear (Figs. 6.3, 6.4), suggesting that the ear-mandible attachment is variable. The middle ears likely had disparate morphological patterns in the shape of the angular region of the dentary, the posterior opening of the postdentary trough, the presence or absence of grooves and fossae in the postdentary trough, as well as the curvature of Meckel's sulcus. For example, the postdentary trough and Meckel's sulcus for attachment of middle ear elements on the mandible have been retained in all non-mammalian mammaliaforms (Figs. 6.3, 6.4), but other plesiomorphic features, such as the medial ridge above the postdentary trough for reinforcing the middle ear bones, are more prominent in mammaliaforms than in cynodonts (Kermack et al. 1973; Allin 1975).

The middle ear bones are not preserved in most mammaliaform and other stem mammal fossils owing to their delicate nature and detachment from the dentary. In such cases, the postdentary trough, its grooves and fossae, and the Meckel's sulcus are considered osteological proxies for inferring the presence of the MMEC for cynodonts and mammaliaforms (Luo et al. 2015a). A previous study (Luo et al. 2001) reported that the derived mammaliaform *Hadrocodium* lacked the postdentary trough, and thus it had a DMME based on conventional fossil preparation. However, re-examination of μCT scans of the fossil, as first noticed by Dr.

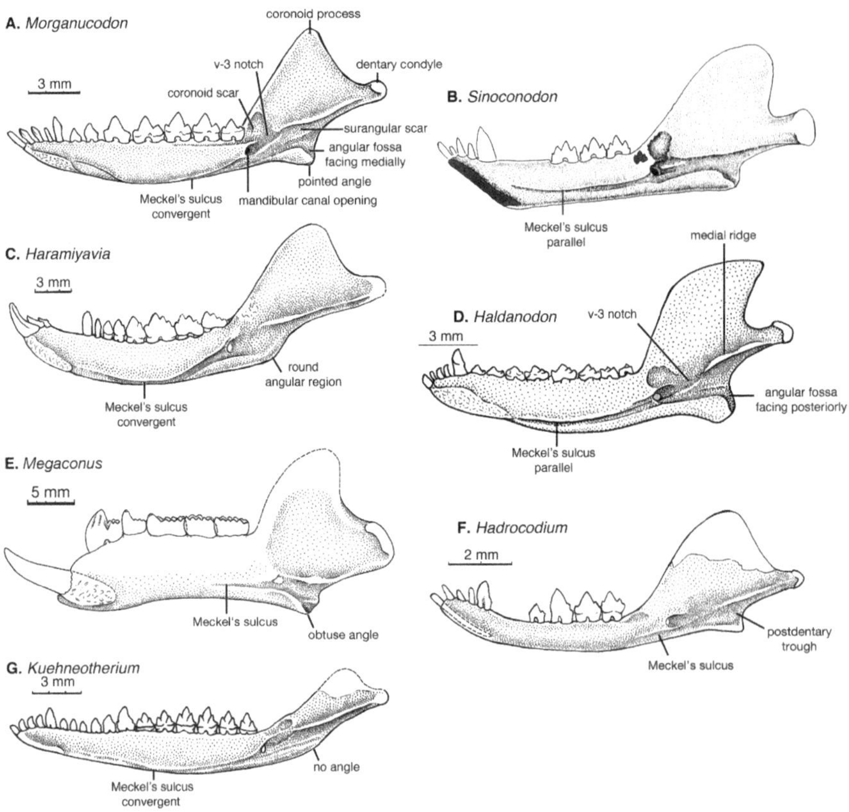

Fig. 6.4 Disparity of middle ear structures of mandibles among mammaliaforms: **(A)** *Morganucodon*, **(B)** *Sinoconodon*, **(C)** *Haramiyavia*, **(D)** *Haldanodon*, **(E)** *Megaconus*, **(F)** *Hadrocodium*, **(G)** *Kuehneotherium*. *V-3 notch*: the notch anterior to the medial ridge and conduit for the trigeminal mandibular nerve (*V-3*) to enter the mandibular canal. (**A** original drawing of authors; **B** modified from Luo 2011; **C** from Luo et al. 2015; **D** modified from Luo 2011; **E** modified from Zhou et al. 2013; **F** new drawing based on CT scanning of the holotype of authors and corrected from Luo et al. 2001; **G** new drawing based on Gill et al. 2014)

Bhart-Anjan Bhullar (B.-A. Bhullar, personal communication) and further reconstruction of the CT dataset confirmed that the *Hadrocodium* fossil possesses the postdentary trough and Meckel's sulcus, although the structures are shallower than those of other stem mammaliaforms (Fig. 6.3F). Thus, *Hadrocodium* had the MMEC. Likewise, a recent re-study of the mammaliaform *Haramiyavia* (Luo et al. 2015a) indicates that almost all of the mandibular features common to other mammaliaforms are also preserved in *Haramiyavia*, including the postdentary trough, which was previously questioned (e.g., Bi et al. 2014).

Stem mammaliaforms exhibit a wide range of variation in the shape of their middle ear bones and their attachment to the dentary. Most mammaliaforms have pointed mandibular angles, and the angular process has a concavity associated with a widening of the posterior opening of the postdentary trough (e.g., *Morganucodon*, Fig. 6.3A). The curved ectotympanic (= reflected lamina of angular) is nestled in this angular concavity on the dentary in *Sinoconodon*, *Morganucodon*, and *Hadrocodium* (Figs. 6.3, 6.4). Docodonts such as *Haldanodon* are unique in that the angular concavity faces posteriorly and is deeply incised into the angular region of the dentary (Fig. 6.4D) (Ji et al. 2006; Luo et al. 2015b). *Hadrocodium* is notable in that its angular process is slightly inflected medially (Fig. 6.4F). In contrast, *Kuehneotherium* and *Megazostrodon* lack a pointed mandibular angle, and the "angular region" of the mandible is only represented by a shallow convexing contour with a narrower opening of the postdentary trough (Fig. 6.4G) (Gow 1986; Gill et al. 2014). The narrow opening would correspond to a more acute angle between the reflected lamina (= ventral limb) and the dorsal limb of the ectotympanic, morphologically comparable to the angle of the refected lamina of the cynodonts *Cynognathus* and *Probainognathus* (Kermack et al. 1973; Allin and Hopson 1992; Kemp, Chap. 5). Thus we speculate that the ectotympanic is likely smaller and not ring-like in *Kuehneotherium* and *Megazostrodon*, as seen in *Sinoconodon*, *Morganucodon*, and *Agilodocodon* (Meng et al. 2015). *Haramiyavia* shows a discernible angular region that is morphologically intermediate between *Kuehneotherium* with no angle and other mammaliaforms with a pointed angle (Fig. 6.4).

6.2.2 Ectotympanic Structure and Function

Mammaliaforms have several neomorphic characters of the ectotympanic bone that are not present in earlier cynodonts. For example, the ventral limb of this bone (homologous to the curved ectotympanic ring of extant mammals; see Ekdale, Chap. 7) assumes a more mammalian appearance in that it is rod-like, curved, and tapered toward its apex (Figs. 6.2, 6.3). In contrast, the reflected lamina in cynodonts is plate-like, broader at the apical part than its base, and is relatively straight, as exemplified by *Probainognathus*, *Diademodon*, and *Cynognathus* (Allin 1975; Hopson and Kitching 2001). The more gracile and curved ectotympanic of mammaliaforms suggests that a relatively larger tympanic membrane was developed in these taxa, which would enable greater direct sound reception by the membrane, than in the more basal cynodonts in which the plate-like reflected lamina still played a major (if not a dominating) role in sound reception (Allin 1975).

This functional change of the ectotympanic between cynodonts and mammaliaforms is likely to have occurred through some transitional states in phylogenetically intermediate taxa. For example, the reflected lamina of the angular is straight (plesiomorphic), thinner, and morphologically similar to the lamina of mammaliaforms in the tritylodontid *Kayentatherium* (Sues 1986). However, the ventral limb of the mammaliaform ectotympanic is quite short compared to extant mammals. In fact, it is shorter than half of the dorsal limb of the angular in the derived mammaliaforms *Sinoconodon*, *Morganucodon*, and docodonts (Ji et al. 2006; Meng et al. 2015), shorter than that of early mammalian eutriconodonts (Luo et al. 2007; Meng et al. 2011), and certainly shorter than that of extant monotremes (platypus and echidnas) and therian mammals (marsupials and placentals) (Doran 1879; Fleischer 1973).

6.2.3 Malleus Structure and Function

The malleus (articular) of *Sinoconodon* and *Morganucodon* has a short retroarticular process with a pointed ventral apex (Figs. 6.2, 6.3) (Crompton and Luo 1993; Luo 2011). This short retroarticular process represents the base for the manubrium of the malleus but not the manubrium itself as developed in extant mammals (Figs. 6.2, 6.3) (Luo 2011). The manubrium is not preserved in these two mammaliaforms, suggesting that it is either cartilaginous and not ossified or that it is not developed at all, based on the well-preserved periosteal surface of the retroarticular process in *Sinoconodon*. A gracile manubrium-like structure has never been confirmed in *Morganucodon* (Crompton and Luo 1993). The absence of a malleal manubrium in basal mammaliaforms is consistent with independent evidence from developmental morphogenesis. Cell lineage tracing of the retroarticular process (also known as the short process) of the malleus of *Mus musculus* (house/laboratory mouse) shows that it is derived from neural crest cells of hyoid arch origin in the mouse (O'Gorman 2005; but see Maier and Ruf 2016), which is the same as the retroarticular process of the quadrate in non-mammalian vertebrates. Because non-mammalian vertebrates do not have a malleal manubrium, the earlier reconstruction of a gracile manubrium extending from the retroarticular process of the malleus (e.g., Kermack et al. 1981) lacked support and is likely incorrect.

In contrast, a manubrium of the malleus, demarcated by a distinctive bend from the base of the malleus, is known for docodonts (Ji et al. 2006; Meng et al. 2015). This morphology suggests that manubrial support in the central part of the tympanic membrane, as seen in extant mammals, is already developed in docodonts (Luo 2011). The haramiyidan *Megaconus* also appears to have a manubrium (Zhou et al. 2013). Thus, the manubrium of the malleus evolved in some (although not all) mammaliaforms preceding the ancestor of crown Mammalia.

The retroarticular process is present in pre-mammaliaform cynodonts, but is commonly small (e.g., the tritylodontid *Kayentatherium*) (Fig. 16 in Sues 1986). This structure is short in other cynodonts, or more likely, it is incompletely ossified, except in *Diademodon* for which a large and hook-like retroarticular process has been described (Brink 1963). An elongate, extant mammal-like manubrium of the malleus first appeared in docodonts (Ji et al. 2006; Meng et al. 2015) and likely also in some haramiyidans (Zhou et al. 2013). Several studies reconstructed the retroarticular process and the ectotympanic to be longer, more curved, and like extant mammals in the earlier diverging *Morganucodon* (e.g., Allin 1975; Kermack and Mussett 1983), but examination of better-preserved fossils indicates that the extant mammal-like ectotympanic and manubrium of *Morganucodon* as reconstructed by Allin (1975) and Kermack et al. (1981) is incorrect (Crompton and Luo 1993; Luo 2011). Rather, the retroarticular process of *Morganucodon* is small, as originally described (Figs. 6.3A, 6.4A) (Kermack et al. 1973), and the reflected lamina of the angular is also short. The incorrect reconstruction of these two elements in *Morganucodon* (e.g., Kermack and Mussett 1983) influenced several studies that interpreted inaccurately the form and function (hearing capacity) of the middle ear of *Morganucodon* and other mammaliaforms (e.g., Lillegraven and Krusat 1991; Rosowski 1992).

The middle ear bones of mammaliaforms are similar to those of mammalian eutriconodonts in both size and proportion. For example, the ectotympanic is curved to less than a 60° arc in both groups (Fig. 6.3) (Luo et al. 2007; Meng et al. 2011). The ectotympanic ring of less than 60° arc curvature is likely the ancestral condition of Mammalia (Fig. 6.5). However, the ectotympanic is curved to a greater degree in extant mammal clades (e.g., about 90° in the duck-billed platypus *Ornithorhynchus* and 120° in the short-beaked echidna *Tachyglossus*; even greater curvatures are observed in placental and marsupial mammals), which gives the bone the iconic "ring-like" morphology in living mammals (Fig. 6.2B). The ectotympanic curvature beyond 90° probably evolved separately in monotremes and in crown therians, given that the ectotympanic is curved to a lesser degree in eutriconodonts. An alternative interpretation is that the lesser curvature of the ectotympanic in eutriconodonts would be an atavistical reversal if the over 90° of ectotympanic ring of extant monotremes represented the ancestral condition of the crown Mammalia. The latter hypothesis would gain support if spalacotherioids and multituberculates turned out to have a monotreme-like ectotympanic curvature and greater curvature than the ectotympanics of eutriconodonts (Luo et al. 2007; Meng et al. 2011). Currently, the full ectotympanic is not preserved in latter Mesozoic mammals, but this alternative hypothesis can be tested if better fossils are discovered in the future. Overall, the variation in ectotympanic arc curvature within mammaliaform clades is evident from the differences of extant monotremes (Fleischer 1973; Luo 2011) and marsupials (Doran 1879).

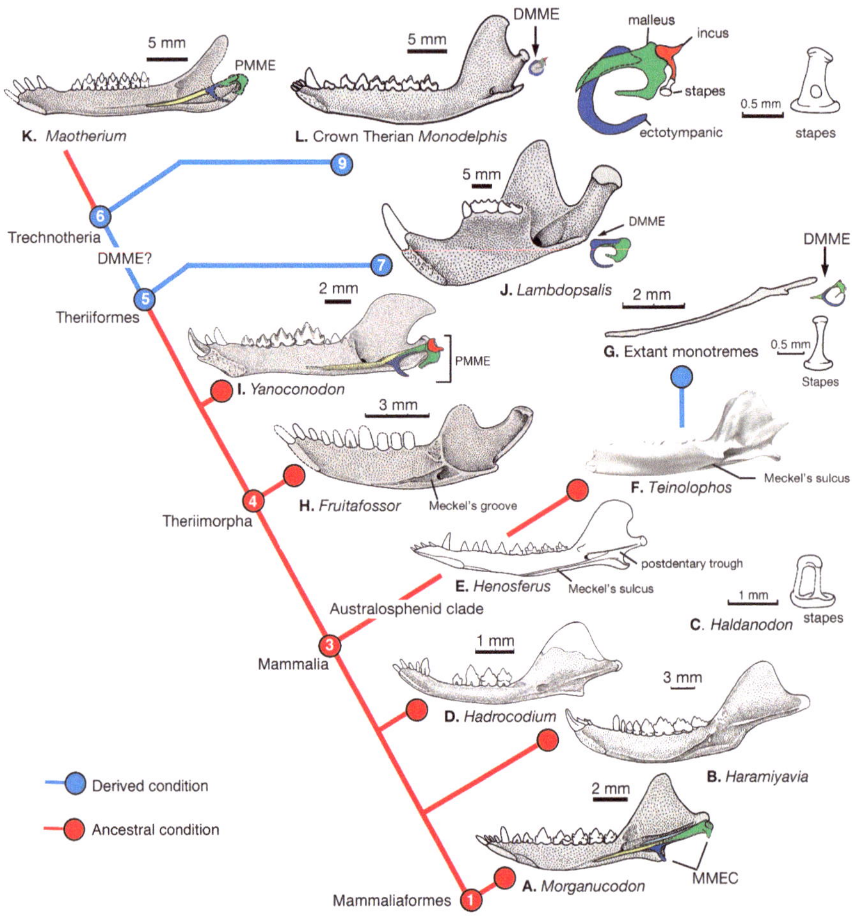

Fig. 6.5 Homoplastic evolution of definitive mammalian middle ear (*DMME*) among mammaliaform clades: **(A)** *Morganucodon*, **(B)** *Haramiyavia*, **(C)** *Haldanodon*, stapes only, **(D)** *Hadrocodium*, **(E)** *Henosferus*, **(F)** Stem monotreme *Teinolophos*, **(G)** extant monotreme *Tachyglossus* (stapes on a different scale), **(H)** *Fruitafossor*, **(I)** eutriconodont *Yanoconodon*, (J) multituberculate *Lambdopsalis*, **(K)** spalacotherioid *Maotherium*, **(L)** marsupial *Monodelphis*: *left*, mandible; *middle*, ear bones; *right*, stapes of an junvenile of *Didelphis* (on different scale). The middle ear elements are not preserved in the majority of Mesozoic theriiforms (*node 5*). A conservative assessment of the disconnection of the ear from the mandible should be based on only the well-preserved fossils with middle ears. If so, it would appear that the DMME is acquired independently in monotremes, in multituberculates, and in crown therians. The *PMME* (partial mammalian middle ear) in spalacotherioids would be either an atavistic reversal, or an intermediate/plesiomorphic condition. *MMEC*, mandibular middle ear of cynodonts. *Cladogram nodes same as in* Fig. 6.1. (**C** new drawing based on Ruf et al. 2013; **D** corrected from Luo et al. 2001; **E** original drawing courtesy of Dr. G. W. Rougier, used by permission, Rougier et al. 2007; **F** original drawing courtesy of Dr. Thomas H. Rich, used with permission, Rich et al. 2005; **G** original drawing; **H** from Luo and Wible 2005; **I** from Luo et al. 2007; **J** drawing by authors, modified from Miao 1988; **K** drawing by Luo, Ji et al. 2009; **L** drawing by Luo)

6.2.4 *Structure of the Incus*

In the cynodont–mammaliaform transition, the quadrate (incus) gained more mobility at the cranio-incus contact, following the loss of the quadrato-jugal in cynodonts (Luo and Crompton 1994; Kemp 2005). Concurrently, the cranial suspension of the incus is also transformed: in the primitive cynodont condition, the incus (quadrate) is suspended primarily by the squamosal; but in the derived mammaliaform condition, the incus is suspended entirely by the petrosal (Luo and Crompton 1994; Luo 2011). These changes of the incus have contributed to a greater mobility of the middle ear bones, which is essential for increased hearing sensitivity (Luo and Crompton 1994).

The stapedial process of the incus (quadrate) is a part of the middle ear lever system for transmitting sound with impedance matching (Durrant and Lovrinic 1995; Clack and Allin 2004). The stapedial process of the incus is absent in most premammaliaform cynodonts except brasilodontids (Bonaparte et al. 2005; Luo 2007). By comparison, stem mammaliaforms have this stapedial process (Kermack et al. 1981), except *Sinoconodon* (Luo 2007). These structures of the incus show incongruent distribution among the transitional mammaliamorphs to mammaliaforms, although they are essential for improved hearing. Thus the inference is that their evolution is homoplastic (Luo 2007, 2011).

6.2.5 *Structure of the Stapes*

Most cynodonts have a columnar stapes with a large stapedial foramen with the proximal (fenestra) end and distal (incus) end of approximately the same size (Allin and Hopson 1992). The derived brasilodontid cynodonts, which are placed very close to mammaliaforms and are among the mammaliamorphs phylogenetically (Fig. 6.1) (Bonaparte et al. 2005; Liu and Olsen 2010), have retained this condition (see Fig. 7 in Rodrigues et al. 2013). This "columelliform-perforate" stapes (*sensu* Novacek and Wyss 1986) is an ancestral condition from which mammaliaforms and crown mammals evolved the extremely diverse structure of the stapes.

Precise fitting of the fenestra vestibuli rim with the stapedial footplate is already present in the derived mammaliamorph cynodont *Brasilitherium* (Rodrigues et al. 2013), possibly also in tritylodontids (Luo 2001). In contrast to the oval outline of the footplate in the majority of placentals, the stapedial footplate of mammaliaforms has an almost completely circular outline. The footplate shape can be slightly convex as in *Haldanodon* (Ruf et al. 2013) or flat as in *Sinoconodon* (Fig. 6.2A) or *Morganucodon* (Kermack et al. 1981).

A notable, neomorphic feature of the stapes in phylogenetically derived mammaliaforms is the substantially smaller size of the lateral (incus) end of the stapes compared to the larger proximal stapedial footplate that fits into the fenestra vestibuli. This is exemplified by *Sinoconodon* and *Haldanodon* (Figs. 6.2A, 6.5C)

(Luo 2007; Ruf et al. 2013). Although an earlier study reconstructed a cynodont-like columelliform stapes for *Morganucodon* (Kermack et al. 1981; also accepted by Novacek and Wyss 1986), the later, improved reconstruction shows a substantially smaller stapedial head than footplate for *Morganucodon* (Allin and Hopson 1992). It is believed that stem mammaliaforms ancestrally have a smaller stapedial head than the footplate, as seen in *Sinoconodon* and *Haldanodon* (Figs. 6.2A, 6.5C), which is a derived condition compared to the primitive shape of the stapes in cynodonts (Allin and Hopson 1992). The stapedial head versus footplate size disparity is further increased in the more derived clades in mammal phylogeny (Novacek and Wyss 1986).

For crown mammals, monotremes have columelliform stapes (Fig. 6.5G), which is known only in one multituberculate (Meng 1992). Marsupials show both columelliform and bicrurate stapes (Novacek and Wyss 1986; Schmelzle et al. 2005), whereas placentals have mostly bicrurate stapes, although they also have a vast range of other forms that have been documented extensively in the literature (Doran 1879; Novacek and Wyss 1986). The evolutionary pattern in basal crown mammals is unclear because of the lack of well-preserved stapes among Mesozoic clades of mammals and the wide range of variation in placental mammals.

6.2.6 *Independent Evolution of Definitive Middle Ear in Monotremes*

The evolution and morphological diversity of the middle and inner ears of crown mammals, which include the common ancestor of living monotremes, marsupials, and placentals, and all of their descendants (Rowe 1988), is discussed elsewhere in this volume (Ekdale, Chap. 7). Due to the recent discovery that *Hadrocodium* has a postdentary trough (Fig. 6.4F), it is now clear that all stem mammaliaforms have the MMEC condition. We can now posit that the origin of crown Mammalia is accompanied by the evolutionary innovation of the partial mammalian middle ear PMME (Figs. 6.3, 6.4 6.5).

However, recognition of the disparate morphologies of middle ears in the Mesozoic mammal clades is necessary for the proper polarization of the ear features within the entire Mammaliaformes. For example, extant monotremes (platypuses and echidnas) possess a DMME that is like their marsupial and placental mammal relatives. However, the DMME of extant monotremes may represent an independent evolution given that the monotreme clade is phylogenetically bracketed by mammaliaforms with the MMEC and the several Mesozoic theriimorph mammal clades that have either an ossified Meckel's cartilage or a Meckel's sulcus indicating the presence of the cartilage in adulthood (Fig. 6.5).

The earliest fossils of the monotreme clade that preserve some parts of the posterior region of the dentary are assigned to the genus *Steropodon* (Archer et al. 1985) and the genus *Teinolophos* (Rich et al. 2001; Philips et al. 2009), both from the Early Cretaceous of Australia. Both *Steropodon* and *Teinolophos* have preserved wide Meckel's sulci.

A postdentary trough was described for *Teinolophos*, which, if in fact is present, would contain the postdentary bones of the middle ear (Rich et al. 2005). However, the presence of the trough has been debated, and an alternative suggestion is that the trough was an enlarged groove for a large mandibular nerve similar to the hypertrophied mandibular nerve of extant *Ornithorhynchus* (Rougier et al. 2005). Further analysis of the fossils using μCT data has challenged this (Rowe et al. 2008). Regardless, Early Jurassic fossil mammals such *Henosferus* and *Asfaltomylos* possess a postdentary trough and a Meckel's sulcus. These fossils likely belong to a clade along the monotreme stem (Fig. 6.4) (O'Meara and Thompson 2014). Thus, an independent acquisition of the DMME must have occurred during monotreme evolution.

6.2.7 Independent Evolution of Definitive Midde Ear in Theriimorph Mammals

Theriimorpha is defined as the common ancestor of eutriconodonts, multituberculates, spalacotherioids, dryolestoids, through living placentals and marsupials (Theria) (Fig. 6.1) (Rowe 1988; see Fig. 7.1 in Ekdale, Chap. 7). Among theriimorphs, an ossified Meckel's cartilage has been described in eutriconodonts (Luo et al. 2007; Meng et al. 2011) and Mesozoic "acute-triangles" symmetrodonts called spalacotherioids, which are stem relatives of extant marsupials and placentals (Meng et al. 2003; Ji et al. 2009). In both eutriconodont and spalacotherioid clades, Meckel's cartilage is curved at mid-length (Fig. 6.3). The curvature of Meckel's cartilage allows for the middle ear bones to be displaced mediolaterally and rotated away from the mandible while retaining a full connection of the cartilage in its mandibular sulcus, which defines the PMME condition (Fig. 6.3). The curvature of this element is slightly variable among eutriconodonts and is especially accentuated in those specimens with more massive Meckel's cartilages (Meng et al. 2011; Martin et al. 2015).

Fossilized Meckel's cartilage in spalacotherioids is almost identical to, albeit more gracile than, those of eutriconodonts (Li et al. 2003; Luo et al. 2007). Because spalacotherioids are a crownward clade that is phylogenetically separated from eutriconodonts (Fig. 6.1; see Fig. 7.1 in Ekdale, Chap. 7), it adds weight to the hypothesis that, prior to the disconnection of Meckel's cartilage from the mandible by resorption in evolution, the PMME was already medially displaced from the mandible (Fig. 6.3). If there were sufficient evidence that the DMME represents the ancestral condition of theriiforms or the ancestral condition of the clade containing spalacotherioids and crown therians, then the PMME in spalacotherioids could be an evolutionary reversal among theriiform clades. Although this is an unlikely scenario, it cannot be rejected at this time (Fig. 6.5: node 5, Theriiformes; node 6, Trechnotheria).

Occurrence of the ossified Meckel's cartilage in mammal fossils may seem unlikely based on classic comparative anatomical studies that focus on extant vertebrates (e.g., Maier and Ruf 2016). Indeed, fossilized Meckel's cartilages in Mesozoic mammals are different in size and shape from those in extant mammal embryos and neonates. Thus, it was only natural that ossified Meckel's cartilage in extinct mammals was questioned when it was first proposed (Li et al. 2001). The discovery of an ossified Meckel's cartilage in situ on the mandible of the spalacotherioid *Maotherium* (Fig. 6.5K) indicates that it is indeed connected to the mandible in spalacotherioids as interpreted for gobiconodontids by Meng et al. (2003). Maier and Ruf (2016) questioned the hypothesis, noting that the massive Meckel's element in the eutriconodont *Repenomamus* seems to be too large when compared to the relatively small size of its embryonically transient equivalent in extant mammals. However, the ossified Meckel's cartilage in eutriconodonts appears to be a structure of adults or sub-adults and not embryos. For example, the eutriconodont *Spinolestes* possesses both an ossified Meckel's cartilage and substantial tooth replacement (Martin et al. 2015), suggesting an advanced stage of maturity beyond the fetal stage.

Although Meckel's cartilage does not normally ossify in extant mammals, extant species have conserved the genetic and developmental capacity to retain a large Meckel's cartilage into adulthood. Fate mapping studies show that Meckel's cartilage chondrocytes have the capacity for transformation into osteoblasts, fibroblasts, and osteocyte-like cells (Harada and Ishizeki 1998; Anthwal et al. 2013). Thus, the potential for ossification of the cartilage is retained at the cellular and tissue levels, at least in *Mus musculus*. The enhanced signaling of *Bmp2* and *Bmp7* of the BMP's family involved in chondrogenesis can generate larger Meckel's cartilages under certain experimental conditions (Wang et al. 2013). Thus, developmental experiments support the idea that Meckel's cartilage can be retained and ossified (Anthwal et al. 2013) and may even become enlarged enough to overlap the size range of this element as observed in some eutriconodonts (Wang et al. 2013).

There are now over 20 distinctive clades of Mesozoic mammaliaforms recognized compared to only three extant mammal clades (Kielan-Jaworowska et al. 2004; Close et al. 2015). Given this great taxonomic diversity of extinct animals, Mesozoic mammals and their close relatives exhibit a greater morphological disparity than the extant descendants of the most recent common ancestor of all Mammalia. The difference between the Meckel's cartilage of Mesozoic mammaliaforms and the embryonic counterpart in extant mammals is a reflection of the great disparity exhibited by Mesozoic mammals. Overall, the weight of the fossil evidence strongly supports the presence of an ossified Meckel's cartilage in eutriconodont and spalatherioid groups and, therefore, a PMME rather than a fully developed DMME.

6.3 Inner Ear Evolution

The cochlear part of the inner ear underwent major transformation in evolution from premammalian cynodonts to modern mammals (Figs. 6.1, 6.6, 6.7). In extant mammals, the osseous structures (also known as the bony labyrinths) are intricately associated with the membranous labyrinths directly supporting the sensory and the neural structures. The tight association of the membranous and the osseous labyrinths has made it possible to re-trace the evolution of at least some of the sensory and neural "soft tissue" structures of the inner ear by examining their osteological correlates in the inner ear by μCT scans (e.g., Rodrigues et al. 2013; Ruf et al. 2013). Now it is also feasible to attempt a more accurate estimate of hearing function by morphometric studies of inner ear endocasts (e.g., Kirk et al. 2014).

6.3.1 Inner Ear Bony Housing

The cavities and bony housing of the inner ears of mammaliaforms and cynodonts differ (Figs. 6.6, 6.7). The bony housing of the inner ear in cynodonts is a composite of several bones that are sutured in adulthood, namely the prootic and the opisthotic (collectively known as the periotics) and the basisphenoid wing that overlaps the external surface of the prootic (see Clack, Chap. 1). For example, in the cynodont *Thrinaxodon*, this wing forms part of a thickened and elevated rim that defines the fenestra vestibuli for the stapedial footplate (Fig. 6.5A), and the cochlear recess is formed by both the prootic and the basisphenoid. The vestibule and semicircular canals are housed mostly by the opisthotic, the exoccipital, and a part of the posterior semicircular canal is enclosed by the supraoccipital (Fourie 1974; Luo et al. 1995).

By comparison, the inner ear bony housing of mammaliaforms shows an apomorphic pattern of ossification. The prootic and opisthotic are fused to become a single petrosal bone that encloses all parts of the inner ear (Fig. 6.6) (Luo 2001; Walsh et al. 2013). The pars cochlearis is enlarged to form a pronounced ventral eminence known as the promontorium, which accommodates the elongate cochlear canal internally. The basisphenoid wing that extends to contribute a part of the fenestra vestibuli in cynodonts is completely lost in mammaliaforms and no longer a part of the cochlear housing (Fig. 6.6C, D). These anatomical changes are key apomorphies of all mammaliaforms including extant mammals (Wible and Hopson 1993; Rougier and Wible 2006).

Other aspects of the inner ear bony housing are also transformed. The pars cochlearis forms a complete medial bony wall to separate the inner ear space from the brain cavity, which in turn defines the internal auditory meatus with an ossified floor for the vestibulo-cochlear nerve (= vestibulo-acoustic cranial nerve VIII). This differs from the incomplete ossification of the medial wall of the periotics in basal cynodonts (Rowe et al. 1995; Kemp 2007; Kemp, Chap. 5). The inner ear bony

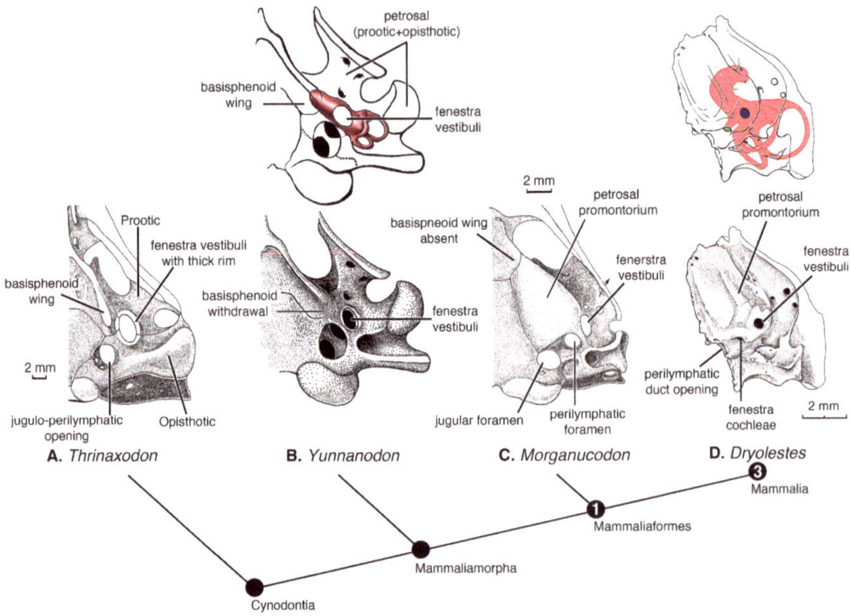

Fig. 6.6 Evolution of the inner ear bony housing in mammaliaforms: (**A**) cynodont outgroup, *Thrinaxodon,* (**B**) mammaliamorph outgroup, tritylodontid *Yunnanodon*, (**C**) *Morganucodon*, (**D**) crown mammal *Dryolestes*. Cynodonts show separate prootic and opisthotic, and the basisphenoid wing participates in the fenestra vestibuli. Mammaliamorphs are derived in that the prootic and opisthotic are fused to form the petrosal. The petrosal inner ear housing partially displaced the basisphenoid wing from the fenestra vestibuli. Mammaliaforms are further transformed in having the bony cochlear housing formed by the promontorium. The derived cladotherians show the most derived condition in which the perilymphatic canaliculus is separated from the fenestra cochleae. *Cladogram nodes as in* Fig. 6.1. (**A**, **C** drawing by Luo, modified from Luo et al. 1995; **B** from Luo 2001; **D** drawing by Luo, modified from Luo et al. 2012)

housing is ossified well enough that the division of the utricle of the vestibule can be recognized in many taxa as well (Luo et al. 1995; Ruf et al. 2013).

The inner ear evolution of mammaliaforms is now better informed by the intermediate and precursor conditions in mammaliamorphs (definition modified from Rowe 1988 to include tritheledontids and brasilodontids) (Liu and Olsen 2010). The petrosals and bony labyrinths of the inner ear of these mammaliamorphs are already quite similar to those of mammaliaforms in several ways (Fig. 6.7). For example, the medial (cranial) wall is fully ossified and forms the internal auditory meatus (Luo 2001; Rodrigues et al. 2013). In tritylodontids and brasilodontids, an incipient promontorium of the petrosal is expanded at the expense of the basisphenoid wing in ventral view (Luo 2001; Bonaparte et al. 2005). The basisphenoid wing is reduced (more so in smaller taxa than in larger taxa) and withdrawn from the margin of the fenestra vestibuli (Fig. 6.6). The fenestra has a

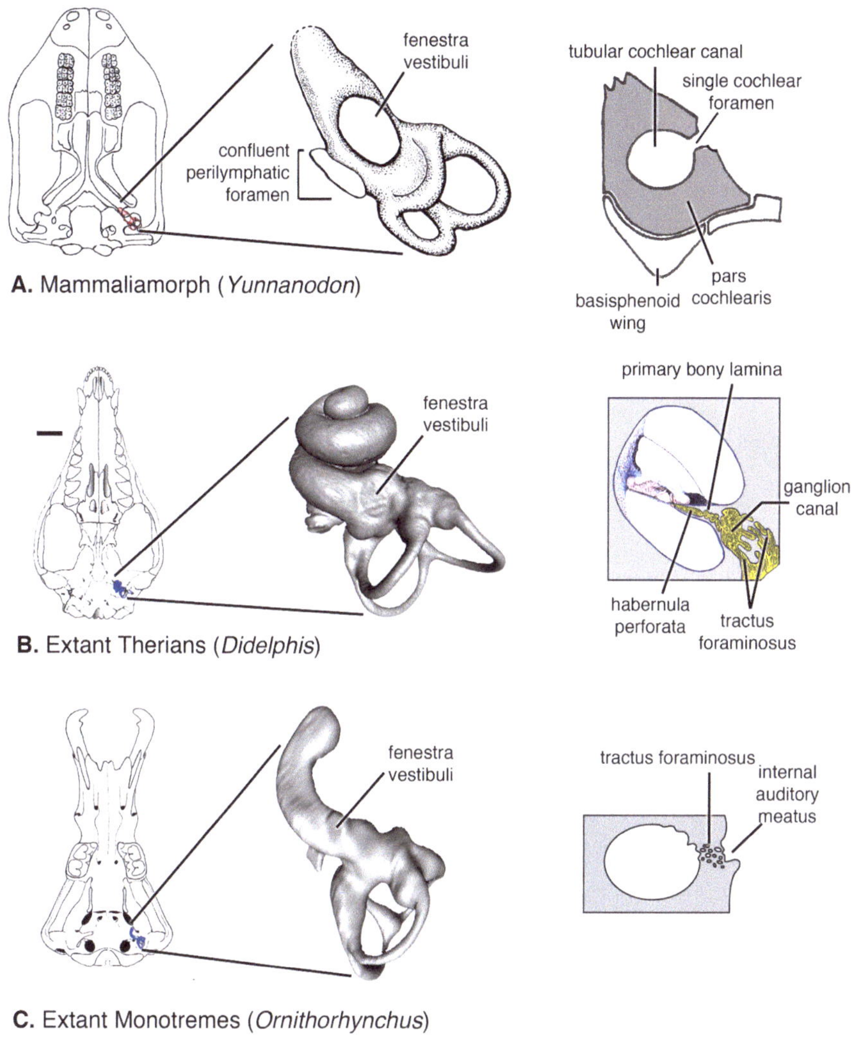

Fig. 6.7 Comparison of osteological structures for cochlear nerve of cynodonts and extant mammals. **(A)** Mammaliamorph condition, exemplified by *Yunnanodon:* skull (*left*) and the inner ear endocast (*middle*) in ventral view and schematic cross section at the internal auditory meatus (*right*). **(B)** Therian condition, exemplified by the marsupial *Didelphis*: skull (*left*) and inner ear endocast (*middle*) in ventral view, schematic cross section at the internal auditory meatus (*right*). **(C)** Monotreme condition, exemplified by *Ornithorhynchus*: skull (*left*) and inner ear endocast (*middle*) in ventral view, and schematic cross section (*right*). (**A** redrawn from Luo 2001; **B** modified from Luo et al. 2011; **C** from Luo et al. 2011 and this study)

sharper rim for a more precise fit with the stapedial footplate. Derived cynodont groups of tritylodontids and brasilodontids lack the thickened rim of this fenestra typical of the primitive cynodonts (Luo 2001; Rodrigues et al. 2013).

6.3.2 Cochlear Canals of Mammaliaforms

In cynodonts (Fig. 6.6), the bony cochlear canal (or recess) is too short to give an indication of the canal curvature, but the derived mammaliaform cochlea is elongate, tubular, and shows variable curvature (Figs. 6.7, 6.8, 6.9, 6.10) (Luo et al. 1995; Ruf et al. 2013), which curves in the opposite direction to that of extant diapsid reptiles and birds (Wever 1978; Hurum 1998; Sobral, Reisz, Neenan,

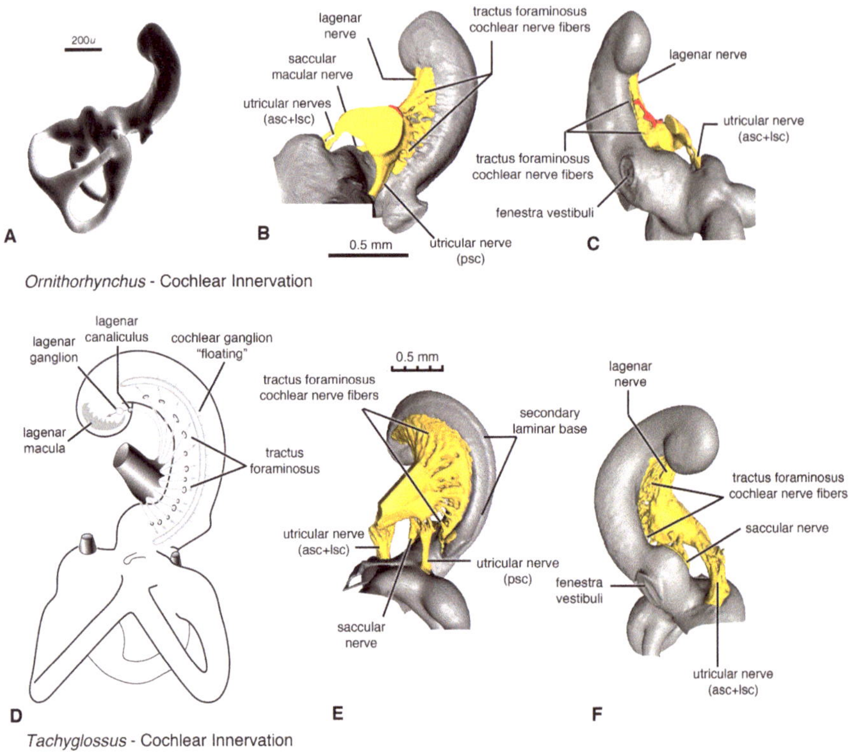

Fig. 6.8 Monotreme cochlear innervation. **(A–C)** *Ornithorhynchus*, entire inner ear. **(A)** Innervation of the vestibulo-cochlear nerve (VIII) to the cochlea, the saccule, and the utricle, shown as endocasts of channels for nerve branches and fibers in dorsal view **(B)** and in lateral view **(C)**. **(D–F)** *Tachyglossus*. **(D)** Dorsal view schematic pattern of innervation to the cochlear ganglion and the lagenar ganglion. **(E)** Innervation of the vestibulo-cochlear nerve in dorsal (endocranial) view. **(F)** Innervation of the vestibule-cochlear nerve in ventro-lateral view of the inner ear. The bony channels and foramina of branches of the vestibule-cochlear nerve are represented by endocasts from μCT rendering. *asc*, anterior semicircular canal; *lsc*, lateral semicircular canal; *psc*, posterior semicircular canal. (New line drawing by Schultz; CT rendering by Schultz from authors' data sets)

Müller, and Scheyer, Chap. 8; Evans, Chap. 9; Sobral and Müller, Chap. 10). In addition, the ratio of cochlear length relative to body mass is higher in most mammaliaforms than cynodonts (Rodrigues et al. 2013; Kirk et al. 2014).

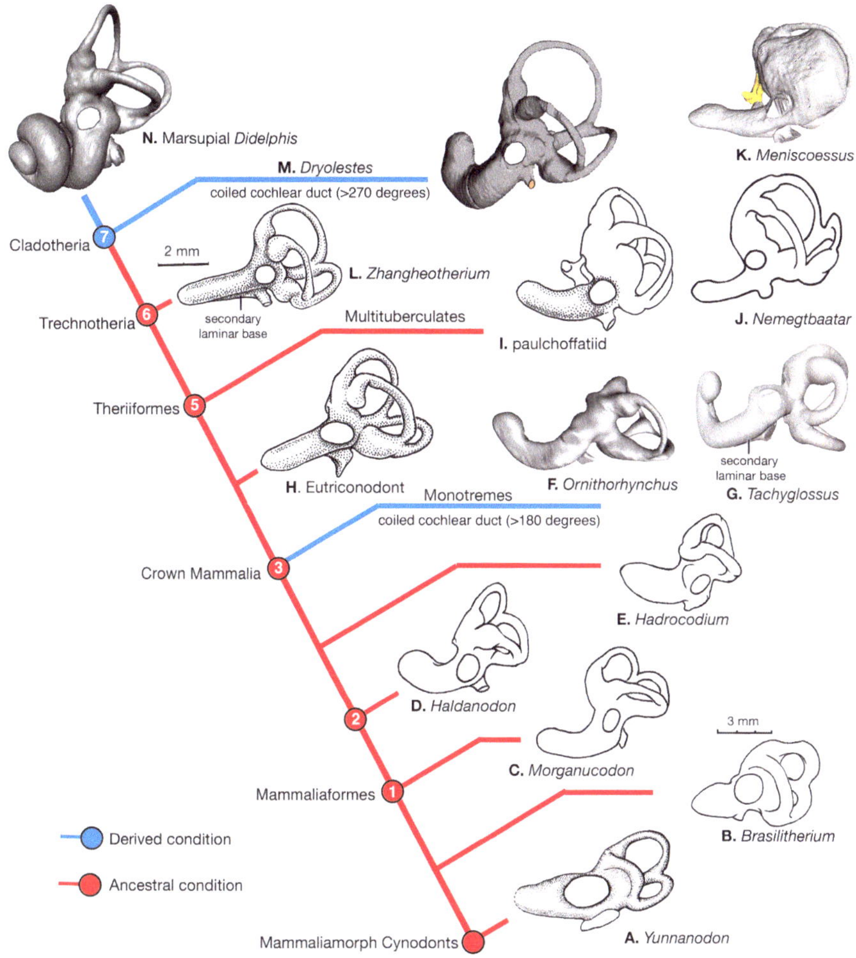

Fig. 6.9 Mammaliaform inner ear evolution: cochlear canal curvature and coiling (inner ear endocasts in ventral view). **(A)** Mammaliamorph *Yunnanodon*, **(B)** Mammaliamorph *Brasilitherium*, **(C)** *Morganucodon*, **(D)** docodont *Haldanodon*, **(E)** *Hadrocodium*, **(F)** *Ornithorhynchus*, **(G)** *Tachyglossus*, **(H)** generalized inner ear of eutriconodont (based on gobiconodontids), **(I)** generalized paulchoffatiid multituberculate, **(J)** *Nemegtbaatar*, **(K)** taeno-labidoid *Meniscoessus*; *yellow* represents vestibulo-cochlear nerve (*viii*) to the cochlea, the saccule, and the utricle, **(L)** Spalacotherioid *Zhangheotherium*, **(M)** Cladotherian *Dryolestes* (dryolestoid), **(N)** Marsupial *Didelphis*. CT rendering based on authors' data sets. *Cladogram nodes: (2) unnamed clade (docodonts+Mammalia); (7) Cladotheria; for names of other cladogram nodes see* Fig. 6.1. (**B** redrawn from Rodrigues et al. 2014; **D** redrawn from Ruf et al. 2013; **F** modified from Luo et al. 2011; **H**, **I**, **K**, **L** original from authors; **J** redrawn from Hurum 1998; **M** from Luo et al. 2011; **N** from Luo et al. 2011)

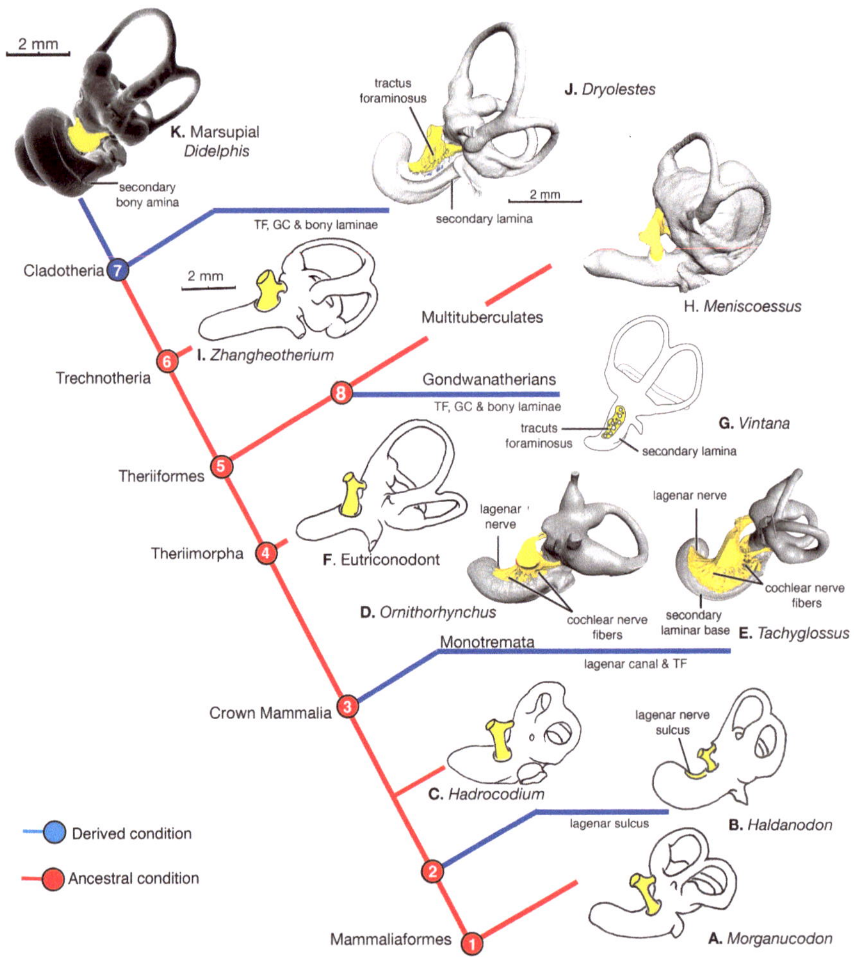

Fig. 6.10 Mammaliaform inner ear evolution: pattern of innervation of cochlear duct and canal curvature (inner ear endocasts in dorsal or endocranial view). **(A)** *Morganucodon*, **(B)** *Haldanodon*, **(C)** *Hadrocodium*, **(D)** *Ornithorhynchus* (the anterior and posterior semicircular canals are incomplete as broken in the original petrosal bone), **(E)** *Tachyglossus*, **(F)** generalized eutriconodont (based on gobiconodontids), **(G)** gondwanatherian *Vintana*, **(H)** taeniolabidoid multituberculate *Meniscoessus*, **(I)** spalacotherioid *Zhangheotherium*, **(J)** cladotherian *Dryolestes*, **(K)** marsupial *Didelphis*. *GC*, ganglion of the cochlea; *TF*, tractus foraminosus ("cribriform plate"); *for clades see* Fig. 6.1. (**B** redrawn from Ruf et al. 2013; **G** redrawn from Hoffmann et al. 2014; all others are new illustrations by the authors)

The interior (axial) wall of the cochlear canal is uniformly simple in contrast to the inner cochlear wall of extant mammals (see Ekdale, Chap. 7). Ancestrally, cochlear innervation was transmitted through a single and large cochlear foramen in early mammaliaforms (Figs. 6.7, 6.10, 6.11). The cochlear nerve was probably connected to a cochlear ganglion surrounded by the soft spiral ligaments to innervate the organ of Corti as in monotremes and non-mammalian, extant amniotes (Alexander 1904; Weston 1939), or embedded in similar membranous labyrinth tissues to innervate the basilar papilla as in non-mammalian amniotes (Weston 1939; Wever 1978).

Haldanodon is more derived among mammaliaforms in showing a more pronounced cochlear curvature up to 180° (Ruf et al. 2013). The cochlea of *Haldanodon* also possesses a distinctive lagenar sulcus on the interior surface of the cochlear canal wall, extending from a distinctive notch of the cochlear foramen (Ruf et al. 2013). The lagenar sulcus and its related notch at the cochlear foramen unequivocally indicate the presence of a lagenar macula (Figs. 6.10, 6.11). The lagenar macula is a sensory area in the apex of the membranous labyrinth of the cochlear duct that is plesiomorphic for amniotes (Wever 1978; Manley and Clack 2004). Using an extant phylogenetic bracket (this extends between the nodes bracketing the extant groups above and below the fossil in question) (Witmer 1995), the presence of the lagenar macula in non-mammalian amniotes and extant monotremes implies that a lagenar macula was most likely present in extinct mammaliaforms, despite the soft-tissue feature not always having an osteological correlate in mammaliaforms, except in *Haldanodon*. However, given the placement of *Haldanodon* on the mammaliaform phylogeny, the bony channel (sulcus) for its lagenar nerve is best hypothesized to be an evolutionary convergence to the lagenar canaliculus of extant monotremes (Figs. 6.10, 6.11).

6.3.3 Ancestral Morphotype of Crown Mammalia

The fully coiled cochlea is one of the most distinctive features of extant therians. The cochlea of extant monotremes is curved, but it does not complete a full coil (Fig. 6.8). In addition to the differences in internal osseous structures of monotremes and therians, the organ of Corti of monotremes is also different from those of therians, in the number of inner sensory hair cell rows and outer hair cell rows (Vater and Kössl 2011; Fritzsch et al. 2015). From these differences, it appears that coiling of the membranous cochlear duct is not homologous between monotremes and therians (Zeller 1989; Luo et al. 2011). This hypothesis is now corroborated by recent phylogenies in which the simple, uncoiled, and tubular cochlear canal is the typical condition along the entire cladogram backbone from mammaliaforms through stem therians (Fig. 6.9).

On the whole, cochlear curvature is variable across the mammalian phylogeny. Among stem mammaliaforms, the cochlear canal ranges from a strongly curved canal of *Haldanodon* (Ruf et al. 2013) to weakly curved canals in *Morganucodon* and *Hadrocodium* and to relatively straight canals of eutriconodonts, spalacotherioids, and some multituberculates (Figs. 6.9, 6.10) (Hurum 1998; Ladevèze et al. 2010). Further discussion of the variation of cochlear curvature among mammal clades is provided by Ekdale (Chap. 7).

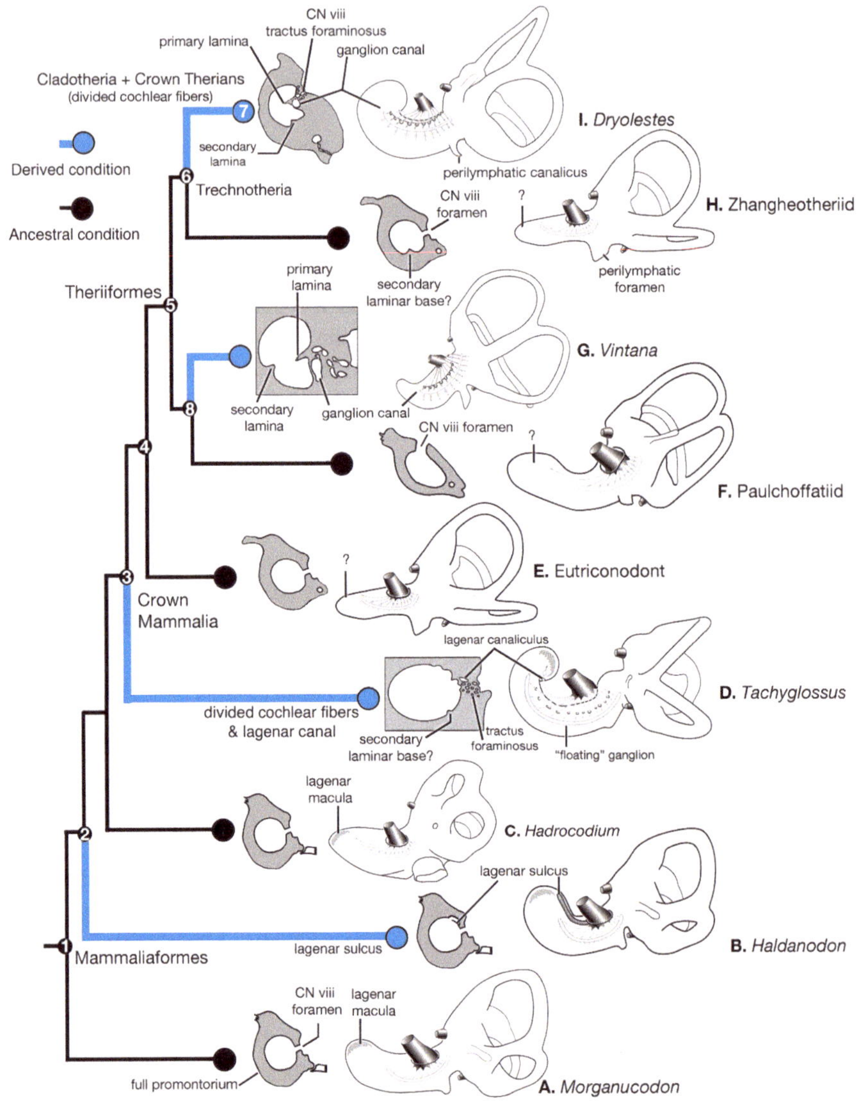

Abundant fossils suggest that the common ancestor of extant mammals (Fig. 6.11: node 3 Mammalia) retained all the plesiomorphic characters of the inner ear of mammaliaforms (Fig. 6.11: node 1): simple, tubular cochlear canal without interior osseous structures, and a single large opening for a single (or single bundled) cochlear nerve. Also, the cochlea in most Mesozoic crown clades is not further elongated beyond the range of stem mammaliaforms (Figs. 6.7, 6.8, 6.9, 6.10) (see Table 2 in Kirk et al. 2014) nor is the cochlear canal particularly curved. Cochlear canals of eutriconodonts, spalacotherioids, and some multituberculates are

◂**Fig. 6.11** Evolutionary pattern of interior structures of cochlear canals among mammaliaforms. **(A)** *Morganucodon*: cochlear canal elongate, housed by a full promontorium; but its simple canal has a single large cochlear foramen (plesiomorphic), indicating that a cochlear nerve entered the cochlear foramen, connecting to an intra-oticganglion before ganglionic nerve fibers connect to sensory structures (plesiomorphic). **(B)** Docodont *Haldanodon*: the same plesiomorphic condition of a simple cochlear canal and a single cochlear foramen but derived in having a lagenar sulcus on the cochlear canal endocast, likely convergent to the lagenar canal in monotremes (Ruf et al. 2013). **(C)** *Hadrocodium*: the same plesiomorphic condition. **(D)** Monotreme *Tachyglossus*: tractus foraminosus with foramina for cochlear nerve fibers that are divided before entering the cochlear canal, apomorphic as compared to mammaliaforms. The cochlear ganglion is not supported by an internal bony structure within the cochlear canal, thus "floating." Monotremes have a macula lagena separated from the organ of Corti and a lagenar nerve in a lagenar canaliculus separated from cochlear nerve fibers in the tractus foraminosus. **(E)** Eutriconodonts: the same plesiomorphic condition. **(F)** Paulchoffatiid multituberculate: the same plesiomorphic condition. **(G)** Gondwanatherian *Vintana*, derived condition of tractus foraminosus, complete cochlear ganglion canal, and both primary and secondary bony laminae. **(H)** Spalacotherioid, same as mammaliaform condition, except for an apomorphic bony base of the secondary lamina. **(I)** *Dryolestes* and its cladotherian clade: apomorphies of tractus foraminosus indicating the cochlear nerve fibers are divided before entering the cochlear fiber foramina (convergent to monotremes and gondwanatherians); bony Rosenthal's canal enclosing the cochlear (spiral) ganglion in the base of primary bony lamina, a derived feature of cladotherian (including marsupials and placentals) but convergent to that of *Vintana*. *For clade names see* Figs. 6.1 *and* 6.5 *captions; CN viii*, acoustic cranial nerve (= vestibilo-cochlear nerve, VIII). (Cross sections modified from Luo et al. (2012) and Hoffmann et al. (2014); outlines of inner ear features are original renderings on the published descriptions on respective taxa)

completely straight; other multituberculates show a slightly curved cochlear canal (Figs. 6.8, 6.9, 6.10).

In extant monotremes, the cochlear duct has two distinctive sensory areas: the apical lagenar macula (ancestral retention from non-mammalian amniote ancestors) and an organ of Corti that is shared with extant marsupial and placental (therian) mammals (Alexander 1904). The lagenar macula in the apex is similar to the vestibular sensory epithelium of therians in its sensory hair cell characteristics (Ladhams and Pickles 1996) but different from the hair cells of the organ of Corti. The apically positioned lagenar macula is separated from the organ of Corti in the more basal part of the cochlear duct (Jørgensen and Locket 1995). In addition to structural and cellular differences, the lagena is innervated by a separate ganglion (Alexander, 1904) from the spiral cochlear ganglion that connects to the organ of Corti (Ladhams and Pickles 1996). The apex of the cochlear canal is perforated by small canaliculi of the lagenar nerve fiber(s), which are distinct from, but may be anastomosed locally with, channels for the main cochlear nerve fibers (Fig. 6.8).

The cochlear nerve fibers of all extant mammals enter the cochlea through densely packed foramina that perforate the bony junction of the cochlear canal with the internal auditory meatus (Fox and Meng 1997; Luo et al. 2011). The tract of these foramina is termed the tractus spiralis foraminosus (*sensu* Williams et al. 1989) or simplified as the tractus foraminosus (*sensu* Hoffmann et al. 2014). The same structure is commonly called the cribriform plate for its sieve-like morphology of the axial region of the cochlea. The foramina in the tractus foraminosus are connected with a network of braided channels that are a conduit of cochlear nerve fibers to connect with the cochlear ganglion in the cochlear canal (Figs. 6.7, 6.9, 6.10) (see Fig. 1 in Luo et al. 2011). Although presence of a cribriform plate characterizes all three extant mammal groups, the structure is absent in such major extinct clades as eutriconodonts, multituberculates, and spalacotherioids (Fig. 6.10) (also see Ekdale, Chap. 7).

The cochlear ganglion of monotremes is not supported by interior bony structures of the canal, and the ganglion "floats" within the primary and secondary spiral ligaments and the rest of the membranous tissues (Fig. 6.9D) (Alexander 1904; Zeller 1989). In contrast, osseous primary and secondary laminae on the inner and outer walls of the cochlear canal respectively, are present in extant therians. The primary bony lamina and Rosenthal's canal for the spiral nerve ganglion are key apomorphies of extant therians (Fleischer 1973; Williams et al. 1989), and these structures have been documented for all stem eutherians and metatherians for which the inner ears have been studied (Meng and Fox 1995; Ekdale and Rowe 2011).

The primary and secondary bony laminae of the basilar membrane have been extensively examined in placental mammals, especially those with exceptional hearing capabilities, such as cetaceans and bats (Pye 1970; Fleischer 1976). The primary lamina provides a stiff bony support for the modiolar (inner) edge of the basilar membrane along the entire length of cochlea, which is fundamental to overall cochlear mechanics (Echteler et al. 1994). Development of robust primary and secondary bony laminae in cladotherians (therians + dryolestoids) signifies a broadening of hearing capabilities exhibited by crown therians.

The secondary bony lamina for the basilar membrane is absent in all non-mammalian mammaliaforms, most Mesozoic mammal clades, and in the extant duck-billed platypus *Ornithorhynchus*. A bony base for the secondary lamina is present in some (but not all) specimens of *Tachyglossus* (Figs. 6.8, 6.11), and the distribution of secondary bony laminae for the basilar membrane is not universal among extant therians (Ruf et al. 2009; Ekdale 2013).

Among living monotremes, *Tachyglossus* has a distinctive sulcus on the cochlear endocast (Figs. 6.8E, G) in a position that corresponds to the secondary lamina of therians (Fig. 6.10K). It is proposed that *Tachyglossus* possesses a base for a lamina but not a fully developed secondary bony lamina itself. Furthermore, this feature is variable and may be absent in other specimens of *Tachyglossus*. *Ornithorhynchus* does not have a bony ridge of the secondary lamina in histological sections or on endocasts (Figs. 6.7C, 6.8B), suggesting a lability of the structure in early mammal evolution. Similarly, the stem therian spalacotherioids have a groove in the homologous position of the secondary lamina, but no crest to indicate a full bony lamina. Thus, it is suggested that neither *Tachyglossus* nor spalacotherioids should be considered to have possessed a "secondary bony lamina."

6.3.4 Vestibular Features of Mammaliaforms

Mammaliaforms show a trend of increased differentiation of the cochlea from the vestibule within the inner ear labyrinth, as well as in the separation of the perilymphatic channel from the membranous labyrinth to the braincase. The result of the differentiation is better structural support by bone for the base of the cochlea, which is the region responsible for higher-frequency hearing and also pressure release mechanisms via the perilymphatic foramen between the middle and inner ear cavities (Fig. 6.6). The full separation of the base of the cochlear canal from the jugular foramen, which transmits cranial nerves and veins, evolved in derived cynodonts, particularly in tritylodontids and basilodontids, as evidenced in their inner ear endocasts. Both groups show clear bony separation of the basal cochlear canal from the vestibule (Luo 2001; Rodrigues et al. 2013). *Brasilitherium* further shows a bony separation of the saccular space from the utricular space within the vestibule (Rodrigues et al. 2013). However, information is not available for many of the cynodonts, such as *Massetognathus* and *Probainognathus* (Quiroga 1979; Allin and Hopson 1992), to assess fully the taxonomic distribution of cochlea-vestibule differentiation.

Concurrent with changes in the bony junction between the cochlear base and the vestibule in mammaliaforms, changes in ossification of the otic capsule also modified the perilymphatic conduit to the braincase—development of the pressure release window of the inner ear (Rougier and Wible, 2006). As in other primitive reptiles (see Sobral, Reisz, Neenan, Müller, and Scheyer, Chap. 8; Evans, Chap. 9; Sobral and Müller, Chap. 10), the primitive cynodont condition is that the perilymphatic foramen is confluent with the jugular foramen (Fig. 6.5A, B) (Wible and Hopson 1993; Luo 1994). In derived mammaliaforms, however, the perilymphatic foramen has separated from the jugular foramen to form its own opening (Fig. 6.5C–D) (Rougier and Wible 2006). This represents the convergent evolution of a pressure relief mechanism, as seen in many later, derived reptiles.

Variation in the construction of the perilymphatic foramen in crown mammals is discussed elsewhere (Ekdale, Chap. 7), but in short, a bony perilymphatic recess for the perilymphatic sac is present external to the perilymphatic foramen as in extant monotremes, which likely is the case for the earliest mammals (Hoffmann et al. 2014). It is hypothesized that this perilymphatic recess was covered by the secondary tympanic membrane in these Mesozoic mammals, as in extant *Ornithorhynchus* (Zeller 1989; Rougier and Wible 2006). In cladotherians (Fig. 6.6C), the open perilymphatic recess becomes enclosed by a structure known as the processus recessus of the petrosal and is converted into the perilymphatic canaliculus (= aqueductus perilymphaticus), separated from the fenestra cochleae or round window (Rougier et al. 1996; Luo et al. 2011).

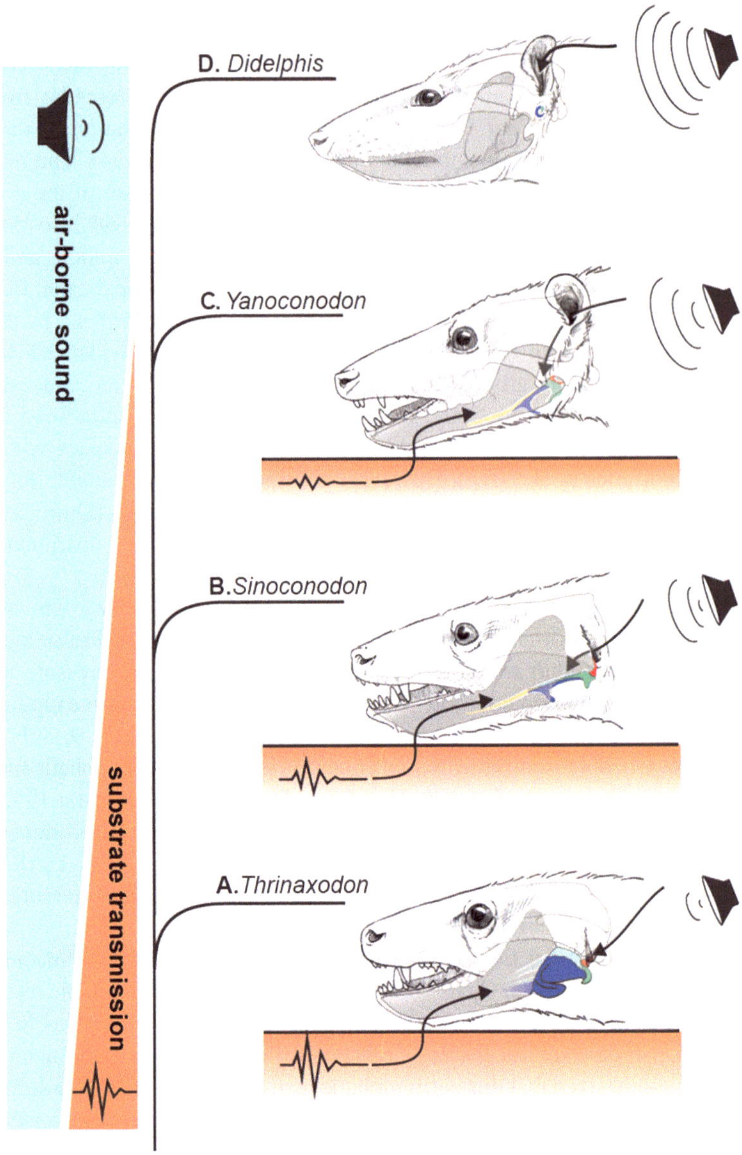

6.4 Functional Evolution of Mammaliaform Ears

Hearing is a major functional adaptation of extant mammals as a whole, by comparison with other extant non-mammalian amniotes (Manley 2010, 2012), but it has long been a major challenge for evolutionary morphologists to estimate the hearing frequency ranges and hearing sensitivity. Thanks to the rapidly growing

◂**Fig. 6.12** Hypotheses on routes of sound conduction and transmission in cynodont–mammal evolution. **(A)** Cynodont *Thrinaxodon* already had developed hearing of airborne sound by the tympanic membrane suspended by the angular bone (ectotympanic) and an air-filled middle ear cavity; but it simultaneously had the capacity to hear substrate-transmitted sound through the mandible to the angular and surangular, to the articular (malleus), and the quadrate (incus), then to the stapes. **(B)** *Sinoconodon* had thinner middle ear bones and a larger tympanic membrane with a great capacity to receive airborne sound, but hearing via substrate-transmission was still fully functional with the mandible-attached middle ear. **(C)** *Yanoconodon* with its middle ear partially displaced from the mandible (PMME) could still hear the substrate-transmitted sound through the ossified Meckel's element to the middle ear, while mainly relying on the tympanic membrane to hear airborne sound; first development of external pinna in eutriconodonts, as indicated by *Spinolestes* (Martin et al. 2015). **(D)** *Didelphis* with DMME exclusively relied on the tympanic membrane for hearing, although various fossorial extant mammals can develop secondary adaptations to hear substrate-transmitted sound. *blue*, ectotympanic; *green*, malleus; *red*, incus; *yellow*, Meckel's cartilage. (Illustration by April I. Neander of the Luo Lab, University of Chicago)

information about the inner ear due to discoveries of better fossils and μCT scanning (Luo et al. 2011; Hoffmann et al. 2014), the hearing function of extinct mammaliaforms can now be estimated by morphometric analyses of the cochlear canal (e.g., Kirk et al. 2014). Based on allometrical analysis by regression of cochlear canal length on body mass, the latest consensus is that stem mammaliaforms have a shorter cochlear canal, relative to body mass, than monotremes. They also have shorter cochlear canals in absolute measurement than monotremes (Ladhams and Pickles 1996; Kirk et al. 2014). Therefore, it is suggested that *Sinoconodon*, *Morganucodon*, *Hadrocodium*, and *Haldanodon* (with the longest cochlear canal known among mammaliaforms) would have lower limits of upper frequency and narrower hearing ranges (fewer octaves) than extant monotremes.

Although it is generally difficult to estimate function for fossil vertebrates, estimates of hearing ranges and sensitivities of Mesozoic mammaliaforms based on their middle ears have been particularly susceptible to the first-hand interpretation of fossil morphologies (see previous clarification on the ectotympanic and malleus size of *Sinoconodon* and *Morganucodon*). Even with the best-preserved fossils (e.g., Luo et al. 2007; Meng et al. 2011), the middle ears (MMEC and PMME) of Mesozoic mammaliaforms have no structural analogues in the ears of living vertebrates for functional inference. The following section discusses just one aspect of the Mesozoic mammaliaform hearing function—the route of sound conduction.

6.4.1 Routes of Sound Conduction to Inner Ear

A tympanic membrane supported by the reflected lamina of the angular and the retroarticular process was already functioning to receive airborne sound during the cynodont–mammal transition (Fig. 6.12) (Allin 1975; Kermack and Mussett 1983). Basal cynodonts already had an air-filled middle ear tympanic cavity (Allin and Hopson 1992; however, see the alternative view of Kemp 2007; Kemp, Chap. 5).

The inferred presence of an air-filled tympanic cavity is consistent with reconstruction of the tympanic membrane as a neomorphic structure that transmitted the airborne sound in extinct cynodonts.

In addition to the tympanic membrane, the plate-like reflected lamina of the angular likely received airbone sound in cynodonts (e.g., Allin 1975) and even in some pre-cynodont therapsids (Allin 1975; Kemp 2007). The reflected lamina of the angular, once having received airborne sound vibrations, would further transmit vibration via bone conduction. Implicit in this hypothesis is that the angular itself transmitted sound independent of, or parallel to, to the tympanic membrane through bone conduction. Bone conduction, either by direct route through the dentary to the postdentary bones (including the angular), or by indirect route through the rest of skull, is a major transmission route of sound waves to the inner ear in synapsids (Tumarkin 1968; Kermack and Mussett 1983; Laaß 2015). Overall, it is still a primary mechanism in most therapsids (Maier and van den Heever 2002; Laaß 2015), while the neomorphic tympanic membrane may have received the airborne sound in parallel.

It is also generally agreed that with large postdentary bones, the sensitivity to higher frequencies would be more limited in basal cynodonts than in derived mammaliaforms. The attachment of the postdentary bones to the mandible was extensive, and these bones are presumably stiffer in cynodonts than mammaliaforms (Kermack et al. 1973). Cranial anchoring of the quadrate and quadrato-jugal complex had not yet achieved the same agility as in later mammaliaforms (Luo and Crompton 1994). This also would have limited hearing sensitivity. As long as the angular and its adjacent surangular were still extensively attached to the mandible, bone conduction most likely would have coexisted with the more direct reception of airborne sound by the neomorphic tympanic membrane (Fig. 6.12A). Most therapsids, including cynodonts, were terrestrial animals with generalized skeletal features, although some were arboreal (Fröbisch and Reisz 2009) or even fossorial in subterranean habitats (Kemp 2005; Fernandez et al. 2013). A mandibular route for bone conduction would convey a functional advantage for fossorial animals, and it is also feasible for hearing in terrestrial species. Thus, in premammaliaforms there were dual routes for sound conduction to the inner ear.

The ossified Meckel's cartilage in stem mammaliaforms and some mammals (e.g., eutriconodonts and spalacotherioids) remained connected by Meckel's element to the mandible (the PMME). In that situation, airborne or substrate transmitted sound vibrations received by the mandible would be transmitted via the ossified Meckel's cartilage to the middle ear. Thus, sound reception via bone conduction would coexist with sound reception by the tympanic membrane (Fig. 6.12B, C). Bone conduction is reduced in successively more crownward mammaliaforms as the tympanic membrane gained a larger surface area for airborne sound reception coupled with better hearing sensitivity and expanded frequency ranges (e.g., Kermack and Mussett 1983; Rosowski 1992). Because the middle ear bones and the ossified Meckel's cartilage, together as a unit, have a larger mass than without the Meckel's cartilage, it is likely that eutriconodonts and spalacotherioids with the PMME almost certainly did not attain the same upper frequency range as those of extant mammals.

The loss of mandibular sound conduction and the shift to hearing airborne sound primarily via the tympanic membrane did not occur until the DMME evolved (Fig. 6.12D), perhaps independently in monotremes and in therians. However, after the achievement of tympanic membrane sound reception as the primary pathway, many fossorial mammals, including the monotreme *Tachyglossus* (Aitkin and Johnstone 1972), independently evolved secondary bone conduction that correlates with their habitat preference and locomotor specialization (Mason and Narins 2001; Laaß 2014).

6.4.2 External Ear Pinna

The external ear pinna is a universal feature of a majority of extant therians that is absent in monotremes and many marine mammals. A pinna has a larger funnel surface to collect sound power from the environment, and it is important in enhancing the sense of direction of the sound source and in steering sound collection toward the source to enhance it (Rosowski 1992; Heffner and Heffner 2014). Several mammaliaforms are now known for skulls fossilized with fur and skin outlines (Bi et al. 2014; Meng et al. 2015), but none has been preserved with an external pinna. An earlier reconstruction of an external pinna for the Jurassic mammaliaform *Castorocauda* (Ji et al. 2006) is now considered to be inaccurate. The earliest known case of an external pinna is documented in *Spinolestes*, a gobiconodontid of the theriimorph clade from the Cretaceous (Martin et al. 2015). In theory, therians are more capable of detecting directionality of the sound source than monotremes due to the pinna. The pinna probably evolved in theriimorph mammals, based on evidence from *Spinolestes* (Martin et al. 2015), after the split of theriimorphs from monotremes that lack the pinna (Fig. 6.12).

6.5 Summary and Remarks

Evolution of middle ear structures underwent a structural transformation from the middle ear bones fully attached to the mandibles (the plesiomorphic "mandibular ear of cynodonts") to an apomorphic structure in which the middle ear as a whole is still connected to the mandible by an ossified Meckel's cartilage, but the ectotympanic and malleus are already displaced ajar from the mandible, as in Mesozoic eutriconodont and spalacotherioid mammals. Recent morphogenetic studies show that the extant mammals (at least the model *Mus*) have conserved the developmental potential for the Meckel's element to ossify. By extrapolation, it is expected that this developmental potential could generate an ossified Meckel's cartilage in early mammalian history, as evidenced by the fossil examples of stem mammaliaforms, eutriconodonts, and spalacotherioids. The new morphologies in fossils helped to reveal that this actually evolved in mammaliaform phylogeny and also that the disconnection of the ear from the mandible occurred independently in monotremes, in therians, and in multituberculate mammals.

A key transformation of the inner ear in the cynodont–mammaliaform transition is the evolution of a single petrosal bone enclosing the entire inner ear, with an enlarged petrosal promontorium to accommodate an elongate cochlear canal in all of the mammaliaforms. This is coiled even further in the more derived cladotherian mammals. Mammaliaforms and the major Mesozoic clades of crown mammals show ancestral features of a simple cochlear canal with a single cochlear nerve foramen. This ancestral cochlear canal lacked any interior bony laminae to support the basilar membrane and its organ of Corti, and the cochlear canal lacked a Rosenthal's canal for the cochlear ganglion.

With the new and more accurate reconstruction of the lagenar nerve channel in extant monotremes (Fig. 6.8) and in some mammaliaform fossils (e.g., Ruf et al. 2013), it is evident that stem mammaliaforms likely possessed the lagenar macula as in extant monotremes (Fig. 6.11). It can now be ascertained that the cladotherians all lacked a lagena; however, it is still equivocal whether or not the Mesozoic clades nested in Mammalia—eutriconodonts, multituberculates, spalacotherioids—would have had a lagena. The sieve-like foramina of the cochlear nerve fibers evolved independently in three Mesozoic mammal lineages: the first time without a canal for the cochlear ganglion in extant monotremes; and the second and third times with Rosenthal's canal for the cochlear ganglion in extinct gondwanatherians (e.g., *Vintana*) and in cladotherians, including crown therians and their proximal relatives (Fig. 6.11). Cochlear canal curvature is homoplastic among various mammaliaform groups. A curvature beyond 270° only evolved in cladotherians, accompanied by the Rosenthal's ganglion canal.

The homoplasies of cochlear structures in early mammal evolution, although seemingly complex, are consistent with a new understanding of labile morphogenesis of the mammalian cochlea regulated by a complex developmental genetic network (Appler and Goodrich 2011; Fritzsch et al. 2015). The completion of cochlear canal coiling, as seen in *Mus*, requires the full activation of this complex network (Bok et al. 2007; Fritzsch et al. 2015). Perturbation of the many pathways of this genetic network can result in phenotypic variations of the cochlear canal, which is under-coiled or has less curvature (Bok et al. 2007; Luo et al. 2011).

New information from the development of the ear in extant mammals and the paleontological record of extinct relatives have added to a growing body of evidence that the seemingly labile evolution of middle ears in Mesozoic mammals was influenced, in a major way, by developmental heterochrony and gene patterning. Extrapolating these common developmental patterns of extant mammals into the fossil record can illuminate the driving mechanism for a similar (convergent) evolution to occur multiple times. On the other hand, the fossil record can provide some concrete examples for morphogenetic potentials that exist in extant mammals but are rarely expressed in normal development. The absence of an ossified Meckel's cartilage in the normal adults of extant monotremes, marsupials, and placentals represents a more canalized development of the middle ear for these living lineages. In contrast, more labile evolutionary development of middle ear features among mammaliaform groups was made possible by a vast range of evolutionary experimentations in their diverse Mesozoic clades (Kielan-Jaworowska et al. 2004; Luo 2007). This appears to be a case of development impacting phylogenetic evolution.

Compliance with Ethics Requirements Zhe-Xi Luo, Julia A. Schultz, and Eric G. Ekdale have declared that they have no conflicts of interest for this publication.

References

Aitkin, L. M., & Johnstone, B. M. (1972). Middle ear function in a monotreme: The echidna (*Tachyglossus aculeatus*). *Journal of Experimental Zoology*, 180, 245–250.

Alexander, G. (1904). Entwicklung und Bau des inneren Gehörorganes von *Echidna aculeata*. *Denkschriften der Medizinisch–Naturwissenschaftlichen Gesellschaft Jena*, 6(2), 1–118.

Allin, E. F. (1975). Evolution of the mammalian middle ear. *Journal of Morphology*, 147, 403–438.

Allin, E. F., & Hopson, J. A. (1992). Evolution of the auditory system in Synapsida ("mammal-like reptiles" and primitive mammals) as seen in the fossil record. In D. B. Webster, R. R. Fay, & A. N. Popper (Eds.), *The evolutionary biology of hearing* (pp. 587–614). New York: Springer-Verlag.

Anthwal, N., Joshi, L., & Tucker, A. S. (2013). Evolution of the mammalian middle ear and jaw: Adaptations and novel structures. *Journal of Anatomy*, 222, 147–160. doi:10.1111/j.1469-7580.2012.01526.x

Appler, J. M., & Goodrich, L. V. (2011). Connecting the ear to the brain: Molecular mechanisms of auditory circuit assembly. *Progress in Neurobiology*, 93, 488–508. doi:10.1016/j.pneurobio.2011.01.004

Archer, M., Flannery, T. F., Ritchie, A., & Molnar, R. (1985). First Mesozoic mammal from Australia—an Early Cretaceous monotreme. *Nature*, 318, 363–366.

Bi, S., Wang, Y.-Q., Guan, J., Sheng, X., & Meng, J. (2014). Three new Jurassic euharamiyidan species reinforce early divergence of mammals. *Nature*, 514, 579–584. doi:10.1038/nature13718

Bok, J., Chang, W., & Wu, D. K. (2007). Patterning and morphogenesis of the vertebrate inner ear. *International Journal of Developmental Biology*, 51, 521–533. doi:10.1387/ijdb.072381jb

Bonaparte, J. F., Martinelli, A. G., & Schultz, C. L. (2005). New information on *Brazilodon* and *Brasilitherium* (Cynodontia, Progainognathia) from the Late Triassic of Southern Brazil. *Revista Brasileira de Paleontologia*, 8, 25–46.

Brink, A. S. (1963). Two cynodonts from the Ntawere Formation in the Luangwa Valley of Northern Rhodesia. *Palaeontologica Africana*, 8, 77–96.

Clack, J. A., & Allin, E. F. (2004) The evolution of single- and multiple-ossicle ears in fishes and tetrapods. In G. A. Manley, A. N. Popper, & R. R. Fay (Eds.), *Evolution of the vertebrate auditory system* (pp. 128–163). New York: Springer Science+Business Media.

Close, R. A., Friedman, M., Lloyd, G. T., & Benson, R. B. J. (2015). Evidence for a Mid-Jurassic adaptive radiation in Mammals. *Current Biology*, 25, 2137–2142. doi:10.1016/j.cub.2015.06.047

Crompton, A. W., & Luo, Z.-X. (1993). Relationships of the Liassic mammals *Sinoconodon*, *Morganucodon*, and *Dinnetherium*. In F. S. Szalay, M. J. Novacek, & M. C. McKenna (Eds.), *Mammal phylogeny: Mesozoic differentiation, multituberculates, monotremes, early therians, and marsupials, vol. 1* (pp. 30–44). New York: Springer-Verlag.

Doran, A. H. G. (1879). Morphology of the mammalian ossicula auditus. *Transactions of the Linnean Society*, 1, 271–498.

Durrant, J. D., & Lovrinic, J. H. (1995). *Bases of hearing science, 3rd ed.* Baltimore: Williams and Wilkins.

Echteler, S. M., Fay, R. R., & Popper, A. N. (1994). Structure of the mammalian cochlea. In R. R. Fay & A. N. Popper (Eds.), *Comparative hearing: Mammals* (pp. 134–171). New York: Springer-Verlag.

Ekdale, E. G. (2013). Comparative anatomy of the bony labyrinth (inner ear) of placental mammals. *PLoS ONE* 8(6), e66624. doi:10.1371/journal.pone.0066624

Ekdale, E. G., & Rowe, T. (2011). Morphology and variation within the bony labyrinth of zhelestids (Mammalia, Eutheria) and other therian mammals. *Journal of Vertebrate Paleontology*, 31, 658–675. doi:10.1080/02724634.2011.557284

Fernandez, V., Abdala, F., Carlson, K. J., Cook, D. C., et al. (2013). Synchrotron reveals Early Triassic odd couple: Injured amphibian and aestivating therapsid share burrow. *PLoS ONE*, doi:10.1371/journal.pone.0064978.

Fleischer, G. (1973). Studien am Skelett des Gehörorgans der Säugetiere, einschließlich des Menschen. *Säugetierkundliche Mitteilungen*, 21, 131–239.

Fleischer, G. (1976). Hearing in extinct cetaceans as determined by cochlear structure. *Journal of Palaeontology*, 50, 133–152.

Fourie, S. (1974). The cranial morphology of *Thrinaxodon liorhinus* Seeley. *Annals of the South African Museum*, 65, 337–400.

Fox, R. C., & Meng, J. (1997). An X-radiographic and SEM study of the osseous inner ear of multituberculates and monotremes (Mammalia): Implications for mammalian phylogeny and evolution of hearing. *Zoological Journal of the Linnean Society*, 121, 249–291.

Fritzsch, B., Pan, N., Jahn, I., & Elliott, K. L. (2015). Inner ear development: Building a spiral ganglion and an organ of Corti out of unspecified ectoderm. *Cell and Tissue Research*, 361, 7–24. doi:10.1007/s00441–014–2031–5

Fröbisch, J., & Reisz, R. R. (2009). The Late Permian herbivore *Suminia* and the early evolution of arboreality in terrestrial vertebrate ecosystems. *Proceedings of the Royal Society of London B: Biological Sciences*, 276, 3611–3618. doi:10.1098/rspb.2009.0911

Gill, P. G., Purnell, M. A., Crumpton, N., Robson Brown, K., et al. (2014). Dietary specializations and diversity in feeding ecology of the earliest stem mammals. *Nature*, 512, 303–306. doi:10. 1038/nature13622

Gow, C. E. (1986). A new skull of *Megazostrodon* (Mammalia: Triconodonta) from the Elliot Formation (Lower Jurassic) of southern Africa. *Palaeontologia Africana*, 26, 13–23.

Harada, Y., & Ishizeki, K. (1998). Evidence for transformation of chondrocytes and site-specific resorption during the degradation of Meckel's cartilage. *Anatomy and Embryology*, 197, 439–450.

Heffner, H. E., & Heffner, R. S. (2014). The behavioral study of mammalian hearing. In A. N. Popper & R. R. Fay (Eds.), *Perspectives on auditory research* (pp. 269–285). New York: Springer Science+Business Media.

Hoffmann, S., O'Connor, P. M., Kirk, E. C., Wible, J. R., & Krause, D. W. (2014). Endocranial and inner ear morphology of *Vintana sertichi* (Mammalia, Gondwanatheria) from the Late Cretaceous of Madagascar. *Journal of Vertebrate Paleontology*, 34(supplement 1), 110–137. doi:10.1080/02724634.2014.956878

Hopson, J. A., & Kitching, J. W. (2001). A probainognathian cynodont from South Africa and the phylogeny of non-mammalian cynodonts. *Bulletin of the Museum of Comparative Zoology*, 156, 5–35.

Hurum, J. H. (1998). The inner ear of two Late Cretaceous multituberculate mammals, and its implications for multituberculate hearing. *Journal of Mammalian Evolution*, 5, 65–94.

Ji, Q., Luo, Z.-X., Yuan, C.-X., & Tabrum, A. R. (2006). A swimming mammaliaform from the Middle Jurassic and ecomorphological diversification of early mammals. *Science*, 311, 1123–1127. doi:10.1126/science.1123026

Ji, Q., Luo, Z.-X., Zhang, X.-L., Yuan, C.-X., & Xu, L. (2009). Evolutionary development of the middle ear in Mesozoic therian mammals. *Science*, 326, 278–231. doi:10.1126/science. 1178501

Jørgensen, J. M., & Locket, N. A. (1995). The inner ear of the echidna *Tachyglossus aculeatus*: The vestibular sensory organs. *Proceedings of the Royal Society of London B: Biological Sciences*, 260, 183–189.
Kemp, T. S. (2005). *The origin and evolution of mammals*. Oxford: Oxford University Press.
Kemp, T. S. (2007). Acoustic transformer function of the postdentary bones and quadrate of a non-mammalian cynodont. *Journal of Vertebrate Paleontology*, 27, 431–441. doi:10.1671/0272–4634(2007)27[431:ATFOTP]2.0.CO;2
Kermack, K. A., & Mussett, F. (1983). The ear in mammal-like reptiles and early mammals. *Acta Palaeontologica Polonica*, 28, 147–158.
Kermack, K. A., Mussett, F., & Rigney, H. W. (1973). The lower jaw of *Morganucodon*. *Zoological Journal of the Linnean Society*, 53, 87–175.
Kermack, K. A., Mussett., F., & Rigney, H. W. (1981). The skull of *Morganucodon*. *Zoological Journal of the Linnean Society*, 71, 1–158.
Kielan-Jaworowska, Z., Cifelli, R. L., & Luo, Z.-X. (2004). *Mammals from the age of dinosaurs: Origins, evolution, and structure*. New York: Columbia University Press.
Kirk, E. C., Hoffmann, S., Kemp, A. D., Krause, D. W., & O'Connor, P. M. (2014). Sensory anatomy and sensory ecology of *Vintana sertichi* (Mammalia, Gondwanatheria) from the Late Cretaceous of Madagascar. *Journal of Vertebrate Paleontology*, 34 (supplement 1), 203–222. doi:10.1080/02724634.2014.963232
Krause, D. W., Hoffmann, S., Wible, J. R., Kirk, E. C., et al. (2014). First cranial remains of a gondwanatherian mammal reveal remarkable mosaicism. *Nature*, 515, 512–517. doi:10.1038/nature13922
Laaß, M. (2014). Bone conduction hearing and seismic sensitivity of the Late Permian anomodont *Kawingasaurus* fossilis. *Journal of Morphology*, 276, 121–143. doi:10.1002/jmor.20325
Laaß, M. (2015) The origins of the cochlear and impedance matching hearing in synapsids. *Acta Palaeontologica Polonica*. doi:10.4202/app.00140.2014.
Ladevèze, S., de Muizon, C., Colbert, M., & Smith, T. (2010). 3D computational imaging of the petrosal of a new multituberculate mammal from the Late Cretaceous of China and its paleobiologic inferences. *Comptes Rendus Palevol*, 9, 319–330. doi:10.1016/j.crpv.2010.07.008
Ladhams, A., & Pickles, J. O. (1996). Morphology of the monotreme organ of Corti and macula lagena. *Journal of Comparative Neurology*, 366, 335–347.
Li, C.-K., Wang, Y.-Q., Hu, Y.-M., & Meng, J. (2003). A new species of *Gobiconodon* (Triconodonta, Mammalia) and its implication for age of Jehol Biota. *Chinese Science Bulletin (English Edition)*, 48, 1129–1134.
Li, J.-L, Wang, Y., Wang, Y.-Q., & Li, C.-K. (2001). A new family of primitive mammals from the Mesozoic of western Liaoning, China. *Chinese Science Bulletin (English Edition)*, 46, 782–785.
Lillegraven, J. A., & Krusat, G. (1991). Cranio-mandibular anatomy of *Haldanodon exspectatus* (Docodonta; Mammalia) from the Late Jurassic of Portugal and its implications to the evolution of mammalian characters. *Contributions to Geology, University of Wyoming*, 28, 39–138.
Liu, J., & Olsen, P. (2010). The phylogenetic relationships of Eucynodontia (Amniota: Synapsida). *Journal of Mammalian Evolution*, 17, 151–176. doi:10.1007/s10914–010–9136–8
Luo, Z.-X. (1994). Sister taxon relationships of mammals and the transformations of the diagnostic mammalian characters. In N. C. Fraser & H.-D. Sues (Eds.), *In the shadow of dinosaurs—Early Mesozoic tetrapods* (pp. 98–128). Cambridge (UK): Cambridge University Press.
Luo, Z.-X. (2001). Inner ear and its bony housing in tritylodonts and implications for evolution of mammalian ear. *Bulletin of the Museum of Comparative Zoology*, 156, 81–97.
Luo, Z.-X. (2007). Transformation and diversification in the early mammalian evolution. *Nature*, 450, 1011–1019. doi:10.1038/nature06277
Luo, Z.-X. (2011). Developmental patterns in Mesozoic evolution of mammal ears. *Annual Review of Ecology, Evolution and Systematics*, 42, 355–380. doi:10.1146/annurev–ecolsys–032511–142302

Luo, Z.-X., & Crompton, A. W. (1994). Transformations of the quadrate (incus) through the transition from non–mammalian cynodonts to mammals. *Journal of Vertebrate Paleontology*, 14, 341–374.

Luo, Z.-X., Crompton, A. W., & Lucas, S. G. (1995). Evolutionary origins of the mammalian promontorium and cochlea. *Journal of Vertebrate Paleontology*, 15, 113–121.

Luo, Z.-X., Crompton, A. W., & Sun, A. L. (2001). A new mammaliaform from the Early Jurassic of China and evolution of mammalian characteristics. *Science*, 292, 1535–1540.

Luo, Z.-X., Kielan-Jaworowska, Z., & Cifelli, R. L. (2002). In quest for a phylogeny of Mesozoic mammals. *Acta Palaeontologica Polonica*, 47, 1–78.

Luo, Z.-X., Chen, P.-J., Li, G., & Chen, M. (2007). A new eutriconodont mammal and evolutionary development of early mammals. *Nature*, 446, 288–293. doi:10.1038/nature05627

Luo, Z.-X., Ruf, I., Schultz, J. A, & Martin, T. (2011). Fossil evidence on evolution of inner ear cochlea in Jurassic mammals. *Proceedings of Royal Society of London B: Biological Sciences*, 278, 28–34. doi:10.1073/pnas.1519387112

Luo, Z.-X., Ruf, I., & Martin, T. (2012). The petrosal and inner ear of the Late Jurassic cladotherian mammal *Dryolestes leiriensis* and implications for evolution of the ear in therian mammals. *Zoological Journal of the Linnean Society*, 166, 433–463. doi:10.1111/j.1096–3642.2012.00852.x

Luo, Z.-X., Gatesy, S. M., Jenkins, F. A., & Amaral, A. A. (2015a). Mandibular and dental characteristics of Late Triassic mammaliaform *Haramiyavia* and their ramifications for basal mammal evolution. *Proceedings of National Academy of Sciences of the USA*, 112, E7101–E7109. doi:10.1073/pnas.1519387112

Luo, Z.-X., Meng, Q.-J., Ji, Q., Liu, D., Zhang, Y.-G., & Neander, A. I. (2015b). Evolutionary development in basal mammaliaforms as revealed by a docodontan. *Science*, 347, 760–764. doi:10.1126/science.1260880

Maier, W., & van den Heever, J. (2002). Middle ear structures in Permian *Glanosuchus sp.* (Therocephalia, Therapsida), based on thin sections. *Mitteilungen aus dem Museum für Naturkunde in Berlin (Geowissenschaften Eihe)*, 5, 309–318.

Maier, W., & Ruf, I. (2016). Evolution of the mammalian middle ear: A historical review. *Journal of Anatomy*, 228, 270–283. doi:10.1111/joa.12379

Manley, G. A. (2010). An evolutionary perspective on middle ears. *Hearing Research*, 263, 3–8. doi:10.1016/j.heares.2009.09.004

Manley, G. A. (2012). Evolutionary paths to mammalian cochleae. *Journal of Association of Research on Otolaryngology*, 13, 733–743. doi:10.1007/s10162–012–349–9

Manley, G. A., & Clack, J. A. (2004). An outline of evolution of vertebrate hearing organs. In G. A. Manley, A. N. Popper, & R. R. Fay (Eds.), *Evolution of the vertebrate auditory system* (pp. 1–26). New York: Springer Science+Business Media.

Martin, T., Marugán–Lobón, J., Vullo, R., Martín–Abad, H., Luo, Z.-X., & Buscalioni, A. D. (2015). A Cretaceous eutriconodont and integument evolution in early mammals. *Nature*, 526, 380–384. doi:10.1038/nature14905

Mason, M. J., & Narins, P. M. (2001). Seismic signal use by fossorial mammals. *American Zoologist*, 41(5), 1171–1184.

Meng, J. (1992) The stapes of *Lambdopsalis bulla* (Multituberculata) and transformational analyses on some stapedial features in Mammaliaformes. *Journal of Vertebrate Paleontology*, 12, 459–471.

Meng, J., & Fox, R. C. (1995). Osseous inner ear structures and hearing in early marsupials and placentals. *Zoological Journal of the Linnean Society*, 115, 47-71.

Meng, J., Hu, Y.-M., Wang, Y.-Q., & Li, C.-K. (2003). The ossified Meckel's cartilage and internal groove in Mesozoic mammaliaforms: Implications to origin of the definitive mammalian middle ear. *Zoological Journal of the Linnean Society*, 138, 431–448.

Meng, J., Wang, Y.-Q., & Li, C. (2011). Transitional mammalian middle ear from a new Cretaceous Jehol eutriconodont. *Nature*, 472, 181–185. doi:10.1038/nature09921

Meng, Q.-J., Ji, Q., Zhang, Y.-G., Liu, D., et al. (2015). An arboreal docodont from the Jurassic and mammaliaform ecological diversification. *Science*, 347, 764–768. doi:10.1126/science.1260879

Miao, D. (1988). Skull morphology of *Lambdopsalis bulla* (Mammalia, Multituberculata). *Contributions in Geology, University of Wyoming, Special Papers*, 4, 1–104.

Novacek, M. J., & Wyss, A. (1986) Origin and transformation of the mammalian stapes. *Contributions to Geology, University of Wyoming, Special Papers*, 3, 35–53.

O'Gorman, S. (2005). Second branchial arch lineages of the middle ear of wild-type and Hoxa2 mutant mice. *Developmental Dynamics*, 234(1), 124–131. doi:10.1002/dvdy.20402

O'Meara, R. N., & Thompson, R. S. (2014). Were there Miocene meridolestidans? Assessing the phylogenetic placement of *Necrolestes paragonenesis* and the presence of a 40 million year meridiolestidan ghost lineage. *Journal of Mammalian Evolution*, 21, 271–284. doi:10.1007/s10914–013–9252–3

Phillips, M. J., Bennett, T. H., & Lee, M. S. Y. (2009). Molecules, morphology, and ecology indicate a recent, amphibious ancestry for echidnas. *Proceedings of the National Academy of Sciences of the USA*, 106, 17089–17094. doi:10.1073/pnas.0904649106

Pye, A. (1970). The structure of the cochlea in Chiroptera—a selection of Microchiroptera from Africa. *Journal of Zoology (London)*, 162, 335–343.

Quiroga, J. C. (1979). The inner ear of two cynodonts (Repitlia–Therapsida) and some comments on the evolution of the inner ear from pelycosaurs to mammals. *Gegenbauers Morphologishces Jahrbuch*, 125, 178–190.

Rich, T. H., Vickers-Rich, P., Trusler, P., Flannery, T. F., et al. (2001). Monotreme nature of the Australian Early Cretaceous mammal *Teinolophos trusleri*. *Acta Palaeontologica Polonica*, 46, 113–118.

Rich, T. H., Hopson, J. A., Musser, A. M., Flannery, T. F., & Vickers, P. (2005). Independent origins of middle ear bones in monotremes and therians. *Science*, 307, 910–914. doi:10.1126/science.1105717

Rodrigues, G. P., Ruf, I., & Schultz, C. L. (2013). Digital reconstruction of the otic region and inner ear of the non–mammalian cynodont *Brasilitherium riograndensis* (Late Triassic, Brazil) and its relevance to the evolution of the mammalian ear. *Journal of Mammalian Evolution*, 20, 291–307. doi:10.1007/s10914–012–9221–2

Rosowski, J. J. (1992). Hearing in transitional mammals: Predictions from the middle-ear anatomy and hearing capabilities of extant mammals. In D. B. Webster, R. R. Fay, & A. N. Popper (Eds.), *The evolutionary biology of hearing* (pp. 615–631). New York: Springer-Verlag.

Rougier, G. W., & Wible, J. R. (2006). Major changes in the mammalian ear region and basicranium. In M. T. Carrano, T. J. Gaudin, R. W. Blob, & J. R. Wible (Eds.), *Amniote paleobiology: Perspectives on the evolution of mammals, birds, and reptiles* (pp. 269–311). Chicago: University of Chicago Press.

Rougier, G. W., Wible, J. R., & Hopson, J. A. (1996). Basicranial anatomy of *Priacodon fruitaensis* (Triconodontidae, Mammalia) from the Late Jurassic of Colorado, and a reappraisal of mammaliaforms interrelationships. *American Museum Novitates*, 3183, 1–38.

Rougier, G. W., Forasiepi, A. M., & Martinelli, A. G. (2005). Comments on "Independent origins of the middle ear bones in monotremes and therians" (II). *Science*, 309, 1492E.

Rougier, G. W., Martinelli, A. G., Forasiepi, A. M., & Novacek, M. J. (2007). New Jurassic mammals from Patagonia, Argentina: A reappraisal of australosphenidan morphology and interrelationship. *American Museum Novitates*, 3566, 1–54. doi:10.1206/0003–0082(2007)3580[1:FJTFSA]2.0.CO;2

Rowe, T. (1988). Definition, diagnosis, and origin of Mammalia. *Journal of Vertebrate Paleontology*, 8, 241–264.

Rowe, T. B. (1993). Phylogenetic systematics and the early history of mammals. In F. S. Szalay, M. J. Novacek, & M. C. McKenna (Eds.), *Mammal phylogeny: Mesozoic differentiation, multituberculates, monotremes, early therians, and marsupials* (pp. 129–145). New York: Springer-Verlag.

Rowe, T., Carlson, W., & Bottorff, W. (1995). *Thrinaxodon* – Digital Atlas of the skull. Austin (TX): University of Texas Press.

Rowe, T. B., Rich, T. H., Vickers-Rich, P., Woodburne, M. O., & Springer, M. (2008). The oldest platypus and its bearing on divergence timing of the platypus and echidna clades. *Proceedings of the National Academy of Sciences of the USA*, 105, 1238–1342. doi:10.1073/pnas.0706385105

Ruf, I., Luo, Z.-X., Wible, J. R., & Martin, T. (2009). Petrosal anatomy and inner ear structure of the Late Jurassic mammal *Henkelotherium* and the ear region characters of basal therian mammals. *Journal of Anatomy*, 214, 679–693. doi:10.1111/j.1469-7580.2009.01059.x

Ruf, I., Luo, Z.-X., & Martin, T. (2013). Re–investigation of the basicranium of *Haldanodon exspectatus* (Docodonta, Mammaliaformes). *Journal of Vertebrate Paleontology*, 33, 382–400. doi:10.1080/02724634.2013.722575

Schmelzle, T. M. R., Nummela, S., & Sanchez–Villagra, M. R. (2005). Phylogenetic transformations of the ear ossicles in marsupial mammals, with special reference to diprotodontians: A character analysis. *Annals of the Carnegie Museum*, 74, 189–200.

Sidor, C. A. (2001). Simplification as a trend in synapsid cranial evolution. *Evolution*, 55, 1419–1442.

Sues, H.-D. (1986). The skull and dentition of two tritylodontid synapsids from the Lower Jurassic of western North America. *Bulletin of the Museum of Comparative Zoology*, 151, 217–268.

Takechi, M., & Kuratani, S. (2010). History of studies on mammalian middle ear evolution: A comparative morphological and developmental biology perspective. *Journal of Experimental Zoology Part B: Molecular and Developmental Evolution*, 314B, 417–433. doi:10.1002/jez.b.21347

Tumarkin, A. (1968). Evolution of the auditory conducting apparatus in terrestrial vertebrates. In A. S. de Reuck & J. Knight (Eds.), *Cib foundation symposium on hearing mechanisms in vertebrates* (pp. 18–37). London: Churchill.

Vater, M., & Kössl, M. (2011). Comparative aspects of cochlear functional organization in mammals. *Hearing Research*, 273, 89–99. doi:10.1016/j.heares.2010.05.018

Walsh, S. A., Luo, Z.-X., & Barrett, P. M. (2013). Modern imaging techniques as a window to prehistoric auditory worlds. In C. Köppel, G. A. Manley, A. N. Popper, & R. R. Fay (Eds.), *Insights from comparative hearing research.* (pp. 227–261). New York: Springer Science +Business Media.

Wang, Y., Zheng, Y., Chen, D., & Chen, Y.-P. (2013). Enhanced BMP signaling prevents degeneration and leads to endochondral ossification of Meckel's cartilage in mice. *Developmental Biology*, 381, 301–311. doi:10.1016/j.ydbio.2013.07.016

Weston, J. K. (1939). Notes on comparative anatomy of the ganglion cells associated with the vertebrate inner ear sensory areas. *Journal of Anatomy*, 73, 263–288.

Wever, E. G. (1978). *The reptile ear.* Princeton: Princeton University Press.

Wible, J. R., & Hopson, J. A. (1993). Basicranial evidence for early mammal phylogeny. In F. S. Szalay, M. J. Novacek, & M. C. McKenna (Eds.), *Mammal phylogeny: Mesozoic differentiation, multituberculates, monotremes, early therians, and marsupials* (pp. 45–62). New York: Springer-Verlag.

Williams, P. L., Williams, R., Dyson, M., & Bannister, L. H. (1989). *Gray's Anatomy, 37th ed.* New York: Churchill Livingstone.

Witmer, L. M. (1995). The Extant Phylogenetic Bracket and the importance of reconstructing soft tissues in fossils. In J. Thomason (Ed.), *Functional morphology in vertebrate paleontology* (pp. 19–33). Cambridge (UK): Cambridge University Press.

Zeller, U. (1989). Die Enwicklung und Morphologie des Schädels von *Ornithorhynchus anatinus* (Mammalia: Prototheria: Monotremata). *Abhandlungen der Senckenbergischen Naturforschenden Gesellschaft*, 545, 1–188.

Zhou, C.-F., Wu, S., Martin, T., & Luo, Z.-X. (2013). A Jurassic mammaliaform and the earliest mammalian evolutionary adaptations. *Nature*, 500, 163–167. doi:10.1038/nature12429

Chapter 7
The Ear of Mammals: From Monotremes to Humans

Eric G. Ekdale

> *"With malleus*
> *Aforethought*
> *Mammals*
> *Got an earful*
> *of their ancestors'*
> *Jaw."*
>
> – *Gould* (1990*: 23)*

Abstract Mammals hear across a greater range of auditory frequencies than other vertebrates, which was made possible through major modifications of sound transmission and processing pathways within the middle and inner ears. The first step in the evolution of the ear of mammals is a combination of the prootic and opisthotic bones to form a single petrosal bone that fully encapsulates the organs of the inner ear. Elongation of the cochlea occurred simultaneously with the development of the petrosal. The curvature of the cochlea itself is variable among Mesozoic mammals, and extreme coiling of the cochlea is coincident with acquisition of major morphogenetic genes within the mammalian genome. A cribriform plate that is penetrated by multiple branches of the cochlear nerve in all extant mammals, internal cochlear structures such as the auditory nerve ganglion canal, and primary and secondary bony laminae for supporting the basilar membrane further broadened the bandwidth of audible frequencies in placental and marsupial mammals. Although often considered to be a major diagnostic feature of mammals, the definitive mammalian middle ear, which consists of a complete separation of middle ear elements from the mandible, appears to have evolved multiple times over the course of mammal evolution. As new important fossils are discovered, described, and included in phylogenetic analyses, the accuracy in ascertaining when and how many times the major features of the mammalian ear were acquired will increase.

E.G. Ekdale (✉)
Department of Biology, San Diego State University, 5500 Campanile Drive,
San Diego, CA 92182, USA
e-mail: eekdale@mail.sdsu.edu

E.G. Ekdale
Department of Paleontology, San Diego Natural History Museum,
San Diego, CA 92182, USA

J.A. Clack et al. (eds.), *Evolution of the Vertebrate Ear—Evidence from the Fossil Record*, Springer Handbook of Auditory Research 59,
DOI 10.1007/978-3-319-46661-3_7

Keywords Auditory bulla • Cochlea • Inner ear • Marsupial • Middle ear • Monotreme • Multituberculate • Outer ear • Petrosal • Placental mammal • Semicircular canals • Vestibule

7.1 Introduction

Mammalia consists of three major extant clades – Monotremata (extant monotreme mammals, including platypuses and echidnas), Metatheria (includes extant marsupials such as opossums, kangaroos, koalas, and extinct kin), and Eutheria (includes extant placental mammals, which contribute most living mammal species). Together, metatherians and eutherians contribute to the clade Theria, and several additional clades of extinct mammals are recognized that likely are more closely related to living marsupial and placental mammals than to monotremes (Fig. 7.1).

The auditory system of most mammals rises above that of many other tetrapods in the ability to process sound vibrations above 10 kHz (Fay 1988; Manley 2000), which is considered to be high-frequency sensitivity in this chapter, and "ultrasonic" ranges of 20 kHz to 150 kHz or above, which are processed by bats and toothed whales that use biosonar (Nachtigall 2007; Heffner et al. 2013). On the other hand, some mammals are sensitive to "infrasonic" sounds below 20 Hz, including elephants and possibly baleen whales (Heffner and Heffner 1980; Parks et al. 2007). The achievement of the broad frequency range across which mammals are sensitive was made possible through modifications of the sound transmission pathway of the middle ear and through the processing of sounds within the elongated cochlea of the inner ear (Manley 1972, 2000). Such transformations were important events during mammalian evolution, given that the earliest mammals likely were nocturnal and had to rely on nonvisual sensory cues to survive (Gerkema et al. 2013).

Beyond physiological differences, the ear region of mammals diverges anatomically from all other vertebrates in two significant ways: the presence of three middle ear ossicles (malleus, incus, and stapes); and a single bone (petrosal) that fully encapsulates the auditory and vestibular end organs of the inner ear. In addition, living mammals have a cochlea that is either slightly curved (membranous portion of the cochlea in monotremes) or completes at least one full turn (in marsupials and placentals). The petrosal and middle ear ossicles preserve readily in the fossil record and, from a paleontological perspective, the mammalian ear region has been studied intensively. As a result, the attainment of the unique features of the mammalian ear is well documented. Furthermore, the bony constituents of the ear are important for the interpretation of the evolutionary relationships among and within extinct and extant mammal clades (e.g., Ladevèze 2007; Ekdale et al. 2011).

A foundation in the general anatomy of the auditory region of mammals is needed to investigate the rich fossil record of the mammalian ear and reveal what it can tell us about the phylogenetic relationships of mammals and the evolution of their hearing physiologies. The generalized mammalian ear can be divided into

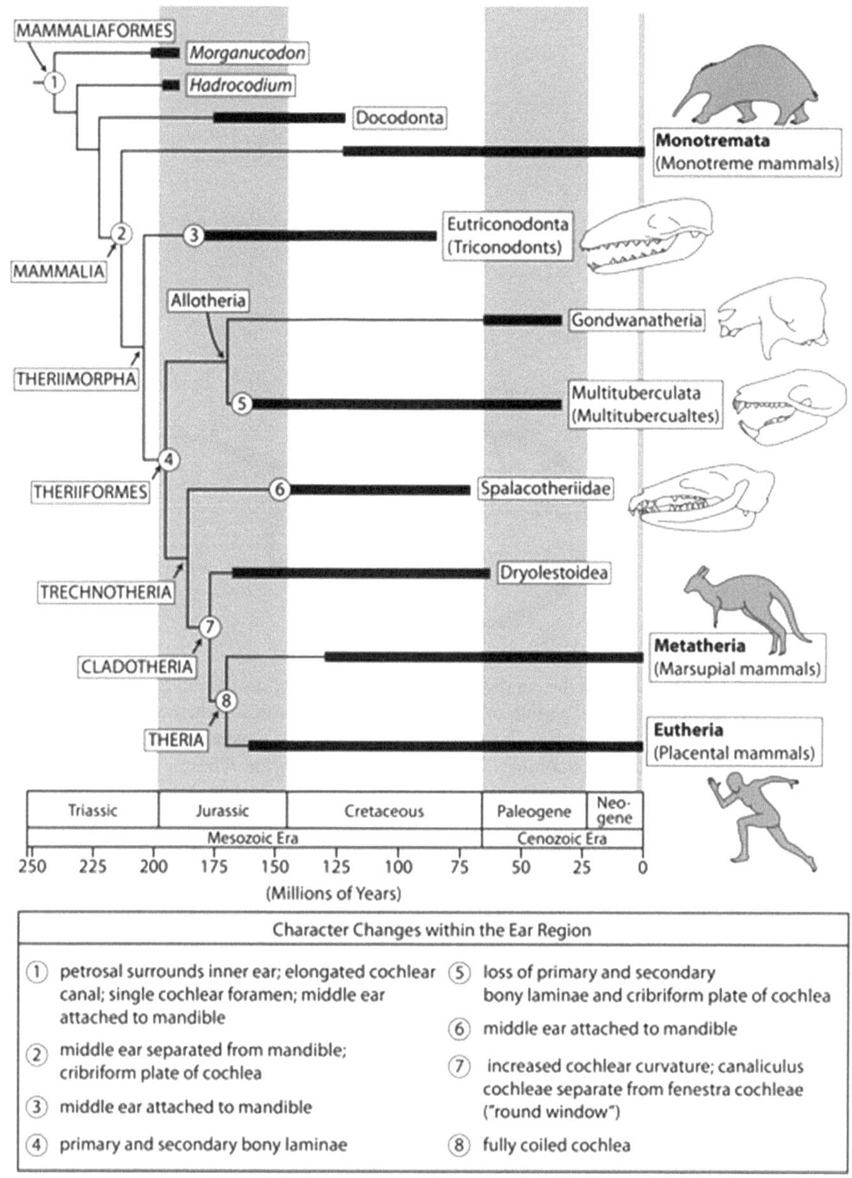

Fig. 7.1 Phylogenetic relationships of mammals (Mammalia) and close extinct relatives discussed in the chapter. Relationships follow Luo (2011) and Krause et al. (2014). Skull reconstructions for extinct mammals include *Jeholodens jenkinsi* representing Eutriconodonta (drawn after Ji et al. 1999), *Vintana sertichi* representing Gondwanatheria (drawn after Krause et al., 2014), *Rugosodon eurasiaticus* representing Multituberculata (drawn after Yuan et al. 2013), and *Maotherium asiaticus* representing Spalacotheriidae (drawn after Ji et al., 2009). Stratigraphic ranges were downloaded from the Paleobiology Database (paleobiodb.org) on 22 September 2015, using the taxonomic name search form for each terminal taxon named on the cladogram

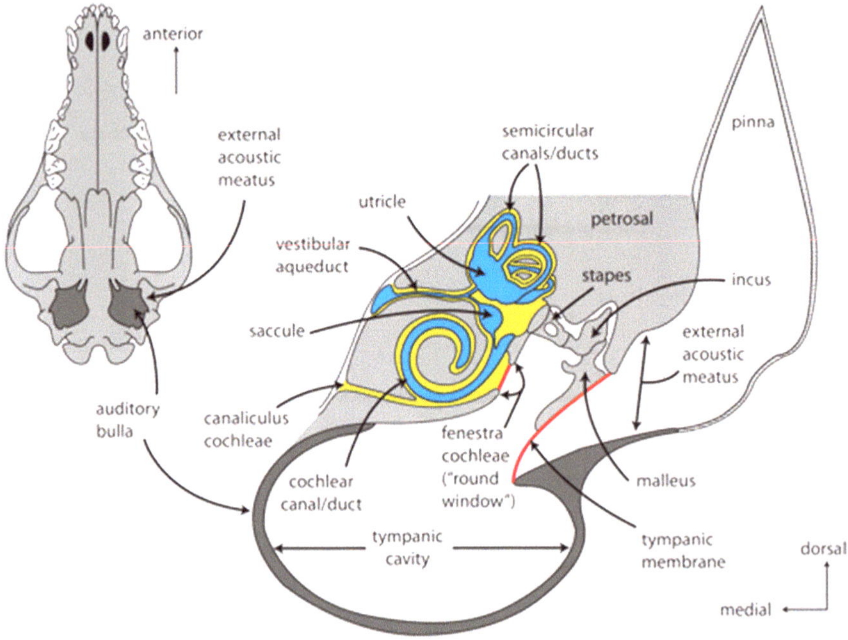

Fig. 7.2 Cross section through the ear region of a domestic dog (*Canis familiaris*). *Blue* structures contribute to the membranous labyrinth of the inner ear, *yellow* indicates the bony labyrinth of the inner ear, and *red* indicates the primary (eardrum) and secondary (spanning fenestra cochleae, or "round window") tympanic membranes. Stapes articulated with the fenestra vestibuli (or "oval window"). (Images modified and reprinted with permission from Ekdale 2015)

three regions: outer ear, middle ear, and inner ear (Fig. 7.2). The outer ear consists of the pinna and external acoustic (auditory) meatus. The middle ear refers to a cavity that is bordered dorsally by elements of the braincase and ventrally by the auditory bulla, which is either bony or cartilaginous dependent on the species. The inner ear contains the auditory and vestibular end organs, including the cochlea, saccule, utricle, and semicircular canals.

7.1.1 Outer Ear

The outer ear of mammals is constructed mostly of soft tissues and thus it generally is not preserved in the fossil record. In particular, the pinna consists of flesh and hair surrounding a cartilaginous skeleton. The pinna is a unique structure in therian mammals (although timing of the first appearance of the pinna cannot be determined in the fossil record) that funnels airborne sound waves from the external environment to the middle ear. The size and shape of the mammalian pinna varies among mammals: from small and simple flaps in the California sea lion (*Zalophus californicus*), to the elongate structures in jackrabbits (*Lepus californicus*) and fennec foxes (*Vulpes zerda*), to the large and complex auditory receivers in ghost bats (*Macroderma gigas*). However, external pinnae are absent in many marine mammals, including true seals (Phocidae), manatees (Sirenia), and whales (Cetacea). Some whales have structures often referred to as "internal acoustic pinnae" that serve the same function as their external counterparts (Cranford et al. 2010; Cranford and Krysl 2015), but these are not homologous to the external pinnae of terrestrial mammals.

The external acoustic meatus often has a bony component in mammals, and so a portion of this structure is easily preserved in the fossil record. In general, the external acoustic meatus extends from the external environment to the eardrum (tympanic membrane) (Fig. 7.2). In some species of mammals, the external acoustic meatus is short (e.g., kangaroo rat, *Dipodomys* sp.), but the meatus is developed into an elongated auditory tube that is completely surrounded by bone in some mammals (e.g., mountain beaver, *Aplodontia rufa*) or only enclosed in bone dorsally (e.g., baleen whale, *Balaenoptera borealis*) (Fig. 7.3).

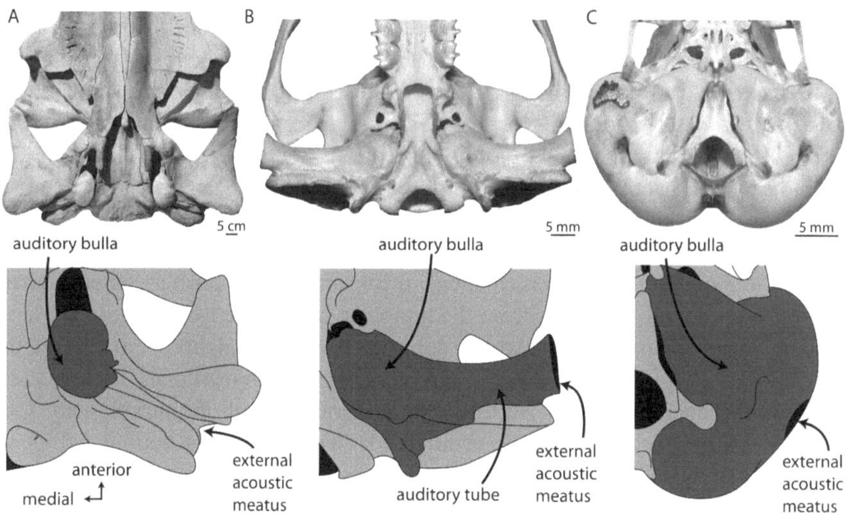

Fig. 7.3 Basicrania (*photographs*) and left ear regions (*line drawings*) of extant mammals in ventral view. Auditory bulla rendered in *dark gray* in line drawings. (**A**) Eden's whale, *Balaenoptera edeni*, and sei whale, *Balaenoptera borealis*, with external acoustic meatus unossified ventrally. (**B**) Mountain beaver, *Aplodontia rufa*, with an elongate and ossified auditory tube. (**C**) Kangaroo rat, *Dipodomys deserti*, without an ossified auditory tube. (**A**, upper photograph from Project 470 at morphobank.org, lower line drawing modified from Ekdale et al. 2011; **B** photograph and line drawing of SDSU-S-665 by the author; **C** photograph and line drawing of SDSU-S-1075 by the author; *SDSU*, San Diego State University Museum of Biodiversity, San Diego, CA)

7.1.2 Middle Ear

The eardrum spans an annular ring of the ectotympanic bone (homologous to the angular of the mandible of non-mammalian synapsids; Allin 1975) and separates the outer ear from the middle ear (tympanic) cavity (Fig. 7.2). The ventral floor of the tympanic cavity is known as the auditory bulla. In many mammals, the auditory bulla is developed as a bulbous outgrowth of the basicranium (Fig. 7.3C). However, the bulla is less distinct in other species (including humans). Developmentally, the auditory bulla is preceded by a fibrous membrane that floors the embryonic tympanic cavity, and the bony and cartilaginous elements of the auditory bulla develop within or adjacent to the membrane (MacPhee 1981).

Multiple bones can contribute to the auditory bulla in different extant and extinct mammals (van der Klaauw 1931; Novacek 1977), although a membranous or cartilaginous bulla likely is the ancestral state for mammals (Novacek 1993). Bony bullae are not preserved and presumably were not present in several Mesozoic mammals (Kielan-Jaworowska 1981; Wible et al. 2009), and bony components of the bulla are either absent or weakly developed in some extant taxa, such as monotremes and shrews (Novacek 1977). Among those mammals that possess osseous bullae, the element is constructed from the ectotympanic bone in most species, although the major component of the bulla is the alisphenoid in marsupials, the basisphenoid in hedgehogs and tenrecs, the entotympanic in tree shrews, hyraxes, some rodents, and some carnivorans, and the petrosal in primates (Novacek 1977, 1993). The bulla is a mosaic of various elements in elephant shrews (MacPhee 1981), and the relative contributions of the alisphenoid and squamosal vary intraspecifically among populations of the grey or northern common cuscus, *Phalanger orientalis breviceps* (Norris 1993).

The three middle ear ossicles (malleus, incus, and stapes) form an osseous chain that transmits sound vibrations from the tympanic membrane to the inner ear via an opening in the petrosal bone, which contributes the dorsal border (roof) of the tympanic cavity (Fig. 7.2). The petrosal is a bone unique to mammals. It completely encloses the inner ear organs, and it is homologous to both the prootic and opisthotic bones of reptiles (Kermack et al. 1981). The cochlea is housed within an anterior swelling of the petrosal known as the promontorium, and the semicircular canals are contained within the posterior portion of the bone (Fig. 7.4). The middle and inner ear cavities communicate through openings within the promontorium, primarily through the fenestra vestibuli (often referred to as the "oval window") that accommodates the footplate of the stapes (Figs. 7.2, 7.4). Posterior and medial to the fenestra vestibuli is the fenestra cochleae (often referred to as the "round window") in marsupials and placentals, and the perilymphatic foramen in monotremes and most extinct mammal clades (Zeller 1993). Those openings are covered by a secondary tympanic membrane and serve to equalize pressure within the membranous ducts of the inner ear.

Endocranially, cranial nerves VII (facial) and VIII (vestibulo-cochlear or vestibulo-acoustic) penetrate into the petrosal via a pair of openings within an

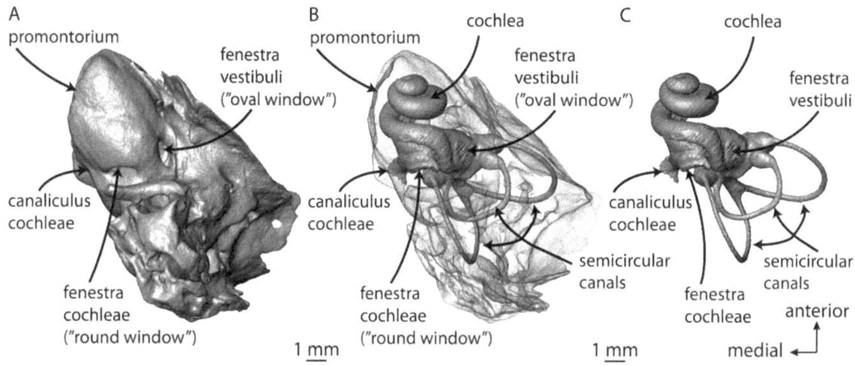

Fig. 7.4 Left petrosal bone and bony labyrinth of the extant nine-banded armadillo, *Dasypus novemcintus*, rendered from CT scan data. (**A**) Ventral view of the petrosal. (**B**) Petrosal bone rendered semitransparent to reveal the bony labyrinth within. (**C**) Digital endocast of the bony labyrinth rendered from CT data. (Images modified and reprinted with permission from Ekdale 2013)

internal acoustic meatus. The facial nerve passes completely through the petrosal before ultimately exiting on the tympanic (ventral) surface of the petrosal lateral to the promontorium. A large depression known as the subarcuate fossa for the parafloccular lobe of the cerebellum is located posterior to the internal acoustic meatus in many mammals. A well-developed fossa has been observed in early eutherians (MacIntyre 1972; Cifelli 1982), but it is secondarily lost in many taxa (e.g., some artiodactyls) (O'Leary 2010).

7.1.3 Inner Ear

The inner ear of mammals consists of a series of continuous osseous chambers known as the bony labyrinth within the petrosal bone (Fig. 7.4B, C). The bony labyrinth in turn contains a network of membranous ducts and sacs referred to as the membranous labyrinth (Fig. 7.2). The membranous labyrinth includes the cochlear duct and saccule inferiorly and three semicircular ducts and the utricle of the vestibule superiorly. The membranous vestibule and semicircular ducts are involved with linear and rotational motion sensations. The sense of hearing is contained within the cochlea and specifically is associated with the spiral organ of hearing (or organ of Corti) upon the basilar membrane within the membranous cochlear duct. The basilar membrane is anchored on either side of the bony cochlear canal by dense and unossified connective tissue in monotremes (Luo et al. 2011) and by a pair of bony laminae that extend toward the center of the cochlear canal lumen in therian mammals (Fig. 7.5) (Fleischer 1973). The primary bony lamina emerges from the axial (inner) wall of the cochlea and surrounds a canal, sometimes referred to as Rosenthal's canal, for the auditory nerve ganglion, a branch of cranial nerve VIII. Individual nerve fibers pass through small foramina within a cribriform plate along the axial wall of the cochlea and into the ganglion canal in therians and into the cochlea in monotremes (Luo et al. 2012). The secondary bony lamina emerges from the radial (outer) wall of the cochlear canal in most extant and extinct therian mammals (Fig. 7.5), although a subsequent loss of the secondary bony lamina occurred in some lineages (Ekdale 2013; Hoffmann et al. 2014).

The basilar membrane is narrow at its proximal end near the base of the cochlea, and it widens toward the apex (Webster and Webster 1980). In addition, the stiffness and thickness of the membrane decrease from the proximal to distal ends of the cochlear spiral (Naidu and Mountain 2007). Variations in the width, thickness, and stiffness of the basilar membrane along its length correlate with tonotopic differences across the cochlea (Wever et al. 1971; Echteler et al. 1994), in which high frequencies are detected at the base of the cochlea and low frequencies are detected at the apex (Liberman 1982; Müller 1996). Although the anatomy of the basilar membrane in mammals is informative for studies of hearing physiology (e.g., Ketten and Wartzok, 1990; Ketten, 1997), the soft tissues of the membranous labyrinth do not preserve in the fossil record. Thus, direct measurements of the stiffness and dimensions of the basilar membrane of extinct taxa, which would provide auditory function information, cannot be ascertained. However, the primary and secondary bony laminae that support the membrane in therians are often preserved, and their robust character and relative distances have been used to estimate the hearing capabilities of extinct mammals (Fleischer 1976).

The distance between the primary and secondary bony laminae ("basilar gap" of Fleischer 1976, or "laminar gap" of Geisler and Luo 1996) is positively correlated with basilar membrane stiffness. Narrowly separated bony laminae are observed in mammals sensitive to high frequencies, such as bats and toothed whales (Wever et al. 1971; Ramprashad et al. 1979). Likewise, mammals sensitive to higher

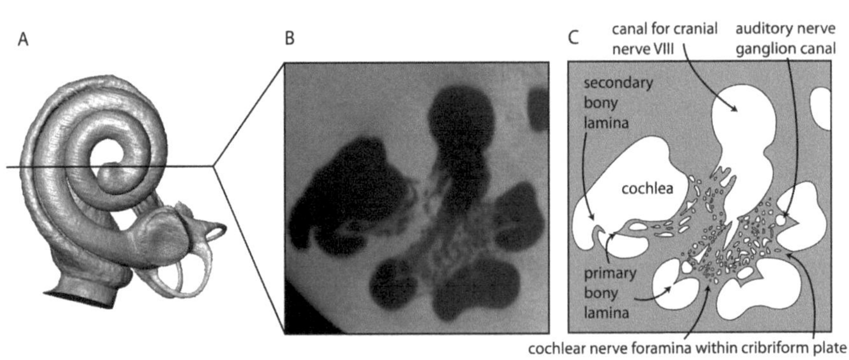

Fig. 7.5 Internal structure of the cochlea of mammals, based on the extant humpback whale, *Megaptera novaengliae*. (**A**) Digital endocast of the left bony labyrinth rendered from CT data. (**B**) Original CT slice through the cochlea. (**C**) Labeled illustration of original CT slice. (Images modified and reprinted with permission from Ekdale and Racicot 2015)

frequencies have secondary bony laminae that extend for a greater proportion of the cochlear length (Ketten 2000; Ekdale and Racicot 2015). Although an extensive secondary bony lamina may indicate high-frequency sensitivity in extinct mammals, some extant taxa, such as baleen whales that likely are sensitive to infrasonic frequencies, have relatively better-developed secondary bony laminae than species that have higher low-frequency thresholds, such as humans and pigs (Fleischer 1973; Ekdale 2013).

A final component of the inner ear is the lagena, an end organ of the inner ear that is present in non-mammalian vertebrates (Khorevin 2008). The lagena is absent in therian mammals (although the extreme apex of the cochlear canal often is called the "lagena" in humans) (Gray 1977), but the organ is present and presumably functional at the apex of the cochlea of monotremes (Pritchard 1881; Ladhams and Pickles 1996). In non-mammalian vertebrates, the lagena is associated with the vestibule and functions in a similar fashion as the utricle and saccule. Although the lagena of monotremes is located at the apex of the cochlea, histological studies reveal that the lagenar hair cells are more vestibular than cochlear in nature (Ladhams and Pickles 1996; Khorevin 2008).

7.2 Monotremata and Early Mammalia

There are only five living species of monotreme—one species of duck-billed platypus (*Ornithorhynchus anatinus*) and four species of echidna or spiny anteater (*Tachyglossus aculeatus* and three *Zaglossus* species). Although monotremes are clearly mammals (Rowe 1988), they retain many ancestral characteristics inherited from the shared common ancestor between mammals and reptiles (e.g., laying shelled eggs) (Jenkins 1990). The cochlea of monotremes has been described as bird-like in its shape and size (Pritchard, 1881), given that the cochlea is elongate and slightly curved in some birds (Gray 1908; Turkewitsch 1934). However, the organ of Corti is distinctly mammalian (Ladhams and Pickles 1996), and monotremes are sensitive to higher frequency sounds than most birds and other reptiles, but lower than many placental and marsupial mammals (upper limit around 14–18 kHz) (Krubitzer 1998; Mills and Shepherd 2001).

Extant monotremes and their extinct allies possess fully developed petrosal bones, which is a primitive retention from mammaliaform ancestors, such as *Morganucodon* (Fig. 7.1) (Rowe 1988). The evolutionary development of the promontorium coincides with an elongation of the cochlear canal in the mammaliaforms (Luo et al. 1995). The cochlea is short and straight in early diverging non-mammalian cynodonts (Luo et al. 1995; Rodrigues et al. 2013), but the degree of coiling varies among Mesozoic stem mammals (see review by Hoffmann et al. 2014). For example, the cochlea of *Morganucodon* and docodonts such as *Haldanodon* (close extinct relatives of mammals) was elongate and slightly curved, a condition that is retained in extant monotremes (Fig. 7.6). Note that although the

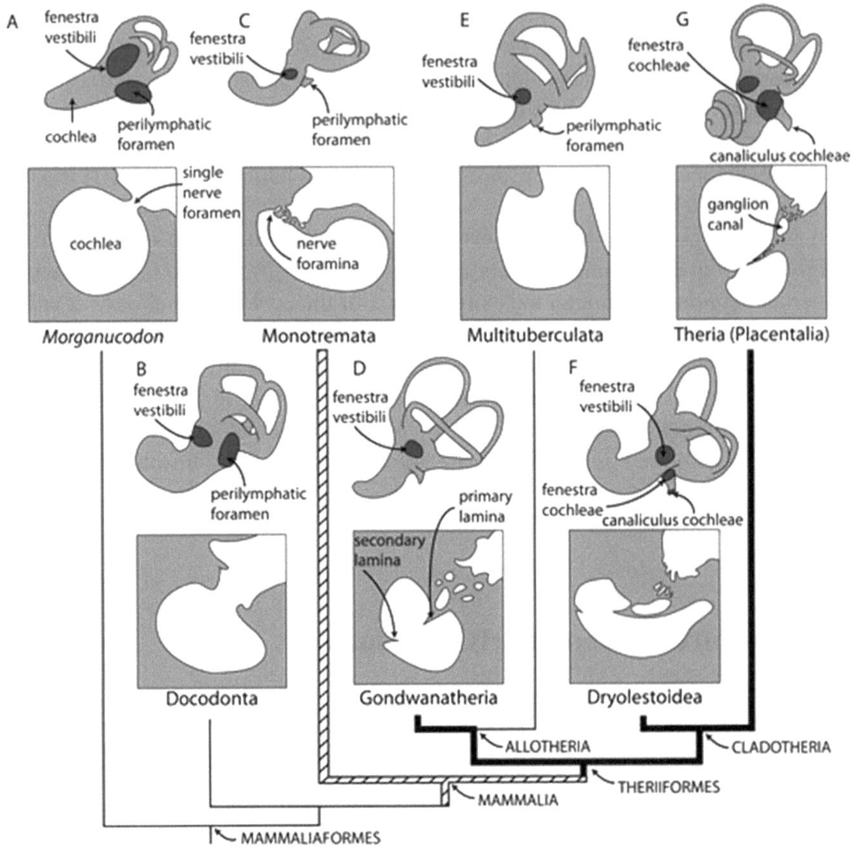

Fig. 7.6 Bony labyrinths and cross sections through the bony cochlea of extinct and extant mammaliaforms. Anatomical structures drawn following published CT images, digital inner ear endocasts, and anatomical descriptions. *Hatched branches* indicate presence of multiple cochlear nerve foramina through a cribriform plate within the cochlea; *thick lines* indicate presence of the cribriform plate, as well as primary and secondary bony laminae; relationships follow Luo (2011) and Krause et al. (2014). **(A)** *Morganucodon* sp. **(B)** Docodonta based on *Haldanodon exspectatus*. **(C)** Monotremata based on *Ornithorhynchus anatinus*. **(D)** Gondwanatheria based on *Vintana sertichi*. **(E)** Multituberculata based on *Chulsanbaatar vulgaris* (*cross section*) and *Nemegtbaatar gobiensis* (*labyrinth*). **(F)** Dryolestoidea based on *Dryolestes leiriensis*. **(G)** Theria (Placentalia) based on the nine-banded armadillo, *Dasypus novemcinctus* (*cross section*), and domestic pig, *Sus scrofa* (*labyrinth*). (**A** modified from Graybeal et al. 1989; **B** modified from Ruf et al. 2013; **C** modified from Luo et al. 2012; **D** modified from Hoffmann et al. 2014; **E** modified from Hurum 1998; **F** modified from Luo et al. 2011; **G** modified from Ekdale, 2013)

bony cochlear canal is essentially straight in the platypus, the membranous cochlear duct is curved at its apex (Luo et al. 2011; Luo, Schultz, and Ekdale, Chap. 6).

The cochlear nerve fibers of monotremes pass through foramina within a cribriform plate, forming the axial wall of the cochlea (Fig. 7.5), but the primitive mammaliaform nervous supply to the cochlea is transmitted through a single cochlear foramen (Fig. 7.6). Conversely, the nerve fibers pass through foramina within a cribriform plate in monotremes, but the auditory nerve ganglion canal is not developed. Similar foramina are observed in Theria, along with the auditory nerve ganglion canal and associated primary bony lamina, but the foramina likely were acquired independently in stem therians.

Monotremes possess the three ossicles of the middle ear as do therians, and that condition is one of the diagnostic hallmarks of Mammalia. The homologies among the middle ear bones and those of the upper and lower jaws of non-mammalian vertebrates have been ascertained embryologically (Zeller 1987; Anthwal et al. 2013), and the evolutionary transition to three middle ear ossicles is well-documented in the fossil record (Kemp, Chap. 5; Luo, Schultz, and Ekdale, Chap. 6). The stapes is homologous to the columella in the ear of non-mammalian tetrapods (Novacek and Wyss 1986), the malleus is homologous to the articular of the lower jaw in reptiles, and the incus is homologous to the quadrate of the basicranium (Reichert 1837; Goodrich 1930). During the evolutionary development of the definitive mammalian middle ear, the malleus and incus were incorporated into the otic region simultaneously (Allin 1975; Rowe 1988). Coincidentally, the lower jaw of mammals consists of a single bone (dentary), whereas the lower jaw of non-mammalian vertebrates consists of multiple bones in addition to the dentary.

Although the transformation of the mammalian ear ossicles is well known, the timing of the evolution of the definitive mammalian middle ear is difficult to determine. Reabsorption of a specific structure known as Meckel's cartilage during embryonic development of the extant mammal ear is the final step in the separation between the middle ear ossicles and the lower jaw (Luo et al. 2007). A groove for Meckel's cartilage is preserved in fossils of non-mammalian cynodonts, indicating attachment of the middle ear bones to the dentary in adulthood in those taxa (Luo 2011). *Morganucodon* (Late Triassic) retained an attachment of the middle ear to the dentary in adulthood, although modified from earlier non-mammalian cynodonts (Fig. 7.7). The lower jaw of *Hadrocodium* (Early Jurassic) was originally reconstructed as lacking a trough for postdentary bones and Meckel's cartilage, suggesting attainment of the definitive mammalian middle ear (bones detached from the dentary) before the divergence of crown Mammalia in the Middle Jurassic (Luo 2011). However, that original reconstruction was inaccurate, and reanalysis of the dentary of *Hadrocodium* indicated that the taxon in fact did have a postdentary trough for attachment of middle ear elements (Luo, Schultz, and Ekdale, Chap. 6), although no middle ear elements are preserved. Docodonts, such as *Agiloconodon* (Middle Jurassic), retain middle ear elements attached to the mandible as well, indicating that the definitive mammalian middle ear is a synapomorphy for Mammalia (Fig. 7.7) (Luo, Schultz, and Ekdale, Chap. 6).

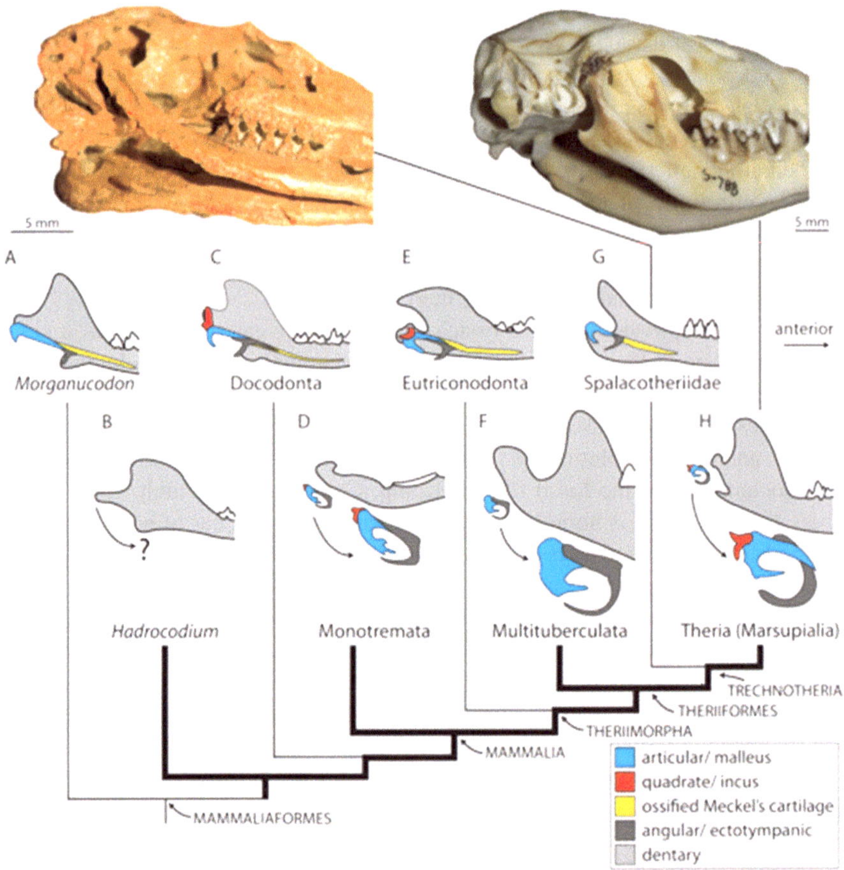

Fig. 7.7 Posterior regions of mandibles and middle ear elements of extinct and extant mammaliaforms in medial view. *Thick black branches* indicate detachment of middle ear bones from the mandible; relationships follow Luo (2011). **(A)** *Morganucodon* sp. **(B)** *Hadrocodium wui* (middle ear bones not preserved). **(C)** Docodonta based on *Agilodocodon scansorius*. **(D)** Monotremata based on the platypus, *Ornithorhynchus anatinus*. **(E)** Eutriconodonta based on *Yanoconodon allini*. **(F)** Multituberculata based on *Lambdopsalis bulla*. **(G)** Spalacotheriidae based on *Maotherium asiaticus*. **(H)** Theria (Marsupialia) based on American opossum *Didelphis virginiana* (*photograph*) and gray short-tailed opossum *Monodelphis domestica* (*line drawing*). (**A** modified from Ji et al. 2009; **B** modified from Luo, Schultz, and Ekdale, Chap. 6; **C** modified from Meng et al. 2015; **D** modified from Meng and Wyss 1995, Musser and Archer 1998; **E** modified from Luo et al. 2007; **F** modified from Miao 1988, Meng and Wyss 1995; **G** line drawing modified and photograph reprinted with permission from Ji et al. 2009; **H** photograph of SDSU-S-788, San Diego State University Museum of Biodiversity, San Diego, CA, and line drawing modified from Ji et al. 2009)

Precise determination of the evolution of the definitive mammalian middle ear hinges on the middle ear anatomy of extinct mammal clades. A trough for postdentary bones was described in a fossil attributed to the extinct monotreme *Teinolophos trusleri* (Rich et al. 2005), which proved to be controversial (see Bever et al. 2005; Rougier et al. 2005). If *Teinolophos* was an early monotreme with postdentary bones, then the definitive mammalian middle ear would have originated independently in monotremes and therians (Rich et al. 2005). Reanalysis of the phylogenetic relationships of *Teinolophos* and other early mammals confirms that *Teinolophos* is a monotreme, but high resolution CT data revealed that the trough for postdentary bones that originally was described for the specimen was misinterpreted (Rowe et al. 2008). Thus, *Teinolophos* likely possessed three middle ear ossicles, although as with *Hadrocodium*, fossil mallei and incudes for *Teinolophos* are unknown.

Although controversial, multiple origins of the definitive mammalian middle ear are supported by an ossified Meckel's cartilage within the mandible of several eutriconodont and spalacotheriid fossils (Fig. 7.7G). A connection between the ossicular chain and the dentary via an ossified Meckel's cartilage in extinct clades of mammals suggests either secondary losses of the definitive mammalian middle ear in eutriconodonts and spalacotheriids, or perhaps more likely, a paedomorphic retention of Meckel's cartilage into adulthood and, therefore, multiple origins for the definitive mammalian middle ear (Luo et al. 2007; Meng et al. 2011).

7.3 Theriimorpha

Fossilized ear regions of theriimorphs (the sister-taxon to monotremes) have been recovered from Jurassic strata, including fossils attributed to the eutriconodonts (Hu et al. 1997; Rougier et al. 1996). The inner ears of eutriconodonts (Middle to Late Jurassic) have yet to be described (Hughes et al. 2015), but the promontorium is elongate, which may suggest a straight cochlea or, at most, a cochlea that is slightly curved. The cochleae of early diverging theriiforms are curved 180° or less, and the same may be likely for eutriconodonts given the curved nature in monotremes, docodonts, and *Morganucodon* (Fig. 7.6).

7.3.1 Allotheria (Multituberculata and Gondwanatheria)

Multituberculates are one of the longest lived lineages of mammals with a stratigraphic range extending from the Upper Jurassic into the Lower Eocene—a duration of approximately 120 million years (Simpson 1937; Ostrander 1984). The petrosal of some multituberculate species is inflated and reminiscent of the auditory bulla of more recent placental mammals, which is a result of extreme expansion of the vestibule of the inner ear (Chow and Qi 1978; Miao 1988). The auditory bulla itself is unossified in multituberculates, although an annular ectotympanic and middle ear ossicles are known, all of which are completely detached from the lower jaw (Fig. 7.7). Some multituberculate species were reconstructed as fossorial (burrowing) animals sensitive to low-frequency noises based, in part, on expansion of the vestibule (Miao and Lillegraven 1986; Miao 1988) as compared to the expanded vestibular cavities in several non-mammalian tetrapods that also are known to burrow (Bramble 1982; Olori 2010).

The promontorium of many Late Cretaceous multituberculates tends to be large relative to the length of the dentary when compared to contemporaneous mammals (Hughes et al. 2015). The morphology of the promontorium often is reflective of the shape of the cochlea, which varies from straight and finger-like to slightly bent among Cretaceous multituberculates, (Fig. 7.6) (Luo and Ketten 1991; Ladevèze et al. 2010). Because the cochlea is curved in monotremes and the closest relatives of mammals, a truly straight cochlea indicates a reversal to an earlier cynodont morphology in those multituberculates. As previously discussed, the bony cochlea of the extant platypus is straight but the membranous duct is curved, and a similar situation may be a more likely case for multituberculates rather than an evolutionary reversal. An expanded region of the cochlear apex in some, but not all, multituberculates may have accommodated an apical curvature of the cochlea. The expanded region has been described as a lagena (Fox and Meng 1997; Hurum 1998); however, the excavation lacks evidence of a lagenar nerve, thereby indicating that a lagena was not present in multituberculates after all (Ruf et al. 2013).

The Mesozoic mammal *Vintana sertichi* is the best known member of the enigmatic mammal clade Gondwanatheria (sister-clade of Multituberculata with a stratigraphic range of Upper Cretaceous to Paleogene; Fig. 7.1) in that it is the only gondwanatherian for which cranial remains are known. The inner ear of *Vintana* is a mosaic. The cochlea of *Vintana* is short and only slightly curved, but both primary and secondary bony spiral laminae are present, which are observed in the later diverging Cladotheria (including therian mammals and closely related dryolestoids; Fig. 7.6). With regard to length, the cochlea of *Vintana* is short relative to the promontorium as compared to cladotherians, but it is long when compared with many earlier diverging non-mammalian taxa (Hoffmann et al. 2014; Kirk et al. 2014). A generalized correlation between a long cochlea and low-frequency hearing sensitivity has been proposed (West 1985), and if the correlation applies to extinct mammals, then gondwanatherians would have been sensitive to higher frequencies than other contemporaneous Mesozoic mammals.

The presence of the bony spiral laminae within the cochlea of *Vintana*, along with the associated cribriform plate and auditory nerve ganglion canal, suggests an independent derivation of the modern and sophisticated cochlear nervous supply in gondwanatherians and cladotherians (Kirk et al. 2014). The secondary bony lamina is absent in multituberculates (Fig. 7.6), but it is particularly well-developed in *Vintana* in that it extends for a greater length of the cochlear canal than in some Mesozoic eutherians (Hoffmann et al. 2014). Such development of the secondary bony lamina supports a hypothesis for frequency sensitivity above 20 kHz, which is within the range of extant therian physiology. However, the cochlea is shorter than that of therians, which would have limited the frequency bandwidths across which *Vintana* was sensitive. That suggests that *Vintana,* and perhaps other gondwanatherians, had either an extremely narrow range of frequency sensitivity or poor frequency resolution (Kirk et al. 2014). Those observations indicate that the evolution of the inner ear form and function in mammals is complicated, and more research is needed among early mammals to clarify the picture.

7.4 Cladotheria

Extreme cochlear curvature, bordering on coiling at approximately 270°, characterizes the basal cladotherians Dryolestoidea (Middle Jurassic to early Neogene; Figs. 7.1, 7.6) (Ruf et al. 2009). Along with increased curvature came significant development to the primary and secondary laminae and the cochlear nervous supply, including the cribriform plate for passage of auditory nerve fibers (Fig. 7.6). Such neural advancements coincided with high-frequency sensitivity (above 10–20 kHz) as well as associated morphogenetic genes, all of which were acquired relatively early in the evolutionary history of mammals and before development of the fully coiled therian cochlea (Ruf et al. 2009; Luo et al. 2011).

7.4.1 Theria (Marsupials and Placentals)

Theria include the most recent common ancestor of Eutheria (including placental mammals) and Metatheria (including marsupial mammals) and all descendants of that ancestor (see Rowe 1988). Coiling of the cochlea varies from approximately a single turn in Mesozoic therians to four or more turns in more recent mammals (Ekdale and Rowe 2011; Ekdale 2013). Significance of the coiled cochlea is not completely understood. A large number of turns completed by the cochlea, and in particular when the number of turns is multiplied by the length of the cochlear canal, has been correlated with low-frequency hearing sensitivity (West 1985). In addition, the graded curvature (or diametric difference) between successive turns of the coiled cochlea is related to the optimal frequencies that are detected by the cochlea (Manoussaki et al. 2008).

In non-cladotherians, the membranous perilymphatic duct exits the inner ear through the perilymphatic foramen and is transmitted in a groove that extends toward the jugular notch of the basicranium (Fig. 7.8A, B). In cladotherians, the perilymphatic duct is confined to its own bony channel called the canaliculus cochleae, which is separated physically from the fenestra cochleae (Figs. 7.6, 7.8). The separation between the two openings is the result of a structure known as the processus recessus that develops during the isolation of the inner ear chambers from the tympanic cavity of the middle ear (Zeller 1985). The bony canaliculus cochleae for the membranous perilymphatic duct is a neomorphic structure in Cladotheria

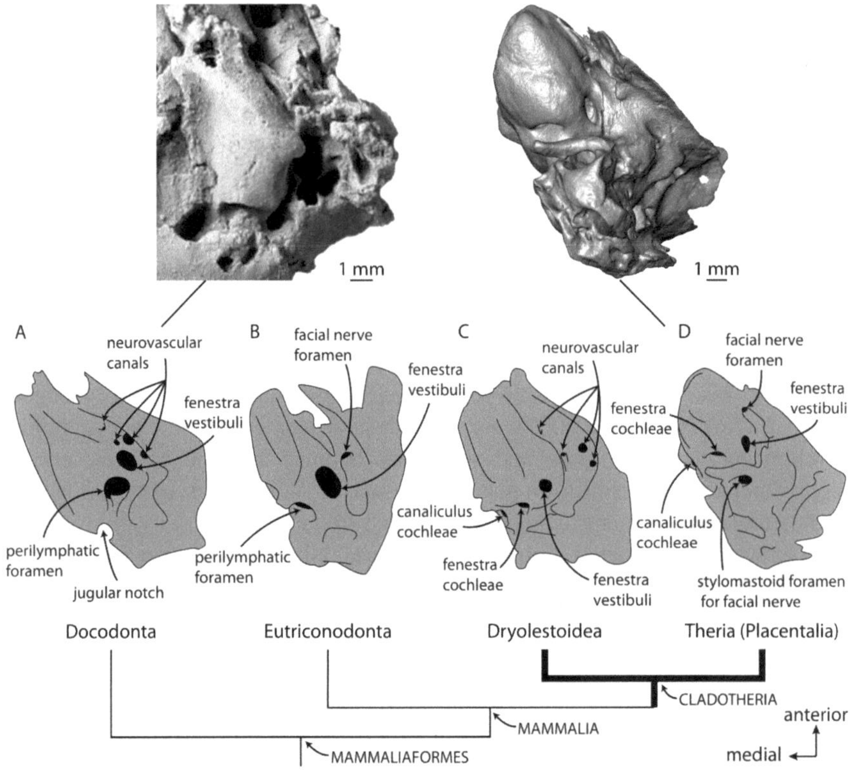

Fig. 7.8 Left petrosal bones of extinct and extant mammaliaforms in ventral view. *Thick black branches* indicate separation of fenestra cochleae and canaliculus cochleae for the perilymphatic duct; relationships follow Luo (2011). **(A)** Docodonta based on *Haldanodon exspectatus*. **(B)** Eutriconodonta based on *Priacodon fruitaensis*. **(C)** Dryolestoidea based on *Dryolestes leiriensis*. **(D)** Theria (Placentalia) based on the nine-banded armadillo, *Dasypus novemcinctus*. (**A** reprinted and modified with permission from Ruf et al. 2013; **B** modified from Rougier et al. 1996; **C** modified from Luo et al. 2012; **D** digital rendering from CT data reprinted and modified with permission from Ekdale 2013)

(Novacek 1993; Hughes et al. 2015), although the passage coalesces with the fenestra cochleae independently in several placental mammal lineages, including manatees, elephants, and baleen whales (see Sect. 7.5.2).

Fossils of isolated therian petrosals can be difficult to identify as either eutherian or metatherian because of the broad morphological variations found in each clade. Only a handful of features have been proposed as potentially diagnostic, but those often are problematic. For example, the footplate of the stapes tends to be more oval in extant and extinct placental mammals compared with marsupials (Segall 1970). A circular stapedial footplate characterizes monotremes and non-therian theriiforms (e.g., multituberculates; Meng and Wyss 1995). However, there are numerous exceptions to the rule, and the shape of the stapedial footplate (often estimated using dimensions of the fenestra vestibuli) varies intraspecifically (Ekdale 2011).

Neurovascular structures that pass through the middle ear region have been a common source of characters used to determine the evolutionary relationships among mammals (MacIntyre 1972; Wible 1986). For example, sulci and canals on the petrosals of stem therians indicate the presence of the stapedial artery, which is a branch of the internal carotid artery, and the lateral head vein, which is a branch diverging from the prootic sinus (Wible 1990). The prootic sinus and associated lateral head vein are absent in adulthood in eutherians, although similar vascular structures have been reconstructed for some Cretaceous eutherians (Ekdale et al. 2004). Rather, blood is diverted to the capsuloparietal emissary vein in placentals (Wible 1990). In contrast, the portion of the lateral head vein that passes through the tympanic cavity is reduced during ontogeny in extant marsupials, and blood is diverted to the neomorphic sphenoparietal emissary vein. Furthermore, marsupials lack the stapedial artery and, therefore, any osteological correlates of the vessel.

Late Cretaceous petrosals from Montana and Wyoming that were identified as metatherian lack the sulci for the stapedial artery, but they possess short prootic canals for the prootic sinus and its distributaries (Wible 1990). The similarity of the basicranial vasculature reconstructed for these and other Cretaceous metatherians (Rougier et al. 1998), along with observations in extant marsupials such as the opossum *Didelphis virginiana*, suggest that the pattern is diagnostic for Metatheria. However, the optimization of phylogenetic characters across therian phylogeny implies that the vascular pattern may be characteristic of Theria as a whole, which subsequently is reversed in Eutheria (O'Leary et al. 2013; Williamson et al. 2014). Unfortunately, the ear regions of basal metatherians (Szalay and Trofimov, 1996; Luo et al. 2003) and closest therian outgroups (Ji et al. 2002) are poorly preserved as fossils, thereby obscuring our understanding of the polarities of many features of the therian ear.

7.5 Placentalia

Placental mammals contribute over 90% of all extant mammal species. Along with great taxonomic diversity come great morphological and ecological diversity (Nowak 1999). Likewise, the organs of the inner ear vary in absolute size, ranging in volume from less than one cubic millimeter to over one cubic centimeter, and the cochlea coils from around one turn in shrews to over four in guinea pigs (Ekdale 2013). The ear regions of extinct placentals are well described, and the remaining sections of this chapter are provided as a small sampling offered to highlight the array of forms and functions preserved in the fossil record of placental mammals.

7.5.1 Archaic Ungulates ("Condylarthra")

Archaic ungulates (sometimes referred to as condylarths) form a paraphyletic "waste basket" group of placentals that are thought to have given rise to extant hoofed mammals (Prothero et al. 1988; Archibald 1998). Taken together, the various clades of archaic ungulates contributed nearly seventy percent of the known North American mammal species in the early part of the Paleogene (Van Valen 1978; Prothero 1999). One of the most diverse groups of endemic mammals from South America throughout the Cenozoic Era is a clade of archaic ungulate called Notoungulata (Patterson and Pascual 1968; Simpson 1980).

The nature of cochlear coiling in notoungulates suggests that some taxa were adapted for low-frequency to infrasonic hearing (Macrini et al. 2013) as has been determined for living elephants. In contrast, North American archaic ungulates may have been sensitive to much higher frequencies (e.g., ultrasonic ranges above 20 kHz). For example, the Eocene archaic ungulate *Hyopsodus lepidus* may have been sensitive to high frequencies, as indicated by features of the endocranium that relate to specific regions of the brain (Orliac et al. 2012a), although the shape of the cochlea and interaural distance (distance between the ears) do not indicate that the species was particularly specialized for ultrasonic sound reception (Ravel and Orliac 2015). On the other hand, the earliest artiodactyls (even-toed ungulates) were specialized for high-frequency hearing based on cochlear structure (Orliac et al. 2012b). Even if sensitivity to high-frequency vibrations was an ancestral condition for artiodactyls, low-frequency thresholds subsequently were lowered in more recent species such as cows, pigs, and goats (Heffner and Heffner 1983, 1990).

7.5.2 Cetacea (Whales and Dolphins)

The ear region has played an important role in the interpretation of the evolution of whales (Cetacea) due to the robust nature of the cetacean petrosal bone and auditory bulla. In addition, the ear bones of extinct and extant cetaceans hold significant diagnostic power (Geisler and Luo 1996; Ekdale et al. 2011). Features of the ear region were used to support a relationship between cetaceans and a group of archaic ungulates known as mesonychids (Geisler and Luo 1998; Luo and Gingerich 1999), although many of the same morphologies are present in extant artiodactyls such as hippopotamatids (O'Leary 2010). Most recent phylogenetic analyses recover a close relationship between whales and hippos rather than mesonychids, thereby nesting Cetacea within Artiodactyla (Spaulding et al. 2009). One anatomical example uniting whales and terrestrial artiodactyls is an absence of a distinct pit on the endocranial surface of the petrosal known as the subarcuate fossa for the parafloccular lobe of the cerebellum. The fossa is absent in many extinct and extant species of artiodactyls, including whales (O'Leary 2010), but the depression is deep and distinct in early artiodactyls (Fig. 7.9). Thus, absence of a subarcuate fossa is derived in artiodactyls and plesiomorphic for whales. In fact, deep subarcuate fossae are lost in extant members of many artiodactyl clades, including camelids, despite the distinct fossa in early clade members (Fig. 7.9B) (O'Leary 2010). Although there is a direct correlation between the absence of both a parafloccular lobe and subarcuate fossa in elephants (Shoshani et al. 2006; Ekdale 2011), cetaceans lack subarcuate fossae (O'Leary 2010; Ekdale et al. 2011) despite the presence of a well-developed parafloccular lobe (Ries and Langworthy 1937; Voogd and Glickstein 1998). Therefore, absence of the osseous pit does not necessarily indicate absence of the cerebellar structure.

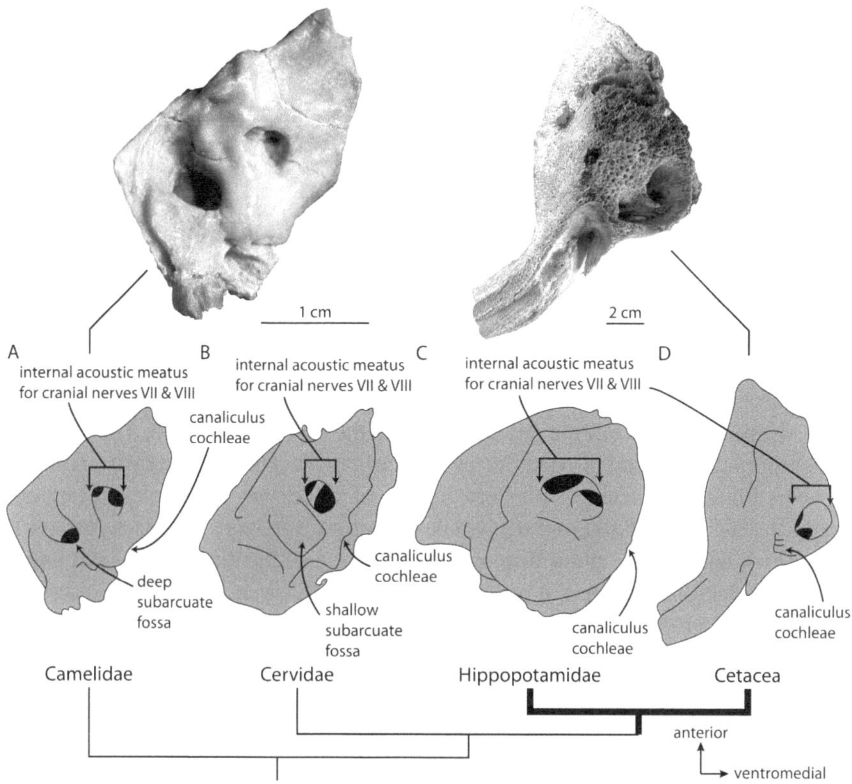

Fig. 7.9 Left petrosal bones of extinct and extant artiodactyls in dorsomedial (cerebellar) view; relationships follow Spaulding et al. (2009). *Thick black branches* indicate loss of the subarcuate fossa. (**A**) Extinct Camelidae (camel) based on *Poebrotherium* sp.. (**B**) extant Cervidae (deer) based on the white-tailed deer, *Odocoileus virginianus*. (**C**) Extant Hippopotamidae (hippopotamuses) based on the pygmy hippopotamus, *Choeropsis liberiensis*. (**D**) Extant Cetacea (whales and dolphins) based on Eden's whale, *Balaenoptera edeni*. (**A** upper photograph reprinted with permission from M. O'Leary, Project 48, morphobank.org; **A**, **B**, **C** line drawings modified from O'Leary 2010; **D** upper photograph modified and reprinted with permission from Ekdale et al. 2011)

Cetaceans also are interesting from an auditory perspective because they occupy both extremes of the overall sensitivity spectrum of mammals. Odontocetes (toothed whales) are sensitive to high-frequency and ultrasonic vibrations, whereas mysticetes (baleen whales) are sensitive to low-frequency and infrasonic vibrations (Ketten 2000). The earliest odontocetes from the Oligocene were capable of biosonar and presumably were sensitive to high-frequency sounds (Geisler et al. 2014), but sensitivity to low-frequency sounds in mysticetes appears to be an ancestral retention from their early whale ancestors (Ekdale and Racicot 2015).

Sound reception in odontocetes occurs via the mandible and soft tissues of the gular region (Norris 1968; Cranford et al. 2008), and reception of high-frequency sound waves has been interpreted in the fossil record through measurement of the dentary (Bianucci and Gingerich 2011) and cranial asymmetry (Fahlke et al. 2011; Geisler et al. 2014). Sound reception in mysticetes has been difficult to model (e.g., Nummela et al. 1999), but recent evidence from CT data suggest that bone conduction is the primary method, with some transmission through associated soft tissues (Cranford and Krysl 2015; Yamato and Pyenson 2015). Hearing through bone conduction has been reconstructed for early whale ancestors as well (Nummela et al. 2004, 2007).

In mysticetes, the auditory bulla is fused to the petrosal by two stout processes. This fusion, and the solid attachment of the petrosal to the skull, facilitates sound reception through bone conduction (Cranford and Krysl 2015). Conversely, the petrosal of odontocetes is completely separate from the basicranium and surrounded by peribullary sinuses (Cranford et al. 2010). Likewise, the odontocete bulla is not firmly attached to the petrosal as it is in mysticetes. In the fossil record, whale ancestors have a mosaic of odontocete and mysticete traits—the petrosal is firmly fixed within the skull, but the articulation with the bulla is not fused. The bulla and petrosal remain unfused in odontocetes and early mysticetes, but extensive fusion of the two bones characterizes the ear region of all extant mysticetes (Marx and Fordyce 2015). The bulla of early whale ancestors was large and pachyosteosclerotic (thick and dense), and it displays an expansion of the dorsomedial rim known as the involucrum, which is a neomorphic structure common to all cetaceans and their closest terrestrial relative *Indohyus* (Thewissen et al. 2007; Uhen 2010). Presence of the involucrum has been cited as evidence for low-frequency sensitivity in early cetaceans (Luo 1998). Subsequent fusion of the bulla and petrosal in extant mysticetes and their closest extinct relatives would have further facilitated low-frequency sound reception via bone conduction.

Within the cetacean inner ear, the cochlea is disproportionately larger than the vestibule and semicircular canals when compared with terrestrial mammals. Although a voluminous cochlea is associated with a lowering of high and low-frequency thresholds in primates (Kirk and Gosselin-Ildari 2009), regression analyses over body size revealed that the vestibule and semicircular canals were reduced during the evolution of early whales, and cochlear size was not increased by auditory pressures (Spoor et al. 2002). The gradual reduction of the semicircular canals is well documented in the cetacean fossil record, and that may be a result of adaptation to a fully aquatic lifestyle (Spoor et al. 2002; however, see Kandel and Hullar 2010 for an opposing view).

Within the cochlea, the auditory nerve ganglion canal is large in echolocating odontocetes, but it is narrower in non-echolocating mysticetes (Fleischer 1976). Likewise, the bony laminae are better developed in odontocetes compared to mysticetes, which corresponds to the physiological differences measured in the two clades of whales (Fleischer 1976). The overall internal structure of the cochlea of the extinct whale ancestor *Zygorhiza kochii* from the Eocene supports the hypothesis that low-frequency sensitivity is the ancestral physiology for whales, and toothed whales are derived in their high-frequency sensitivity (Ekdale and Racicot 2015).

7.5.3 Chiroptera (Bats)

The forelimbs of the Eocene bat *Onychonycteris finneyi* were relatively short, but the phalanges were elongated, and the animal certainly was volant (Simmons et al. 2008). However, each digit bore a claw, which is a unique condition among bats, and *Onychonycteris* may have been proficient at arboreal locomotion as well. Nevertheless, fossil specimens of the oldest bats are preserved with well-developed wings, and so it is clear that bats have been capable of powered flight for their entire evolutionary history (Simmons and Geisler 1998).

One of the most pervasive questions surrounding bat evolution is the relative timing of flight and the acquisition of biosonar. As with whales, only some bats are capable of echolocation and have high-frequency sensitivity, but others are not. Unlike whales, the physiological differences are not concordant with phylogeny (Teeling et al. 2002). The ability to produce and receive high-frequency sounds has been reconstructed for several early bats through gross measurements of the cochlea (Novacek 1987; Simmons and Geisler 1998). In contrast, the cochlea of *Onychonycteris* is small, and in regression analyses it clusters with the cochleae of non-echolocating species (Simmons et al. 2008). Thus, it appears that flight preceded echolocation. However, recent investigations of the positional relationship between the stylohyal and auditory bulla of *Onychonycteris* indicate that the species was capable of producing the same "tongue clicks" used by extant bats during echolocation (Veselka et al. 2010). In particular, the stylohyal contacts the auditory bulla directly in *Onychonycteris* and in all bats that utilize biosonar. Although there is a possibility that *Onychonycteris* was capable of producing echolocation clicks but could not perceive sounds of comparable frequency, that scenario seems unlikely. The holotype of the species is flattened and distorted, which might provide an artificial articulation between the stylohyal and auditory bulla (Veselka et al. 2010), thereby leaving the earliest evidence for the attainment of biosonar in bats unresolved.

7.5.4 Proboscidea (Elephants)

Proboscideans (elephants, mammoths, and mastodons) are well represented in the fossil record, primarily by teeth and cranial elements, but relatively little is known about the ear regions of these giant mammals. Living elephants possess a secondarily undivided perilymphatic foramen in the absence of the canaliculus cochleae and fenestra cochleae (Fischer 1990), reminiscent of the opening in non-cladotherians, (Fig. 7.8), some extant and extinct mysticetes (e.g., gray whales) (Ekdale et al. 2011), and extant sirenians (dugongs and manatees) (Fischer 1990). The secondarily undivided perilymphatic foramen certainly was derived independently in elephants and gray whales, but sirenians are the extant sister-clade of proboscideans, and the foramen is optimized as a synapomorphy for the pair when only extant taxa are considered (Ekdale 2013). However, the undivided perilymphatic foramen was independently derived in sirenians and proboscideans based on a divided foramen in early members of both clades (Fig. 7.10) (Court and Jaeger 1991; Court 1992a).

Living elephants vocalize in low frequencies (Payne et al. 1986; Poole et al. 1988), and their inner ears are equipped to perceive the same range of sounds (Heffner and Heffner 1980). In particular, extinct and extant elephantimorphs lack a secondary bony lamina (Ekdale 2011; Benoit et al. 2013a), which corresponds to low-frequency sensitivity. Closely related sirenians and arsinotheres also lack the secondary bony lamina (Ekdale 2013; Benoit et al. 2013a). The secondary bony lamina is present in the early proboscidean *Numidotherium koholense* (Court 1992b), which implies that low-frequency sensitivity may be a recent acquisition for elephants. However, additional detailed descriptions of the inner ears of extinct proboscideans and allied taxa will shed new light on the evolution of low-frequency hearing sensitivity in Proboscidea.

7.5.5 Human Evolution

As in many mammalian clades, the middle and inner ear regions of extinct and extant hominins (modern humans and all extinct species more closely related to humans than to any other great ape lineage) are phylogenetically informative. For example, the mallei of *Australopithecus africanus* and *Paranthropus robustus* are distinctly human in form, but the incudes and stapedes resemble those of living great apes (Quam et al. 2013). Additionally, the postglenoid and styloid processes of the basicranium, as well as the posterior semicircular canals within the inner ear, are distinctly smaller in *Homo erectus* than in modern humans (Martínez and Arsuaga 1997). The posterior semicircular canal assumes a more ventral position relative to the plane of the lateral canal in Neanderthals than it does in extant *Homo* as well (Hublin et al. 1996; Spoor et al. 2003).

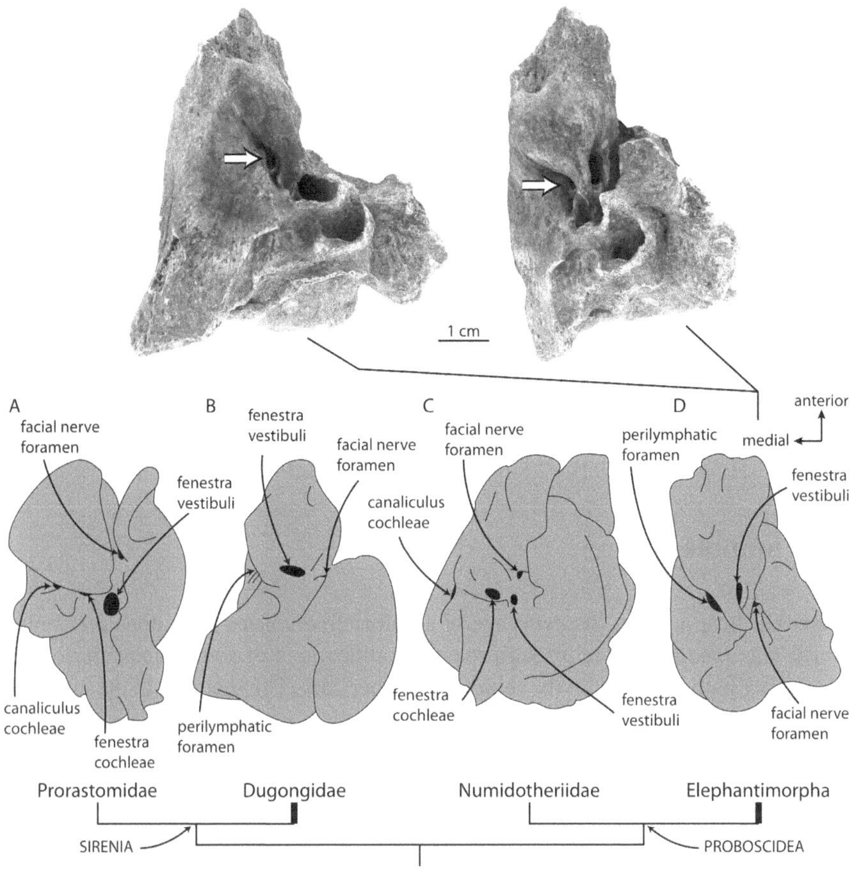

Fig. 7.10 Convergent evolution of the secondarily undivided perilymphatic foramen between Proboscidea (elephants) and Sirenia (manatees). Left petrosals in ventral (tympanic) view; *black branches* indicate coalescence of the canaliculus cochleae and fenestra cochleae; relationships follow Benoit et al (2013a). **(A)** Extinct early sirenian Prorastomidae based on *Prorastomus sirenoides*. **(B)** Extinct derived sirenian based on *Metaxotherium albifontanum*. **(C)** Extinct early proboscidean Numidotheriidae based on *Numidotherium koholense*. **(D)** Extinct derived proboscidean based on unidentified Elephantimorpha, either a mammoth or mastodon. *Left photograph* in medial view and *right photograph* in ventral view, both from TMM 933-1389, (Vertebrate Paleontology Laboratory of the Texas Natural Science Center, Austin, TX); *white arrows* indicate undivided perilymphatic foramen in photographs. (**A** modified from Benoit et al., 2013b; **B** modified from Vélez-Juarbe & Domning 2014; **C** modified from Benoit et al. 2013a; **D** line drawing modified from Ekdale 2011)

Much attention has been paid to the semicircular canals of human ancestors, primarily focusing on the emergence of bipedalism and other locomotor behaviors (e.g., Spoor 2003). However, recent evidence has revealed that changes within the cochlea occurred between humans and our closest non-human ape relatives as well. In particular, hominids (great apes and humans) have longer cochlear canals relative to the number of cochlear turns and larger fenestrae cochleae areas than that predicted for catarrhine primates of similar body size. Early members of *Homo* exhibit a further expansion of the cochlear canal by as much as 30% over earlier australopithecine ancestors, which relates to a downward shift in frequency thresholds within the ear (Braga et al. 2015). The physiological transition is a partial result of increased body size within the hominid lineage, for example, *Homo erectus* is roughly 30% larger in brain size and 10% larger in overall body size than *Australopithecus* species (Antón et al. 2014), but other factors may be in play, including phylogenetic signal and pressures of natural selection. The shift toward lower frequency sensitivity in humans may have coincided with long distance communication in early members of *Homo* (Braga et al. 2015), but the exact interplay among phylogeny, allometry, and other evolutionary pressures associated with the lengthening of the cochlear canal during human evolution are not understood completely at this time.

7.6 Summary

Mammals hear across a greater range of auditory frequencies than other vertebrates, which was made possible through major modifications of sound transmission and processing pathways within the middle and inner ears. The first step in the evolution of the ear of mammals is a combination of the prootic and opisthotic to form a single petrosal bone that fully encapsulates the organs of the inner ear. The development of the petrosal occurred simultaneously with an elongation of the cochlea. Given the tonotopic distribution of the basilar membrane of the cochlea, with low-frequency receptors toward the apex, an elongation of the cochlea across mammalian evolution likely facilitated the expansion of frequency sensitivities in mammals.

The cochlea itself is variably curved among Mesozoic mammals, and extreme curvature of the cochlea coincides with acquisition of major morphogenetic genes within the mammalian genome. A cribriform plate for multiple branches of the cochlear nerve, as well as associated structures including the auditory nerve ganglion canal and primary and secondary bony laminae, further broadened the bandwidth of audible frequencies. Future investigations of the morphology of the inner ear of early mammals, primarily eutriconodonts, gondwanatheres, and spalacotheriids, will help to resolve many issues surrounding the evolution of auditory physiology in living mammals.

Although often considered to be a major diagnostic feature of mammals, the definitive mammalian middle ear, which consists of a complete separation of middle ear elements from the mandible, appears to have evolved multiple times over the course of mammal evolution. As new important fossils are discovered, described, and included in phylogenetic analyses, the accuracy in ascertaining when and how many times the mammalian ear was acquired will increase.

One aspect of evolutionary anatomy that is difficult to address is that of intraspecific variation, in large part due to the nature of the fossil record. Additional work is needed to explore variation within the inner ear, such as the degree and nature of cochlear coiling and the connections between interspecific and intraspecific form-function relationships. The latter may alter understanding of the evolution of the anatomy and physiology of extinct organisms. In addition, further explorations into the frontiers of evolutionary developmental biology and evolutionary genomics will inform us on the ontogenetic history of significant transitions, such as the development of the definitive mammalian middle ear and innervations of the cochlear end organ.

Acknowledgements The author thanks M. O'Leary for access to photographs and M. Van Patten for access to specimens within the San Diego State University Museum of Biodiversity. Development of the chapter was much improved by discussions with A. Ekdale and Z.-X. Luo. Further thanks go to the editors of the volume, J. A. Clack, R. R. Fay, and A. N. Popper.

Compliance with Ethics Requirements Eric G. Ekdale had no conflicts of interest.

References

Allin, E. F. (1975). Evolution of the mammalian middle ear. *Journal of Morphology*, 147, 403–438.

Anthwal, N., Joshi, L., & Tucker, A. S. (2013). Evolution of the mammalian middle ear and jaw: Adaptations and novel structures. *Journal of Anatomy*, 222, 147–160.

Antón, S. C., Potts, R., & Aiello, L. C. (2014). Evolution of early *Homo*: An integrated biological perspective. *Science*, 345, 1236828/1–1236828/13.

Archibald, J. D. (1998). Archaic ungulates ("Condylarthra"). In C. M. Janis, K. M. Scott, & L. L. Jacobs (Eds.), *Terrestrial carnivores, ungulates, and ungulatelike mammals. Evolution of Tertiary mammals of North America, vol. 1* (pp. 292–329). Cambridge (UK): Cambridge University Press.

Benoit, J., Merigeaud, S., & Tabuce, R. (2013a). Homoplasy in the ear region of Tethytheria and the systematic position of Embrithopoda (Mammalia, Afrotheria). *Geobios*, 46, 357–370.

Benoit, J., Adnet, S., Mabrouk, E. E., Khayati, H., et al. (2013b). Cranial remain from Tunisia provides new clues for the origin an evolution of Sirenia (Mammalia, Afrotheria) in Africa. *PLoS ONE*, 8(1), e54307.

Bever, G. S., Rowe, T., Ekdale, E. G., Macrini, T. E., Colbert, M. W., & Balanoff, A. M. (2005). Comment on "Independent origins of middle ear bones in monotremes and therians" (1). *Science*, 309, 1492a.

Bianucci, G., & Gingerich, P. D. (2011). *Aegyptocetus tarfa*, n. gen. et sp. (Mammalia, Cetacea), from the middle Eocene of Egypt: Clinorhynchy, olfaction, and hearing in a protocetid whale. *Journal of Vertebrate Paleontology*, 31, 1173–1188.

Braga, J., Loubes, J.–M., Descouens, D., Dumoncel, J., et al. (2015). Disproportionate cochlear length in genus *Homo* shows a high phylogenetic signal during apes' hearing evolution. *PLoS ONE*, 10(6), e0127780.

Bramble, D. M. (1982). *Scaptochelys*: Generic revision and evolution of gopher tortoises. *Copeia*, 1982, 852–867.

Chow, M., & Qi, T. (1978). Paleocene mammalian fossils from Nomogen Formation of inner Mongolia. *Vertebrata PalAsiatica*, 16, 77–85.
Cifelli, R. L. (1982). The petrosal structure of *Hyopsodus* with respect to that of some other ungulates, and its phylogenetic implications. *Journal of Paleontology*, 56, 795–805.
Court, N. (1992a). The skull of *Arsinotherium* (Mammalia, Embrithopoda) and the higher order interrelationships of ungulates. *Palaeovertebrata* 22, 1–43.
Court, N. (1992b). Cochlea anatomy of *Numidotherium koholense*: Auditory acuity in the oldest known proboscidean. *Lethaia*, 25, 211–215.
Court, N., & Jaeger, J.–J. (1991). Anatomy of the periotic bone in the Eocene proboscidean *Numidotherium koholense*: An example of parallel evolution in the inner ear of tethytheres. *Compte Rendu de l'Académie des Sciences*, 312, 559–565.
Cranford, T. W., & Krysl, P. (2015). Fin whale sound reception mechanisms: Skull vibration enables low-frequency hearing. *PLoS ONE*, 10(1), e0116222.
Cranford, T. W., Krysl, P., & Amundin, M. (2010). A new acoustic portal into the odontocete ear and vibrational analysis of the tympanoperiotic complex. *PLoS ONE*, 5(8), e11927.
Cranford, T. W., McKenna, M. C., Soldevilla, M. S., Wiggins, S. M., et al. (2008). Anatomic geometry of sound transmission and reception in Cuvier's beaked whale (*Ziphius cavirostris*). *Anatomical Record*, 291, 353–378.
Echteler, S. M., Fay, R. R., & Popper, A. N. (1994). Structure of the mammalian cochlea. In R. R. Fay & A. N. Popper (Eds.), *Comparative hearing: Mammals* (pp. 134–171). New York: Springer-Verlag.
Ekdale, E. G. (2011). Morphological variation in the ear region of Pleistocene Elephantimorpha (Mammalia, Proboscidea) from central Texas. *Journal of Morphology*, 272, 452–464.
Ekdale, E. G. (2013). Comparative anatomy of the bony labyrinth (inner ear) of placental mammals. *PLoS ONE*, 8(6), e66624.
Ekdale, E. G. (2015). Form and function of the mammalian inner ear. *Journal of Anatomy*, doi:10.1111/joa.12308.
Ekdale, E. G., & Rowe, T. (2011). Morphology and variation within the bony labyrinth of zhelestids (Mammalia, Eutheria) and other therian mammals. *Journal of Vertebrate Paleontology*, 31, 658–675.
Ekdale, E. G., & Racicot, R. A. (2015). Anatomical evidence for low frequency sensitivity in an archaeocete whale: Comparison of the inner ear of *Zygorhiza kochii* with that of crown Mysticeti. *Journal of Anatomy*, 226, 22–39.
Ekdale, E. G., Archibald, J. D., & Averianov, A. O. (2004). Petrosal bones of placental mammals from the Late Cretaceous of Uzbekistan. *Acta Palaeontologica Polonica*, 49, 161–176.
Ekdale, E. G., Berta, A., & Deméré, T. A. (2011). The comparative osteology of the petrotympanic complex (ear region) of extant baleen whales (Cetacea: Mysticeti). *PLoS ONE*, 6(6), e21311.
Fahlke, J., Gingerich, P. D., Welsh, R. C., & Wood, A. R. (2011). Cranial asymmetry in Eocene archaeocete whales and the evolution of directional hearing in water. *Proceedings of the National Academy of Sciences of the USA*, 108, 14545–14548.
Fay, R. R. (1988). *Hearing in vertebrates: A psychophysics databook*. Winnetka (IL): Hill-Fay Associates.
Fischer, M. (1990). Un trait unique de l'oreille des elephants et des Siréniens (Mammalia): un paradoxe phylogénétique. *Compte Rendus de l'Académie des Sciences, Paris*, 311, 157–162.
Fleischer, G. (1973). Studien am Skellet des Gehörorgans der Säugetiere, einschließlich des Menschen. *Säugtierkundliche Mitteilungen*, 21, 131–239.
Fleischer, G. (1976). Hearing in extinct cetaceans as determined by cochlear structure. *Journal of Paleontology*, 50, 133–152.
Fox, R. C., & Meng, J. (1997). An X–radiographic and SEM study of the osseous inner ear of multituberculates and monotremes (Mammalia): Impressions for mammalian phylogeny and evolution of hearing. *Zoological Journal of the Linnean Society*, 121, 249–291.
Geisler, J. H., & Luo, Z.–X. (1996). The petrosal and inner ear of *Herpetocetus* sp. (Mammalia: Cetacea) and their implications for the phylogeny and hearing of archaic mysticetes. *Journal of Paleontology*, 70, 1045–1066.

Geisler, J. H., & Luo, Z.–X. (1998). Relationships of Cetacea to terrestrial ungulates and the evolution of cranial vasculature in Cete. In J. G. M. Thewissen (Ed.), *The emergence of whales* (pp. 163–212). New York: Plenum Press.

Geisler, J. H., Colbert, M. W., & Carew, J. L. (2014). A new fossil species supports an early origin for toothed whale echolocation. *Nature*, 508, 383–386.

Gerkema, M. P., Davies, W. I. L., Foster, R. G., Menaker, M., & Hut, R. A. (2013). The nocturnal bottleneck and the evolution of activity patterns in mammals. *Proceedings of the Royal Society of London B: Biological Sciences*, 280, 20130508.

Goodrich, E. S. (1930). *Studies on the structure and development of vertebrates*. London: Macmillan.

Gould, S. J. (1990). An earful of jaw. *Natural History*, 99(3), 12–23.

Gray, A. A. (1908). *The labyrinth of animals, vol. 2*. London: J. and A. Churchill.

Gray, H. (1977). The organs of special sense. In Pick, T. P., & Howden, R. (Eds.), *Anatomy* (pp. 819–868). New York: Portland House.

Graybeal, A., Rosowski, J. J., Ketten, D. R., & Crompton, A. W. (1989). Inner-ear structure in *Morganucodon*, an early Jurassic mammal. *Zoological Journal of the Linnean Society*, 96, 107–117.

Heffner, R. S., & Heffner, H. E. (1980). Hearing in the elephant (*Elephas maximus*). *Science*, 208, 518–520.

Heffner, R. S., & Heffner, H. E. (1983). Hearing in large mammals: Horses (*Equus caballus*) and cattle (*Bos taurus*). *Behavioral Neuroscience*, 97, 299–309.

Heffner, R. S., & Heffner, H. E. (1990). Hearing in domestic pigs (*Sus scrofa*) and goats (*Capra hircus*). *Hearing Research*, 48, 231–240.

Heffner, R. S., Koay, G., & Heffner, H. E. (2013). Hearing in American leaf–nosed bats. IV. The common vampire bat, *Desmodus rotundus*. *Hearing Research*, 296, 42–50.

Hoffmann, S., O'Connor, P. M., Kirk, E. C., Wible, J. R., & Krause, D. W. (2014). Endocranial and inner ear morphology of *Vintana sertichi* (Mammalia, Gondwanatheria) from the Late Cretaceous of Madagascar. *Society of Vertebrate Paleontology Memoir 14. Journal of Vertebrate Paleontology*, 34 (Supplement to 6), 110–136.

Hu, Y., Wang, Y., Luo, Z.–X., & Li, C. (1997). A new symmetrodont mammal from China and its implications for mammalian evolution. *Nature*, 390, 137–142.

Hublin, J.–J., Spoor, F., Braun, M., Zonneveld, F., & Condemi, S. (1996). A late Neanderthal associated with Upper Palaeolithic artifacts. *Nature*, 381, 224–226.

Hughes, E. M., Wible, J. R., Spaulding, M., & Luo, Z.–X. (2015). Mammalian petrosal from the Upper Jurassic Morrison Formation of Fruita, Colorado. *Annals of Carnegie Museum*, 83, 1–17.

Hurum, J. H. (1998). The inner ear of two Late Cretaceous multituberculate mammals, and its implications for multituberculate hearing. *Journal of Mammalian Evolution*, 5, 65–93.

Jenkins, F. A., Jr. (1990). Monotremes and the biology of Mesozoic mammals. *Netherlands Journal of Zoology*, 40, 5–31.

Ji, Q., Luo, Z.–X., Yuan, C.–X., Wible, J. R., Zhang, J.–P., & Georgi, J. A. (2002). The earliest known eutherian mammal. *Nature*, 416, 816–822.

Ji, Q., Luo, Z.–X., Zhang, & Ji, S.–A. (1999). A Chinese triconodont mammal and mosaic evolution of the mammalian skeleton. *Nature*, 398, 326–330.

Ji, Q., Luo, Z.–X., Zhang, X., Yuan, C.–X., & Xu, L. (2009). Evolutionary development of the middle ear in Mesozoic therian mammals. *Science*, 326, 278–281.

Kandel, B. M., & Hullar, T. E. (2010). The relationship of head movements to semicircular canal size in cetaceans. *The Journal of Experimental Biology*, 213, 1175–1181.

Kermack, K. A., Mussett, F., & Rigney, H. W. (1981). The skull of *Morganucodon*. *Zoological Journal of the Linnean Society*, 71, 1–158.

Ketten, D. R. (1997). Structure and function in whale ears. *Bioacoustics*, 8, 103–135.

Ketten, D. R. (2000). Cetacean ears. In W. W. L. Au, A. N. Popper, & R. R. Fay (Eds.), *Hearing by whales and dolphins* (pp. 43–108). New York: Springer-Verlag.

Ketten, D. R., & Wartzok, D. (1990). Three-dimensional reconstructions of the dolphin ear. In J. Thomas & R. Kastelein (Eds.), *Sensory abilities of cetaceans* (pp. 81–105). New York: Plenum Press.

Khorevin, V. I. (2008). The lagena (the third otolith end organ in vertebrates). *Neurophysiology*, 40, 142–159.

Kielan–Jaworowska, Z. (1981). Evolution of the therian mammals in the Late Cretaceous of Asia. Part IV. Skull structure in *Kennalestes* and *Asioryctes*. *Palaeontologia Polonica*, 42, 25–78.

Kirk, E. C., & Gosselin-Ildari, A. D. (2009). Cochlear labyrinth volume and hearing abilities in primates. *Anatomical Record*, 292, 765–776.

Kirk, E. C., Hoffmann, S., Kemp, A. D., Krause, D. W., & O'Connor, P. M. (2014). Sensory anatomy and sensory ecology of *Vintana sertichi* (Mammalia, Gondwanatheria) from the Late Cretaceous of Madagascar. *Society of Vertebrate Paleontology Memoir 14. Journal of Vertebrate Paleontology*, 34 (Supplement to 6), 203–222.

Krause, D. W., Wible, J. R., Hoffmann, S., Groenke, J. R., et al. (2014). Craniofacial morphology of *Vintana sertichi* (Mammalia, Gondwanatheria) from the Late Cretaceous of Madagascar. *Society of Vertebrate Paleontology Memoir 14. Journal of Vertebrate Paleontology*, 34 (Supplement to 6), 14–109.

Krubitzer, L. (1998). What can monotremes tell us about brain evolution? *Philosophical Transactions of the Royal Society of London B: Biological Sciences*, 353, 1127–1146.

Ladevèze, S. (2007). Petrosal bones of metatherian mammals from the Late Palaeocene of Itaboraí (Brazil), and a cladistic analysis of petrosal features in metatherians. *Zoological Journal of the Linnean Society*, 150, 85–115.

Ladevèze, S., de Muizon, C., Colbert, M., & Smith, T. (2010). 3D computational imaging of the petrosal of a new multituberculate mammal from the Late Cretaceous of China and its paleobiologic inferences. *Comptes Rendus Palevol*, 9, 319–330.

Ladhams, A., & Pickles, J. O. (1996). Morphology of the monotreme organ of Corti and macula lagena. *Journal of Comparative Neurology*, 366, 335–347.

Liberman, M. C. (1982). The cochlear frequency map for the cat: Labeling auditory-nerve fibers of known characteristic frequency. *The Journal of the Acoustic Society of America*, 72, 1441–1449.

Luo, Z.–X. (1998). Homology and transformation of cetacean ectotympanic structures. In J. G. M. Thewissen (Ed.), *The emergence of whales* (pp. 269–301). New York: Plenum Press.

Luo, Z.–X. (2011). Developmental patterns in Mesozoic evolution of mammal ears. *Annual Review of Ecology, Evolution, and Systematics*, 42, 355–380.

Luo, Z.–X., & Ketten, D. R. (1991). CT scanning and computerized reconstructions of the inner ear of multituberculate mammals. *Journal of Vertebrate Paleontology*, 11, 220–228.

Luo, Z.–X., Crompton, A. W., & Lucas, S. G. (1995). Evolutionary origins of the mammalian promontorium and cochlea. *Journal of Vertebrate Paleontology*, 15, 113–121.

Luo, Z.–X., & Gingerich, P. D. (1999). Terrestrial Mesonychia to aquatic Cetacea: Transformation of the basicranium and evolution of hearing in whales. *University of Michigan Papers on Paleontology*, 31, 1–98.

Luo, Z.–X., Ji, Q., Wible, J. R., & Yuan, C.–X. (2003). An Early Cretaceous tribosphenic mammal and metatherian evolution. *Science*, 302, 1934–1940.

Luo, Z.–X., Chen, P., Li, Gang, & Chen, M. (2007). A new eutriconodont mammal and evolutionary development in early mammals. *Nature*, 446, 288–293.

Luo, Z.–X., Ruf, I., Schultz, J. A., & Martin, T. (2011). Fossil evidence on evolution of inner ear cochlea in Jurassic mammals. *Proceedings of the Royal Society of London B: Biological Sciences*, 278, 28–34.

Luo, Z.–X., Ruf, I., & Martin, T. (2012). The petrosal and inner ear of the Late Jurassic cladotherian mammal *Dryolestes leiriensis* and implications for ear evolution in therian mammals. *Zoological Journal of the Linnean Society*, 166, 433–463.

MacIntyre, G. T. (1972). The trisulcate petrosal pattern of mammals. In T. Dobzhansky, M. K. Hecht, & W. C. Steere (Eds.), *Evolutionary biology, vol. 6* (pp. 275–303). New York: Appleton-Century-Crofts.

MacPhee, R. D. E. (1981). Auditory regions of primates and eutherian insectivores: Morphology, ontogeny, and character analysis. In F. S. Szalay (Ed.), *Contributions to primatology* (pp. 1–282). Basel: S. Karger.

Macrini, T. E., Flynn, J. J., Ni, X., Croft, D. A., & Wyss, A. R. (2013). Comparative study of notoungulate (Placentalia, Mammalia) bony labyrinths and new phylogenetically informative inner ear characters. *Journal of Anatomy*, 223, 442–461.

Manley, G. A. (1972). A review of some current concepts of the functional evolution of the ear in terrestrial vertebrates. *Evolution*, 26, 608–621.

Manley, G. A. (2000). Cochlear mechanisms from a phylogenetic viewpoint. *Proceedings of the National Academy of Sciences of the USA*, 97, 11736–11743.

Manoussaki, D., Chadwick, R. S., Ketten, D. R., Arruda, J., Dimitriadis, E. K., & O'Malley, J. T. (2008). The influence of cochlear shape on low-frequency hearing. *Proceedings of the National Academy of Sciences of the USA*, 105, 6162–6166.

Martínez, I., & Arsuaga, J. L. (1997). The temporal bones from Sima de los Huesos Middle Pleistocene site (Sierra de Atapuerca, Spain). A phylogenetic approach. *Journal of Human Evolution*, 33, 283–318.

Marx, F. G., & Fordyce, R. E. (2015). Baleen boom and bust: A synthesis of mysticete phylogeny, diversity and disparity. *Royal Society Open Science*, 2, 140434.

Meng, J., & Wyss, A. (1995). Monotreme affinities and low-frequency hearing suggested by multituberculate ear. *Nature*, 377, 141–144.

Meng, J., Wang, Y., & Li, C. (2011). Transitional mammalian middle ear from a new Cretaceous Jehol eutriconodont. *Nature*, 472, 181–185.

Meng, Q.–J., Ji, Q., Zhang, Y.–G., Liu, D., et al. (2015). An arboreal docodont from the Jurassic and mammaliaform ecological diversification. *Science*, 347, 764–768.

Miao, D. (1988). Skull morphology of *Lambdopsalis bulla* (Mammalia, Multituberculata) and its implication to mammalian evolution. *Contributions to Geology, The University of Wyoming*, 4, 1–104.

Miao, D., & Lillegraven, J. A. (1986). Discovery of three ear ossicles in a multituberculate mammal. *National Geographic Research*, 2, 500–507.

Mills, D. M., & Shepherd, R. K. (2001). Distortion product otoacoustic emission and auditory brainstem responses in the echidna (*Tachyglossus aculeatus*). *Journal of the Association for Research in Otolaryngology*, 2, 130–146.

Musser, A. M., & Archer M. (1998). New information about the skull and dentary of the Miocene platypus *Obdurodon dicksoni*, and a discussion of ornithorhynchid relationships. *Philosophical Transactions of the Royal Society of London B: Biological Sciences,* 353, 1063–1079.

Müller, M. (1996). The cochlear place-frequency map of the adult developing Mongolian gerbil. *Hearing Research*, 94, 148–156.

Nachtigall, P. E., Mooney, T. A., Taylor, K. A., & Yuen, M. M. L. (2007). Hearing and auditory evoked potential methods applied to odontocete cetaceans. *Aquatic Mammals*, 33, 6–13.

Naidu, R. C., & Mountain, D. C. (2007). Basilar membrane tension calculations for the gerbil cochlea. *Journal of the Acoustical Society of America*, 121, 994–1002.

Norris, C. A. (1993). Changes in the composition of the auditory bulla in southern Solomon Islands populations of the grey cuscus, *Phalanger orientalis breviceps* (Marsupialia, Phalangeridae). *Zoological Journal of the Linnean Society*, 1077, 93–106.

Norris, K. S. (1968). The evolution of acoustic mechanisms in odontocete cetaceans. In E. T. Drake (Ed.), *Evolution and environment* (pp. 297–324). New Haven: Yale University Press.

Novacek, M. J. (1977). Aspects of the problem of variation, origin, and evolution of the eutherian auditory bulla. *Mammal Review*, 7, 131–149.

Novacek, M. J. (1987). Auditory features and affinities of the Eocene bats *Icaronycteris* and *Palaeochiropteryx* (Microchiroptera, incertae sedis). *American Museum Novitates*, 2877, 1–18.

Novacek, M. J. (1993). Patterns and diversity in the mammalian skull. In J. Hanken & B. K. Hall (Eds.), *Patterns of structural and systematic diversity. The skull, vol. 2* (pp. 438–545). Chicago: The University of Chicago Press.

Novacek, M. J., & Wyss, A. (1986). Origin and transformation of the mammalian stapes. *Contributions to Geology, University of Wyoming Special Paper*, 3, 35–53.
Nowak, R. M. (1999). *Walker's mammals of the world*. Baltimore (MD): Johns Hopkins University Press.
Nummela, S., Wäger, T., Hemilä, S., & Reuter, T. (1999). Scaling of the cetacean middle ear. *Hearing Research*, 133, 71–81.
Nummela, S., Thewissen, J. G. M., Bajpai, S., Hussain, S. T., & Kumar, K. (2004). Eocene evolution of whale hearing. *Nature*, 430, 776–778.
Nummela, S., Thewissen, J. G. M., Bajpai, S., Hussain, T., & Kumar, K. (2007). Sound transmission in archaic and modern whales: Anatomical adaptations for underwater hearing. *Anatomical Record*, 290, 716–733.
O'Leary, M. A. (2010). An anatomical and phylogenetic study of the osteology of the petrosal of extant and extinct artiodactylans (Mammalia) and relatives. *Bulletin of the American Museum of Natural History*, 335, 1–206.
O'Leary, M. A., Bloch, J. I., Flynn, J. J., Gaudin, T. J., et al. (2013). The placental mammal ancestor and the post–K–Pg radiation of placentals. *Science*, 339, 662–667.
Olori, J. C. (2010). Digital endocasts of the cranial cavity and osseous labyrinth of the burrowing snake *Uropeltis woodmasoni* (Alethinophidia: Uropeltidae). *Copeia*, 2010, 14–26.
Orliac, M. J., Argot, C., & Gilissen, E. (2012a). Digital cranial endocast of *Hyopsodus* (Mammalia, "Condylarthra"): A case of Paleogene terrestrial echolocation? *PLoS ONE*, 7(2), e30000.
Orliac, M. J., Benoit, J., & O'Leary, M. A. (2012b). The inner ear of *Diacodexis*, the oldest artiodactyl mammal. *Journal of Anatomy*, 221, 417–426.
Ostrander, G. E. (1984). The early Oligocene (Chadronian) Raben Ranch Local Fauna, northwest Nebraska: Multituberculata, with comments on the extinction of the Allotheria. *Transactions of the Nebraska Academy of Science*, 7, 71–80.
Parks, S. E., Ketten, D. R., O'Malley, J. T., & Arruda, J. (2007). Anatomical predictions of hearing in the North Atlantic right whale. *Anatomical Record*, 290, 734–744.
Patterson, B., & Pascual, R. (1968). The fossil mammal fauna of South America. *Quarterly Review of Biology*, 43, 409–451.
Payne, K. B., Langbauer, W. R., Jr., & Thomas, E. M. (1986). Infrasonic calls of the Asian elephant (*Elephas maximus*). *Behavioral Ecology and Sociobiology*, 18, 297–301.
Poole, J. H., Payne, K., Langbauer, W. R., Jr., & Moss, C. J. (1988). The social contexts of some very low frequency calls of African elephants. *Behavioral Ecology and Sociobiology*, 22, 385–392.
Pritchard, U. (1881). The cochlea of the *Ornithorhynchus platypus* compared with that of ordinary mammals and of birds. *Philosophical Transactions of the Royal Society of London B: Biological Sciences*, 172, 267–282.
Prothero, D. R. (1999). Ungulates. In R. Singer (Ed.), *Encyclopedia of paleontology* (pp. 1303–1305). Chicago: Fitzroy Dearborn Publishers.
Prothero, D. R., Manning, E. M., & Fisher, M. (1988). The phylogeny of the ungulates. In M. J. Benton (Ed.), *Mammals. The phylogeny and classification of the tetrapods, vol. 2* (pp. 201–234). Oxford: Clarendon Press.
Quam, R. M., de Ruiter, D. J., Masali, M., Arsuaga, J.–L., et al. (2013). Early hominin auditory ossicles from South Africa. *Proceedings of the National Academy of Sciences of the USA*, 110, 8847–8851.
Ramprashad, F., Landolt, J. P., Money, K. E., Clark, D., & Laufer, J. (1979). A morphometric study of the cochlea of the little brown bat (*Myotis lucifugus*). *Journal of Morphology*, 160, 345–358.
Ravel, A., & Orliac, M. J. (2015) The inner ear morphology of the 'condylarthran' *Hyopsodus lepidus*. *Historical Biology*, 27, 957–969.
Reichert, C. (1837). Ueber die Visceralbogen der Wirbelthiere im Allgemeinen und deren Metamorphosen bei den Vögeln und Säugethieren. *Archiv für Anatomie, Physiologie, und wissenschafliche Medizin, Leipzig*, 1837, 120–222.

Rich, T. H., Hopson, J. A., Musser, A. M., Flannery, T. F., & Vickers-Rich, P. (2005). Independent origins of middle ear bones in monotremes and therians. *Science*, 307, 910–914.
Ries, F. A., & Langworthy, O. R. (1937). A study of the surface structure of the brain of the whale (*Balaenoptera physalus* and *Physeter catodon*). *Journal of Comparative Neurology*, 68, 1–36.
Rodrigues, P. G., Ruf, I., & Schultz, C. L. (2013). Digital reconstruction of the otic region and inner ear of the non-mammalian cynodont *Brasilitherium riograndensis* (Late Triassic, Brazil) and its relevance to the evolution of the mammalian ear. *Journal of Mammalian Evolution*, 20, 291–307.
Rougier, G. W., Wible, J. R., & Hopson, J. A. (1996). Basicranial anatomy of *Priacodon fruitaensis* (Triconodontidae, Mammalia) from the Late Jurassic of Colorado, and a reappraisal of mammaliaform interrelationships. *American Museum Novitates*, 3183, 1–38.
Rougier, G. W., Wible, J. R., & Novacek, M. J. (1998). Implications of *Deltatheridium* specimens for early marsupial history. *Nature*, 396, 459–463.
Rougier, G. W., Forasiepi, A. M., & Martinelli, A. G. (2005). Comment on "Independent origins of middle ear bones in monotremes and therians (II). *Science*, 309, 1492b.
Rowe, T. (1988). Definition, diagnosis, and origin of Mammalia. *Journal of Vertebrate Paleontology*, 8, 241–264.
Rowe, T., Rich, T. H., Vickers-Rich, P., Springer, M., & Woodburne, M. O. (2008). The oldest platypus and its bearing on divergence timing of the platypus and echidna clades. *Proceedings of the National Academy of Sciences of the USA*, 105, 1238–1242.
Ruf, I., Luo, Z.-X., Wible, J. R., & Martin, T. (2009). Petrosal anatomy and inner ear structures of the Late Jurassic *Henkelotherium* (Mammalia, Cladotheria, Dryolestoidea): Insight into the early evolution of the ear region in cladotherian mammals. *Journal of Anatomy*, 214, 679–693.
Ruf, I., Luo, Z.-X., & Martin, T. (2013). Reinvestigation of the basicranium of *Haldanodon exspectatus* (Mammaliaformes, Docodonta). *Journal of Vertebrate Paleontology*, 33, 382–400.
Segall, W. (1970). Morphological parallelisms of the bulla and auditory ossicles in some insectivores and marsupials. *Fieldiana Zoology*, 51, 169–205.
Shoshani, J., Kupsky, W. J., & Marchant, G. H. (2006). Elephant brain part I: Gross morphology, functions, comparative anatomy, and evolution. *Brain Research Bulletin*, 70, 124–157.
Simmons, N. B., & Geisler, J. H. (1998). Phylogenetic relationships of *Icaronycteris*, *Archaeonycteris*, *Hassianycteris*, and *Palaeochiropteryx* to extant bat lineages, with comments on the evolution of echolocation and foraging strategies in Microchiroptera. *Bulletin of the American Museum of Natural History*, 235, 1–181.
Simmons, N. B., Seymour, K. L., Habersetzer, J., & Gunnell, G. F. (2008). Primitive early Eocene bat from Wyoming and the evolution of flight and echolocation. *Nature*, 451, 818–821.
Simpson, G. G. (1937). Skull structure of the Multituberculata. *Bulletin of the American Museum of Natural History*, 73, 727–763.
Simpson, G. G. (1980). *Splendid isolation: The curious history of South American mammals*. New Haven: Yale University Press.
Spaulding, M., O'Leary, M. A., & Gatesy, J. (2009). Relationships of Cetacea (Artiodactyla) among mammals: Increased taxon sampling alters interpretations of key fossils and character evolution. *PLoS ONE*, 4(9), e7062.
Spoor, F. (2003). The semicircular canal system and locomotor behaviour, with special reference to hominin evolution. *Courier Forschungsinstitut Senckenberg*, 243, 93–104.
Spoor, F., Bajpai, S., Hussain, S. T., Kumar, K., & Thewissen, J. G. M. (2002). Vestibular evidence for the evolution of aquatic behaviour in early cetaceans. *Nature*, 417, 163–166.
Spoor, F., Hublin, J.-J., Braum, M., & Zonneveld, F. (2003). The bony labyrinth of Neanderthals. *Journal of Human Evolution*, 44, 141–165.
Szalay, F. S., & Trofimov, B. A. (1996). The Mongolian Late Cretaceous *Asiatherium*, and the early phylogeny and paleobiogeography of Metatheria. *Journal of Vertebrate Paleontology*, 16, 474–509.
Teeling, E. C., Madsen, O., Van Den Bussche, R. A., de Jong, W. W., Stanhope, M. J., & Springer, M. S. (2002). Microbat paraphyly and the convergent evolution of key innovation in

Old World rhinolophoid microbats. *Proceedings of the National Academy of Sciences of the USA*, 99, 1431–1436.
Thewissen, J. G. M., Cooper, L. N., Clementz, M. T., Bajpai, S., & Tiwari, B. N. (2007). Whales originated from aquatic artiodactyls in the Eocene epoch of India. *Nature*, 450, 1190–1195.
Turkewitsch, B. G. (1934). Zur Anatomie des Gehörorgans der Vögel (Canales semicirculares). *Zeitschrift für Anatomie und Entwicklungsgeschischte*, 103, 551–608.
Uhen, M. D. (2010). The origin(s) of whales. *Annual Review of Earth and Planetary Science*, 38, 189–219.
van der Klaauw, C. J. (1931). The auditory bulla in some fossil mammals, with a general introduction to this region of the skull. *Bulletin of the American Museum of Natural History*, 62, 1–352.
Van Valen, L. (1978). The beginning of the age of mammals. *Evolutionary Theory*, 4, 45–80.
Vélez–Juarbe, J., & Domning, D. P. (2014). Fossil Sirenia of the west Atlantic and Caribbean region. IX. *Metaxytherium albifontanum*, sp. nov. *Journal of Vertebrate Paleontology*, 32, 444–464.
Veselka, N., McErlain, D. D., Holdsworth, D. W., Eger, J. L., et al. (2010). A bony connection signals laryngeal echolocation in bats. *Nature*, 463, 939–942.
Voogd, J., & Glickstein, M. (1998). The anatomy of the cerebellum. *Trends in Neuroscience*, 21, 370–375.
Webster, D. B., & Webster, M. (1980). Morphological adaptations of the ear in the rodent family Heteromyidae. *American Zoologist*, 20, 247–254.
West, C. D. (1985). The relationship of the spiral turns of the cochlea and the length of the basilar membrane to the range of audible frequencies in ground dwelling mammals. *The Journal of the Acoustical Society of America*, 77, 1091–1101.
Wever, E. G., McCormick, J. G., Palin, J., & Ridgway, S. H. (1971). Cochlea of the dolphin, *Tursiops truncatus*: The basilar membrane. *Proceedings of the National Academy of Sciences of the USA*, 11, 2708–2711.
Wible, J. R. (1986). Transformations in the extracranial course of the internal carotid artery in mammalian phylogeny. *Journal of Vertebrate Paleontology*, 6, 313–325.
Wible, J. R. (1990). Petrosals of Late Cretaceous marsupials from North America and a cladistic analysis of the petrosal in therian mammals. *Journal of Vertebrate Paleontology*, 10, 183–205.
Wible, J. R., Novacek, M. J., Rougier, G. W., & Asher, R. J. (2009). The eutherian mammal *Maelestes gobiensis* from the Late Cretaceous of Mongolia and the phylogeny of Cretaceous Eutheria. *Bulletin of the American Museum of Natural History*, 327, 1–123.
Williamson, T. E., Brusatte, S. L., & Wilson, G. P. (2014). The origin and early evolution of metatherian mammals: The Cretaceous record. *ZooKeys*, 465, 1–76.
Yamato, M., & Pyenson, N. D. (2015). Early development and orientation of the acoustic funnel provides insight into the evolution of sound reception pathways in cetaceans. *PLoS ONE*, 10 (3), e0118582.
Yuan, C.–X., Ji, Q., Meng, Q.–J., Tabrum, A. R., & Luo, Z.–X. (2013). Earliest evolution of multituberculate mammals revealed by a new Jurassic fossil. *Science*, 341, 779–783.
Zeller, U. (1985). The morphogenesis of the fenestra rotunda in mammals. *Fortschritte der Zoologie*, 30, 153–157.
Zeller, U. (1987). Morphogenesis of the mammalian skull with special reference to *Tupaia*. In H.–J. Kuhn, & U. Zeller (Eds.), *Morphogenesis of the mammalian skull* (pp. 54–98). Hamburg: Verlag Paul Parey.
Zeller, U. (1993). Ontogenetic evidence for cranial homologies in monotremes and therians, with special reference to *Ornithorhynchus*. In F. S. Szalay, M. J. Novacek, & M. C. McKenna (Eds.), *Mammal phylogeny: Mesozoic differentiation, multituberculates, monotremes, early therians, and marsupials* (pp. 95–107). New York: Springer-Verlag.

Chapter 8
Basal Reptilians, Marine Diapsids, and Turtles: The Flowering of Reptile Diversity

Gabriela Sobral, Robert Reisz, James M. Neenan, Johannes Müller, and Torsten M. Scheyer

Abstract Reptiles form the most diverse clade of living land vertebrates. They include lizards, snakes, crocodiles, birds, and turtles, as well as many fossil groups. In this chapter we revise the otic anatomy of early reptilians, including some aquatic groups and turtles. Basal members possessed a stout stapes that still retained its ancestral bracing function, and they lacked a tympanic membrane. The acquisition of tympanic hearing did not happen until later in the evolution of the clade and occurred independently in both parareptiles and diapsids. Parareptiles also show additional otic modifications that are convergent with much later reptilians, which are potentially related to the evolution of more terrestrial habits. In contrast, in aquatic reptiles, such as ichthyosaurs, thalattosaurs, and sauropterygians, the otic anatomy and hearing capacities are adapted to an aquatic medium, resulting in many convergences in their otic anatomy. In turtles, however, there are differences in the configuration and morphology among Triassic and modern taxa.

G. Sobral (✉)
Departamento de Ecologia e Zoologia, Universidade Federal de Santa Catarina, Campus Universitário s/n, Florianópolis 88040-970, Brazil
e-mail: gabisobral@gmail.com

G. Sobral · J. Müller
Museum für Naturkunde Berlin, Invalidenstraße 43, 10115 Berlin, Germany
e-mail: Johannes.Mueller@mfn-berlin.de

R. Reisz
University of Toronto at Mississauga, 3359 Mississauga Road, Mississauga, ON L5L 1C6, Canada
e-mail: robert.reisz@utoronto.ca

J.M. Neenan
Oxford University Museum of Natural History, Parks Road, Oxford OX1 3PW, UK
e-mail: james.neenan@oum.ox.ac.uk

T.M. Scheyer
Paläontologisches Institut und Museum der Universität Zürich, Karl Schmid-Strasse 4, 8006 Zurich, Switzerland
e-mail: tscheyer@pim.uzh.ch

J.A. Clack et al. (eds.), *Evolution of the Vertebrate Ear—Evidence from the Fossil Record*, Springer Handbook of Auditory Research 59,
DOI 10.1007/978-3-319-46661-3_8

Keywords Aquatic adaptations · Diapsida · Marine reptiles · Otic region · Parareptilia · Reptilia · Testudines · Tympanic hearing

8.1 Introduction

Reptilia forms the largest and most diverse group of tetrapods today. In our current understanding of amniote phylogeny, Reptilia is the sister-taxon of Synapsida, to which mammals belong (Kemp, Chap. 5; Luo, Schultz, and Ekdale, Chap. 6; and Ekdale, Chap. 7), and includes living lizards, snakes, turtles, crocodilians and birds, as well as a number of extinct groups. The technical definition of the group is that it has "the most inclusive clade containing *Lacerta agilis* Linnaeus 1758 and *Crocodylus niloticus* Laurenti 1768, but not *Homo sapiens* Linnaeus 1758" (Modesto and Anderson 2004).

This chapter discusses a range of early reptiles, many of which have no especially close relationship to each other or whose placement among reptiles is controversial. Some of the groups are totally extinct with no living relatives, but many are members of the basal radiation of diapsid reptiles, which today includes all of the living reptiles and birds. Because of the uncertainty in their relationships, it is difficult to understand their patterns of otic evolution. Therefore, the groups will be treated somewhat separately in the following chapter. A cladogram of relationships as accepted here is given in Fig. 8.1.

As it currently stands, the first clade to branch off the reptilian tree is Parareptilia, which used to include a group previously known as "Anapsida." These were forms in which there were no holes or vacuities in the posterior skull roof but which had complete bony cheeks behind the eyes, a condition found in non-amniote fossil tetrapods from earlier periods. In contrast, diapsids have two vacuities in the posterior cheek, as it is seen today (e.g., crocodiles) (Sobral and Müller, Chap. 10). The anapsids included groups such as mesosaurs, millerettids, procolophonoids, pareiasaurs, and also turtles (Modesto and Anderson 2004). Many of these are Permian in age and became extinct during the end-Permian mass extinction event. By contrast, the earliest unequivocal turtles with a rigid shell do not appear until in the Late Triassic.

Unfortunately, there is as yet no recent, detailed analysis tackling early reptilian phylogenetic relationships, but the group Parareptilia is still accepted as monophyletic if turtles are excluded from it. Turtles are diapsid reptiles (see Sect. 8.4) (Joyce 2015), but they are grouped together with other early reptile clades in this volume for space issues only.

The unique development of an external layer in the amniote egg, which may be flexible or consolidated with calcium and which prevents dehydration of the embryo, means that amniotes are fully terrestrial clades. Terrestrial habits for Paleozoic amniotes, and reptilians in particular, are indicated by a series of natural traps in Nova Scotia and Oklahoma. The aquatic habits of some, especially fossil amniote groups, are thus considered secondary adaptations. These taxa are,

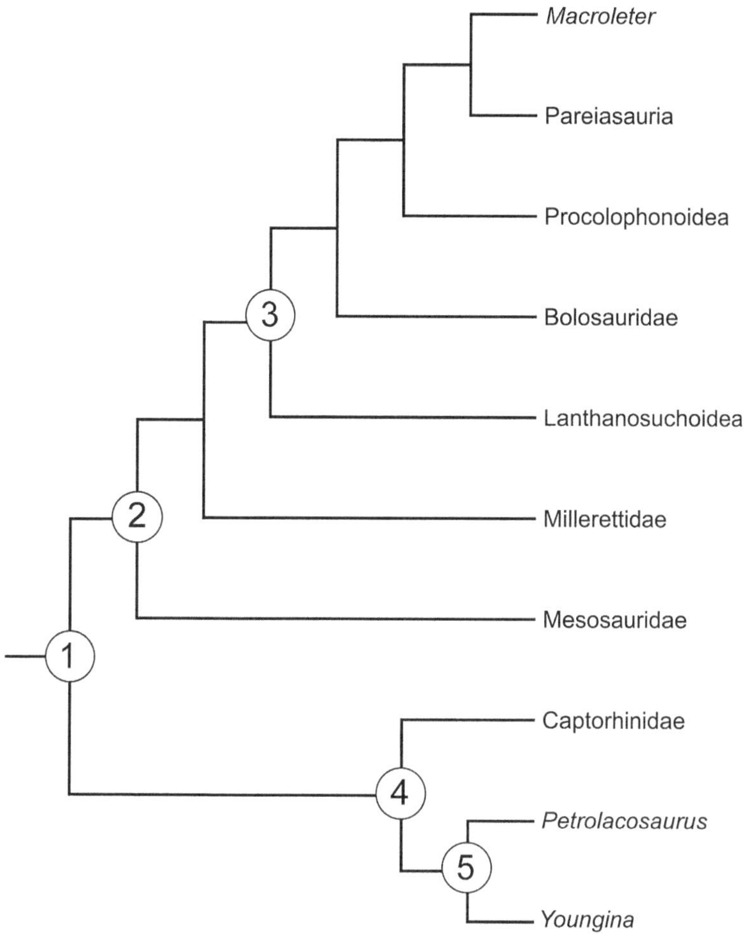

Fig. 8.1 Phylogenetic relationships of early reptilians. Numbered nodes correspond to: (1) Reptilia, (2) Parareptilia, (3) Ankyramorpha, (4) Eureptilia, (5) Diapsida. (Based on Tsuji and Müller 2009; Sobral et al. 2015)

therefore, often excluded from phylogenetic analyses, making inferences about their phylogenetic relationships and about ancestral Reptilia ecology difficult (Sobral et al. 2015). Mesosaurs were, for instance, fully aquatic. Among the earliest amniotes, they are sometimes considered the basalmost parareptilian clade or even the basalmost reptilian group (Piñero et al. 2012). Likewise, amphibious habits have been suggested for captorhinids (Canoville and Laurin 2010). The paraphyletic nature of the so-called "marine reptiles" and their highly controversial phylogenetic position indicate that aquatic habits were widespread among early diapsids (Sobral et al. 2015). Marine reptiles, such as the popularly known ichthyosaurs and plesiosaurs, are also included in this chapter.

Fig. 8.2 *Captorhinus*: **(A)** Reconstruction; **(B)** Photograph of the left half of the skull in posterior view with stapes stippled out. Braincase in **(C)** posterior, **(D)** lateral, **(E)** dorsal, and **(F)** ventral views. **(G)** Opishotic in anterior (*above*) and posterior (*below*) views. **(H)** Left stapes in dorsal (*above*) and posterior (*below*) views. *dp*, dorsal process; *fo*, fenestra ovalis; *fp* footplate of the stapes; *hym*, hyomandibular branch; *lc*, lagenar crest; *lr*, lagenar recess; *lsc*, lateral semicircular canal; *mf*, metotic foramen; *op*, opisthotic; *par*, posterior ampullary recess; *pp*, paroccipital process; *pr*, prootic; *rst*, recessus scala tympani; *sf*, shaft of the stapes; *so*, supraoccipital; *st*, stapes. *Roman numerals* indicate corresponding cranial nerves. (**A**, **B** provided by J. A. Clack; **C–H** reproduced with permission of the Oklahoma Geological Society from Heaton 1979)

8.2 Basal Reptilians

8.2.1 Captorhinidae

Captorhinids are more closely related to diapsids than to parareptiles and, together with diapsids, they form the clade Eureptilia (Tsuji and Müller 2009). Captorhinids are mainly small animals from the Lower Permian of North America, and they give us the best preserved non-diapsid reptilian braincases. The braincase of *Captorhinus* has been extensively studied by Price (1935) and Heaton (1979), who provided not only a very detailed morphological account of the skull of *Captorhinus* but also a careful comparison with the available developmental and physiological record of extant clades available at the time. The braincase of captorhinids is the best known among early reptiles and traditionally has been used on a comparative basis in reptilian braincase studies. Therefore, it is described first.

The presence of tympanic hearing in *Captorhinus* is unlikely because it lacks an otic conch or notch on the quadratojugal that could have supported a tympanic membrane (Fig. 8.2A, B). Cranial nerves (CN) IX–XI exit the brain via the occiput (Fig. 8.2C). The exoccipital forms the posterior border of the metotic foramen, which bears a channel that indicates the exit of the nerves (Heaton 1979). The basioccipital formed the anterior portion of the floor of the foramen. The basioccipital also contributed to the posterior portion of the cochlear recess and to the posteroventral border of the fenestra ovalis (FO).

The stapes of *Captorhinus* is a massive and complex bone (Fig. 8.2C–G). The footplate is much expanded. It has extensive contacts to the opisthotic and the basioccipital posteriorly and articulates with the prootic and the basisphenoid anteriorly. The posterolaterally directed shaft is triangular in cross section with a stout, medially directed dorsal process. This contacts the ventral portion of the prootic at the anterodorsal margin of the FO. A groove ventral to the dorsal process marks the course of the hyomandibular branch of the facial nerve (CN VII). A foramen for the stapedial artery pierces the anteroproximal part of the shaft. The distal tip is oval and sits in a shallow groove on the quadrate. It is unfinished, suggesting that it continued in cartilage or connected to a cartilaginous extrastapes. The paroccipital process of *Captorhinus* is relatively broad and stout and approaches the quadrate closely. However, the process is somewhat different in more basal

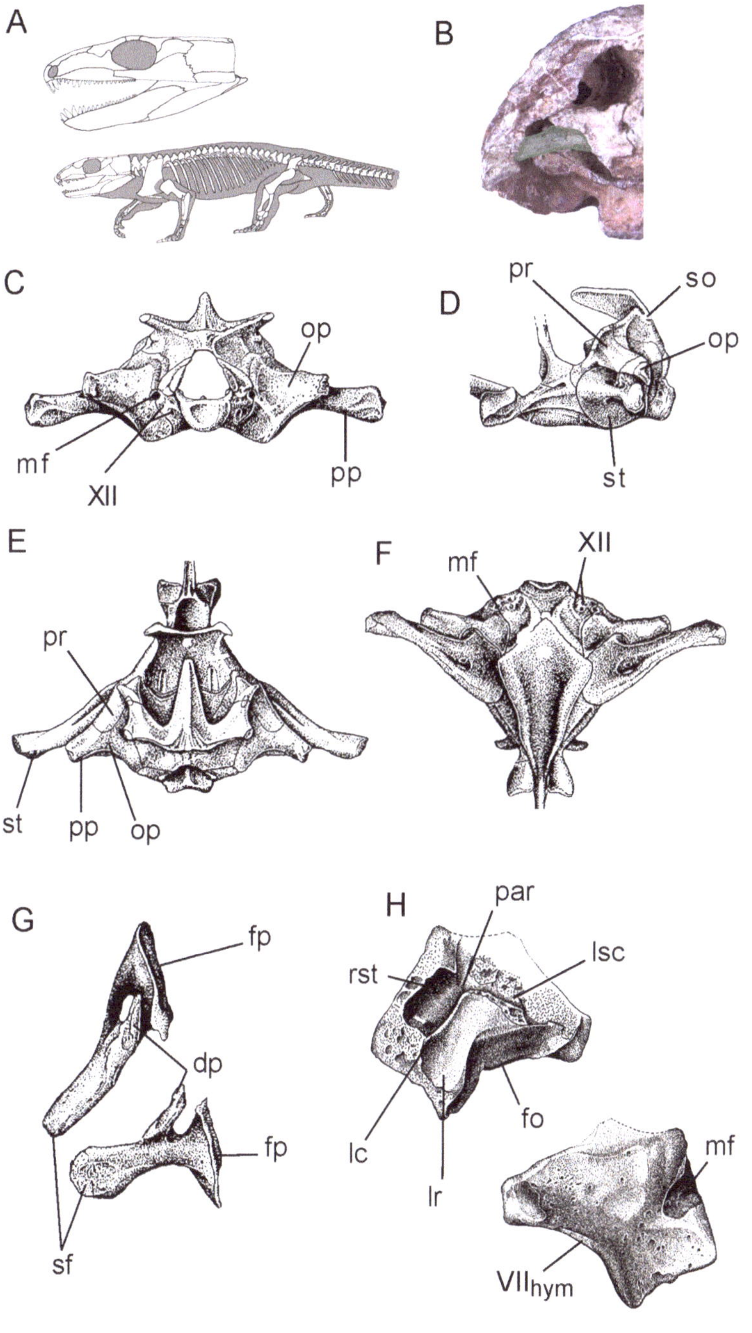
A
B
C
op
mf
XII
pp
D
pr
so
op
st
E
pr
st
pp
op
F
mf
XII
G
fp
dp
fp
sf
H
par
rst
lsc
lc
lr
fo
mf
VIIhym

captorhinids in that it is poorly developed. In these, the shaft of the stapes is long and narrow (Heaton 1979).

According to Heaton (1979), the lagenar recess was located posterodorsal to the FO, while the recessus scalae tympani was posterior to it (Fig. 8.2H). There is, however, no indication of a connection between the recessus and the metotic foramen, which contradicts the extra-capsular definition of the recessus made by de Beer (1937). There is also no indication of the presence or location of the perilymphatic foramen or the perilymphatic duct. The structures of the inner ear left shallow impressions on the medial surface of the opisthotic, indicating the generally poor ossification of the otic capsule. However, although not figured, Heaton (1979) mentioned a thin, vertical medial wall formed by the opisthotic, suggesting that there was some ossification of the medial wall of the capsule. The border of the FO was formed anteroventrally by the basisphenoid and anterodorsally by the prootic, where it also contacted the footplate of the stapes.

8.2.2 Parareptilia

The oldest known record of Parareptilia dates from the latest Carboniferous (late Pennsylvanian) (Modesto et al. 2015). By the Middle Permian, parareptiles had become quite diverse, and by the end of the Permian, the group could be found worldwide (Tsuji and Müller 2009). However, in the Triassic, the only parareptilian clade left after the extinction event was the Procolophonidae. Parareptilia includes groups that were among the first reptilians to evolve herbivory and associated modified feeding mechanisms (Tsuji and Müller 2009). Compared to *Captorhinus*, there are major differences in the construction of the otic region of Parareptilia. This group shows many evolutionary novelties that parallel and predate those seen in other amniote groups. Among those novelties are the independent acquisition of tympanic hearing and impedance-matching hearing (e.g., see Sect. 8.2.2.3, pareiasauromorphs).

8.2.2.1 Mesosaurs

The phylogenetic position of mesosaurs is uncertain. They are usually considered the basalmost parareptile group, but analyses have also placed them as stem reptilians (Piñero et al. 2012) or even as the basalmost amniotes (Hill 2005). Mesosaurs were marine animals, and they show many adaptations to a fully aquatic lifestyle, such as heavy, extremely ossified trunk ribs and limbs modified into paddles (Modesto 2006). They have a geographically very restricted distribution and are found only in the Lower Permian sediments of eastern South America and western southern Africa. Historically, they were among the first indicators that Africa and South America were once joined and that the continental land masses might have moved, giving rise to the theory of “continental drift” (Du Toit 1927).

Unfortunately, given the environment in which they lived and the types of sediments in which they are found, most mesosaur fossils are dorsoventrally compressed, hampering a detailed analysis of their braincase, and thus their otic region is poorly known. The morphology is generally described as similar to that of *Captorhinus*, with an occipital metotic foramen and no contact between the paroccipital process and the skull cheek (Modesto 2006).

8.2.2.2 Millerettids

Millerettids are small animals that have an overall "generalist" morphology (Fig. 8.3A, B) and are often considered to be more basal than the sister-clades Procolophonoidea and Pareiasauria. There is good three-dimensional millerettid material available, but the articulated nature of most of them makes a conventional approach to their otic morphology difficult. They also have not yet been the subject of computed tomography (CT) scanning techniques. The braincase of *Milleretta*, however, provides a good opportunity to study not only the individual braincase bones but also the medial morphology of the brain cavity and inner ear (Gow 1972). As currently understood, the braincase of *Milleretta* is very similar to that of *Captorhinus*, with the metotic foramen directed posteriorly and a very large but poorly defined lateral FO (Fig. 8.3C–E). The parabasisphenoid participated extensively in the ventral and anteroventral portions of the FO. The metotic foramen appears to have been narrow and elongate. The groove marking the exits of cranial nerves IX–XI lies close to the dorsal margin of the foramen. As depicted in Gow (1972), the medial surface of both the prootic and opisthotic is more strongly ossified than that of *Captorhinus*. They have marked recesses for the anterior and posterior ampullae (Fig. 8.3D). The posterior ampullary recess was separated from the area of the metotic foramen by a small medial growth of the opisthotic. The lateral semicircular canal seems to have left only grooves on the opisthotic and prootic and probably did not have a large radius of curvature.

The stapes of *Milleretta* is very different from that of *Captorhinus* (Fig. 8.3F). It is stout and bears a rather narrow footplate and a very short shaft (Gow 1972). The shaft expands significantly distally to become wider than the footplate. It has a large stapedial foramen and either no dorsal process at all or only a very small one. The stapes does not contact the quadrate. The stapes of "*Millerosaurus*" *nuffieldi*, however, is different. The shaft is straight and there is a clearer separation between the dorsal and hyoid processes (Watson 1957). There is a weak contact between the paroccipital process and the squamosal, which contributes to the dorsal half of the otic notch for the attachment of the tympanic membrane. The ventral half of the notch is formed by the quadratojugal, and the notch itself occupies only the ventral half of the posterior skull margin.

The phylogeny of millerettids is poorly understood. If they are found not to be closely related to other parareptiles, the presence of an otic notch might indicate a separate development of a tympanic membrane convergent with that of ankyramorphs (see Sect. 8.2.2.3).

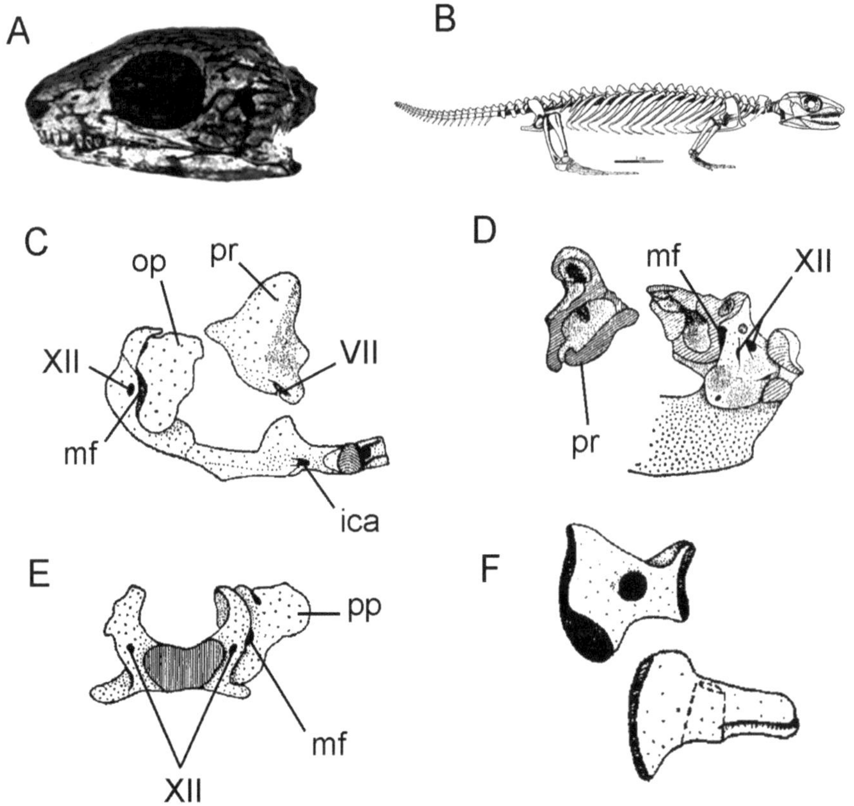

Fig. 8.3 *Milleretta* (**A**) fossil and (**B**) reconstruction. Braincase details: (**C**) right-lateral, (**D**) medial, and (**E**) posterior views. (**F**) Right stapes in anterior (*above*) and dorsal (*below*) views. *ica*, internal carotid artery; *mf*, metotic foramen; *op*, opisthotic; *pp*, paroccipital process; *pr*, prootic. *Roman numerals* indicate corresponding cranial nerves. (Reproduced with permission from Gow 1972)

8.2.2.3 Ankyramorphs: Bolosaurids, Pareiasauromorphs, and Procolophonoids

Ankyramorpha is a recently named clade of derived parareptilians that includes bolosaurids, pareiasauromorphs, procolophonoids, and some other more basal groups. The fossils rarely include complete and articulated braincases, so their otic morphology is not well understood. In the basal groups, the overall morphology of the posterolateral part of the skull shows little emargination that would unambiguously support a tympanic membrane. In some clades, however, there is a slight emargination and in pareiasauromorphs, this emargination can be extreme (Müller and Tsuji 2007).

Basal ankyramorph braincases are represented by *Delorhynchus* and *Bolosaurus*. Most of the braincase of *Delorhynchus* is embedded in sediment, and only the

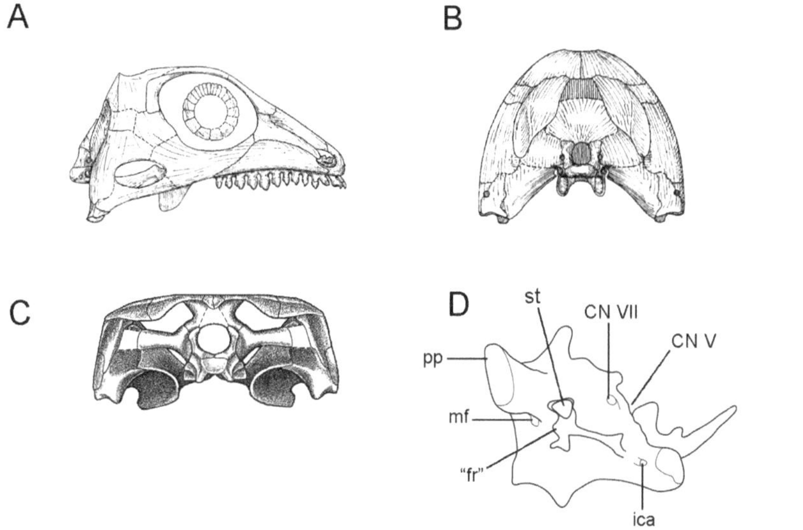

Fig. 8.4 Skull of *Bolosaurus*: (**A**) right-lateral and (**B**) posterior views. (**C**) Skull of *Owenetta* in posterior view. (**D**) Braincase of *Procolophon* in right-lateral view. *ica*, internal carotid artery; "*fr*", fenestra "rotunda"; *mf*, metotic foramen; *pp*, paroccipital process; *st*, stapes. *Roman numerals* indicate corresponding cranial nerves. (**A**, **B** reproduced with permission of the Museum of Comparative Zoology and Harvard University from Watson 1954; **C** reproduced with permission of the Society of Vertebrate Paleontology from Reisz and Scott 2002; **D** re-drawn from Carroll and Lindsay 1985)

stapes is visible. It resembles closely that of *Capthorhinus* (Reisz et al. 2014). The braincase morphology of *Bolosaurus* (Fig. 8.4A, B) is better preserved but has not been much explored since Watson (1954). The paroccipital process is expanded dorsoventrally and seems to have had an extensive contact with the side of the skull. In posterior view, the paroccipital process is positioned more ventrally and extends straight laterally.

The procolophonoid *Owenetta* (Fig. 8.4C) has slender paroccipital processes that extend straight laterally to contact the squamosal (Reisz and Scott 2002). The FO appears to have been small, and there is a more obvious emargination on the posterolateral side of the skull. In the braincase of *Procolophon*, a very small opening anterior to the metotic foramen may indicate the independent exit of CN IX (Carroll and Lindsay 1985). The reconstruction of the posterior margin of the FO of *Procolophon* shows a small process at about its midheight, subdividing the FO into dorsal and ventral spaces (Fig. 8.4D). The proximal part of the stapes would have articulated with the narrow space dorsal to the process. The ventral portion was termed "fenestra rotunda" and a similar structure was identified by Müller and Tsuji (2007) in *Macroleter*. However, the reconstruction of *Procolophon* by Kemp (1974) does not show such a regionalization of the FO. As preserved, the stapes is very short, but the distance of the distal end from the deep, well-developed otic

Fig. 8.5 *Macroleter*: **(A)** Fossil (*top*) and reconstruction showing position of the tympanic ▶ membrane (*bottom*); **(B)** braincase in ventral view; **(C)** skull in posterior view. **(D)** Braincase of *Pareiasuchus* in left-lateral view. **(E)** Braincase of *Deltavjatia* in left (*above*) and right (*below*) lateral views. *ap*, anterior process; *es*, extrastapes; *fo*, fenestra ovalis; *fp*, fenestra pseudorotunda; "*fr*", fenestra "rotunda"; *pp*, paroccipital process; *pr*, prootic; *st*, stapes; *tm*, tympanic membrane. *Roman numerals* indicate corresponding cranial nerves. (**A–C** modified from Müller and Tsuji 2007; **D** reproduced with permission of Wiley from Lee et al. 1997; **E** reproduced with permission of The Royal Society of Edinburgh from Tsuji 2013)

notch may indicate that it was much longer. The paroccipital process is well-developed, extending dorsally to contact the skull table.

In general, pareiasauromorphs have very prominent otic notches (Fig. 8.5A), indicating the undoubted presence of a large tympanic membrane. In their case, the otic notch is supported by the squamosal and quadratojugal, in contrast to diapsids in which it is supported by the quadrate. This was suggested by Müller and Tsuji (2007) to indicate the independent development of tympanic hearing in the group (See Sect. 8.4). The braincase of *Macroleter* conforms largely to that of *Procolophon* in that the ventral extension of the opisthotic bears an anteriorly directed process that separates the FO into dorsal and ventral regions (Müller and Tsuji 2007). In *Macroleter*, however, the prootic also seems to bear a posteriorly directed process to contact the opisthotic, completely separating the regions of the FO (Fig. 8.5B). The metotic foramen of *Macroleter* is enlarged. Although no sign of the inner ear structures can be seen on the medial surface of the brain cavity, swellings on its anterior region indicate extensive ossification of the otic capsule. The stapes bears a small footplate. Although the shaft is long, it would have had no close contact with the lateral side of the skull (Fig. 8.5C). The shaft is also very slender. Proximally, close to the footplate, it bears a small foramen interpreted as pneumatic (Müller and Tsuji 2007).

The braincase of *Pareiasuchus* must be among the best known pareiasaur material. The paroccipital process is very well-developed, and its distal part is directed dorsally to contact the supratemporal. The ventral border of the FO is formed mainly by the basisphenoid with some contribution from the basioccipital posteroventrally (Fig. 8.5D). The posterior border of the FO of *Pareiasuchus* was described by Lee et al. (1997) as incomplete and confluent with a large metotic foramen. The opisthotic would appear to have lacked a ventral ramus, being restricted to the posterodorsal margin of the FO and the anterodorsal margin of the metotic foramen. In contrast, the ventral ramus of the opisthotic of *Deltavjatia* is complete and fully separates the FO from the metotic foramen (Tsuji 2013). In *Deltavjatia* (Fig. 8.5E), the FO is larger than the metotic foramen. There is a subtle indication that a dorsoventral subdivision of the FO does exist in this taxon, in which case there may have been a "fenestra rotunda."

Contrasting with procolophonoids, the medial wall of the otic capsule of *Deltavjatia* is not well-ossified. The medial surface of the opisthotic is marked with pits indicating the location of the inner ear structures (Tsuji 2013). Among pareiasaurs, an unossified medial braincase wall is considered plesiomorphic, whereas

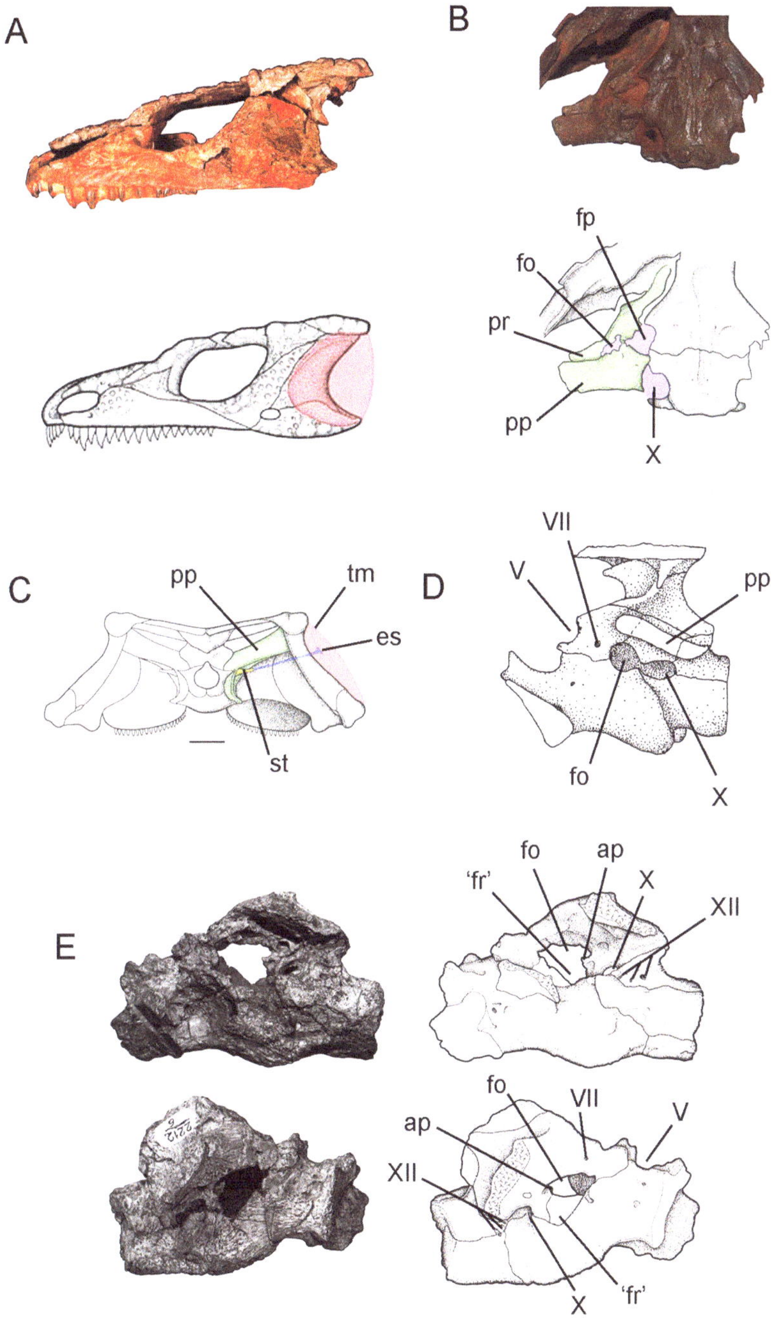
A
B
fp
fo
pr
pp
X
C
pp
tm
es
st
D
VII
V
pp
fo
X
E
fo
ap
'fr'
X
XII
fo
VII
V
ap
XII
X
'fr'

the ossified condition would be derived (Lee 1997). Tsuji (2013) further suggested it was an ontogenetic feature in taxa with the derived state. The preserved part of the stapes of *Deltavjatia* is very short, and the footplate is formed by two articular surfaces separated by a sulcus. On the proximal part of the shaft, close to the footplate, there is a large stapedial foramen. In contrast to other pareiasaurs, the otic notch of *Pareiasuchus* is either much reduced or absent.

8.2.3 Basal Diapsida

The phylogenetic relationships of basal diapsid clades are still controversial, and their early evolutionary history remains poorly understood (Sobral et al. 2015). The earliest diapsids, *Petrolacosaurus* (Reisz 1981) and *Araeoscelis* (Vaughn 1955), show otic morphologies in some ways similar to that of *Captorhinus*. There is no evidence of a tympanic ear in these early diapsids. They lack an otic notch and have a laterally or ventrally oriented stapes with a dorsal process (Fig. 8.6A–D). In fact, the stapes seems to have functioned more as a support for the jaw joint, directly or indirectly. The paroccipital processes contacted the quadrate, the squamosal, or the tabular.

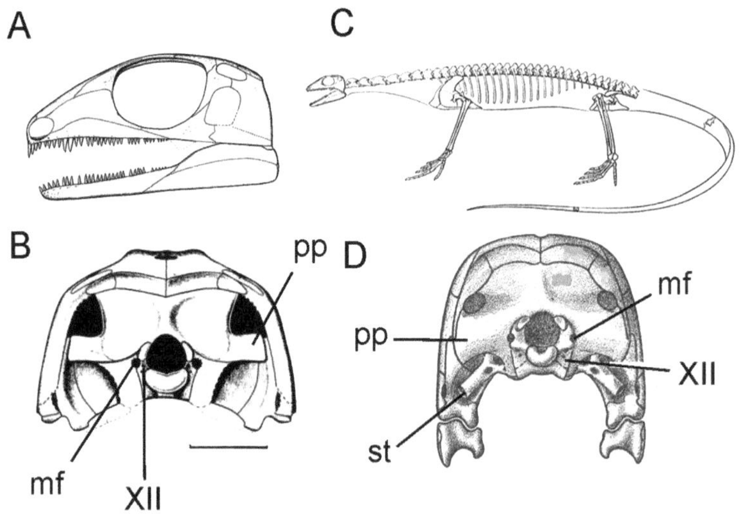

Fig. 8.6 Basal diapsids. **(A)** Skull reconstruction of *Petrolacosaurus*, left-lateral view and **(B)** posterior view. **(C)** Reconstruction of *Araeoscelis*, left-lateral view, and **(D)** skull in posterior view. *mf*, metotic foramen; *pp*, paroccipital process; *st*, stapes. *Roman numerals* indicate corresponding cranial nerves. (**A**, **B** used with permission of the University of Kansas and the Museum of Natural History from Reisz 1981; **C**, **D** reproduced with permission of the Museum of Comparative Zoology and Harvard University from Vaughn 1955)

The stapes of *Araeoscelis* has a ventrally directed footplate. Compared to *Captorhinus*, it is small and narrow. The dorsal process probably articulated with the braincase. The shaft of the stapes is oval in cross section, and its distal part is directed posteroventrally to a recess on the quadrate, though not contacting it. The foramen for the stapedial artery is large and lies close to the junction of the dorsal process with the footplate.

A large amount of cranial material exists for *Youngina capensis*, including a complete and disarticulated braincase (Evans 1987). Also, an articulated specimen has recently been μCT scanned (Gardner et al. 2010). *Youngina* is probably the best-known stem diapsid to date. In the past it was thought to be a basal lepidosauromorph and played an important role in discussions on the origin of this clade (Rieppel et al. 2003). It is now, however, considered to fall outside the lepidosauromorph–archosauromorph split, but its exact position is uncertain (Sobral et al. 2015).

As in other basal reptilians, the posterolaterally placed metotic foramen is larger than that in *Captorhinus*. The margins of the FO are not well-ossified (Fig. 8.7A). The paroccipital processes are well-developed, posterolaterally directed, and in extensive contact with the quadrates (Fig. 8.7A–C). In contrast to *Captorhinus* and other basal taxa, the ventral ramus of the opisthotic is thick, not plate-like, although not yet club-shaped like that of *Prolacerta* and other archosauromorphs (see Sobral and Müller, Chap. 10).

Youngina is the basalmost diapsid for which a floccular fossa has been described (Fig. 8.7D). Most of the medial wall of the otic capsule is unossified. As in *Captorhinus*, shallow grooves mark the inner ear structures on the medial surfaces of the braincase and are separated from one another by shallow ridges. There is no indication of the location and direction of the perilymphatic foramen on the medial side of the ventral ramus of the opisthotic. In the inner ear, the lateral semicircular canal is the longest, although the anterior canal has the greatest radius of curvature. The posterior canal is the shortest. The anterior and the posterior semicircular canals meet at the short common crus, which is well-marked off from the vestibule, similar to that in archosauromorphs (see Sobral and Müller, Chap. 10). The cochlear recess seems short and does not excavate the dorsal surface of the basisphenoid.

The stapes is very long and robust. There is no sign of an osseous dorsal process and the shaft is perforated by a large foramen for the stapedial artery (Fig. 8.7E). The footplate is separated from the shaft by a poorly defined neck. It is not much larger than the shaft itself. The shaft is long and slender and appears to extend laterally toward a slight emargination of the squamosal-quadrate complex. An imperforate, ossified extrastapes has also been identified (Gardner et al. 2010). A tympanic membrane was probably absent because the posterior margin of the skull lacks an otic notch, although a subtle emargination is present (Gow 1972).

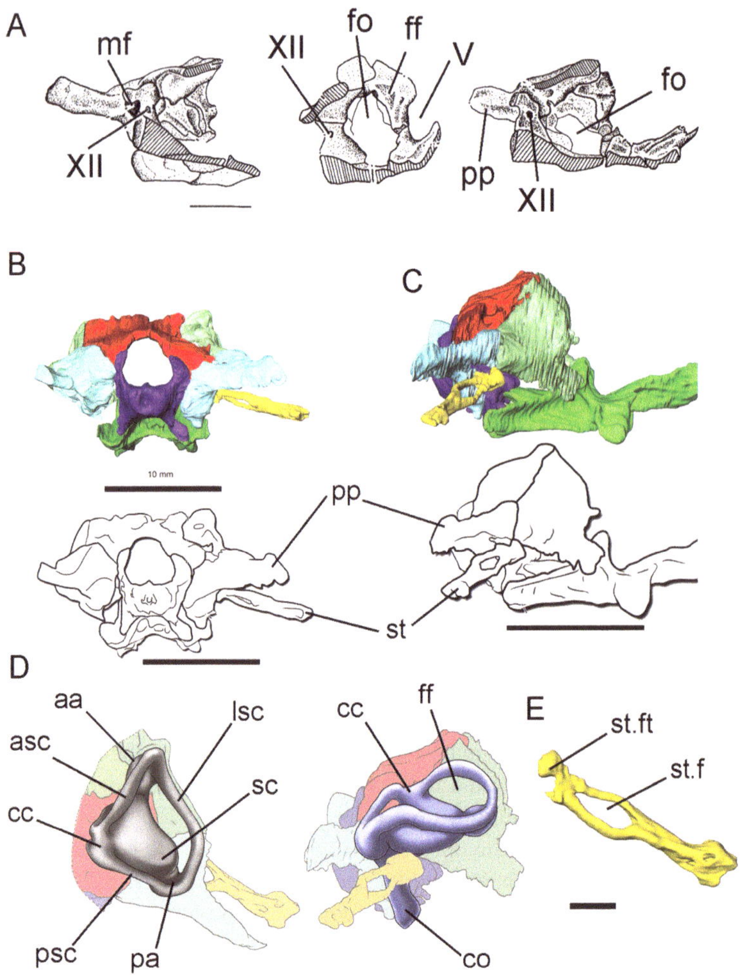

Fig. 8.7 *Youngina* braincases. **(A)** Disong National Museum of Natural History, former Transvaal Museum, Pretoria, South Africa specimen TM 3603 in different medial views. **(B–E)** American Museum of Natural History (AMNH) 5561 in posterior view **(B)** and right-lateral view **(C)**; **(D)** reconstruction of the inner ear relative to briancase bones; **(E)** stapes. *aa*, anterior ampula; *asc*, anterior semicircular canal; *cc*, common crus; *co*, cochlea; *fo*, fenestra ovalis; *ff*, floccular fossa; *lsc*, lateral semicircular canal; *mf*, metotic foramen; *pa*, posterior ampulla; *pp*, paroccipital process; *psc*, posterior semicircular canal; *sc*, saccule; *st*, stapes; *st. ft*, stapes footplate; *st. f*, stapes foramen. *Scale bar in A = 5 mm; B, C = 10 mm; E = 2 mm.* (**A** reproduced with permission of E. Schweizerbart'sche Verlagsbuchhandlung from Evans 1987; **B–E** reproduced with permission of the Palaeontological Association from Gardner et al. 2010)

8.3 Marine Diapsid Reptiles

There are basically three dominant lineages of secondarily marine reptiles that appeared shortly after the Permo-Triassic transition 250 million years ago, following in the aftermath of the largest and most severe mass-extinction event in earth history (Scheyer et al. 2014). The systematic position of these three major marine groups among Neodiapsida is still debated, mainly because of the high degree of morphological convergences among aquatic taxa (Neenan et al. 2013; Kelley and Pyenson 2015). All three lineages, the Ichthyosauromorpha, the Sauropterygia, and the Thalattosauriformes (Fig. 8.8), appeared in the Early Triassic with overall body plans already adapted in various degrees to aquatic life, without hinting at their terrestrial ancestors. Whereas thalattosauriforms became extinct by the end of the Triassic, the other two lineages thrived throughout most of the Mesozoic, becoming extinct only at the beginning (ichthyosauromorphs) or the end (sauropterygians) of the Late Cretaceous (Bardet et al. 2014).

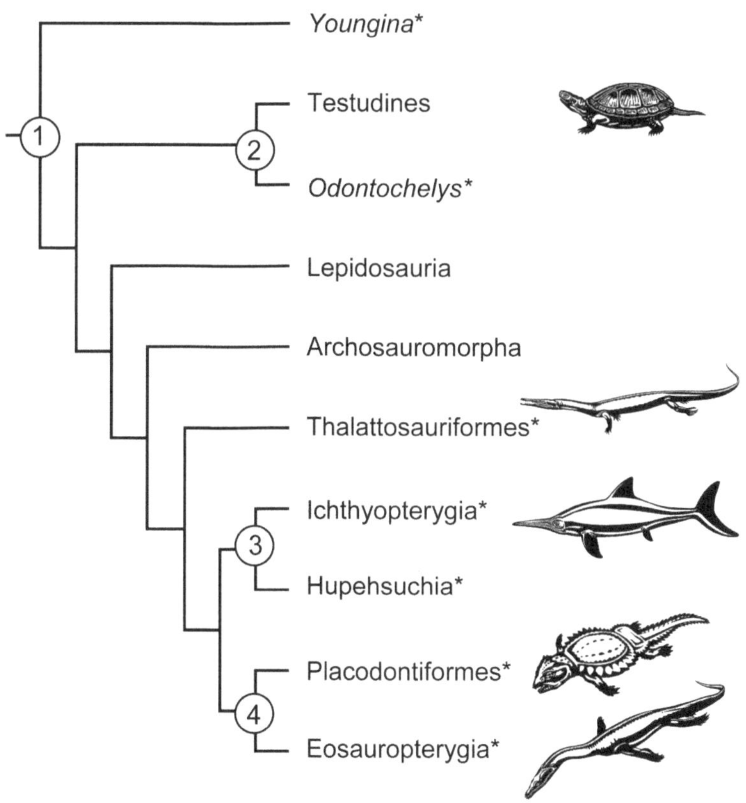

Fig. 8.8 Phylogenetic relationships of neodiapsid marine reptile lineages discussed in the chapter. (1) Neodiapsida, (2) Testudinata, (3) Ichthyosauromorpha, (4) Sauropterygia. *Extinct taxa/clades are marked by an asterisk.* (Original artwork by T. Scheyer based on Neenan et al. 2015; sketches from Beat Scheffold, the Palaeontological Institute and Museum University of Zurich)

Of the three lineages, the Ichthyosauromorpha, with their streamlined body shapes and flipper-like appendages, most strongly resemble fish (Motani et al. 2014) or dolphins, whereas the other two lineages retained more features typical of their terrestrial ancestors. As such, thalattosauriforms and early sauropterygians were probably able to leave the water, at least temporarily (e.g., for oviposition) (Sander 2012).

As in modern secondarily aquatic animals, external ears were probably absent in Mesozoic marine reptiles (Fig. 8.9) because sound could be transmitted directly

Fig. 8.9 (**A**) Skull of the small ichthyosaur *Mixosaurus* cf. *cornalianus* (PIMUZ T4848) in right-lateral view. (**B**) Skull of the medium-sized thalattosaur *Askeptosaurus italicus* (PIMUZ T 4831) in dorsal view. (**C**) Skull of the medium-sized nothosaur *Ceresiosaurus lanzi* (PIMUZ T2464, holotype) in dorsal view. (**D**) Skull of the small-sized pachypleurosaur *Serpianosaurus mirigiolensis* (PIMUZ 3931, holotype) in dorsal view. Note that although overall quality and completeness of the fossils is excellent, preservation on thick slabs of sedimentary rock often hinders the detailed study of the otic region in many specimens. All fossil material comes from the Middle Triassic of Monte San Giorgio, Canton Ticino, Switzerland and is stored at the Palaeontological Institute and Museum University of Zurich (PIMUZ), Zurich, Switzerland. (All images taken by T. Scheyer)

from the water to body tissue, obviating the need to channel sound into an ear canal (Hetherington 2008). At the same time, adaptations similar to modern marine reptiles, such as sea snakes and sea turtles, may have evolved in these extinct groups as well, such as hearing sensitivity to low-frequency wave lengths or a sealed-off outer ear in which a conspicuous tympanic membrane is absent.

8.3.1 Ichthyosauromorpha

The term Ichthyosauromorpha was recently introduced to combine the sister-clades of fish-shaped reptiles, the Hupehsuchia and the Ichthyosauriformes (Motani et al. 2014). The former are a group of poorly understood small animals restricted to the Early Triassic of China, often showing a strong degree of ossification and body armour, but little is known about their braincases or ear regions.

Ichthyosauriformes represent a well-studied, cosmopolitan group of predatory reptiles including both small forms, such as *Mixosaurus*, and also whale-sized forms that reached 20 metres or more in total body length (McGowan and Motani 2003). All ichthyosauriforms have comparatively huge eyes, larger than those of any other vertebrate group (Motani et al. 1999). This indicates that vision was an important, if not the most important, sense for these hunting and often deep-diving animals (McGowan 1973).

Little has been published on the morphology of the ear and hearing capacity of ichthyosauromorphs, although the topic has been addressed in a few well-preserved examples (Maisch et al. 2006; Marek et al. 2015). Most of these studies are traditional osteological descriptions (McGowan 1973; Maisch et al. 2006), whereas some studies, such as on the Lower Cretaceous ichthyosaur *Platypterygius longmani* from Australia (Kear 2005) and the Lower Jurassic ichthyosaur '*Hauffiopteryx*' *typicus* from England (Marek et al. 2015), also used CT (see Clack, Chap. 1 for the various forms of CT scans) to elucidate internal braincase, brain endocast, and ear-related structures.

The otic capsules of ichthyosaurs, as in other reptiles, are formed mainly by the prootics and opisthotics, where they are firmly integrated between other braincase elements (i.e., the squamosals, the supraoccipital, exoccipitals, the basioccipital, and the stapes). The form and shape of the otic capsules differ substantially between Triassic ichthyosaurs such as *Mixosaurus* (Middle Triassic, Switzerland), and the later Jurassic and Cretaceous forms (McGowan 1973; Maisch et al. 2006). The otic capsule of *Mixosaurus* was well ossified (Fig. 8.10). The lateral wall was closed off by the sutured opisthotic and prootic, and a crista interfenestralis, usually ossified, formed the posterior border of the FO and the anterior border of the metotic foramen. The latter thus was encased completely by the exoccipital and opisthotic (Maisch et al. 2006). The paroccipital process was pronounced and elongated. In contrast, the otic capsules were to a large degree cartilaginous in post-Triassic ichthyosaurs such as *Ichthyosaurus* (McGowan 1973).

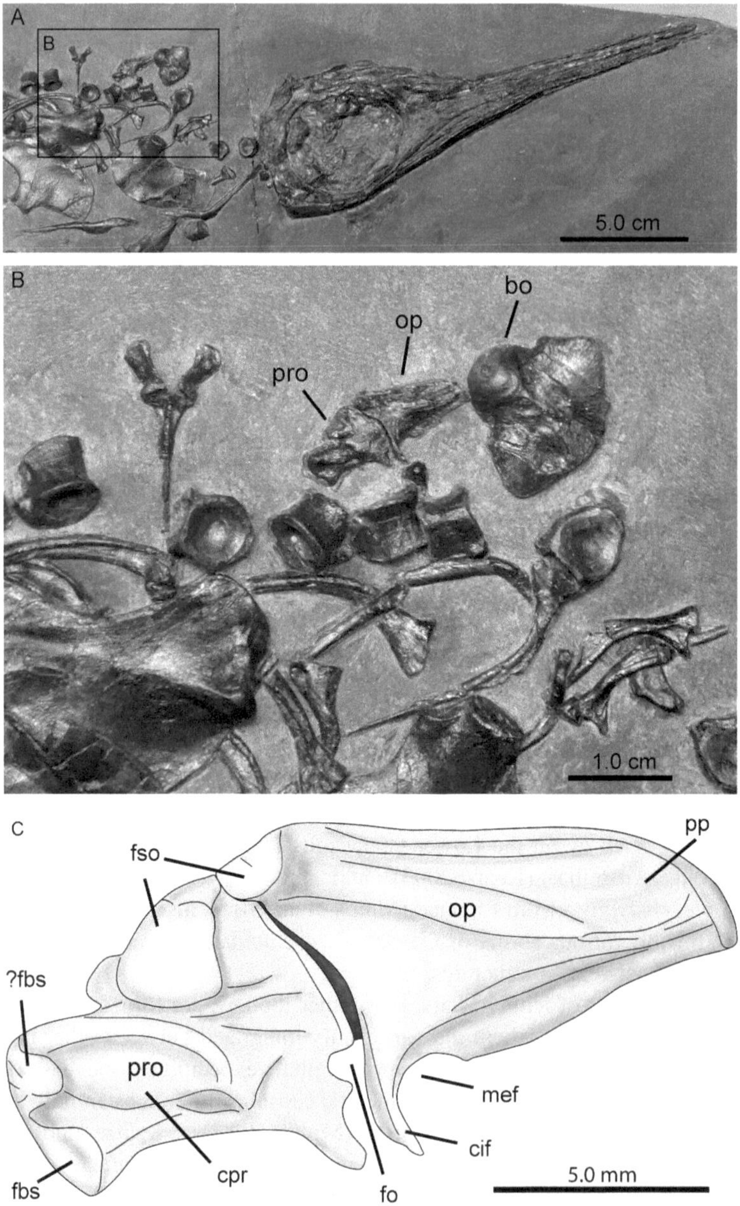

Fig. 8.10 (**A**) Head and neck region of the ichthyosaur *Mixosaurus* cf. *cornalianus* (PIMUZ T4848). (**B**) Detail of isolated braincase elements (basioccipital, opisthotic, and prootic) next to postcranial bone of the axial and appendicular skeleton. (**C**) Interpretative drawing of the otic capsule. *bo*, basioccipital; *cif* crista interfenestralis; *cpr*, crista prootica lateralis; *fbs*, facet for basisphenoidal portion of parabasisphenoid; *fso*, facet for supraoccipital; *fo*, fenestra ovalis; *mef*, metotic foramen; *op*, opisthotic; *pp*, paroccipital process; *pro*, prootic. (**A**, **B** by T. Scheyer; **C** modified from Maisch et al. 2006 and reproduced with permission of the Swiss Geological Society)

Fischer et al. (2012, 2014) provided an overview of the semicircular canals, saccule, ventricle, and ampullae on isolated opisthotics and prootics of a number of ophthalmosaurid ichthyosaurs from the Jurassic and Cretaceous. Imprints of the dorsal surface of the anterior and posterior semicircular canals and utricle on the supraoccipital have been investigated in the Early Cretaceous ophthalmosaurids *Acamptonectes* and *Sisteronia* (Fischer et al. 2012, 2014). The canals form a 170° angle, and the utricular impression is small and round in these taxa. A juvenile specimen of '*Hauffiopteryx*' *typicus* (Lower Jurassic, UK) also preserved parts of the vestibule on the opisthotics, prootics, and supraoccipital bone as revealed by CT scanning and virtual reconstruction (Marek et al. 2015).

The stapes of most ichthyosaurs consists of a dorsomedially broadened footplate (sometimes referred to as the occipital head) (e.g., Fischer et al. 2012) and a narrower ventrolateral shaft. In some cases, such as in the Jurassic *Temnodontosaurus*, thickness is more consistent throughout the shaft, resulting in overall stout stapedial morphology. The stapes structurally supports the braincase by articulating with the basioccipital and the basisphenoid medially and the quadrate laterally (McGowan 1973; Kear 2005). A massive stapes does not necessarily imply lack of hearing, however, because it could still transduce low-frequency sound waves via a bony connection directly to the fluids of the inner ear (Hetherington 2008). Additionally, the ventral surface of the stapes of some specimens of *Acamptonectes* was described as having a small triangular articulation facet for the pterygoid, further increasing the integration of the stapes with the palate (Fischer et al. 2012). The basioccipital and basisphenoid bear large, laterally facing facets for a cartilaginous articulation with the stapes, whereas the quadrate exhibits only a small medially positioned facet (Marek et al. 2015).

In general, there seems to be no free stapedial movement that would allow transmission of vibrations from a tympanic membrane to the FO. Instead, the lack of an otic notch together with the bony contact between the otic capsules and the suspensorium led to the proposition that a tympanic membrane was completely lost in ichthyosaurs and that they were not capable of acute, directional hearing (McGowan 1973).

8.3.2 *Thalattosauriformes*

Thalattosauriformes is a cosmopolitan lineage of marine reptiles restricted to the Triassic, with fossil finds from North America, Europe, and Asia (Bardet et al. 2014; Scheyer et al. 2014). Although in the past few decades an increasing number of thalattosauriforms have been recovered from black shale deposits of southern China (Wang et al. 2008), they usually constitute rare faunal elements in the fossil communities, compared to ichthyosaurs and sauropterygians.

Thalattosauriforms exhibit a huge variety of body and skull shapes, ranging from forms with tapering, edentulous snouts, downturned snouts, to large crocodile-like, predatory morphotypes, reflecting a diversity of lifestyles and feeding ecologies

(Müller 2005; Rieppel et al. 2005). The sensory array of thalattosauriforms is still underexplored, and the hearing capacity of these animals is no exception. The braincase elements are well-described for a series of taxa, including *Askeptosaurus* (Müller 2005), the most abundant taxon in the Middle Triassic of Europe, and some Chinese taxa (e.g., Liu 2013).

In thalattosaurs, the supraoccipital has a median crest and an expanding mediolateral curvature housing part of the vestibular system (Müller 2005). The supraoccipital meets the exoccipitals and opisthotics posteroventrally. In some taxa, the exoccipitals are not fused to the prootics or opisthotics; whereas in others, they appear to be fused to the paroccipital process of the opisthotics (Rieppel and Liu 2006). The metotic foramen comprises only a single, undivided opening, which reflects the plesiomorphic condition seen in early diapsid reptiles.

The quadrate in thalattosaurs is a massive element with a convex anterior margin and an emarginated and curving posterior margin. The convex margin of the quadrate forms a mobile articulation with the cranium called streptostyly. It is found in many reptiles, such as squamates, that have kinetic skulls (Müller 2005; also see Evans, Chap. 9). The lateral concave posterior emargination of the quadrate further argues for the presence of a tympanic membrane. In *Concavispina* (Late Triassic, China), the prominent anterior flange of the quadrate (= pterygoid ramus) contributes to the posterior portion of the lateral wall of the middle ear (Liu et al. 2013). The vestibular system is still unknown in thalattosauriforms.

The margins of the FO in *Askeptosaurus* and some Chinese forms (Müller 2005; Liu and Rieppel 2001) are formed by the prootic anterodorsally, the ventral ramus of the opisthotic posteriorly, and the parabasiphenoid complex ventrally. The paroccipital process of the opisthotic originates posterolateral to the otic capsule in *Askeptosaurus*. Its stapes is an elongated rod-like bone that medially thickens into a two-headed knob reminiscent of the stapedial shapes found in various stem diapsids (also see Sect. 8.2.3) (Müller 2005). *Askeptosaurus* seems to lack the stapedial foramen described for the CT-reconstructed massive stapes of the holotype skull of *Youngina* (Gardner et al. 2010). A partially preserved stapes missing the distal tip was found in situ in the holotype skull of *Miodentosaurus* (Late Triassic, China) (Cheng et al. 2009), but preservation is insufficient to identify details of the footplate articulating with the FO.

8.3.3 Sauropterygia: Placodontiformes and Eosauropterygia

Sauropterygia (Figs. 8.8, 8.9) comprises two groups, the placodontiforms and the eosauropterygians. The latter is a diverse assemblage grouping together the pachypleurosaurs, nothosaurs, pistosaurs, and plesiosaurs (e.g., Neenan et al. 2015). Sauropterygians are one of the most successful groups of extinct marine reptiles, having been both temporally and geographically widespread while inhabiting a wide range of ecological roles (Neenan et al. 2013, 2015). The clade experienced its greatest diversity during the Triassic, with only the plesiosaurs making it past the Triassic–Jurassic boundary to become one of the most iconic marine reptile lineages. However, despite the importance of sauropterygians, relatively few CT studies have been published to date that reveal details of the inner and middle ear anatomy.

Placodonts were a clade of nearshore, mostly durophagous taxa that were often heavily armored—termed Cyamodontoidea (Neenan et al. 2013, 2015). In all placodontiforms, the posterior quadrate margin is concave, which is thought to have supported a tympanic membrane. However, in the highly derived cyamodontoid placodonts, a notch is also present on the anteromedial surface of the quadrate. Presumably, it would have received the stapes, in which case there would have been no tympanic membrane (Rieppel 2000, 2001). Only one stapes is known to be preserved in one specimen of the basal placodont species *Placodus gigas* (Nosotti and Rieppel 2002), indicating that this structure may not have been fully ossified in derived placodonts. The inner ear of *Placodus* was reconstructed in two exceptionally well-preserved specimens with the use of μCT scanning (Neenan and Scheyer 2012), the first time this was achieved for a sauropterygian (Fig. 8.11). The anterior and posterior semicircular canals are dorsoventrally compressed and anteroposteriorly elongate, much like those of some modern aquatic reptiles, including crocodylians and turtles (Georgi and Sipla 2008). Additionally, the orientation of the lateral semicircular canals indicate that, in life, the head of *Placodus* would have dipped by over 20 degrees: an "alert" position ideal for feeding on the sea floor. Thus, despite the relatively basal position of *Placodus* in sauropterygian phylogeny, it seems plausible that these animals were already well adapted to an aquatic lifestyle.

Pachypleurosaurs are small, relatively abundant Triassic sauropterygians that are especially well-known from localities in central Europe and southern China. Much like placodontiforms, the posterior margin of the quadrate in most pachypleurosaurs is concave, possibly supporting a tympanic membrane and indicating the presence of an air-filled middle ear (Rieppel 1989; Klein 2009). Indeed, the tympanic membrane would have been particularly large in pachypleurosaurs as they have the unique feature of a quadrate fossa posterior to the articular facet of the retroarticular process on the lower jaw, which contributed to form part of the middle ear chamber (Lin and Rieppel 1998). The middle ear chamber is also composed of the medial portion of the quadrate, the squamosal, and the paroccipital process. However, the Chinese pachypleurosaur *Keichousaurus* does not have a quadrate with a curved posterior margin, suggesting that this taxon lacked a tympanic membrane (Holmes et al. 2008). In addition, the stapes has been reported in only one pachypleurosaur genus, *Neusticosaurus*, consisting of a short, slender, and cylindrical rod (Carroll and Gaskill 1985).

In the large, predatory nothosaurs, pistosaurs, and plesiosaurs, the tympanic membrane appears to have been completely lost, possibly due to these taxa being increasingly aquatic and not requiring terrestrial hearing (e.g., Rieppel 1998; Sato et al. 2011). However, narrow stapes have been described for the nothosaur

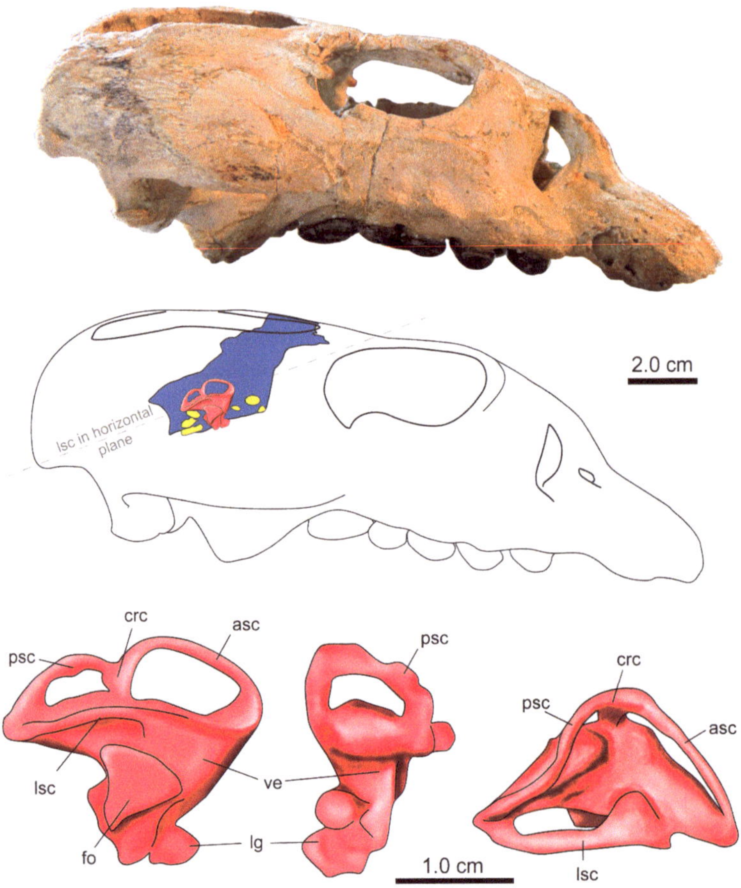

Fig. 8.11 Mirrored image of the left-lateral view of skull and schematic drawing of the placodont reptile *Placodus gigas* (Urwelt-Museum Oberfranken, Bayreuth, Germany specimen UMO BT 13), indicating the position of the reconstructed virtual cranial endocast (*blue*) and endosseous labyrinth (*red*). Cranial nerves are in *yellow*. *asc*, anterior semicircular canal; *crc*, common crus; *fo*, fenestra ovalis; *lg*, lagena; *lsc*, lateral semicircular canal; *psc*, posterior semicircular canal; *ve*, vestibule of inner ear. (Modified from Neenan and Scheyer 2012; reproduced with permission of Taylor and Francis)

Ceresiosaurus (Hänni 2004) and for the pistosaur *Yunguisaurus* (Sato et al. 2014), but it remains unclear whether they were used for hearing.

Plesiosaurs from the Early Jurassic, such as *Plesiosaurus* and *Stratesaurus* are commonly reported to have had narrow stapes (Storrs et al. 1997; Benson et al. 2015). However, from the Middle Jurassic until the clade died out at the end of the Cretaceous, no conclusive examples of plesiosaur stapes have been described (Sato et al. 2011; Benson et al. 2015), indicating the possibility that they probably were not ossified. In addition, little is currently known about the morphology of the plesiosaurian inner ear. The polycotylid *Dolichorhynchops* appears to exhibit a vestibule that is housed within the prootic and fused exoccipital-opisthotic, with preserved semicircular canals that run between the supraoccipital and prootic (anterior semicircular canal), supraoccipital and exoccipital-opisthotic (posterior semicircular canal), and the prootic and exoccipital-opisthotic (lateral semicircular canal) (Sato et al. 2011). The presence of an FO has been debated for many plesiosaur taxa, whereas Sato et al. (2011) argued that it may simply be obscured by deformation in many cases and that in *Dolichorhynchops*, the FO is positioned almost horizontally between the prootic and exoccipital-opisthotic.

8.3.4 Testudinata: Chelonians and Their Kin

The systematic position of turtles among amniotes is one of the most contested mysteries of vertebrate palaeontology/zoology largely due to their unique body plan, including a shell and the anapsid skull configuration shown by modern (and most extinct) turtles (Rieppel 2013). Depending on the data used to reconstruct their relationships, either molecular, morphological, or a combination of the two, turtles were thought to be either the last living representatives of Parareptilia or they were placed within Eureptilia, the group that includes all other living reptiles (Lee 2013; Joyce 2015). Within the Eureptilia, molecular phylogenies have tended to place them among the archosauromorphs (Field et al. 2014), although some have placed them among the lepidosauromorphs (Lyson et al. 2012).

It was only recently that a putative new stem turtle, *Pappochelys* from the Middle Triassic of southern Germany, was described that might shed light on the issue. These small (20 cm) animals not only had broadened ribs and a massive gastral apparatus that may be considered a precursor of a fused shell, but they also had a lizard-like diapsid skull with teeth (Schoch and Sues 2015).

Crown group turtles are usually separated into hidden-necked (Cryptodira) and side-necked (Pleurodira) turtles, depending on the neck retraction system. Pleurodira are also easily recognised by a pelvic girdle that is firmly attached to both the dorsal carapace and the ventral plastron of the shell (Joyce et al. 2013).

The earliest representatives of Testudinata, on the stem to crown-group turtles, were shown to have flexible and mobile necks that co-evolved with the shell. As exemplified by *Proganochelys* from the Late Triassic of Europe, at least some measure of protection by hiding the neck and head was possible (Werneburg et al. 2015). This early turtle appeared with contemporaneous forms during the Late Triassic, by which time they had already spread over several continents, including South and North America, Europe, and Asia. These ancient turtles were probably terrestrial, similar to modern tortoises (Testudinoidea), with one exception. *Odontochelys*, a toothed turtle lacking a well-sutured dorsal carapace, was found in Upper Triassic marine sediments in China; its exact ecology and lifestyle, however, are still contested (Li et al. 2008; Joyce 2015).

In general, turtles exhibit a wide range of ecological niches and lifestyles. This variety, including dietary preferences from herbivory to omnivory to carnivory and habitats from terrestrial, freshwater, and marine environments, is reflected in skull shape (Gaffney 1979) and in the proportional development of cranial sensory organs (Thewissen and Nummela 2008). Among modern turtles, sight, chemoreception, and magnetoreception (especially in marine turtles) are considered to be more important than hearing in both terrestrial and aquatic habitats (Thewissen and Nummela 2008). Similar sensory capabilities have been proposed for fossil turtles as well.

Cranial anatomy, including the braincase, has been described for a large number of fossil turtles including *Proganochelys* (Fig. 8.12) (Gaffney 1979, 1990). Concerning hearing and associated ear structures, only few data are available so far from fossils. This is surprising given the abundance of fossil descriptions and stands in contrast to the growing dataset of extant aquatic and terrestrial turtle species for which CT scans are available (e.g., Georgi and Sipla 2008; Walsh et al. 2009). Studies of structures associated with hearing have been undertaken in terrestrial stem turtles (including *Proganochelys*, Late Triassic, Germany), and marine and non-marine fossil cryptodiran turtles (including, for example, the Late Jurassic *Plesiochelys etalloni* from Switzerland). Furthermore, CT scan data have been published for a few taxa (e.g., Brinkman et al. 2006; Lipka et al. 2006), but it is only in *Plesiochelys etalloni* that an inner ear was reconstructed virtually (Carabajal et al. 2013). In addition, natural or artificial endocasts are known for a small number of fossil species (Carabajal et al. 2013).

Among the Triassic stem turtles, *Proganochelys* is among the best known and most extensively studied taxa (Gaffney 1990; Clack and Allin 2004). The species has a curved, c-shaped quadrate lacking a posterior flange, which differs from the funnel-shaped quadrate of modern turtles. As in the other stem turtles, *Palaeochersis* (Late Triassic, South Africa) and *Australochelys* (Early Jurassic, South Africa), *Proganochelys* lacks a well-developed, deep lateral concavity housing the middle ear and distalmost aspect of the stapes (the so-called cavum tympani referred to as CAT; Fig. 8.12). This is composed of the quadrate and quadratojugal and the more medial chamber of the middle ear, a cone-shaped second cavity formed mainly by the squamosal and, to a lesser degree, by the quadrate (Gaffney 1972; Sterli and Joyce 2007). This more medially situated chamber serves to enlarge the middle ear cavity to increase the range of sound perception (De Iuliis and Pulerà 2011). The older stem turtles such as *Odontochelys* (Late Triassic, China) (Li et al. 2008) show only weakly concave posterior margins of the quadrate. At the lateral wall of the middle ear, the quadrate of *Proganochelys* and other stem turtles exhibits a shallow depression interpreted as the articulation surface with the distal end of the stapes. This bony articulation is lost in all post-Triassic turtles, in which the stapedial shaft articulates with the tympanic membrane via a cartilaginous extrastapes (Sterli et al. 2007). In *Kayentachelys*, the tympanic membrane was attached to a rim formed mostly by the quadrate and to a lesser part by the squamosal (Sterli and Joyce 2007). The stem turtle *Eileanchelys* (Middle Jurassic, Scotland) shows a deep CAT, a more medial chamber of the

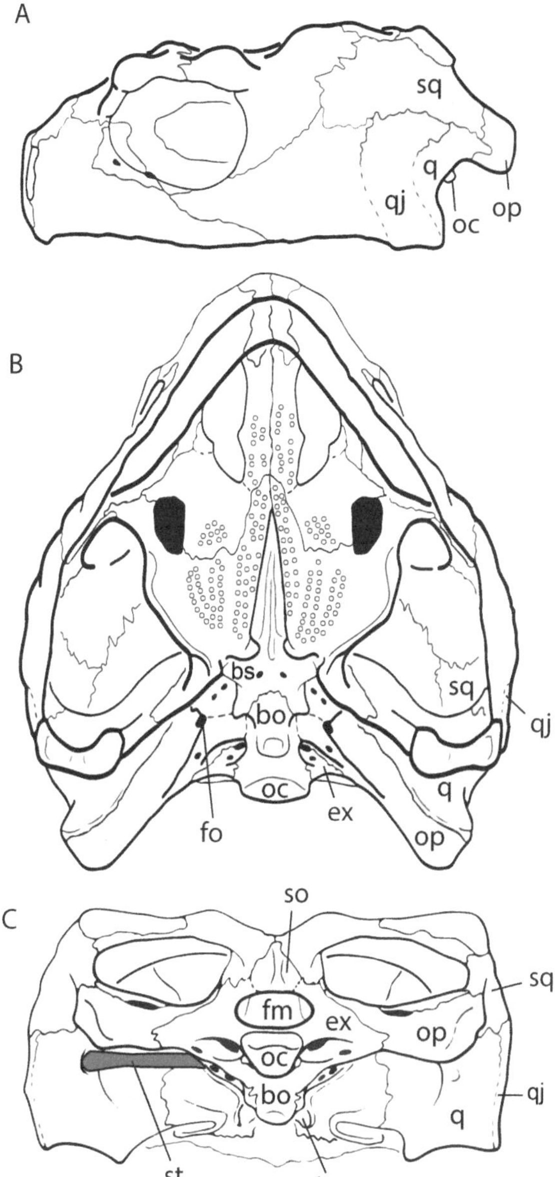

Fig. 8.12 Schematic drawing of the skull of the Late Triassic turtle *Proganochelys*: **(A)** lateral, **(B)** ventral, and **(C)** occipital view. Bones pertaining to the ear and braincase as discussed in the text are highlighted in *light gray*. Note position of stapes (*dark gray*) indicated on left side of skull in **C**. *bo*, basioccipital; *bs*, basisphenoid; *ex*, exoccipital; *fo*, fenestra ovalis; *fm*, foramen magnum; *oc*, occipital condyle; *op*, opisthotic; *q*, quadrate; *qj*, quadratojugal; *so*, supraoccipital; *sq*, squamosal; *st*, stapes. [Modified from Gaffney and Meeker (1983), reproduced with permission of Taylor and Francis; Gaffney (1990) reproduced with permission of the American Museum of Natural History, New York; and Joyce (2007) reproduced with permission of the Yale Peabody Museum of Natural History, New Haven]

middle ear, and an incision housing the stapes that is widely open posteroventrally (Anquetin et al. 2009). The latter is formed by a constriction of the quadrate around the stapes, and a wide incision possibly reflects the plesiomorphic condition for turtles. In some turtles, this incision encloses both the stapes and the Eustachian tube (Gaffney 1972).

Virtual reconstructions of the inner ear and brain endocasts in *Plesiochelys etalloni* (Carabajal et al. 2013) basically reflect the morphology seen in extant turtles, such as in the common snapping turtle, *Chelydra serpentina* (e.g., Walsh et al. 2009), and the red-eared slider, *Trachemys scripta* (Carabajal et al. 2013). As described by the authors, *Plesiochelys* has a globose and extremely short lagena, as well as short, subequal, and oval-shaped semicircular canals. The anterior and posterior semicircular canals are also dorsoventrally flattened. Furthermore, the FO is large and circular, the bony inner ear cavity has a bony floor as is typical in cryptodiran turtles, and the stapes is housed laterally by the CAT and medially by the middle ear cavity between the lateral CAT and the medial part of the bony inner ear cavity. The middle ear cavity houses, among other structures, the proximal part of the stapes and the stapedial artery (Gaffney 1972; Carabajal et al. 2013).

As in *Plesiochelys etalloni*, in modern turtles the FO separates the bony inner ear cavity from the middle ear cavity. The posteromedial section of the middle ear is offset by an interfenestral process of the opisthotic to form the recess of the tympanic duct. The middle ear is typically unfloored in stem turtles, including *Proganochelys* among others (Anquetin et al. 2009). In the skull of *Helochelydra* (Early Cretaceous, Isle of Wight, UK), the cavum labyrinthicum, the cavum acustico-jugulare, and the recess of the tympanic duct were described as forming a single "conjoined fossa" (with predicted soft tissue structures acting as separation), a condition not found in extant turtles. In this taxon, the interfenestral process of the opisthotic appears to be absent, so the posterior portion of the FO is not demarcated by bone (Joyce et al. 2011).

In turtles, sound energy is transmitted through the air-filled middle ear to the inner ear via the stapes. In some squamates, excess energy is dissipated through the fenestra pseudorotunda. Turtles, however, lack a fenestra pseudorotunda, so after the sound energy passes through the inner ear, it dissipates directly back into the large fluid-filled space surrounding the stapes, a structure also called the pericapsular recess. This sound dissipation method is referred to sometimes as a "re-entrant fluid system" (Clack and Allin 2004; Hetherington 2008). In *Proganochelys*, however, the re-entrant fluid system is not present, and instead it is considered that the metotic foramen could have provided pressure relief and energy dissipation similarly to early diapsids (Clack and Allin 2004).

In *Palaeochersis* and *Proganochelys*, the incision around the stapes is weakly developed, the medial footplate of the stout, rod-like stapes articulates with the FO, and the lateral aspect of the stapes fits into an articular pocket on the medial side of the quadrate (e.g., Gaffney 1990; Sterli et al. 2007). On the other hand, many fossil and modern turtles show a strongly developed, funnel-shaped incision, as exemplified by a well-preserved specimen of the pleurodiran turtle *Bairdemys sanchezi* (Gaffney et al. 2008) from the Late Miocene of Urumaco, Venezuela (Fig. 8.13). It has a delicate stapes from which a lateral extension—the extrastapes—articulates with the tympanic membrane. Somewhere on the stem or at the base of the turtle crown, the modification to a true impedance-matching tympanic ear thus must have occurred, and transitional stages might be expected to occur in the fossil record in the time interval between the end of the Late Triassic and the Middle Jurassic (Joyce et al. 2013).

According to Sterli and Joyce (2007), *Kayentachelys* from the Early Jurassic of Arizona, USA, is the oldest turtle in which inner and middle ear structures can be studied. The anatomy of the otic region of *Kayentachelys* is of importance because of the contested phylogenetic position of the taxon: it has been recovered either as a stem turtle (e.g., Sterli and Joyce 2007) or it has been found to be one of the basalmost members of the turtle crown (Gaffney and Jenkins 2010). Sterli and Joyce (2007) did not find remains of a stapes in this species, but they argued for the stapes to be "a slender rod-like element that has an impedance-matching function, as in modern turtles" (p. 689 in Sterli and Joyce 2007), based on other features of the otic region (i.e., small FO, well-developed incision for the stapes, and absence of a stapedial articular pocket on the quadrate). Gaffney and Jenkins (2010) later described a small and slender bone embedded in matrix as the distal part of the stapes.

8.4 Early Reptilian Evolutionary History

The composition and phylogenetic relationships of early reptile clades is far from being resolved. However, it is clear that early reptilians did not possess tympanic hearing (Clack 1997, 2002). Current evidence also does not support a single appearance of this system within reptilians, which independently evolved in ankyramorph parareptiles and in diapsids. Even within Diapsida, it is possible that tympanic hearing evolved more than once (Clack 1997; Evans, Chap. 9; Sobral and Müller, Chap. 10).

Hair cells are the basic sensory receptors of the auditory system of vertebrates. The arrangement of the otic structures can provide a low-impedance (= low-resistance) pathway for transmission of sound waves and displacement of the fluids around the hair cells on the auditory end organ (Lombard and Hetherington 1993). The marked differences in the impedances of the external environment (air) and inner ear (liquid) of terrestrial vertebrates result in significant energy loss at the interface of both during sound transmission. Mechanisms are needed for equalization of wave amplitudes, for amplification of the signal, and to avoid distortions (Manley 1990). A tympanic ear has not only a pressure-transducing function, but also an impedance-matching one. It involves a system that includes an external collector (tympanum = tympanic membrane), a transducing lever (stapes and extrastapes), and an air-filled cavity (middle ear). In extant groups, the tympanum is a superficial membrane located in a frame formed by bone on its anterior, dorsal, and ventral portions, and by soft tissue posteriorly (Wever 1978). The tympanum is connected to the stapes by a short cartilaginous extrastapes and connected to the bony frame by one or two ligaments fixing the extrastapes. The stapes is usually a narrow rod with a cartilaginous covering. Its distal-most part connects to the extrastapes while its

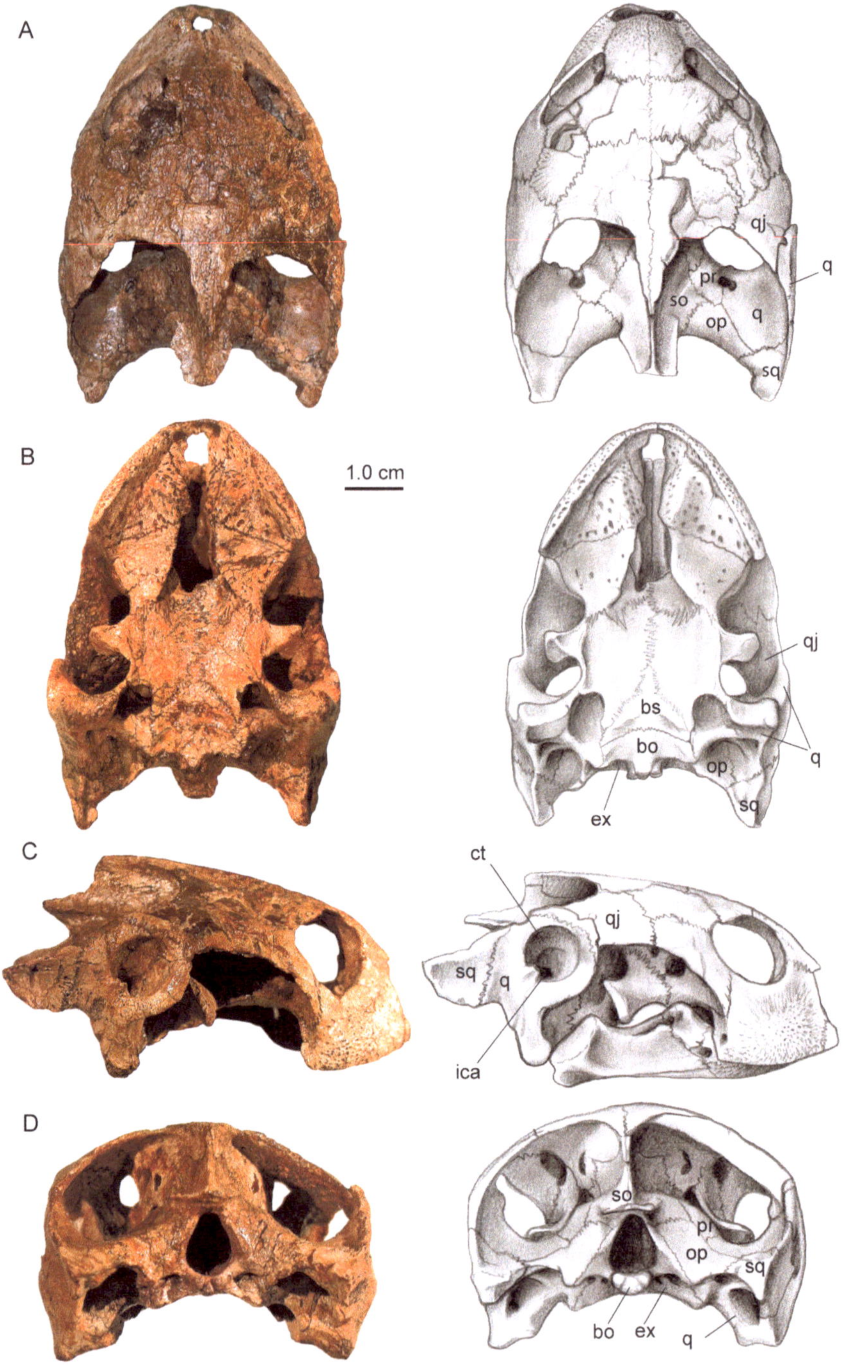
A
qj
q
pr
so
q
op
sq
B
1.0 cm
qj
bs
bo
q
op
ex
sq
C
ct
qj
sq
q
ica
D
so
pr
op
sq
bo
ex
q

◂**Fig. 8.13** Images and interpretative drawings of the holotype skull (Colección de Paleontología de Vertebrados de la Alcaldía de Urumaco, Estado Falcón, Venezuela, specimen AMU-CURS 186) of the fossil pleurodiran turtle *Bairdemys sanchezi*: **(A)** dorsal, **(B)** ventral, **(C)** right-lateral, and **(D)** occipital view. Note the external round lateral concavity housing the middle ear and distal-most aspect of the stapes, which progresses into the small incision accommodating the more proximal part of the stapes in **C**. *bo*, basioccipital; *bs*, basisphenoid; *ct*, cavum tympani; *ex*, exoccipital; *ica*, incision housing the stapes (= incisura columellae auris); *op*, opisthotic; *pr*, prootic; *q*, quadrate; *qj*, quadratojugal; *so*, supraoccipital; *sq*, squamosal. (Modified from Gaffney et al. 2008 with photos by F. Carlini, Argentina, and drawings by J. Gonzalez, Argentina)

proximal footplate inserts into the FO. Ligaments fix the footplate on the FO. The tympanic ear acts as an impedance-matching mechanism in two ways. The first mechanism is through amplification of the energy arriving at the body surface by the tympanic membrane; the second mechanism is based on the specific arrangement of the lever system (Saunders et al. 2000). Features of the middle ear cavity, like its volume and the composition of its walls (such as muscles, connective tissue, cartilage or bone) can heavily influence frequency response (Lombard and Hetherington 1993). At high frequencies, much of the energy is lost due to flexion of the system. Thus, ossification of the cavity and increased stiffness of the extrastapes help the detection of higher frequency sounds.

In early tetrapodomorphs (see Clack and Anderson, Chap. 4) and in stem amniotes, the stapes served an important bracing function. This function is represented by the articulation of the dorsal process with the opisthotic and by the extensive contact between the edges of the footplate and the border of the FO proximally. Distally, the shaft either sat in a groove of the quadrate or contacted it directly. However, during the transition into terrestrial environments, tetrapods seem to have benefited from the increasing sound transmitting function of the stapes.

The paroccipital process gradually replaced the initial bracing function of the stapes. This replacement is characterized by eventual loss of the dorsal process and reduction in the size of the footplate and in the thickness of the shaft. At the same time, an increase in the length of the paroccipital process eventually brought the process and the skull cheek in contact. This was, however, not a straightforward process. In millerettid parareptilians the stapes was a stout element closely related to the skull cheek (Gow 1972). It becomes slender and more independent of the skull cheek in more derived clades. In captorhinids, the stapes retained an articulation with the quadrate. The same is true for the basal diapsid *Youngina* (Gardner et al. 2010). In this taxon, however, there is a loss of the dorsal process and a relative slenderness of the stapes in relation to the paroccipital process. In this scenario, a complete detachment of the stapes from its bracing function occurred at least twice in the evolutionary history of reptilians: in parareptiles and in diapsids.

The transformation of the stapes into a sound conducting element shows a mosaic pattern even within Parareptilia. For example, the basal *Milleretta* had a short stapes with a small footplate, while the more derived *Millerosaurus* had a longer stapes (Gow 1972). The narrow stapes of ankyramorph parareptiles suggests that good hearing function had been acquired by these taxa. The stapes of *Youngina*, while still robust, shows functional modifications in comparison to *Captorhinus*. The presence of an ossified extrastapes in *Youngina*, however, seems to support a less specialized hearing function than that of procolophonoids.

The evolution of a specialized hearing system in parareptiles is also shown by the development of the otic notch. In extant squamates, the absence of a tympanic membrane is related to the absence of a conch on the quadrate bone that supports the membrane (Evans 2008). Therefore, the lack of an otic notch in fossils is usually regarded as evidence for the absence of a tympanic membrane. The posterior portion of the skull of millerettids shows an emargination formed by the squamosal and quadratojugal that is suitable for the attachment of a tympanic membrane (Gow 1972). The otic notch becomes expanded anteriorly and quite deep medially in procolophonoids and much more so in pareiasauromorphs. In contrast, the skulls of early eureptilians show no signs of emargination, indicating the absence of a tympanic membrane in these groups. The absence of an otic notch in early eureptilians and the fact that in parareptiles the notch is formed by the squamosal and quadratojugal (not by the quadrate) may indicate that tympanic hearing appeared independently in these two groups (Müller and Tsuji 2007). The authors also pointed out that this appearance in parareptiles would be the earliest instance of tympanic hearing among amniotes. In the skull of pareiasaurs, the notch is concealed by the dermal roofing bones in lateral view (Tsuji 2013).

Presence of a tympanic membrane in early diapsids is disputed. Some authors interpret the skull of *Youngina* as lacking a notch, but others identify a shallow emargination. The presence of a tympanic membrane in diapsids is thus inconclusive. There is also a chance that the evolution of such a system occurred independently in both clades of crown diapsids, Lepidosauromorpha and Archosauromorpha (Clack 1997, 2002; Evans, Chap. 9; Sobral and Müller, Chap. 10).

Another important piece of evidence for the refinement of terrestrial hearing in procolophonoids is the "fenestra rotunda." In tympanic hearing, the FO is basically both the entrance and the exit of sound waves in and out the inner ear system. The fluid passageway of the inner ear leads the sound waves from the internal boundary of the vestibule to the cochlea in a circular path back to the lateral surface of the footplate (Wever 1978). When the fluid is displaced with incoming sound waves, the movements of the inner ear structures involve the cochlear pathway. However, a reversal in the motion of the footplate causes the entire circuit to move in reverse as well. As it involves the complete circuit back to the footplate, it sets up a churning movement in the system. This movement makes the basilar membrane react not only to the incoming sound waves but also to the system going back to its initial position.

In the crown-groups of Diapsida and Mammalia, a pressure-relief window has evolved independently to compensate for the dampening created by this system (Clack 1997). In these groups, the window represents a subdivision of the metotic foramen, being thus related to the extracapsular space of the recessus scalae tympani. Sobral and Müller (Chap. 10) provide a more detailed explanation of impedance-matching in hearing. In contrast, in procolophonoids and pareiasauromorphs, this window is formed by the regionalization of the FO, a morphology that is unparalleled in other amniotes. On the diapsid side, a specialized pressure-relief mechanism did not evolve before the appearance of the crown-group. On the other hand, the metotic foramen can, to some extent, serve as a pressure-relief mechanism. The foramen of early reptilians was small but was enlarged in *Youngina*. Enlargement of the foramen is further seen in archosauromorphs (Sobral and Müller, Chap. 10).

The development of a pressure-relief window is also associated with the ossification of the otic capsule. One way to enhance acoustic abilities is to force the path of the sound wave to follow the location of the sound-detecting tissues. Ossification of the otic capsule helps to direct the waves along these routes (for further explanations see Sobral and Müller, Chap. 10). The otic capsule of *Captorhinus* was largely unossified. The structures of the vestibule and semicircular canals left only shallow grooves on the medial wall of the opisthotic (Heaton 1979). In addition, the ventral ramus of the opisthotic is plate-like with little medial expansion. The process of ossification of the otic capsule in parareptiles is largely unknown, but the capsule of *Macroleter* seems to be extensively ossified in its canalicular portion (Müller and Tsuji 2007). The otic capsule of *Youngina* is not as ossified, but the medial surface of the opisthotic and prootic shows deep recesses for the ampullae and the cochlea (Evans 1987).

While our knowledge of braincase and otic morphology of early reptilians still remains poor, it is possible to state that procolophonoid and pareiasauromorph parareptiles show extreme morphological convergences with diapsids, possibly related to a more terrestrial mode of life. Changes in the otic region and hearing capacities of procolophonoids and pareiasauromorphs predate by far those seen in diapsids. Similar anatomical transformations are not found prior to the diapsid crown group. The otic specializations of diapsids were acquired in a much slower pace, at least in the archosaur line (Sobral and Müller, Chap. 10). The appearance of a separate pressure relief window, for instance, only takes place in derived crocodylomorphs and in dinosaurs, including birds.

8.5 Summary

Reptiles form the most diverse group of living land vertebrates. They include lizards, snakes, crocodiles, birds, and turtles, as well as many fossil groups. This chapter reviewed the otic anatomy of early reptilian groups and turtles. Basal reptilians possessed a stout stapes that articulated with the skull, retaining a bracing function, and they lacked a tympanic membrane. The development of tympanic hearing did not happen until later in the evolution of the clade, in parareptiles and diapsids independently. Parareptiles also show other modifications that are convergent with much later reptilians. These include the development of the paroccipital process as the bracing structure of the skull, a slender stapes, a well-ossified otic capsule, and the development of a pressure-relief window for the inner ear.

These changes may be related with the acquisition of more terrestrial habits of pareiasauromorph parareptilians. In basal diapsids, the stapes remains relatively stout, albeit shorter. It loses its dorsal process but retains the foramen for the stapedial artery. The ossification of the otic capsule occurs in the area of the semicircular canals, and there is an enlargement of the metotic foramen. In contrast, in aquatic groups such as ichthyosaurs, thalattosaurs, or sauropterygians, the otic anatomy and hearing capacity are adapted to the denser aquatic medium, and convergences in otic anatomy are common in these groups. A lateral concave posterior emargination of the quadrate in many taxa argues for the presence of a tympanic membrane. In ichthyosaurs, the stapes structurally still supports the braincase, but it could transduce low-frequency sounds via a bony connection with the fluids of the inner ear. In turtles, sound energy is transmitted through the air-filled middle ear to the inner ear via the stapes. There are differences, however, in the configuration and morphology of the bones housing the stapes between Triassic stem turtles and modern turtles, in that the re-entrant fluid system is absent in some of the most basal forms. In these taxa, such as *Proganochelys*, the construction is more like that of a basal diapsid.

Acknowledgements G. S. was supported by the DAAD and CAPES program with funding provided by CAPES (BEX 3474/09-7). T. M. S. acknowledges funding by SNF (grant no. 31003A_149506). J. M. N. has a SNF Postdoc-Mobility grant no. P2ZHP3_162102. J. M. was funded by the Deutsche Forschungsgemeinschaft (MU 1760/2-3).

Compliance with Ethics Requirements G. Sobral, R. Reisz, J. M. Neenan, J. Müller, and T. M. Scheyer have declared that they had no conflicts of interest.

References

Anquetin, J., Barrett, P. M., Jones, M. E. H., Moore-Fay, S., & Evans, S. E. (2009). A new stem turtle from the Middle Jurassic of Scotland: New insights into the evolution and palaeoecology of basal turtles. *Proceedings of the Royal Society of London B: Biological Sciences*, 276(1658), 879–886. doi:10.1098/rspb.2008.1429

Bardet, N., Falconnet, J., Fischer, V., Houssaye, A., et al. (2014). Mesozoic marine reptile palaeobiogeography in response to drifting plates. *Gondwana Research*, 26(3-4), 869–887. doi:10.1016/j.gr.2014.1005.1005

Benson, R. B. J., Evans, M., & Taylor, M. A. (2015). The anatomy of *Stratesaurus* (Reptilia, Plesiosauria) from the lowermost Jurassic of Somerset, United Kingdom. *Journal of Vertebrate Paleontology*, 35(4), e933739. doi:10.1080/02724634.2014.933739

Brinkman, D., Hart, M., Jamniczky, H., & Colbert, M. (2006). *Nichollsemys baieri* gen. et sp. nov., a primitive chelonioid turtle from the Late Campanian of North America. *Paludicol*, 5 (4), 111–124.

Canoville, A., & Laurin, M. (2010). Evolution of humeral microanatomy and lifestyle in amniotes, and some comments on palaeobiological inferences. *Biological Journal of the Linnean Society*, 100(2), 384–406.

Carabajal, A. P., Sterli, J., Müller, J., & Hilger, A. (2013). Neuroanatomy of the marine Jurassic turtle *Plesiochelys etalloni* (Testudinata, Plesiochelyidae). *PLoS ONE*, 8(7), e69264. doi:10.1371/journal.pone.0069264

Carroll, R. L., & Gaskill, P. (1985). The nothosaur *Pachypleurosaurus* and the origin of plesiosaurs. *Philosophical Transactions of the Royal Society of London B: Biological Sciences*, 309(1139), 343–393.

Carroll, R. L., & Lindsay, W. (1985). Cranial anatomy of the primitive reptile *Procolophon*. *Canadian Journal of Earth Sciences*, 22, 1571–1587.

Cheng, Y.-N., Wu, X.-C., & Sato, T. (2009). A new thalattosaurian (Reptilia: Diapsida) from the Upper Triassic of Guizhou, China. *Vertebrata PalAsiatica*, 45(3), 246–260.

Clack, J. A. (1997). The evolution of tetrapod ears and the fossil record. *Brain, Behavior and Evolution*, 50, 198–212.

Clack, J. A. (2002). Patterns and processes in the early evolution of the tetrapod ear. *Journal of Neurobiology*, 53, 251–264.

Clack, J. A., & Allin, E. (2004). The evolution of single- and multiple-ossicle ears in fishes and tetrapods. In G. A. Manley, A. N. Popper, & R. R. Fay (Eds.), *Evolution of the vertebrate auditory system* (pp. 128–163). New York: Springer.

de Beer, G. R. (1937). *The development of the vertebrate skull*. London: Oxford University Press.

De Iuliis, G., & Pulerà, D. (2011). *The dissection of vertebrates - A laboratory manual*. 2nd ed. Oxford: Academic Press/Elsevier.

Du Toit, A. L. (1927). A geological comparison of South America with South Africa. *Carnegie Institute Publications*, 381, 1–158.

Evans, S. E. (1987). The braincase of *Youngina capensis* (Reptilia: Diapsida; Permian). *Neues Jahrbuch für Geologie und Paläontologie Monatshefte*, 4, 193–203.

Evans, S. E. (2008). The skull of lizards and tuatara. In C. Gans, A. S. Gaunt, & K. Adler (Eds.), *Biology of the Reptilia: The skull of Lepidosauria, vol. 20. Morphology H* (pp. 1–347). Salt Lake City, UT: Society for the Study of Amphibians and Reptiles.

Field, D. J., Gauthier, J. A., King, B. L., Pisani, D., Lyson, T. R., & Peterson, K. J. (2014). Toward consilience in reptile phylogeny: miRNAs support an archosaur, not lepidosaur, affinity for turtles. *Evolution and Development,* 16, 189-196. doi:10.1111/ede.12081

Fischer, V., Maisch, M. W., Naish, D., Kosma, R. et al. (2012). New ophthalmosaurid ichthyosaurs from the European Lower Cretaceous demonstrate extensive ichthyosaur survival across the Jurassic-Cretaceous boundary. *PLoS ONE*, 7(1), e29234. doi:10.1371/journal.pone.0029234

Fischer, V., Bardet, N., Guiomar, M., & Godefroit, P. (2014). High diversity in Cretaceous ichthyosaurs from Europe prior to their extinction. *PLoS ONE*, 9(1), e84709. doi:10.1371/journal.pone.0084709

Gaffney, E. S. (1972). An illustrated glossary of turtle skull nomenclature. *American Museum Novitates*, 2486, 1–33.

Gaffney, E. S. (1979). Comparative cranial morphology of Recent and fossil turtles. *Bulletin of the American Museum of Natural History*, 164(2), 67–376.

Gaffney, E. S. (1990). The comparative osteology of the Triassic turtle *Proganochelys*. *Bulletin of the American Museum of Natural History*, 194, 1–263.

Gaffney, E. S., & Meeker, L. J. (1983). Skull morphology of the oldest turtles: A preliminary description of *Proganochelys quenstedti*. *Journal of Vertebrate Paleontology*, 3, 25–28.

Gaffney, E. S., & Jenkins, F. A., Jr. (2010). The cranial morphology of *Kayentachelys*, an Early Jurassic cryptodire, and the early history of turtles. *Acta Zoologica (Stockholm)*, 91(3), 335–368. doi:10.1111/j.1463-6395.2009.00439.x

Gaffney, E. S., Scheyer, T. M., Johnson, K. G., Bocquentin Villanueva, J., & Aguilera, O. (2008). Two new species of the side necked turtle genus, *Bairdemys* (Pleurodira, Podocnemididae), from the Miocene of Venezuela. In M. R. Sánchez-Villagra (Ed.), *Contributions on vertebrate paleontology in Venezuela. Paläontologische Zeitschrift*, 82(2), 209–229. doi:10.1007/BF02988411

Gardner, N. M., Holliday, C. M., & O'Keefe, F. R. (2010). The braincase of *Youngina capensis* (Reptilia, Dipsida): New insights from high-resolution CT scanning of the holotype. *Palaeontologia Electronica*, 13(3), 19A:16p.
Georgi, J. A., & Sipla, J. S. (2008). Comparative and functional anatomy of balance in aquatic reptiles and birds. In J. G. M. Thewissen & S. Nummela (Eds.), *Sensory evolution on the threshold. Adaptations in secondarily aquatic vertebrates* (pp. 233–256). Berkeley: University of California Press.
Gow, C. E. (1972). The osteology and relationships of the Millerettidae (Reptilia: Cotylosauria). *Journal of Zoology*, 167, 219–264.
Hänni, K. (2004). *Die Gattung Ceresiosaurus. Ceresiosaurus calcagnii Peyer und Ceresiosaurus lanzi n. sp (Lariosauridae, Sauropterygia).* Zürich: Vdf Hochschulverlag ETH Zürich.
Heaton, M. J. (1979). Cranial anatomy of primitive captorhinid reptiles from the Late Pennsylvanian and Early Permian Oklahoma and Texas. *Bulletin of the Oklahoma Geological Survey*, 127, 1–80.
Hetherington, T. (2008). Comparative anatomy and function of hearing in aquatic amphibians, reptiles, and birds. In J. G. M. Thewissen & S. Nummela (Eds.), *Sensory evolution on the threshold. Adaptations in secondarily aquatic vertebrates* (pp. 183–209). Berkeley: University of California Press.
Hill, R. V. (2005). Integration of morphological data sets for phylogenetic analysis of Amniota: The importance of integumentary characters and increased taxonomic sampling. *Systematic Biology*, 54(4), 530–547.
Holmes, R., Cheng, Y.-N., & Wu, X.-C. (2008). New information on the skull of *Keichousaurus hui* (Reptilia: Sauropterygia) with comments on sauropterygian interrelationships. *Journal of Vertebrate Paleontology*, 28(1), 76–84.
Joyce, W. G. (2007). Phylogenetic relationships of Mesozoic turtles. *Bulletin of the Peabody Museum of Natural History*, 48(1), 3–102.
Joyce, W. G. (2015). The origin of turtles: A paleontological perspective. *Journal of Experimental Zoology Part B: Molecular and Developmental Evolution*, 324B(3), 181–193. doi:10.110. 1002/jez.b.22609
Joyce, W. G., Chapman, S. D., Moody, R. T. J., & Walker, C. A. (2011). The skull of the solemydid turtle *Helochelydra nopcsai* from the Early Cretaceous of the isle of Wight (UK) and a review of Solemydidae. *Special Papers in Palaeontology*, 86, 75–97. doi:10.1111/ j.1475-4983.2011.01075.x
Joyce, W. G., Schoch, R. R., & Lyson, T. R. (2013). The girdles of the oldest fossil turtle, *Proterochersis robusta*, and the age of the turtle crown. *BMC Evolutionary Biology*, 13, 266. doi:10.1186/1471-2148-1113-1266
Kear, B. P. (2005). Cranial morphology of *Platypterygius longmani* Wade, 1990 (Reptilia: Ichthyosauria) from the Lower Cretaceous of Australia. *Zoological Journal of the Linnean Society*, 145, 583–622.
Kelley, N. P., & Pyenson, N. D. (2015). Evolutionary innovation and ecology in marine tetrapods from the Triassic to the Anthropocene. *Science*, 348(6232), aaa3716. doi:10.1126/science. aaa3716
Kemp, T. S. (1974). The braincase and associated structures of the cotylosaur reptile *Procolophon trigoniceps* Owen. *Annals of the South African Museum*, 64, 11–26.
Klein, N. (2009). Skull morphology of *Anarosaurus heterodontus* (Reptilia: Sauropterygia: Pachypleurosauria) from the lower Muschelkalk of the Germanic Basin (Winterswijk, the Netherlands). *Journal of Vertebrate Paleontology*, 29(3), 665–676. doi:10.1671/1039.1029. 0327
Lee, M. S. Y. (1997). Pareiasaur phylogeny and the origin of turtles. *Zoological Journal of the Linnean Society*, 120, 197–280.
Lee, M. S. Y. (2013). Palaeontology: Turtles in transition. *Current Biology*, 23(12), 1113–1119. doi:10.1016/j.cub.2013.1105.1011
Lee, M. S. Y., Gow, C. E., & Kitching, J. W. (1997). Anatomy and relationships of the pareiasaur *Pareiasuchus nasicornis* from the Upper Permian of Zambia. *Palaeontology*, 40, 307–335.

Li, C., Wu, X.-C., Rieppel, O., Wang, L.-T., & Zhao, L.-J. (2008). An ancestral turtle from the Late Triassic of southwestern China. *Nature*, 456(7221), 497–501. doi:10.1038/nature07533

Lin, K., & Rieppel, O. (1998). Functional morphology and ontogeny of *Keichousaurus hui* (Reptilia, Sauropterygia). *Fieldiana (Geology), New Series*, 39, 1–35.

Lipka, T. R., Therrien, F., Weishampel, D. B., Jamniczky, et al. (2006). A new turtle from the Arundel clay facies (Potomac Formation, Early Cretaceous) of Maryland, U.S.A. *Journal of Vertebrate Paleontology*, 26(2), 300–307.

Liu, J. (2013). On the taxonomy of *Xinpusaurus* (Reptilia: Thalattosauria). *Vertebrata PalAsiatica*, 51(1), 17–23.

Liu, J., & Rieppel, O. (2001). The second thalattosaur from the Triassic of Guizhou, China [in English, with Chinese Abstract]. *Vertebrata PalAsiatica*, 39(2), 77–87.

Liu, J., Zhao, L.-J., Li, C., & He, T. (2013). Osteology of *Concavispina biseridens* (Reptilia, Thalattosauria) from the Xiaowa Formation (Carnian), Guanling, Guizhou, China. *Journal of Paleontology*, 87(2), 341–350. doi:10.1666/1612-1059R1661.1661

Lombard, R. E., & Hetherington, T. E. (1993). Structural basis of hearing and sound transmission. In J. Hanken, & B. K Hall (Eds), *The Skull, vol. 3. Functional and evolutionary mechanisms* (pp. 241–302). Chicago: The University of Chicago Press.

Lyson, T. R., Sperling, E. A., Heimberg, A. M., Gauthier, J. A., et al. (2012). MicroRNAs support and turtle + lizard clade. *Biology Letters*, 8, 104–107.

Maisch, M. W., Matzke, A. T., & Brinkmann, W. (2006). The otic capsule of the Middle Triassic ichthyosaur *Mixosaurus* from Monte San Giorgio (Switzerland): New evidence on the braincase structure of basal ichthyosaurs. *Eclogae geologicae Helvetiae*, 99, 205–210.

Manley, G. A. (1990). *Peripheral hearing mechanisms in reptiles and avialans*. Berlin: Springer-Verlag.

Marek, R. D., Moon, B. C., Williams, M., & Benton, M. J. (2015). The skull and endocranium of a Lower Jurassic ichthyosaur based on digital reconstructions. *Palaeontology*, 58(4), 723–742. doi:10.1111/pala.12174

McGowan, C. (1973). The cranial morphology of the Lower Liassic latipinnate ichthyosaurs of England. *Bulletin of the British Museum (Natural History), Geology*, 24(1), 1–109.

McGowan, C., & Motani, R. (2003). Ichthyopterygia. *Handbuch der Paläoherpetologie [Handbook of Paleoherpetology], Part 8*, 1–173.

Modesto, S. P. (2006). The cranial skeleton of the Early Permian aquatic reptile *Mesosaurus tenuidens*: implications for relationships and palaeobiology. *Zoological Journal of the Linnean Society*, 146, 345–368.

Modesto, S. P., & Anderson, J. S. (2004). The phylogenetic definition of Reptilia. *Systematic Biology*, 53, 815–821.

Modesto, S. P., Scott, D. M., MacDougall, M. J., Sues, H.-D., et al. (2015). The oldest parareptile and the early diversification of reptiles. *Proceedings of the Royal Society of London B: Biological Sciences*, 282, 20141912. doi:10.1098/rspb.2014.1912

Motani, R., Rothschild, B. M., & Wahl, Jr., W. (1999). Large eyeballs in diving ichthyosaurs. *Nature*, 402, 747. doi:10.1038/45435

Motani, R., Jiang, D.-Y., Chen, G.-B., Tintori, A., et al. (2014). A basal ichthyosauriform with a short snout from the Lower Triassic of China. *Nature*, 517, 483–488. doi:10.1038/nature13866

Müller, J. (2005). The anatomy of *Askeptosaurus italicus* from the Middle Triassic of Monte San Giorgio and the interrelationships of thalattosaurs (Reptilia, Diapsida). *Canadian Journal of Earth Sciences*, 42(7), 1347–1367. doi:10.1139/e1305-1030

Müller, J., & Tsuji, L. A. (2007). Impedance-matching hearing in Paleozoic reptiles: Evidence of advanced sensory perception at an early stage of amniote evolution. *PLoS ONE*, 2, e889. doi:10.1371/journal.pone.0000889

Neenan, J. M., & Scheyer, T. M. (2012). The braincase and inner ear of *Placodus gigas* (Sauropterygia, Placodontia) - a new reconstruction based on micro-computed tomographic data. *Journal of Vertebrate Paleontology*, 32(6), 1350–1357. doi:10.1080/02724634.2012.695241

Neenan, J. M., Klein, N., & Scheyer, T. M. (2013). European origin of placodont marine reptiles and the evolution of crushing dentition in Placodontia. *Nature Communications*, 4, 1621. doi:10.1038/ncomms2633

Neenan, J. M., Li, C., Rieppel, O., & Scheyer, T. M. (2015). The cranial anatomy of Chinese placodonts and the phylogeny of Placodontia (Diapsida: Sauropterygia). *Zoological Journal of the Linnean Society*, 175(2), 415–428. doi:10.1111/zoj.12277

Nosotti, S., & Rieppel, O. (2002). The braincase of *Placodus* Agassiz, 1833 (Reptilia, Placodontia). *Memoire della Società Italiana di Scienze Naturali e del Museo Civico di Storia Naturale di Milano*, 31(1), 3–18.

Piñero, G., Ferigolo, J., Ramos, A., & Laurin, M. (2012). Cranial morphology of the Early Permian mesosaurid *Mesosaurus tenuidens* and the evolution of the lower temporal fenestration reassessed. *Comptes Rendus Palevol*, 11, 379–391.

Price, L. I. (1935). Notes on the brain case of *Captorhinus*. *Boston Society of Natural History Proceedings*, 40, 377–385.

Reisz, R. R. (1981). A diapsid reptile from the Pennsylvanian of Kansas. *Special Publications of the Museum of Natural History, University of Kansas*, 7, 1–74.

Reisz, R. R., Macdougall, M. J., & Modesto S. P. (2014) A new species of the parareptile genus *Delorhynchus*, based on articulated skeletal remains from Richards Spur, Lower Permian of Oklahoma. *Journal of Vertebrate Paleontology*, 34(5), 1033–1043.

Reisz, R. R., & Scott, D. (2002). *Owenetta kitchingorum*, sp. nov., a small parareptile (Procolophonia: Owenettidae) from the Lower Triassic of South Africa. *Journal of Vertebrate Paleontology*, 22, 244–256.

Rieppel, O. (1989). A new pachypleurosaur (Reptilia: Sauropterygia) from the Middle Triassic of Monte San Giorgio, Switzerland. *Philosophical Transactions of the Royal Society of London B: Biological Sciences*, 323, 1–73.

Rieppel, O. (1998). *Corosaurus alcovensis* Case and the phylogenetic interrelationships of Triassic stem-group Sauropterygia. *Zoological Journal of the Linnean Society*, 124(1), 1–41. doi:10.1111/j.1096-3642.1998.tb00568.x

Rieppel, O. (2000). Sauropterygia I - Placodontia, Pachypleurosauria, Nothosauroidea, Pistosauroidea. In P. Wellnhofer (Ed.), *Encyclopedia of Paleoherpetology, vol. 12A*. Munich: Verlag Dr. Friedrich Pfeil.

Rieppel, O. (2001). The cranial anatomy of *Placochelys placodonta* Jaekel, 1902, and a review of the Cyamodontoidea (Reptilia, Placodonta). *Fieldiana: Geology, New Series*, 45(1514), 1–104.

Rieppel, O. (2013). The evolution of the turtle shell. In D. B. Brinkman, P. A. Holroyd, & J. D. Gardner (Eds.), *Morphology and evolution of turtles. Proceedings of the Gaffney turtle symposium (2009) in honor of Eugene S Gaffney. Vertebrate Paleobiology and Paleoanthropology Series* (pp. 51–62). Dordrecht: Springer.

Rieppel, O., & Liu, J. (2006). On *Xinpusaurus* (Reptilia: Thalattosauria). *Journal of Vertebrate Paleontology*, 26(1), 200–204.

Rieppel, O., Fraser, N. C., & Nosotti, S. (2003). The monophyly of Protorosauria (Reptilia, Archosauromorpha): A preliminary analysis. *Atti della Società Italiana di Scienze Naturali e del Museo Civico di Storia Naturale di Milano*, 144, 359–382.

Rieppel, O., Müller, J., & Liu, J. (2005). Rostral structure in Thalattosauria (Reptilia, Diapsida). *Canadian Journal of Earth Sciences*, 42(12), 2081–2086. doi:10.1139/E2005-2076

Sander, P. M. (2012). Reproduction in early amniotes. *Science*, 337, 806–808.

Sato, T., Wu, X.-C., Tirabasso, A., & Bloskie, P. (2011). Braincase of a polycotylid plesiosaur (Reptilia: Sauropterygia) from the Upper Cretaceous of Manitoba, Canada. *Journal of Vertebrate Paleontology*, 31(2), 313–329.

Sato, T., Zhao, L.-J., Wu, X.-C., & Li, C. (2014). A new specimen of the Triassic pistosauroid *Yunguisaurus*, with implications for the origin of Plesiosauria (Reptilia, Sauropterygia). *Palaeontology*, 57(1), 55–76 doi:10.1111/pala.12048.

Saunders, J.C., Duncan, R. K., Doan, D. E., & Werner, Y. L. (2000). The middle ear of reptiles and avialans. In R. J. Dooling, R. R. Fay, & A. N. Popper (Eds.). *Comparative hearing: Birds and reptiles* (pp. 13–69). New York: Springer-Verlag.

Scheyer, T. M., Romano, C., Jenks, J., & Bucher, H. (2014). Early Triassic marine biotic recovery: The predators' perspective. *PLoS ONE*, 9(3), e88987. doi:10.1371/journal.pone.0088987

Schoch, R. R., & Sues, H.-D. (2015). A Middle Triassic stem-turtle and the evolution of the turtle body plan. *Nature*, 523, 584–587. doi:10.1038/nature14472

Sobral, G., Sues, H.-D., & Müller, J. (2015). Anatomy of the enigmatic reptile *Elachistosuchus huenei* Janensch, 1949 (Reptilia: Diapsida) from the Upper Triassic of Germany and its relevance for the origin of Sauria. *PLoS ONE* 10(9), e0135114. doi:10.1371/journal.pone.0135114

Sterli, J., & Joyce, W. G. (2007). The cranial anatomy of the Early Jurassic turtle *Kayentachelys aprix*. *Acta Palaeontologica Polonica*, 52(4), 675–694.

Sterli, J., de La Fuente, M. S., & Rougier, G. W. (2007). Anatomy and relationships of *Palaeochersis talampayensis*, a Late Triassic turtle from Argentina. *Palaeontographica Abt. A*, 281(1-3), 1–61.

Storrs, G., Callaway, J., & Nicholls, E. (1997). Morphological and taxonomic clarification of the genus *Plesiosaurus*. In J. M. Callaway & E. L. Nicholls (Eds.), *Ancient marine reptiles* (pp. 145–190). San Diego, CA: Academic Press.

Thewissen, J. G. M., & Nummela, S. (Eds). (2008). *Sensory evolution on the threshold. Adaptations in secondarily aquatic vertebrates*. Berkeley: University of California Press.

Tsuji, L. A. (2013). Anatomy, cranial ontogeny and phylogenetic relationships of the pareiasaur *Deltavjatia rossicus* from the Late Permian of central Russia. *Earth and Environmental Science Transactions of the Royal Society of Edinburgh*, 104, 81–122.

Tsuji, L. A., & Müller, J. (2009). Assembling the history of the Parareptilia: Phylogeny, diversification, and a new definition of the clade. *Fossil Record*, 12, 71–81.

Vaughn, P. P. (1955). The Permian reptile *Araeoscelis* restudied. *Bulletin of the Museum of Comparative Zoology*, 113, 305–467.

Walsh, S. A., Barrett, P. M., Milner, A. C., Manley, G., & Witmer, L. M. (2009). Inner ear anatomy is a proxy for deducing auditory capability and behaviour in reptiles and birds. *Proceedings of the Royal Society of London B: Biological Sciences*, 276, 1355–1360. doi:10.1098/rspb.2008.1390

Wang, X., Bachmann, G. H., Hagdorn, H., Sander, P. M., et al. (2008). The Late Triassic black shales of the Guanling area, Guizhou Province, south-west China: A unique marine reptile and pelagic crinoid fossil lagerstätte. *Palaeontology*, 51(1), 27–61. doi:10.1111/j.1475-4983.2007.00735.x

Watson, D. M. S. (1954). On *Bolosaurus* and the origin and classification of reptiles. *Bulletin of the Museum of Comparative Zoology*, 111, 1–449.

Watson, D. M. S. (1957). On *Millerosaurus* and the early history of the sauropsid reptiles. *Philosophical Transactions of the Royal Society of London B: Biological Sciences*, 240, 325–400.

Werneburg, I., Wilson, L. A. B., Parr, W. C. H., & Joyce, W. G. (2015). Evolution of neck vertebral shape and neck retraction at the transition to modern turtles: An integrated geometric morphometric approach. *Systematic Biology*, 64(2), 187–204. doi:10.1093/sysbio/syu1072

Wever, E. G. (1978). *The reptile ear: Its structure and function*. Princeton: Princeton University Press.

Chapter 9
The Lepidosaurian Ear: Variations on a Theme

Susan E. Evans

Abstract Today, Lepidosauria encompasses more than 9,000 species of lizards, snakes, and amphisbaenians (Squamata), as well as the New Zealand Tuatara, *Sphenodon* (Rhynchocephalia). In many lizards, an efficient tympanic middle ear and an effective inner-ear compensatory mechanism permit acute hearing across a range of frequencies. *Sphenodon* lacks a tympanic membrane, but this is the result of secondary loss. Fossils of stem lepidosaurs and early rhynchocephalians indicate that the ancestral lepidosaurian middle ear was tympanic, although the compensatory mechanism was probably rudimentary. Derived rhynchocephalians like *Sphenodon* lost the tympanic ear, possibly in association with feeding specializations, whereas squamates improved it by developing a more efficient compensatory window. However, the timing of this change is uncertain as the earliest lizard fossils are uninformative in this respect. Lizards from the Early Cretaceous onward show the derived condition. Squamates are morphologically and ecologically diverse, and some specialized lifestyles have affected ear anatomy. Among extant squamates, the only obligate marine swimmers are sea snakes, but in the Cretaceous, mosasaurs dominated the marine niche. These aquatic lizards show a middle ear morphology analogous to that of extant marine turtles (bulla-like quadrate, expanded extrastapes, loss of the tympanum?). Loss of the tympanum also occurs in squamate burrowers but in conjunction with the possession of a robust stapes with an enlarged footplate and, frequently, reduction or modification of the compensatory mechanism. Ears of this type are found in the enigmatic Cretaceous *Sineoamphisbaena* and in amphisbaenians from the Eocene to the present day. Where known, the ears of early snakes more closely resemble those of burrowers than swimmers.

Keywords Amphisbaenian · Burrowing · Lepidosauromorph · Lizard · Mosasaur · Quadrate · Recessus scalae tympani · Rhynchocephalian · Snake · Sphenodon · Stem lepidosaur · Tuatara · Vestibular apparatus

S.E. Evans (✉)
Department of Cell and Developmental Biology (Anatomy Building),
University College London, Gower Street, London WC1E 6BT, UK
e-mail: s.e.evans@ucl.ac.uk

J.A. Clack et al. (eds.), *Evolution of the Vertebrate Ear—Evidence from the Fossil Record*, Springer Handbook of Auditory Research 59,
DOI 10.1007/978-3-319-46661-3_9

Figure Label Abbreviations

aam	anterior ampulla
aLRST	apertura lateralis recessus scalae tympani
aMRST	apertura medialis recessus scalae tympani
ar	acoustic recess (entry of CN VIII, nerve foramina)
art.s	articular surface
asc	anterior semicircular canal
as.ip	articular region of internal process
Bo	basioccipital
CCF	crista circumfenestralis
Cd	cochlear duct
cEs	contact surface on stapedial shaft for extrastapes
cif	crista interfenestralis
CN 8	vestibulo-acoustic cranial nerve (VIII)
CN 9	glossopharyngeal cranial nerve (IX)
CN 10	vagus cranial nerve (X)
cp	crista prootica
ctb	crista tuberalis
Cw	compensatory window
Ed	endolymphatic duct
En	endolymphatic sac
Eo	exoccipital
Es	extrastapes
f5	foramen for CN V
f8	foramen for CN VIII
f9	foramen for glossopharyngeal nerve CN IX
f En	foramen for endolymphatic duct
FP	footplate of stapes
fV	fenestra vestibuli
ham	horizontal ampulla
hsc	horizontal semicircular canal
ip	internal process
isp	infrastapedial process
Jr	jugular recess
JSF	juxtastapedial fossa
la	lagena
Mf	Metotic fissure
Mx	maxilla
occ	osseus common crus
O.cp	otic capsule
orb	orbit
pam	posterior ampulla
Pc	periotic cistern
Pd	periotic duct

Pf	periotic foramen
Ppr	paroccipital process of opisthotic
Pro	prootic
Ps	periotic sac
psc	posterior semicircular canal
pt.ip	pit for process internus
Qu	quadrate
Qu.co	quadrate conch
Qu.tc	quadrate tympanic crest
Ra	retroarticular process of jaw
rm	rim around stapedial footplate
RST	recessus scalae tympani
So	supraoccipital
Sp	sphenoid
ssp	suprastapedial process
St	stapes
Stm	meatus for stapes
Sts	stapedial shaft
stt	statolith (=otolith)
Tm	tympanic membrane
Va	vestibular apparatus
Vc	vestibular chamber
vf	vagus foramen

9.1 Introduction

Lepidosauria is the reptile group encompassing lizards, snakes, and their relatives. Today it comprises two monophyletic sister-clades, Squamata and Rhynchocephalia, which are very disparate in terms of species number and biogeography. Squamata includes more than 9,000 species of lizards, snakes (Serpentes), and worm-lizards (Amphisbaenia) and has a global distribution. The group is also morphologically and ecologically diverse. In contrast, Rhynchocephalia is represented by a single living species, *Sphenodon punctatus*, the Tuatara of New Zealand, although the group has a rich Mesozoic history.

Clearly an understanding of lepidosaurian interrelationships is important to our understanding of the polarity (i.e., primitive or advanced) and evolutionary trajectories of different characters, including those of the ear region. Most recent phylogenies (morphological and/or molecular) of squamates recognize the monophyly of the following subgroups: Gekkota (geckos), Dibamidae (rare limbless burrowers), Scincoidea (e.g., skinks, girdled-lizards), Teiioidea (e.g., tegus, whiptails), Lacertidae (e.g., wall lizards), Amphisbaenia (worm lizards), Iguania (e.g., iguanas, chameleons), Anguimorpha (e.g., monitor lizards, slow worms), and Serpentes (snakes). However, there are some major discrepancies in the relative placement of these groups among different researchers. Phylogenetic analyses based solely on morphological characters (e.g., Estes et al. 1988; Conrad 2008; Gauthier et al. 2012) consistently place Iguania as the sister group to all other squamates. However, analyses based on molecular data (e.g., Townsend et al. 2005; Wiens et al. 2012; Pyron et al. 2013) or a combination of molecular and morphological characters (e.g., Wiens et al. 2010; Reeder et al. 2015) always place Iguania as the sister group of Anguimorpha and/or Serpentes, with Gekkota and/or Dibamidae emerging as the sister-taxon to other squamates (Fig. 9.1). These molecular analyses also group Amphisbaenia with Lacertidae and Teiioidea. Snakes are broadly divided between the small-mouthed burrowing Scolecophidia (questionably monophyletic) and Alethinophidia, which encompasses all remaining snakes, most notably the Macrostomata (e.g., boas, pythons, vipers, cobras, sea snakes). For

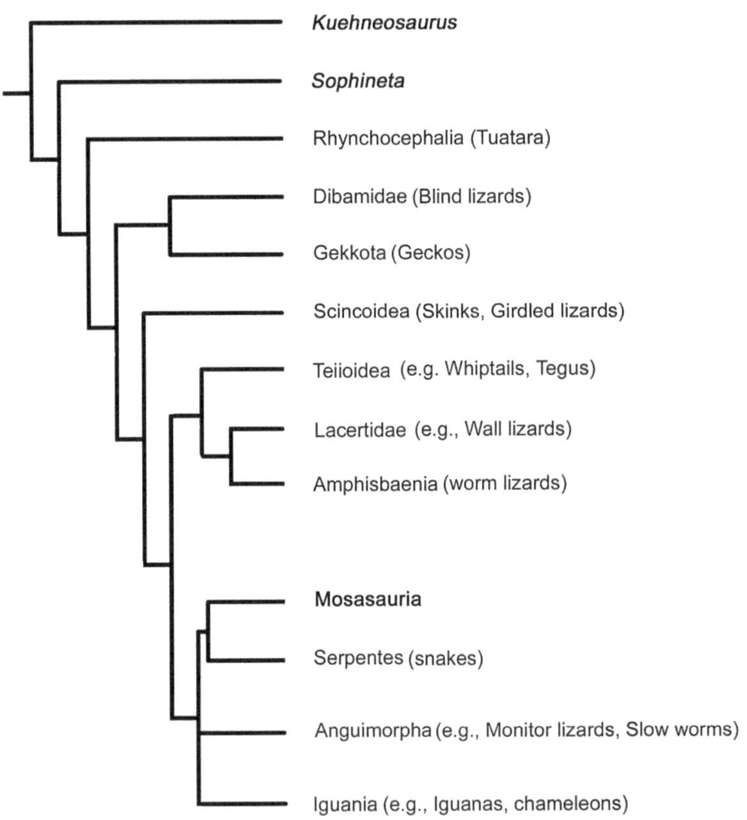

Fig. 9.1 Simplified tree of lepidosaur relationships based mainly on the combined evidence tree of Reeder et al. (2015) with additional data from Evans (2009) and Evans and Borsuk-Białynicka (2009)

obvious reasons, fossil taxa can only be accommodated within morphology-based or combined-evidence analyses, and this can lead to differing opinions as to their precise phylogenetic position.

Note that in this chapter, 'lizard' is used to refer to a squamate that is neither a snake nor an amphisbaenian. In older classifications, the formal name Lacertilia was applied to this grouping, but as lizards do not form a monophyletic group, a formal clade name is inappropriate.

9.2 The Fossil Record of Lepidosauria

Recent molecular divergence estimates (e.g., Jones et al. 2013) place the origin of Lepidosauria in the Early Triassic (~240–250 million years ago, Ma) and the origin of Squamata in the Early Jurassic (~193 Ma). However, most lepidosaurs are small, and this limits both the potential for their skeletons to be preserved after death and the chances of recovering their fossils. As a result, the earliest stages of lepidosaurian history and diversification are documented poorly. The lepidosaurian stem is represented by a small number of Triassic and Jurassic genera, most of which represent survivors of earlier originating lineages. These stem taxa are grouped with lepidosaurs into a more inclusive clade: Lepidosauromorpha. The position of the marine Sauropterygia (nothosaurs, plesiosaurs, placodonts, pliosaurs) and Ichthyosauria in relation to Lepidosauromorpha or Archosauromorpha remains unresolved (e.g., Jiang et al. 2014; Motani et al. 2015), and these taxa are covered in Chap. 8 by Sobral, Reisz, Neenan, Müller and Scheyer. Currently, the first recorded crown-group lepidosaurs are all rhynchocephalians dating from the Middle Triassic onward (Jones et al. 2013). The first uncontentious squamate fossils are from the Middle Jurassic (e.g., Evans 1994, 1998; Evans and Jones 2010), but these are both rare and mostly fragmentary. The squamate record improves substantially from the Cretaceous onward with the first records of specialized marine lizards (Mosasauria) and the first unambiguous snakes in the mid-Cretaceous and the first amphisbaenians in the Paleocene.

Knowledge of lepidosaurian ear evolution is limited because many early fossils come either from microvertebrate localities, where the small bones are three-dimensionally preserved but disarticulated, or from fine-grained deposits (e.g., Solnhofen, Germany; the Yixian Formation, China; Las Hoyas, Spain), where the skeletons are articulated but two-dimensionally compressed. Most data on the lepidosaurian ear, therefore, come from rare three-dimensionally preserved specimens. High resolution X-ray computed tomography (HRXCT) and Synchrotron CT (SRμCT) offer ways to access the morphology of the inner ear but, to date, relatively few lepidosaur fossils have been examined in this way.

9.3 The Lepidosaurian Ear

The ear of a typical lizard provides a good basis from which to consider lepidosaurian ear evolution as a whole. The outline presented here is mainly based on the works of Baird (1970) and Wever (1978) (Fig. 9.2).

There is no external ear in the mammalian sense (Fig. 9.2A). The tympanic membrane is exposed on the side of the head and is usually free of scales (Fig. 9.3A). However, in some lizards the membrane is recessed more deeply below the surface and may be protected by projecting scales and/or by a meatal closure muscle (geckos). The tympanic membrane is supported dorsally, anteriorly, and ventrally by the curved tympanic crest of the quadrate bone and, in part, ventrally by the retroarticular process of the lower jaw (Fig. 9.3B). Posteriorly, it is attached to soft tissue (fascia and muscle). In striking contrast to mammals, the frame supporting the lizard tympanic membrane is not rigid. Tension in the membrane fluctuates during feeding, partly because of the changing angle between quadrate and lower jaw and partly because the quadrate of most lizards is streptostylic (capable of independent movement at its dorsal and ventral articulations). Despite this, lizards can apparently still hear while feeding (Wever, 1978).

Medial to the tympanic membrane, the middle ear cavity is an air-filled diverticulum of the pharynx, usually with little distinction between the two regions, unlike the narrow Eustachian (auditory) tube of mammals (Fig. 9.2A). It is bounded medially by the bony otic capsule, dorsally by muscles and braincase structures such as the paroccipital process, and laterally by the quadrate, quadrate process of the pterygoid, and associated muscles. The stapes of most lizards has a small medial

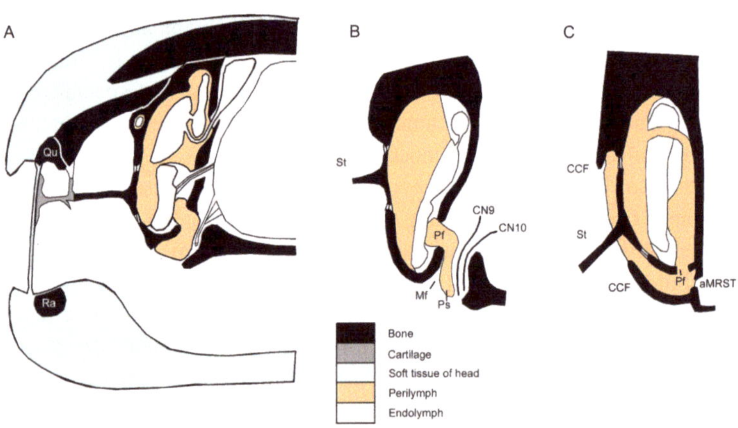

Fig. 9.2 Lepidosaur ear anatomy. **(A)** Schematic coronal section through ear region of a generalized lizard showing features of middle and external ear discussed in the text. **(B)** Schematic coronal section through otic capsule of *Sphenodon* showing the relationships of the periotic system to the metotic fissure. **(C)** As in **B**, for a generalized snake showing the relationship between the periotic sac and the crista circumfenestralis. (**A–C** adapted from Baird 1970)

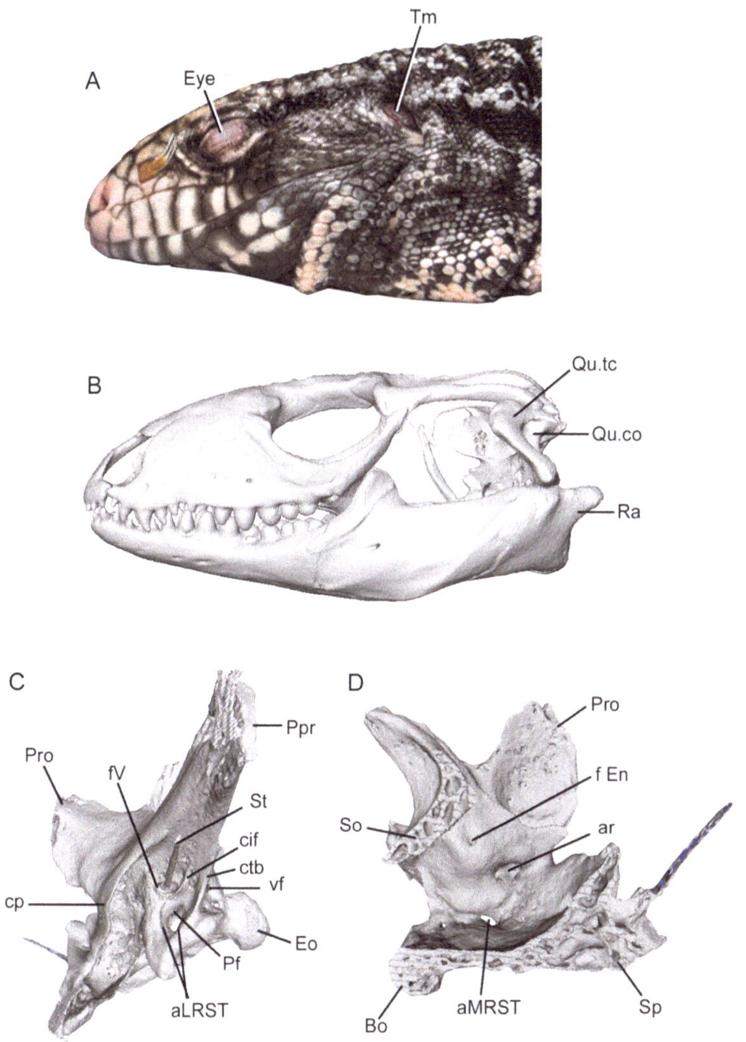

Fig. 9.3 Osteological correlates of ear anatomy in the otic capsule. **(A)** Left-lateral view of the head of the teiid lizard *Tupinambis merianae* showing the external tympanic membrane. **(B)** Lateral view of the skull of *Tupinambis merianae*, based on surface model built from HRXCT data and visualized using Amira 6.3. **(C–D)** Braincase of the lacertid lizard *Gallotia sp.* based on surface models built from HRXCT data and visualized with Meshlab. **(C)** Ventrolateral view showing relationship between fenestra vestibuli, apertura lateralis recessus scalae tympani, and surrounding crests. **(D)** Medial view showing positions of apertura medialis recessus scalae tympani, acoustic recess with foramina for CN8, and endolymphatic foramen. (All figures original)

footplate and a long slender shaft that is extended laterally by a cartilaginous extrastapes (Fig. 9.2A). The extrastapes typically terminates in one or more anterior processes, a posterior process, and an inferior process. The latter two form an oblique bar across part of the tympanic membrane, contributing to its stability and to a lever effect that enhances the transmission of sound pressure. Close to the junction with the bony stapes, two further processes may help to support the ossicle within the middle ear. These are the internal or quadrate process that contacts the quadrate and the dorsal process that meets the paroccipital process. One or more of these processes may be missing.

The otic capsule is composed of the prootic, opisthotic, and supraoccipital (Fig. 9.3C–D). These enclose the cochlear duct (lagena) ventrally and the vestibular apparatus dorsally. The fenestra vestibuli lies at the lateral junction of the prootic and opisthotic (Fig. 9.3C). Early in development, the otic capsule is separate from the basal plate (which will form the sphenoid anteriorly and the basioccipital posteriorly) and occipital arch (which contributes to the exoccipital). Although the occipital arch subsequently contacts the opisthotic posterodorsally, a gap (the metotic fissure) is left between the posterior surface of the otic capsule and the occipital arch/basal plate. In squamates, this metotic fissure is divided into dorsal and ventral portions by a second more ventral contact between the otic capsule and exoccipital. The dorsal passage carries the vagus nerve (cranial nerve, CN 10). The ventral passage is the recessus scalae tympani (RST) (= occipital recess of Oelrich 1956; Rieppel 1985). The RST has a medial opening into the cranial cavity, the apertura medialis recessus scalae tympani (aMRST, Fig. 9.3D), and a lateral opening, the apertura lateralis recessus scalae tympani (aLRST, Fig. 9.3C) into the middle ear cavity. In most lizards, the RST also gives passage to the glossopharyngeal nerve (CN 9) (Figs. 9.2A, 9.3C–D), although some taxa (e.g., gekkotans) subdivide the RST again to produce a separate nerve canal. A periotic (= perilymphatic) foramen opens from the otic capsule into the roof of the RST, providing an exit for the periotic sac (Fig. 9.2A). Some authors (e.g., Wu 1994) have termed this opening the fenestra rotunda (or fenestra cochleae), but this has implications with respect to homology and also causes confusion when the same term is used for the aLRST (e.g., Palci and Caldwell 2014). Fenestra rotunda is better avoided completely in reptiles. In a typical lizard, the periotic sac extends both medially, where it is blind-ending, and laterally, where it attaches to the margins of the aLRST to form a compensatory window (= secondary tympanic membrane) that lies adjacent to the membrane lining the middle ear/pharynx (Fig. 9.2A). This structure is a functional analogue of the mammalian fenestra cochleae (= fenestra rotunda). In many lizards (e.g., *Iguana, Gallotia*), the margins of the aLRST and, therefore, of the compensatory window are expanded by the development of two crests, the crista interfenestralis and crista tuberalis (Fig. 9.3C). The former, as its name suggests, lies between the fenestra vestibuli and the aLRST. The crista tuberalis extends down the edge of the exoccipital and separates the aLRST from the vagus foramen. It may combine with a similar crest on the basioccipital to form a more extensive frame. Medially (Fig. 9.3D), the otic capsule is pierced by anterior and posterior foramina for branches of the vestibulo-acoustic nerve (CN VIII), usually placed within a depression (acoustic recess), and by a more dorsal foramen through which the endolymphatic duct emerges into the cranial cavity from the otic capsule.

The inner ear is enclosed within the otic capsule and comprises a short ventral lagena (= cochlear duct) and the sacs (saccule, utricle), ampullae and semicircular ducts (anterior, posterior, horizontal) of the vestibular apparatus (Baird 1970; Wever 1978). Both the utricle and saccule contain sense organs (maculae) that respond to changes in body position. These are composed of sensory cells overlain by a gel-like matrix into which is embedded calcareous particles. The whole structure sometimes becomes a large otolith that fills the saccular cavity. An endolymphatic duct connects the saccule to a reservoir, the endolymphatic sac, which lies in the cranial cavity (Fig. 9.2A). The endolymph-filled labyrinth is encircled by the perilymph-filled periotic system (Fig. 9.2). This is complex in shape, but the key parts are the periotic cistern, which lies behind the fenestra vestibuli, and the periotic sac, which exits the vestibular cavity at the periotic foramen as described above. The two regions (cistern and sac) are connected by a periotic duct of varying length (Baird 1970; Wever 1978).

The utricle, saccule, and ampullae, as well as the associated periotic labyrinth, are accommodated within the central vestibular chamber (= cavum vestibuli), and the semicircular ducts are within the canals that perforate the more peripheral parts of the capsule. In many lizards there is little separation between the dorsal vestibular portion of the chamber and the ventral cochlear region, but in some taxa (e.g., gekkotans) a variably developed crest (cochlear crest) marks the boundary between the two cavities. The semicircular canals lead off from the vestibular chamber, which also accommodates their ampullae. The cross-sectional diameter of the semicircular ducts varies, as does their angulation in relation to one another and to the vestibular chamber. The sensitivity of the ducts and their ampullary sense organs is reported to increase with both the diameter and length of the ducts, with the latter expressed in terms of radius of curvature (Sipla and Spoor 2008). Burrowers also show a tendency to have semicircular ducts that are more tightly wrapped around the enlarged central chamber than is seen in terrestrial or climbing taxa.

9.4 Osteological Correlates of Ear Function in Lepidosaurs

9.4.1 Middle Ear

Most of the information on the presence and/or size of the tympanic membrane is derived from the quadrate bone (Fig. 9.3B). An expanded lateral quadrate concavity (= conch) bordered by a tympanic crest is usually taken as evidence of a tympanic ear. In the absence of a tympanic membrane, medial rugosities or pits on the quadrate shaft may provide evidence of stapedial abutment. The stapes itself is rarely preserved in fossil lepidosaurs, but where it is, the length and diameter of the shaft and the relative size of the footplate are indicative of hearing ability. Lepidosaurs with good tympanic ears tend to have a long slender stapes with a small footplate; whereas loss of the tympanic membrane generally correlates with an enlarged footplate (sometimes forming almost the entire wall of the otic capsule) and a short, robust shaft (e.g., Wever 1978). Some living squamates that lack the tympanic membrane utilize alternative surfaces for sound reception. These include an unossified region at the posterior end of the pterygoid (e.g., some chameleons) and skin on the mandible (e.g., some amphisbaenians). The former would be evident in a fossil; the latter could be interpreted if a mineralized extrastapedial element were associated with it.

9.4.2 Inner Ear

Osteological correlates for the inner ear comprise external features of the otic capsule and internal casts. The former include the size of the fenestra vestibuli, the position and size of the periotic foramen, the presence and size of the RST and of any crests (e.g., crista interfenestralis, crista tuberalis) that indicate the extent of a lateral compensatory window. In snakes, these crests usually contribute to the formation of a crista circumfenestralis (CCF), associated with the development of a juxtastapedial recess (Sect. 9.5.6.1) and a specialized fluid re-entrant system (Fig. 9.2C).

Internal casts of the inner ear, now possible through segmentation of the vestibular apparatus from HRXCT or SRμCT scan data, provide an approximation of inner ear anatomy that includes the size and shape of the vestibular chamber and lagena, the presence of any otoliths, and the size and orientation of the semicircular canals (note that "duct" is the preferred term for the soft tissue structure and "canal" for the bone cavity that contains it) (Sipla and Spoor 2008). In squamates, as in other groups, there is a correlation between the shape of the vestibular apparatus overall and lifestyle (e.g., Boistel et al. 2011). To date, relatively few fossil lepidosaurs have been studied using HRXCT data, partly because of the rarity of suitably preserved specimens and partly due to limited access to scanning facilities (e.g., Kearney et al. 2005; Daza et al. 2013).

9.5 The Fossil Record of Ear Evolution in Lepidosauria

9.5.1 Overview

The lepidosaurian ear is comparatively well-known in extant taxa (e.g., Wever 1978). The principal questions that can be addressed from an examination of fossil specimens relate to the evolution of the ear in different lineages and ecotypes, notably, the structure of the stem lepidosaurian and early rhynchocephalian ears, the evolutionary history of the derived squamate ear, and the changes that occurred in ear morphology in response to specialized lifestyles.

9.5.2 *Stem Lepidosaurs*

The ancestral sauropsid ear was probably not tympanic (Sobral, Reisz, Müller, Neenan, and Scheyer, Chap. 8) and an impedance matching middle ear arose independently in descendent lineages (Walsh et al. 2013). The fossil record of stem lepidosaurs is very limited, but currently includes *Paliguana whitei* (Permo-Triassic, South Africa), kuehneosaurs (Triassic, Europe and North America: Robinson 1962; Evans 2009), *Sophineta cracoviensis* (Early Triassic, Poland: Evans and Borsuk-Białynicka 2009), and *Marmoretta oxoniensis* (Jurassic, Europe: Evans 1991). With the exception of the latter species, these stem lepidosaurian taxa all have a quadrate with a large lateral conch and a tympanic crest (Fig. 9.4A, B), which are features suggestive of a tympanic ear. However, within the braincase, the metotic fissure is undivided (as determined from the exoccipitals). The stem lepidosaur middle ear may have been a reasonably efficient sound transducer with limitations imposed on sound perception by the lack of an efficient compensatory pathway.

9.5.3 *Rhynchocephalia*

Although *Sphenodon punctatus*, the New Zealand Tuatara, is the sole surviving rhynchocephalian, the group was both geographically and morphologically diverse in the Mesozoic (Jones and Cree 2012). Rhynchocephalians survived in South America into the Paleocene (Apesteguia et al. 2014), and the ancestors of *Sphenodon* are first recorded in the New Zealand Miocene (Jones et al. 2009).

Sphenodon lacks a functional tympanic membrane (and quadrate conch), and the middle ear cavity is filled with fatty tissue. The stapes has a large footplate and a relatively thin shaft that connects to an expanded extrastapes. Between the extrastapes and the overlying depressor mandibulae muscle is a connective tissue sheet that some authors have interpreted as the original tympanic membrane (Gans and Wever 1976). The lateral surface of the head, therefore, may serve as a sound receptor. Medially, the footplate sits in a large fenestra vestibuli. As in stem lepidosaurs, the metotic fissure is undivided. The periotic sac passes via a posteromedial periotic foramen into the cranial cavity and extends laterally to end blindly in the suprapharyngeal region (Fig. 9.2B). The compensatory system is therefore fairly rudimentary. Nonetheless, *Sphenodon* is relatively good at perceiving low-frequency sound (below 2,000 Hz) using both aerial and substrate pathways (Gans and Wever 1976).

In the past, many researchers regarded *Sphenodon* as a primitive reptile ("a living fossil"), leading to the assumption that it represented the primitive lepidosaurian condition in a majority of its morphological features, including those of the ear (Manley 2000). However, the fossil record shows this to be wrong in many respects. Basal rhynchocephalians from the Late Triassic and Early Jurassic (e.g., *Gephyrosaurus bridensis*, Evans 1980; *Diphydontosaurus avonis*, Whiteside 1986;

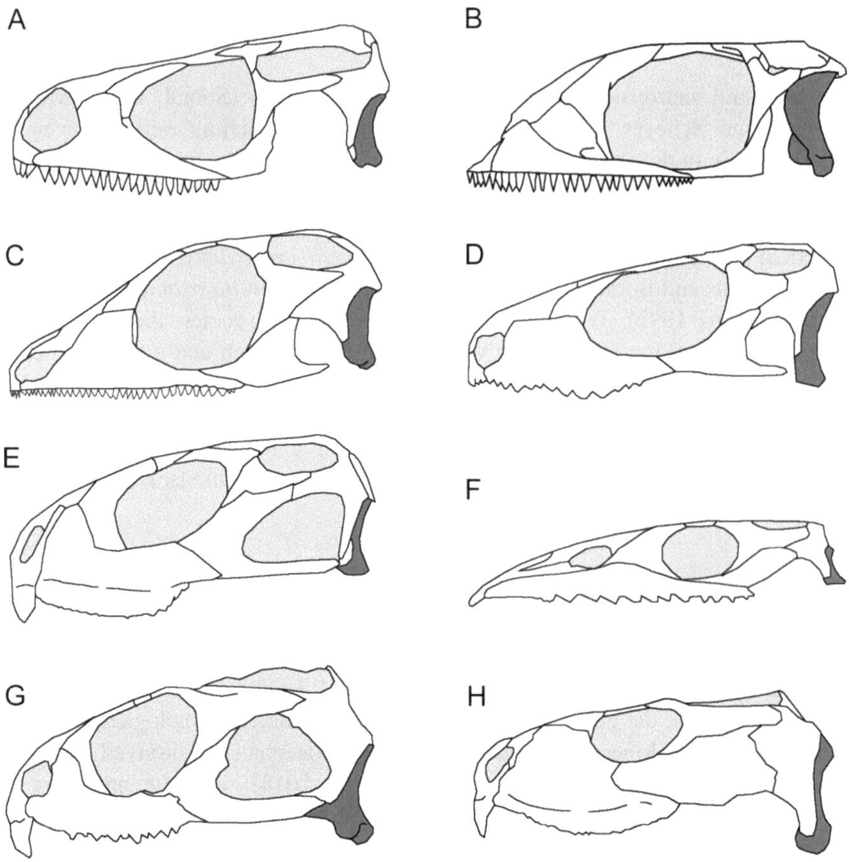

Fig. 9.4 Lateral views of the skull in stem lepidosaurs **(A, B)** and rhynchocephalians **(C–H)** showing differing degrees of quadrate (*dark gray*) embayment in relation to possession of a tympanic membrane. **(A)** *Sophineta cracoviensis* (Early Triassic), **(B)** *Kuehneosaurus latus* (Late Triassic), **(C)** *Gephyrosaurus bridensis* (Early Jurassic), **(D)** *Planocephalosaurus robinsonae* (Late Triassic), **(E)** *Clevosaurus bairdii* (Late Triassic), **(F)** *Pleurosaurus goldfussi* (Late Jurassic—Early Cretaceous), **(G)** *Sphenodon punctatus* (Recent), **(H)** *Priosphenodon avelasi* (Late Cretaceous). (**A** redrawn from Evans and Borsuk-Białynicka 2009; **B**, **C** redrawn from Evans 2009; **D–H** redrawn from Jones 2008)

Planocephalosaurus robinsonae, Fraser 1982), as well as the more derived *Clevosaurus hudsoni* (Robinson 1973; Fraser 1988), resemble stem lepidosaurs and squamates in having a quadrate with a well-developed tympanic crest and lateral conch (Fig. 9.4C–D). They almost certainly had a tympanic middle ear. However, the size of the fenestra vestibuli and, therefore, the size of the stapedial footplate have not been reported in any of these taxa. No fossil rhynchocephalian shows a divided metotic fissure (Jones 2006; contra Wu 1994) or evidence of a compensatory window.

However, if the ears of early rhynchocephalians were tympanic, it raises the question as to why the middle ear degenerated in *Sphenodon*. The quadrate of *Clevosaurus bairdii* (Fig. 9.4E) (Late Triassic, USA) appears less emarginated than that of *C. hudsoni*. A further reduction is seen in many derived rhynchocephalians from the Jurassic onward (e.g., *Palaeopleurosaurus posidoniae* and *Pleurosaurus goldfussi*, Carroll 1985; Carroll and Wild 1994; *Priosphenodon avelasi*, Apesteguia and Novas 2003) (Fig. 9.4F–H). Loss of the tympanic membrane may be related to the refinement of the characteristic fore–aft shearing action by the jaw of derived rhynchocephalians (e.g., Jones 2008; Jones et al. 2012). It is difficult to see how a tympanic membrane stretched between lower jaw and quadrate could have accommodated movement of that kind.

9.5.4 General Squamates

Given the morphology of stem lepidosaurs and rhynchocephalians, it is likely that stem squamates also inherited a tympanic ear that lacked an efficient compensatory mechanism. Development of the latter required the division of the metotic fissure into dorsal and ventral passages as described previously (Sect. 9.3). Again, however, the timing of this change is difficult to determine. Although stem squamates must have existed in the mid-Triassic (due to the presence of their sister clade, Rhynchocephalia), the earliest uncontested squamate fossils are from the Middle Jurassic (Britain, Russia, Central Asia) (Evans 1998; Evans and Jones 2010). This Middle Jurassic material is mostly composed of isolated skeletal elements, but there is currently no braincase material. The squamate record improves in the Late Jurassic, and apart from embayed quadrates (e.g., the Solnhofen lizard, *Eichstaettisaurus schroederi*) (Evans et al. 2004), there are no direct data on the ear in these taxa. However, the anguimorph *Dorsetisaurus* is known from both the Late Jurassic (Portugal: Broschinski 2001; North America: Prothero and Estes, 1980) and Early Cretaceous. Although the Jurassic material of this genus is fragmentary, that from the Early Cretaceous includes a well-preserved braincase and it is reasonable to assume that the Jurassic representatives had a similar morphology.

The squamate fossil record improves significantly in the Cretaceous with well-preserved skulls in many specimens. However, in many cases the only indicator of ear anatomy is the presence of a quadrate that has a lateral conch and tympanic crest (Fig. 9.5A–C). The earliest known squamate braincases are currently from the Early Cretaceous. They include *Dorsetisaurus purbeckensis* (Fig. 9.5D) (UK, Hoffstetter 1967) and other taxa from South Africa (Ross et al. 1999), Mongolia (Conrad and Norell 2006; Conrad and Daza 2015), and Thailand (Fernandez et al. 2015). In all of these braincase specimens, the metotic fissure is fully divided with the aLRST opening defined by flanking crests. Furthermore, where the preservation is sufficiently good (Ross et al. 1999; Conrad and Norell 2006), these Early Cretaceous squamate braincases possess all of the features, internal and external, described in extant lizards (Sect. 9.3). Taken together with

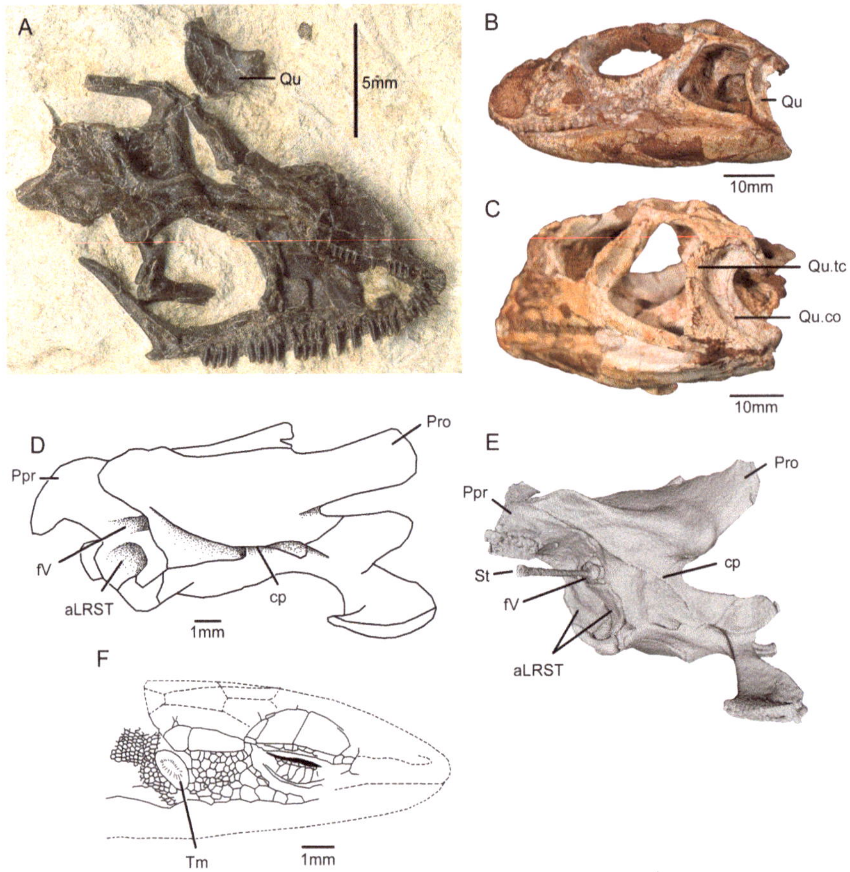

Fig. 9.5 The ear in fossil lizards. **(A)** BGS GSb581, the skull of the Early Cretaceous lacertoid lizard *Purbicella ragei*, Purbeck Limestone Group, UK, with the disarticulated quadrate above it. **(B, C)** Skulls of the Late Cretaceous lizard *Tianyusaurus* sp., Nanxiong Formation, China, showing the typically embayed quadrate associated with a tympanic lizard ear: **(B)** lateral view of an immature skull (NHMG 9316); **(C)** posterolateral view of a mature skull (NHMG 9317) showing the quadrate conch and expanded tympanic crest. **(D)** Lateral view of the braincase of the Early Cretaceous anguimorph *Dorsetisaurus purbeckensis,* showing earliest recorded subdivision of Meckelian fossa to produce a compensatory window in apertura lateralis recessus scalae tympani. **(E)** Lateral view of the braincase of the extant anguimorph lizard *Diploglossus* lessonae for comparison. **(F)** Lateral view of the head of the Eocene *Succinilacerta succinea*, a lacertid lizard preserved in amber. (**A–C** original photographs; **D** redrawn and simplified from Hoffstetter 1967; **E** based on original HRXCT scan, imaged using Meshlab; **F** redrawn and adapted from Borsuk-Białynicka et al. 1999)

quadrate anatomy, this implies that squamate sound perception in the Early Cretaceous was similar to that of extant lizards.

The use of HRXCT permitted a partial reconstruction of the inner ear morphology (semicircular canals, vestibule, and lagenar recess) in the Early Cretaceous,

Mongolian stem gekkotan *Norellius nyctisaurops* (Conrad and Daza 2015). The reconstruction revealed the presence of a constriction between the vestibule and cochlear duct, as well as the presence of a small accessory endolymphatic duct that is connected to the anterior semicircular canal and opens on the dorsal surface of the otic capsule. Further stem gekkotan material was recently described from the mid-Cretaceous (Aptian-Cenomanian) of Burma (Daza et al. 2016). These specimens are three-dimensionally preserved in amber, and they revealed the presence of a stapes perforated by a stapedial foramen as in extant gekkotans, dibamids and some scincids. Remarkably, recently described embryonic lizards from the Early Cretaceous of Thailand (Fernandez et al. 2015) imaged in ovo using SRμCT have preserved large calcified endolymphatic sacs in the cranial cavity. Calcified sacs of this kind are typically (though not exclusively) associated with gekkotan lizards that have hard (rather than leathery) eggshells, and it is interesting that the Thai eggs, though attributed to anguimorphs, also appear to be hard-shelled.

Direct evidence that the concave quadrate supported a tympanum is provided by another Early Cretaceous lizard, *Liushusaurus acanthocaudata* (Yixian Formation, China: Evans and Wang 2010). One specimen (IVPP V14715: Evans and Wang 2010) has traces of cranial soft tissues. As in most extant lizards, the head scales terminate at the anterior border of the quadrate, leaving the outline of the tympanic membrane above the retroarticular process of the lower jaw. Traces of the cartilaginous extrastapes lie adjacent to the tympanic region. The outline of the tympanic membrane and adjacent scalation are also preserved in a younger (Eocene) lizard specimen preserved in amber, the lacertid *Succinilacerta succinea* (Fig. 9.5F) (Borsuk-Białynicka et al. 1999).

Three-dimensionally preserved cranial material is known for many Late Cretaceous lizards, especially from Mongolia and China, but the braincase is only described (often briefly) in a subset of these. They include anguimorphs (e.g., Borsuk-Białynicka 1984; Norell and Gao 1997; Conrad et al. 2011), lacertoids (Borsuk-Białynicka 1988), iguanians (Borsuk-Białynicka and Moody 1984), and a stem gekkotan (Borsuk-Białynicka 1990). The latter was recently redescribed following HRXCT scanning (Daza et al. 2013). Of these examples, the most complete material is that of the anguimorphs *Estesia mongoliensis* and *Gobiderma pulchrum*, the braincases of which are described in detail by Norell and Gao (1997) and Conrad et al. (2011), respectively, with *Gobiderma* examined by HRXCT. Conrad and Norell (2008) also described the braincase and ear region of Neogene anguimorph glyptosaurs. However, in none of these taxa does the ear region depart substantively from that of the typical extant lizard pattern described above, and their hearing is likely to have been similar. To date, however, the inner ear of fossil lizards has received little attention.

Squamates show a diversity of lifestyles, but two in particular are associated with ear modifications that can be recognized in the fossil record. These are burrowing and marine swimming.

9.5.5 Burrowing Squamates

Many squamates burrow in loose soil or leaf litter to find food, escape predators, or avoid excessive heat, but a sizeable minority are adapted more obviously to a fossorial or semi-fossorial existence. These include many limbless or limb-reduced lizards, as well as larger groups like the scolecophidian snakes and amphisbaenians. Burrowing imposes constraints on the ear as it needs to be protected from physical damage. In addition, much of the sound reaching the head comes via the substrate rather than the air. Unlike ears that detect airborne sound, and thus pressure changes, those of burrowers are displacement sensitive (sensu Lombard and Hetherington 1993). Sound vibrations shake the otic capsule creating relative motion between it and the more loosely attached elements like the stapes. Although the footplate may remain still, the relative motion between it and the otic capsule causes displacements at the fenestra vestibuli similar to those that would be caused by movements of a tympanic membrane. This system is improved by enlargement of the stapes and footplate so that they have greater inertia. Alternatively, incoming sound vibrations actually deform the bone of the otic capsule and thus produce displacement of the perilymph directly (Lombard and Hetherington 1993).

Burrowing adaptations have evolved independently in several squamate lineages, but their ears show similar adaptations (e.g., Baird 1970; Wever 1973). The middle ear is typically lost or reduced and may become filled with connective tissue or muscle (e.g., depressor mandibulae). The stapes usually has increased mass, a short robust shaft, and an enlarged footplate that forms much of the lateral wall of the otic capsule. In the absence of a tympanic membrane, the stapes abuts the quadrate, allowing some sound vibrations to be transmitted directly from the surface of the head. The aLRST is often reduced in size, with the periotic sac ending blindly rather than expanding into a compensatory window. However, many burrowers develop a re-entrant fluid circuit (Wever 1973) whereby the periotic sac leaves the vestibular chamber ventrally or medially and finds an alternative pathway to the lateral surface of the stapedial foot plate (Baird 1970; Wever 1973).

Burrowing reptiles are poorly represented in the fossil record. *Tamaulipasaurus morenoi* from the Early Jurassic (Pliensbachian) La Boca Formation of Mexico (Clark and Hernandez 1994) is an enigmatic little burrower represented by skull and neck vertebrae. Originally placed as either a stem lepidosaur or stem squamate (Clark and Hernandez 1994), its phylogenetic position within reptiles is currently unresolved, but it is included here for completeness. The skull is convergent on extant squamate burrowers like amphisbaenians and differs strikingly from the stem lepidosaur taxa described previously (Sect. 9.5.2). The quadrate is narrow and there is a large fenestra vestibuli, presumably accommodating a large stapedial footplate (obscured). A short stapedial shaft projects laterally and contacts a second component (possibly an ossified extrastapes or hyoid element) that abuts the quadrate. The occipital surface of the skull bears a large opening (jugular foramen of Clark and Hernandez 1994) that appears circular in one specimen and ovoid in another. Without visible sutures, the identity of this opening remains uncertain, but it presumably represents all or part of the metotic fissure. This material would benefit from re-examination either with HRXCT or SRμCT.

Other than a few early snakes (Sect. 9.5.6), the only clear example of a Cretaceous squamate burrower is *Sineoamphisbaena hexatabularis* (Fig. 9.6) (Late Cretaceous, China: Wu et al. 1996). Although it was originally classified as a stem amphisbaenian, the relationships of this small lizard are unclear (e.g., Reeder et al. 2015). The skull is rounded and robust (Fig. 9.6A, B) (Wu et al. 1996), and the ear anatomy matches that of extant squamate burrowers as outlined above. The short quadrate lacks an embayment for a tympanic membrane. The stapes has a large footplate and short shaft that abutted the quadrate, and the aLRST is reduced to a small foramen for the glossopharyngeal nerve (Fig. 9.6C). The otic capsules appear inflated compared the rest of the skull (Fig. 9.6B)

Extant amphisbaenians spend most of their lives underground. There is no tympanic membrane, but in many amphisbaenians, the extrastapes forms a cartilage plate or bar that runs along the lower jaw with its tip embedded in the skin—often at a specially enlarged labial scale (Fig. 9.7A). Vibrations pass from the skin to the extrastapes and then along the stapes. The stapedial footplate fits into a fenestra vestibuli that may extend over most of the external surface of the otic capsule

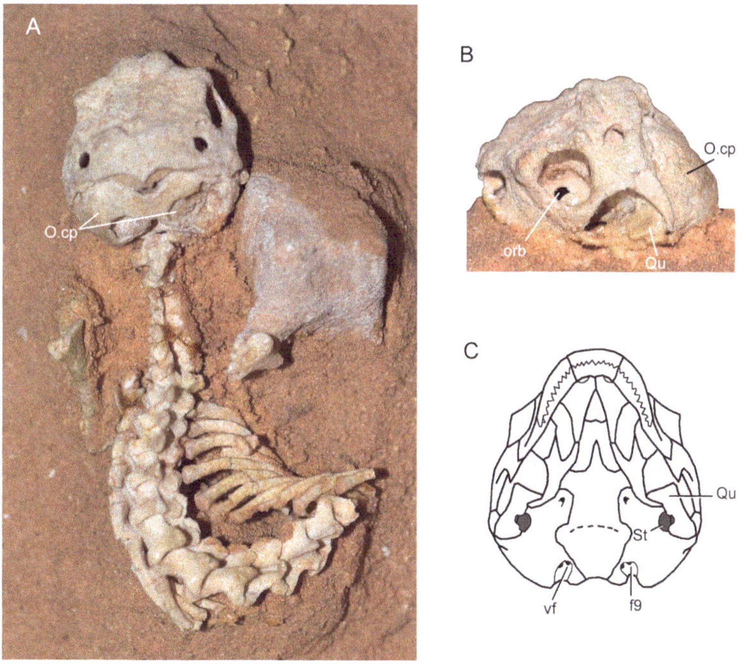

Fig. 9.6 The burrowing squamate ear: *Sineoamphisbaena hexatabularis*. **(A)** Dorsal view of the holotype specimen (IVPP V10593). **(B)** Left-lateral view of the skull of IVPP V10593. **(C)** Reconstructed palatal view showing stapes (dark gray) with short shaft and large footplate. (**A**, **B** original photographs; **C** redrawn from Wu et al. 1996)

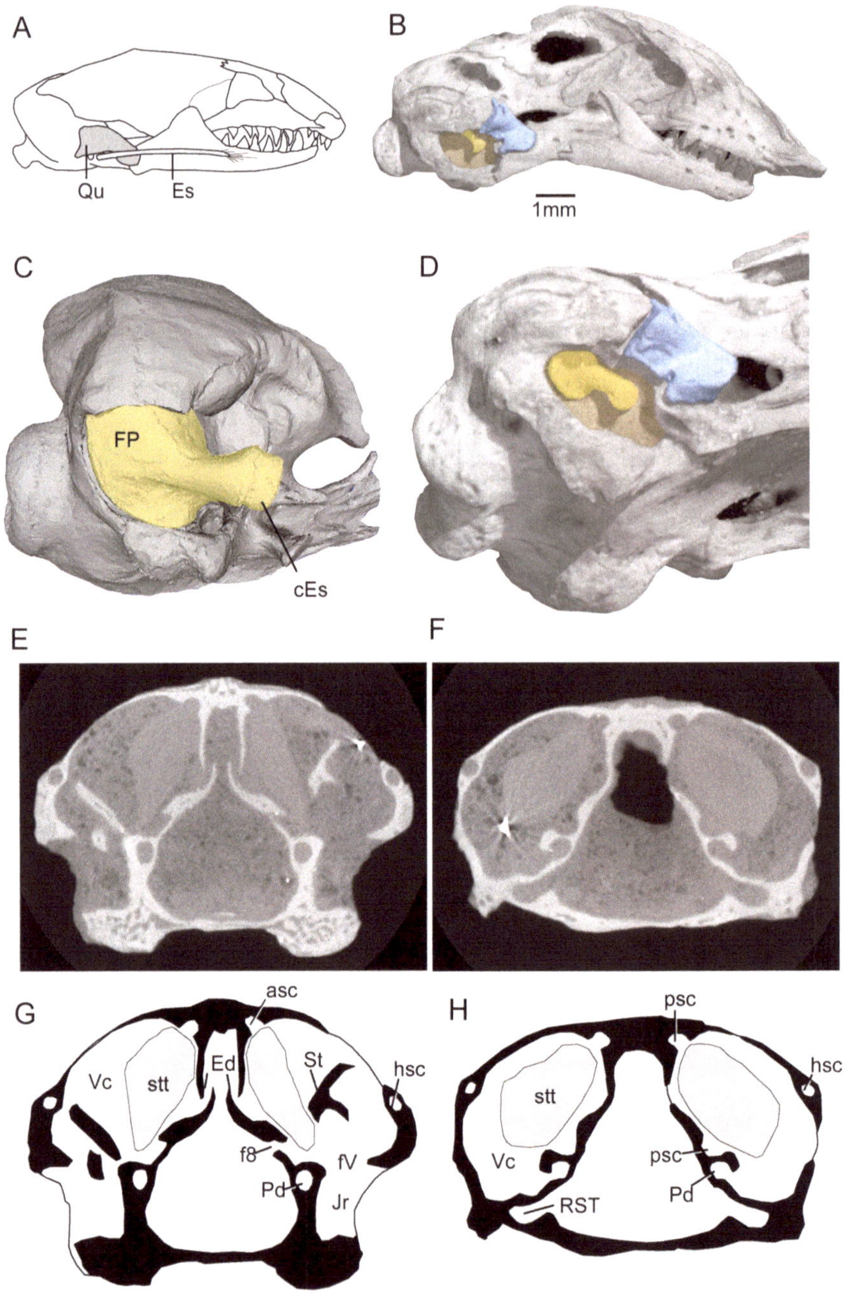
A
Qu
Es
B
1mm
C
FP
cEs
D
E
F
G
asc
Ed
St
hsc
Vc
stt
f8
fV
Pd
Jr
H
psc
hsc
stt
Vc
psc
Pd
RST

◂**Fig. 9.7** The burrowing squamate ear: amphisbaenians. **(A)** Lateral view of the skull of the extant *Amphisbaena alba* showing the relationship of the extrastapes to the quadrate and lower jaw. **(B)** Lateral view of the skull of *Rhineura hatcherii* (BAD 18303), Oligocene, USA, surface model created from HRXCT data. **(C)** The braincase of the extant *Diplometopon zarudnyi*, right-lateral view, to show the size and robusticity of the stapes and its footplate. **(D–H)** *Rhineura hatcherii* (BAD 18303): **(D)** detail of ear region in ventrolateral view; **(E, G)** coronal section through fenestra vestibuli and jugular recess; **(F, H)** coronal section posterior to E showing recessus scala tympani closed laterally. Color scheme: *blue*, quadrate; *yellow*, stapes; *brown*, jugular recess. (**A** redrawn from Gans and Wever 1972; **B**, **D**, **E**, **F**, originals from HRXCT scan data; **C** original based on HRXCT scan, surface visualized with Meshlab; **G**, **H**, original explanatory drawings of **E**, **F**)

(Fig. 9.7C) and is set within a deep jugular recess (Fig. 9.7B, D–G) (Kearney et al. 2005). The stapedial shaft projects laterally to the level of the quadrate where, in most taxa, it abuts the extrastapes (Gans and Wever 1972; Wever and Gans 1973). In the inner ear, the large vestibular chamber generally accommodates an extensive otolith (stt, Fig. 9.7E–G) and, as in many burrowers, the semicircular canals are narrow and compacted, resulting in little separation between them and the central vestibular chamber. As in *Sineoamphisbaena*, the aLRST is restricted to a foramen for the glossopharyngeal nerve and does not accommodate a compensatory window. The periotic cistern fills the space behind the stapedial footplate but then narrows into a long periotic duct that enters a channel (sometimes fully enclosed, e.g., *Rhineura floridana*) that runs first anteriorly and then turns sharply through 180° to run posteromedially within the walls of the otic capsule to reach the periotic foramen. From there, the periotic sac enters the RST and extends medially where it communicates with fluid-filled spaces that run anteromedially through the cranial cavity before passing via the trigeminal foramen to enter the jugular recess and the lateral surface of the stapedial footplate (Gans 1978; Wever 1978). Movements of the stapedial footplate set up a continuous fluid displacement around the circuit (Gans 1978; Wever 1978). Amphisbaenians perceive low-frequency sound (300–700 Hz) and reportedly have better reception than snakes, but this is dependent on the extrastapedial system. Any interruption results in a major loss of sensitivity (Gans and Wever 1972).

Molecular phylogenetic analyses group amphisbaenians with lacertid lizards (e.g., Vidal and Hedges 2005; Pyron et al. 2013), which are predominantly tetrapodal surface dwellers. *Cryptolacerta hassiaca* (Eocene, Germany: Müller et al. 2011) was described as a potential transitional stem amphisbaenian, but some subsequent analyses (Longrich et al. 2015; Tałanda 2016) have placed *Cryptolacerta* with lacertids rather than amphisbaenians. Moreover, the braincase and ear region are crushed. More recently, *Slavoia darevskii* [Late Cretaceous (Campanian), Mongolia] has been proposed as a stem amphisbaenian (Tałanda 2016). Although its morphology is plausible for a burrower (short snout, robust build, reduced limbs), the quadrate appears to be large and embayed, suggesting that the ear was tympanic. However, the braincase has never been described, although many skulls of this taxon exist in collections. HRXCT scanning could be informative.

The earliest uncontested amphisbaenian records are currently from the Paleocene of Europe and North America (Folie et al. 2013; Longrich et al. 2015), but the first three-dimensionally preserved skulls are those of Eocene rhineurids from North America (Berman 1973, 1976). *Spathyrhynchus fossorium* (Middle Eocene, North America: Berman 1973) resembles living amphisbaenians in having a large fenestra vestibuli that lies within a deep jugular recess and is covered by an ovoid stapedial footplate. The stapedial shaft is directed anteroventrally, and it ends in a small knob which, by comparison with extant taxa, would have articulated with the extrastapes (a remnant of which may persist). However, the shaft differs from that of extant taxa in being long and slender rather than short and robust. *Dycticonastis* (Oligocene, North America: Berman 1976) is closely similar, and Berman (1976) correlates the long slender stapedial shaft of these and other early taxa (e.g., *Ototriton* and *Jepsibaena*, Eocene) with the retention of paroccipital processes on the braincase and, therefore, a wider head. In amphisbaenians with narrow skulls (e.g., *Lophocranion*, Miocene, Kenya: Charig and Gans 1990), the stapedial shaft is shorter and more robust. *Rhineura hatcherii* (Oligocene, North America: Kearney et al. 2005) was HRXCT scanned (Fig. 9.7B, D–F). Its ear closely resembles that of the extant *R. floridana*, including an internal channel for the periotic duct that leads to the periotic foramen, a large saccular otolith, a large fenestra vestibuli that opens into the roof of the jugular recess, a stapes with a relatively large footplate and a short shaft bearing a terminal articular region for the extrastapes, and narrow semicircular canals packed close to the vestibular chamber (Fig. 9.7B, D–H).

9.5.6 Marine Squamates, the Mosasaurs

Today, the only squamates that are obligate marine swimmers are sea snakes, although the Galapagos marine iguana, *Amblyrhynchus cristatus,* regularly feeds offshore. However, in the Late Cretaceous, a group of lizards (Mosasauria: Marsh, 1880; Camp 1942) dominated marine ecosystems worldwide for around 30 million years.

Until recently, mosasaurs were widely considered to be relatives of extant varanid lizards (e.g., Conrad 2008; Wiens et al. 2010; Conrad et al. 2011), but their position is now contested (e.g., Gauthier et al. 2012 versus Reeder et al. 2015). Figure 9.1 shows one recent hypothesis (Reeder et al. 2015). Within Mosasauria, three morphological grades are typically recognized: "dolichosaurs", which were small, often long-bodied Albian-Cenomanian lizards (e.g., *Dolichosaurus longicollis:* Caldwell 2001; *Pontosaurus lesinensis*: Caldwell and Dal Sasso 2004; Pierce and Caldwell 2004); "aigialosaurs", which were larger, Cenomanian-Turonian lizards (e.g., *Aigialosaurus dalmaticus,* Carroll and DeBraga 1992; Dutchak and Caldwell 2006); and finally the mainly Turonian-Maastrichtian mosasaurids (Fig. 9.8A) (e.g., Bell 1997). A variety of different within-clade names have been invoked (e.g., Bell 1997), but herein Mosasauridae (mosasaurid) is used for the Late Cretaceous crown clade (Gauthier et al. 2012), Mosasauroidea (mosasauroid) for the group encompassing aigialosaurs and Mosasauridae (e.g., Conrad et al. 2011), and Mosasauria

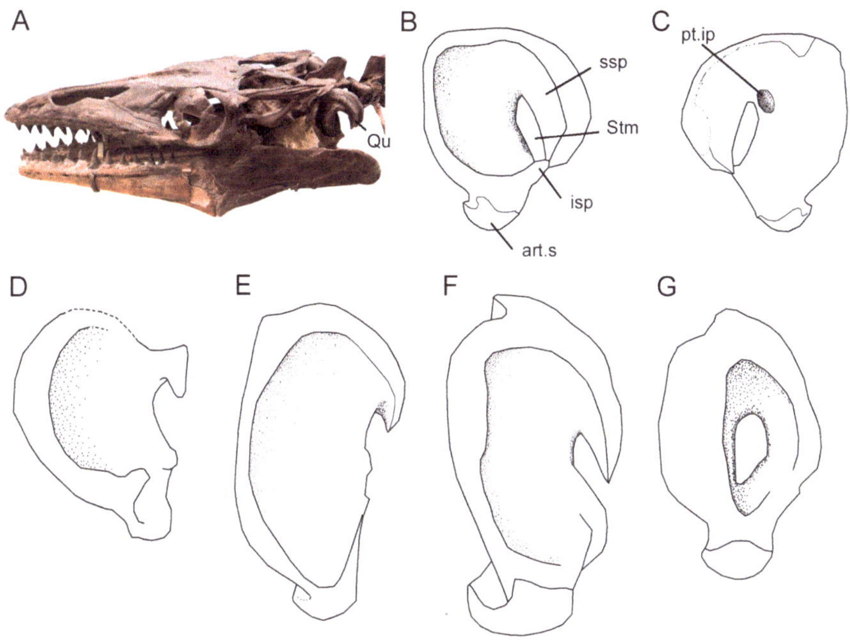

Fig. 9.8 Mosasaur quadrates. **(A)** lateral view of the skull of *Platecarpus tympaniticus* (KUVP 1007) and **(B–G)** isolated right quadrates: **(B, C)** *Platecarpus tympaniticus*, lateral and medial views, respectively; **(D)** *Pontosaurus kornhuberi*; **(E)** *Mosasaurus hoffmannii;* **(F)** *Tylosaurus proriger*; **(G)** *Prognathodon rapax* (not to scale). (**A–C**, **E–G** redrawn and simplified from Russell 1967; **D** redrawn from Caldwell 2006)

(mosasaur) for the total clade (dolichosaurs and Mosasauroidea) (Marsh 1880; Camp 1942). Although Mosasauridae is generally considered as monophyletic (e.g., Gauthier et al. 2012), the other two "groups" may not be. Understanding these relationships is important for understanding the evolutionary trajectory of their ear morphology, particularly in relation to increasing marine specialization.

Dolichosaurs, aigialosaurs, and basal mosasaurids had feet rather than paddles and their pelves were still firmly in contact with the sacrum, suggesting that they were at least facultatively terrestrial (Caldwell and Palci 2007). More derived mosasaurids developed paddles and lost the sacro-pelvic contact, a change that may also mark a shift from a primarily undulatory (anguilliform) locomotor style to one that relied to a greater degree on the tail (Lindgren et al. 2007, 2011). Mosasaurs mainly colonized epicontinental seas, but their global distribution from the Campanian period onward suggests an increasing tendency to move into deeper water as they became more adapted to a pelagic lifestyle (Lindgren et al. 2007, 2011; Houssaye et al. 2013). Based on mosasaur brain endocasts and snout morphology, Polcyn (2010) suggested that the increasing marine specialization in derived mosasaurids was accompanied by a shift from a reliance on olfactory cues to vestibulo-acoustic stimuli, as has been reported for whales.

Sound vibrations in water have greater strength than those in air and no impedance matching is required between the water and soft tissues. Swimming animals also live in a more three-dimensional environment than land animals, and the effects of gravity are reduced by the density and viscosity of the water. Living marine tetrapods (e.g., sea turtles, whales) often show specializations in both the middle and inner ear (Lombard and Hetherington 1993; Hetherington 2008). Some extant taxa continue to use a tympanic route, with a tympanic membrane or plate underlying the skin of the head but overlying an air-filled chamber so that displacements of skin move the stapes. Other taxa rely on bone conduction as in burrowers, with an enlarged stapes or stapedial footplate acting as an inertial element. Mosasaurs seem to fall into the first category.

9.5.6.1 Mosasaur Middle Ears

The mosasauroid quadrate has a distinctive shape (Fig. 9.8). The dorsal margin curves posteriorly or posteroventrally to form a suprastapedial process and this is mirrored to a varying degree by an infrastapedial process (Fig. 9.8B). Where known, the "dolichosaur" quadrate (e.g., *Pontosaurus*: Pierce and Caldwell 2004; Caldwell 2006: Fig. 9.8D) approaches this morphology, but the infrastapedial process is small. Enlargement of this process in more derived taxa creates a deep notch or, frequently, a fully enclosed meatus when the supra-stapedial and infra-stapedial processes make contact (Fig. 9.8B–C, F–G). This meatus forms a passage for the stapedial shaft. Posterodorsal and posteroventral expansion of the tympanic crest may then combine to form a circular or ovoid bulla around the stapedial shaft (Fig. 9.8G). Berman and Regal (1967) speculated that this morphology might protect the middle ear cavity from being collapsed by movements of the quadrate or depressor mandibulae, but this does not occur in lizards with highly mobile quadrates. By comparison with sea turtles, the bulla-like quadrate shape is probably associated with aquatic habits. Variations in mosasauroid quadrate morphology (overall shape, size of the infrastapedial process, depth of the conch, development of the tympanic crests, see Fig. 9.8) have been documented (e.g., Camp 1942; Callison 1967; Konishi and Caldwell 2009; Fernandez and Talevi 2015) and used in phylogenetic analyses (e.g., Bell 1997). However, there have been few attempts to link the variation to hearing and/or lifestyle, other than Polcyn (2010), who suggested that increased depth of the conch in derived plioplatecarpine mosasaurids might have enhanced spatial sensitivity.

A round or ovoid pit in the medial surface of the mosasauroid quadrate (Fig. 9.8C) probably accommodated the equivalent of the internal (= quadrate) process of the extrastapes (McDowell 1967; Rieppel and Zaher 2000). In extant squamates, this process is usually cartilaginous, but mineralization has been recorded in some mosasaurids (*Plioplatecarpus*, McDowell 1967; *Platecarpus*, Callison 1967), suggesting a stiffening of the extrastapedial mechanism to provide additional support (Fig. 9.9C–E). The mosasaurid stapes is usually long and relatively slender [e.g., *Plotosaurus* (as *Kolposaurus*), Camp 1942; *Plioplatecarpus*,

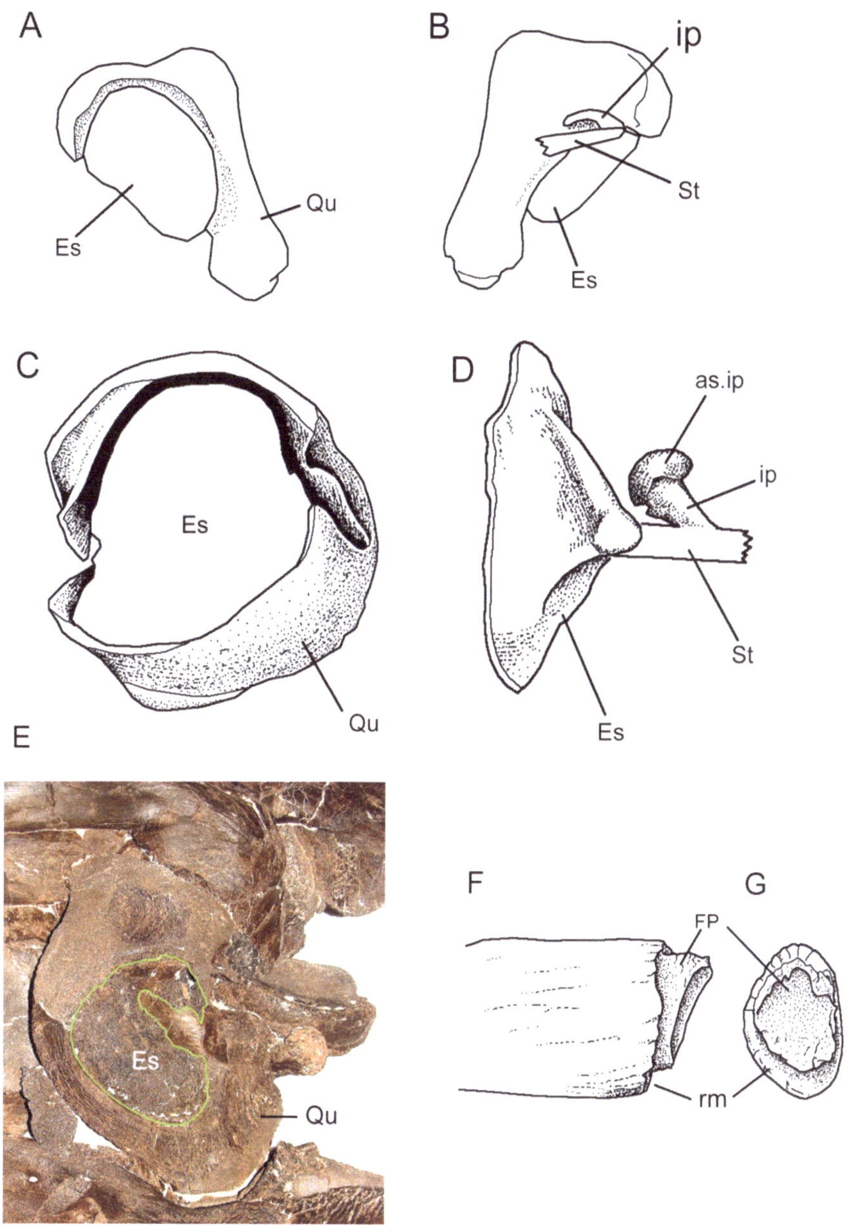

Fig. 9.9 Mosasaur stapes and extrastapes. (**A, B**) Lateral and medial views, respectively, of the quadrate and discoidal extrastapes of the extant lizard *Lanthanotus borneensis*. (**C, D**) *Plioplatecarpus houzeani*: (**C**) lateral view of the right quadrate and mineralized extrastapes; (**D**) reconstruction of the mineralized extrastapes, stapes, and internal process, anterior view. (**E**) Left quadrate of *Tylosaurus proriger* (FFHM-1997-10) showing mineralized extrastapes, outlined in *green*. (**F, G**) Posterior and medial views, respectively, of the footplate region of the stapes in *Platecarpus sp.* (**A**, **B** redrawn and simplified from McDowell 1967; **C**, **D** redrawn from Camp 1942; **F**, **G** redrawn from Callison 1967)

McDowell 1967; *Clidastes, Platecarpus,* and *Tylosaurus*, Callison 1967]. At its medial end, the stapedial shaft bears a terminal footplate that is smaller in diameter than the shaft itself (Fig. 9.9F–G), creating a margin for a thick cartilaginous or fibrous ring that attached the footplate to the rim of the fenestra vestibuli (*Clidastes, Platecarpus*) (Callison 1967). A complementary furrow may be present around the fenestra. This arrangement may have served to resist inward displacement of the footplate due to water pressure (Camp 1942), also aided by a limiting ridge within the fenestra and the medial restriction of the stapedial shaft to a narrow groove (stapedial canal) (Callison 1967) in the otoccipital (e.g., in *Platecarpus, Plioplatecarpus*).

Discs of bone or thick, calcified cartilage have been recorded within the quadrate conch in several mosasaurid specimens (e.g., *Platecarpus* and *Tylosaurus*, Williston 1914; Callison 1967; *Plioplatecarpus*, Camp 1942; *Clidastes*, Callison 1967; *Ectenosaurus*, Russell 1967; *Aigialosaurus*, Carroll and DeBraga 1992; but see Dutchak and Caldwell 2006; *Halisaurus*, Fernandez and Talevi 2015). These discs were originally interpreted as mineralized tympanic membranes that represented an adaptation to deep diving, the pressure of which might rupture a thinner structure. However, a similar discoid cartilage in the extant earless monitor lizard, *Lanthanotus borneensis* (Fig. 9.7A–B) is an extrastapes (McDowell 1967), and subsequent authors have generally accepted this interpretation of the mosasaurid disc (e.g., Konishi and Caldwell 2009). Like *Lanthanotus*, these mosasaurids probably lacked a tympanic membrane, at least of the type that would have responded to airborne sound. Polcyn (2010) proposed that the extrastapes formed a "stiff but functional tympanum" by expansion and calcification of the cartilage, a not unreasonable explanation given that the disc fills the lateral quadrate cavity (Fig. 9.9C, E) and connects with the relatively slender stapes through the stapedial meatus (Fig. 9.9D).

Extant sea turtles provide a partial analogue for the mosasauroid ear (Lombard and Hetherington 1993) in that they have a strongly concave quadrate that forms much of the medial wall of the middle ear, enclosure of the stapedial shaft in a bony meatus, and an enlarged disc-like extrastapes (Lombard and Hetherington 1993; Christensen-Dalsgaard et al. 2012). In sea turtles, a thickened tympanic membrane underlies the skin and scales and is itself underlain by a layer of fatty connective tissue (Wever and Vernon 1956; Hetherington 2008). As in sea turtles, the enclosure of the mosasauroid stapes within a meatus would have limited its movements to a piston-like action with little possibility of rotation or hinging to provide lever action. The relatively slender mosasaurid stapes and small footplate differ strikingly from those of marine ichthyosaurs (Sobral, Reisz, Müller, Neenan, and Scheyer, Chap. 8), in which a massive stapes acts as an inertial element. The mosasaur ear is likely to have been adapted primarily to receive sound passing through water, mainly by a direct "tympanic" route through thick skin and the extrastapedial plate to the stapes and then fenestra vestibuli (Hetherington 2008).

In more basal mosasaurs, where the quadrate is less specialized and the postcranial morphology is suggestive of amphibious habits, the ear may have been able to perceive aerial sounds in a manner similar to that of terrestrial or occasionally amphibious lizards like varanids (e.g., Lombard and Hetherington 1993; Hetherington 2008). A reconstruction of the head of the dolichosaur *Pontosaurus* (Caldwell and Dal Sasso 2004; Caldwell 2006) clearly made that assumption. Although this specimen does preserve traces of head scales, they do not extend to the ear region and, therefore, the restoration of an external tympanum is speculative.

9.5.6.2 The Mosasaur Inner Ear

There are no informative descriptions of the dolichosaur or aigialosaur braincase, but it has been described in several mosasaurids (e.g., Russell 1967; Rieppel and Zaher 2000; Cuthbertson et al. 2015). The morphology is generally similar to that of extant lizards. Taking *Platecarpus* as an example (Camp 1942), the fenestra vestibuli, aLRST, and vagus (= jugular) foramen were described as more broadly exposed than in the extant *Varanus* due to a weaker development of surrounding bony crests (Rieppel and Zaher 2000). However, the aLRST is smaller and forms a rounded aperture or slit. If there was a compensatory window within the aLRST, it would have been small and without an obvious frame.

Camp (1942) prepared an internal cast of the ear region of *Platecarpus* by filling the cavities in the braincase with latex and then removing the surrounding bone on one side (Fig. 9.10A, B). HRXCT permits less destructive access to the inner ear. To date, there has been relatively little work on mosasaur inner ears and much of that is contained in conference abstracts (Polcyn 2008, 2010; Yi et al. 2012) rather than peer-reviewed publications. The exceptions, other than Camp (1942), are Georgi and Sipla (2008) on *Platecarpus* and *Tylosaurus*, and Cuthbertson et al. (2015) on *Plioplatecarpus peckensis*. The latter described *Plioplatecarpus* as having a relatively long, conical lagena, with the fenestra vestibuli placed at its posterodorsal limit.

The vestibular apparatus of *Platecarpus* and *Tylosaurus* has been described as broadly similar to that of *Varanus* (Camp 1942; Georgi and Sipla 2008; Yi et al. 2012), although the central chamber is relatively larger in the living taxon (Fig. 9.10A, D). In *Platecarpus*, the anterior semicircular canal is longer than the posterior one and of slightly greater diameter. The horizontal canal is relatively short (Fig. 9.10A, B). In *Plioplatecarpus* (Cuthbertson et al. 2015), the canals are more symmetrical in their cross-sectional diameter (narrow) and radius of curvature (strong), with the anterior and posterior canals arching up above the level of the central vestibular chamber (as in *Platecarpus*). Taking shape and orientation together, Cuthbertson et al. concluded that the vestibular apparatus of *Plioplatecarpus* was most sensitive in the pitch plane. In their analysis, *Plioplatecarpus* fell within the range of extant lizards with respect to semicircular canal diameter, but the radius of curvature was most like that of arboreal and marine iguanians moving in three-dimensional space (e.g., *Amblyrhynchus*, Fig. 9.10C). Given that most mosasaurs probably retained an essentially lizard-like, undulatory swimming mode (Lindgren et al. 2007, 2011), it is not surprising that their

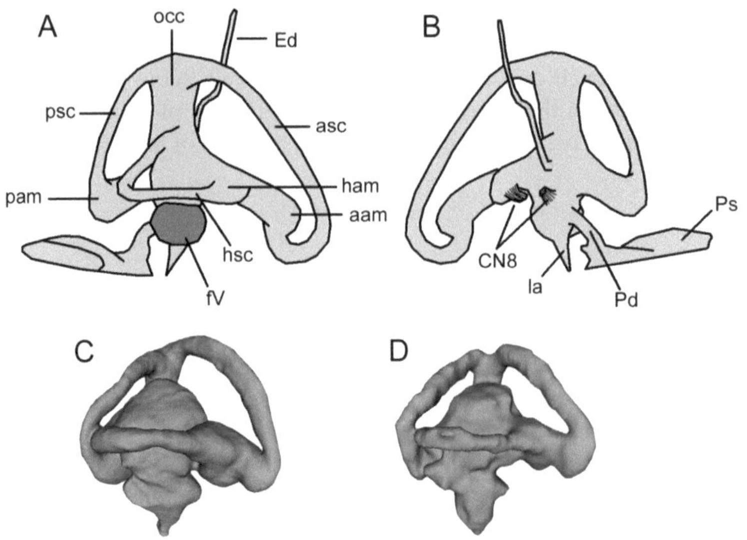

Fig. 9.10 The vestibular apparatus in mosasaurs. (**A, B**) Reconstructions of the vestibular apparatus of *Platecarpus tympaniticus* in (**A**) lateral and (**B**) medial views. (**C, D**) Lateral views of the same region in two extant lizards often used in comparison: (**C**) the marine iguana, *Amblyrhynchus cristatus* and (**D**) a monitor lizard, *Varanus exanthematicus*. (**A, B** redrawn and simplified from Camp 1942; **C**, **D** originals, segmented from HRXCT scans and visualized with Meshlab)

vestibular apparatus generally resembles that of terrestrial lizards that have similar movements. However, some derived mosasaurids may have used locomotor styles more suited to deep water. The mid-Maastrichtian mosasaurid *Plotosaurus* (Lindgren et al. 2007) has been interpreted as having a more tuna-like body shape with a stiff, deep body propelled mainly by the tail. Unfortunately, its inner ear is undescribed.

9.5.7 *Snakes*

The debate as to the origins and relationships of snakes is divided between proponents of a terrestrial ancestry (e.g., Zaher and Scanferla 2012; Hsiang et al. 2015), possibly with a burrowing or semi-burrowing lifestyle (e.g., Camp 1923; Apesteguía and Zaher 2006; Reeder et al. 2015), and advocates of a marine ancestry who contend that snakes evolved directly from early dolichosaur-like mosasaurs (e.g., Caldwell and Lee 1997; Caldwell 2001). The reduced eyes (Walls 1940), reinforced skulls, and body proportions (e.g., Wiens and Slingluff 2001; Wiens et al. 2006) of snakes are consistent with the first hypothesis, as are some features of the ear (Berman and Regal 1967). Like burrowing lizards, snakes lack a tympanic membrane and a middle ear cavity, and are sensitive to ground vibration arriving via the lower jaw and quadrate (Friedel et al. 2008; Christensen-Dalsgaard et al. 2012). The columnar quadrate lacks a conch and the stapedial footplate is generally larger than that of lizards of equivalent size (Baird 1970). Laterally, the stapes is continued by a cartilaginous extension, but this may be homologous to the internal process rather than the main body of the extrastapes as in generalized lizards (Rieppel and Zaher 2000) based on its relationship to the chorda tympani nerve. The process contacts the quadrate (shaft or suprastapedial process) via a hyoid arch derivative (stylohyal) that is fused to the quadrate (Rieppel 1980; Rieppel and Zaher 2000).

In snakes, the periotic sac leaves the otic capsule, passes laterally through the aLRST and continues into a recess (the juxtastapedial recess) (Fig. 9.2C). This recess is created around the aLRST and the fenestra vestibuli (and thus also around the stapedial footplate) by the development of an encircling crista circumfenestralis (CCF) (Figs. 9.2C, 9.11A). Laterally, the periotic sac attaches to the free edges of the CCF so that the juxtastapedial recess is closed off by a thick periotic membrane that is perforated by the stapedial shaft (Fig. 9.2C). The CCF is formed by an amalgamation of the otic crests present in lizards (Rieppel and Zaher 2000): crista prootica, crista interfenestralis, and crista tuberalis (Fig. 9.11A). However, there is variation in the development of component parts of the CCF that lead to disagreement as to the precise definition and homologies (e.g., Rieppel and Zaher 2000; Rieppel et al. 2003a; Palci and Caldwell 2014). The debate is largely beyond the scope of this chapter except that it affects interpretations of the ear region in early fossil snakes.

The Mesozoic fossil record of snakes is dominated by isolated vertebrae, and there is relatively limited information on the early snake skull. Although snakes probably separated from their closest lizard ancestors in the Jurassic (Jones et al. 2013), the earliest informative articulated specimens are from the Late Cretaceous with one possible exception. *Tetrapodophis amplectus* is a remarkable snake-like fossil described recently from the Lower Cretaceous of Brazil (Martill et al. 2015). As its name suggests, it retains both forelimbs and hindlimbs, but the skull is poorly preserved.

9.5.7.1 Terrestrial Fossil Snakes

The Late Cretaceous (Coniacian) snake *Dinilysia patagonica* (Fig. 9.11B–E) is represented by several skull specimens and has been the subject of numerous anatomical (e.g., Estes et al. 1970; McDowell 2008; Zaher and Scanferla 2012) and phylogenetic studies. It has been considered a stem snake (e.g., Scanlon 2006; Maximum Parsimony tree, Gauthier et al. 2012; Longrich et al. 2012), a stem alethinophidian (e.g., Zaher and Rieppel 2002; Wiens et al. 2010), or a crown-group alethinophidian (e.g., Bayesian tree, Gauthier et al. 2012; Zaher and Scanferla 2012). Zaher and Scanferla (2012) provided the most detailed descriptions and illustrations of this taxon. The quadrate is large and somewhat resembles the quadrate of some tympanic lizards (Estes et al. 1970). However, the stapes has an

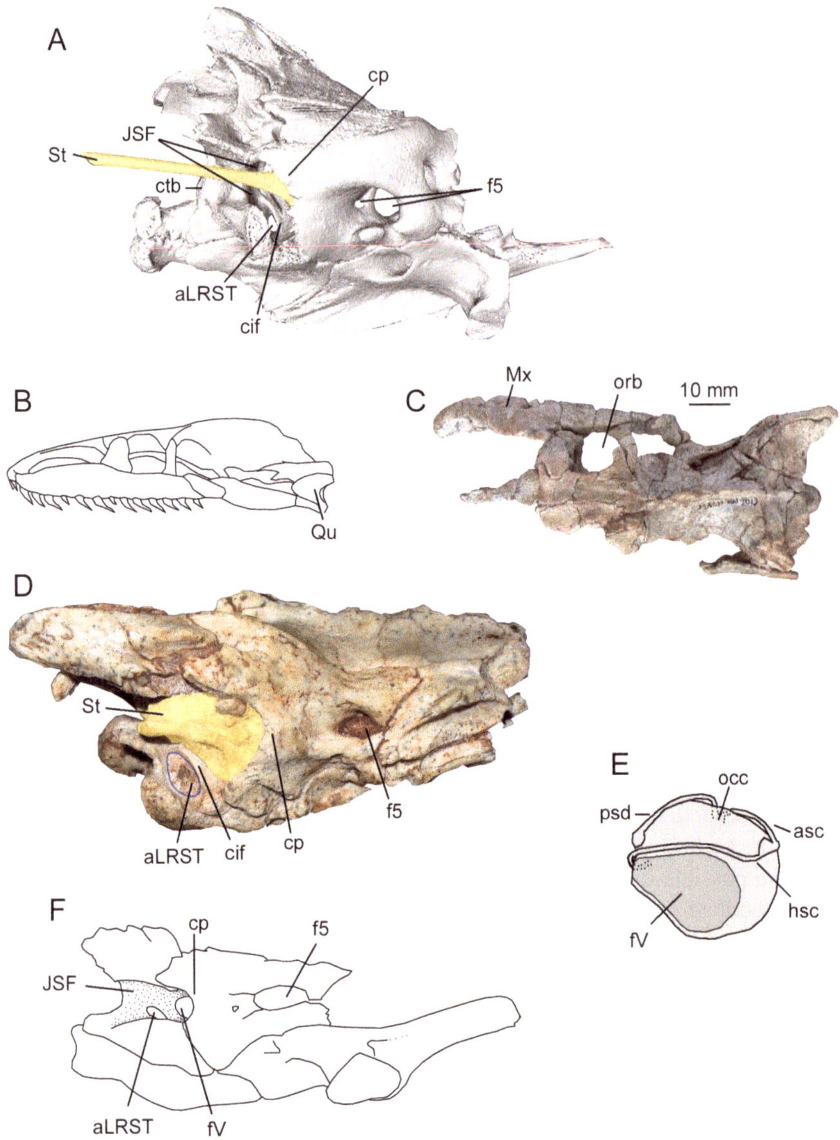

Fig. 9.11 The snake ear. **(A)** Right-lateral view of the braincase of the extant *Python sp.* to explain terminology used in the text. **(B–E)** The Late Cretaceous snake *Dinilysia patagonica*: **(B)** left-lateral view of skull; **(C)** dorsal view of MACN 1013; **(D)** right-lateral view of the braincase, MACN 1014. **(E)** Reconstruction of right side vestibular apparatus of *Dinilysia*. **(F)** The Pleistocene snake *Wonambi*, braincase without stapes in right-lateral view. Stapes is colored *yellow* in **A** and **D**. (**A** original from HRXCT scan, visualized in Meshlab; **B**, **E** redrawn and simplified from Zaher and Scanferla, 2012; **C**, **D** original; **F** redrawn and simplified from Rieppel et al. 2003a)

exceptionally large footplate and a slender shaft directed posterodorsally toward the suprastapedial process of the quadrate (Fig. 9.11D). A small articular facet on the posterodorsal edge of the fenestra vestibuli suggests there may have been a rocking movement of the stapedial footplate about that point (Estes et al. 1970). The aLRST is large, opens ventrolaterally into the juxtastapedial space (sensu Palci and Caldwell 2014: the region adjacent to the stapes irrespective as to whether a recess and CCF are fully developed) that is posterior to the crista interfenestralis (Fig. 9.11D). In contrast, Estes et al. (1970) refer to the aLRST as the fenestra rotunda (but see Sect. 9.3 about use of this terminology). There also is disagreement as to the definition of the CCF and, therefore, as to whether or not a CCF is present in *Dinilysia* (Estes et al. 1970; Zaher and Scanferla 2012; Palci and Caldwell 2014). Regardless of the semantics, *Dinilysia* does not appear to have a functional juxtastapedial recess that could contain an expanded periotic sinus.

Najash rionegrina is a second terrestrial stem snake from the early Late Cretaceous of Argentina (Apesteguía and Zaher 2006; Zaher et al. 2009). It is represented by both cranial and postcranial material, the latter showing the presence of small hind limbs. The ear region is represented by an attributed partial braincase and a small quadrate that lacks a conch (Apesteguía and Zaher 2006; Zaher et al. 2009). The stapedial shaft is broken but, as in *Dinilysia*, a large footplate is preserved in situ within the fenestra vestibuli. Again as in *Dinilysia*, there is a lack of consensus on the presence of a CCF in *Najash* (e.g., Caldwell and Calvo 2008 versus Zaher et al. 2009). The aLRST is smaller than that of *Dinilysia* and opens posteroventrally into a shallow juxtastapedial space.

Thus both *Najash* and *Dinilysia* have large stapedial footplates as found in extant burrowing snakes and lizards, but they lack a deep juxtastapedial recess. It seems unlikely that they had a functional re-entrant fluid system like that of derived snakes. The vestibular apparatus, reconstructed for *Dinilysia* by Zaher et al. (2009), also resembles that of burrowing squamates with regard to the large size of the central chamber compared to the closely packed semicircular canals (Fig. 9.11E). Yi and Norell (2015) came to the same conclusion with respect to burrowing habits in *Dinilysia* based on a new HRXCT scan of the *Dinilysia* skull and a comparison of vestibular morphology in a sample of extant snakes.

Madtsoiids are a group of mostly large snakes known from Gondwana and southern Europe (Rage et al. 2014). They are known primarily from late Cretaceous to Eocene deposits, but they survived into the Pleistocene in Australia. Their phylogenetic position is controversial. Some authors (Scanlon and Lee 2000; Longrich et al. 2012; Palci et al. 2013a) have argued for a basal position, close to *Dinilysia*, whereas others (e.g., Rieppel et al. 2003a; Wiens et al. 2010; Wilson et al. 2010) place madtsoiids within Alethinophidia, either within or on the stem of Macrostomata. The ear region certainly seems more derived than that of *Dinilysia*. Three madtsoiids have well-preserved braincases: the Australian *Wonambi naracortensis* (Scanlon and Lee 2000; Rieppel et al. 2003a; Scanlon 2005); *Yurlunggar camfieldensis* (Scanlon 2006); and the Indian *Sanajeh indicus* (Wilson et al. 2010). The latter species from the Indian Late Cretaceous is the oldest representative with a preserved braincase. Unlike *Najash* and *Dinilysia*, all three madtsoiids possess a distinct juxtastapedial recess framed by a CCF (e.g., *Wonambi*, Fig. 9.11F), although this is relatively low and directed laterally so that the footplate of the stapes, where preserved, is largely visible in lateral view. The aLRST opens posterior to the fenestra vestibuli and is separated from it by a crista interfenestralis that extends ventrally to contribute to the CCF. In *Wonambi*, a slit-like aLRST (= foramen pseudorotunda of Scanlon and Lee 2000) opens into the juxtastapedial recess just behind the crista interfenestralis, but the aMRST is very small. Further details are provided by Scanlon (2005). Individual braincase elements are well illustrated and show the periotic foramen (Scanlon's "foramen cochleae") opening into the RST. The semicircular canals are very small in diameter relative to the large vestibular chamber. The anatomy of *Yurlunggur* appears very similar (Scanlon 2006).

9.5.7.2 Marine Limbed Snakes

Several genera of marine snakes have been described, mostly from Cenomanian deposits in the western Mediterranean region. Their phylogenetic placement varies from basal (e.g., Scanlon and Lee 2000; Scanlon 2006; Palci et al. 2013b) to alethinophidian (e.g., Zaher and Rieppel 2002; Wiens et al. 2010; Gauthier et al. 2012). Although several of these snakes are represented by articulated skeletons, none has an uncrushed braincase. *Pachyrhachis* was described as having a large plate-like quadrate (Lee and Caldwell, 1998), suggesting ear specialization; however, Polcyn et al. (2005), working with HRXCT data, showed that this shape was an artifact of crushing. The only other relevant element is a slender stapedial shaft (Polcyn et al. 2005), but the size of the footplate is unknown. *Haasiophis* also has a slender quadrate but the stapes is not preserved (Tchernov et al. 2000; Rieppel et al. 2003b). It is therefore not possible to determine whether the ear of early marine snakes resembled that of *Najash* and *Dinilysia*.

9.6 Discussion

The combination of an external tympanic membrane held in a supportive quadrate conch/tympanic crest, a gracile stapes, and a compensatory window in the aLRST permits terrestrial lizards to have acute hearing over a range of frequencies. The middle ear of the iguanian *Crotaphytus collaris* was found to increase the reception of aerial sound by around 35 dB over the frequency range of 300–4,000 Hz (Wever and Werner1970), compared to a cat in which the increase was 30 dB over the same range (although the shorter lagena obviously limits the discriminatory ability of the squamate ear). Until recently, the vestibular apparatus had received less attention than the rest of the ear, but this is beginning to change as HRXCT and SRμCT scanning allow visual access to internal structures (e.g., Boistel et al. 2011).

The body of work that exists on extant lepidosaurs permits a degree of confidence in interpretations of fossil taxa. The tympanic crest and lateral conch on the quadrate of stem lepidosaurs and early rhynchocephalians provide compelling evidence that early lepidosaurs had a tympanic middle ear, but this was coupled with a suboptimal compensatory mechanism that probably reduced overall acuity. By subdividing the metotic fissure and expanding the lateral aperture of the recessus scalae tympani (aLRST), squamates developed a more efficient compensatory system and an extended hearing range. Based on current evidence, rhynchocephalians never developed an equivalent mechanism and derived members of the group subsequently lost the tympanic ear, perhaps in conjunction with the evolution of an efficient horizontal (pro-oral) jaw action that allowed shearing (Jones 2008; Jones et al. 2012).

The squamate ear was also adapted in response to specialist lifestyles, although the fossil record of each of the key changes is missing. Arboreal climbers, like chameleons, appear to have sacrificed efficient hearing for a better visual system and feeding strategy, but the earliest known fossil chameleon skulls (e.g., Rieppel et al. 1992) already had an atympanic columnar quadrate and are referable to living genera. The same problem exists for amphisbaenians for which the first known skulls can be assigned to living genera (or their close relatives). Burrowers and derived marine lizards independently lost the tympanic membrane and were presumably reliant mainly on substrateborne/waterborne sound. Nonetheless, they responded in different ways. Burrowers enlarged the stapes and its footplate to form an inertial element. Mosasaurs expanded the quadrate into a bulla and constrained a slender stapes (by the stapedial meatus, soft tissue, fenestral ridge, and stapedial groove) so as to resist inward displacement due to pressure changes during diving.

Understanding how the squamate ear has responded to burrowing versus marine swimming has an impact on current arguments as to the ancestral lifestyle of snakes. What may be called the traditional view is the hypothesis that snakes arose from terrestrial lizards (e.g., Zaher and Scanferla 2012; Hsiang et al. 2015) with a secretive lifestyle (burrowing or semi-burrowing) (e.g., Camp 1923; Apesteguía and Zaher 2006; Reeder et al. 2015; Yi and Norell 2015), which is consistent with their reduced eyes (Walls 1940, but see Simões et al. 2015), reinforced skulls, and trunk-tail proportions (e.g., Wiens and Slingluff 2001; Wiens et al. 2006). Berman and Regal (1967) also considered the ear of living snakes to reflect a burrowing ancestry. A second hypothesis, originally proposed by Cope (1869) but more recently championed by Lee, Caldwell, and collaborators (e.g., Lee 1997; Caldwell and Lee 1997; Caldwell 2001), proposes that snakes evolved directly from early marine mosasaurs ("dolichosaurs"). The dolichosaur quadrate, as exemplified by *Pontosaurus*, suggests these lizards may have had a tympanic middle ear, although confirmation of that would require data on the rest of the ear, which remains unknown. In contrast, where described (*Najash*, *Dinilysia*), the ear of stem snakes (or stem alethinophidians) had the large stapedial footplate of lizard burrowers, although probably not the re-entrant fluid system that characterises the ear of extant snakes. However, the ear of early marine snakes also is unknown.

To a large degree, a fuller understanding of lepidosaurian ear evolution is dependent on the recovery of new and more complete material of stem lepidosaurs, early rhynchocephalians, stem squamates, stem and basal snakes, stem amphisbaenians, and early mosasaurs. However, in the interim there exists a considerable amount of three-dimensional skull material of many of these groups in museum collections. These would benefit from ear-focused (rather than character-scoring) re-study, especially if HRXCT or SRμCT scanning is employed. Moreover, given the large amount of mosasaurid material in collections worldwide, there is clearly scope for a detailed comparative study of the mosasaur ear region (quadrate, braincase, inner ear) in relation to ear function and lifestyle.

9.7 Summary

Fossil evidence provides compelling evidence that stem lepidosaurs had a tympanic ear, and that this ear was inherited by both early rhynchocephalians and early squamates. The tympanic membrane appears to have been lost in more derived rhynchocephalian lineages, perhaps in association with the evolution of a pro-oral shearing bite. In squamates, subdivision of the metotic fissure and the addition of a compensatory window in the apertura lateralis of the recessus scalae tympani would have improved sound perception and extended the frequency range.

The efficient tympanic ear of squamates was modified in burrowing lizards and amphisbaenians on the one hand, and in aquatic mosasaurs on the other, to allow for perception of either substrateborne or waterborne sound. Although these two strikingly different lifestyles show some similarities (e.g., loss of the tympanic membrane and air-filled middle ear), burrowers and swimmers differ in the size of the stapedial footplate and the shape of the quadrate. The ear of snakes more closely resembles that of burrowers than that of swimmers, but the ear of early snakes remains incompletely known.

Reconstruction of the early history of the lepidosaurian ear (stem taxa, extinct rhynchocephalians, early lizards) remains limited by a lack of three-dimensional early fossils and by a tendency to focus descriptions of existing specimens on characters of phylogenetic rather than functional significance. Where three-dimensional fossils exist (e.g. stem lepidosaurs, rhynchocephalians, early lizards, and snakes), greater use of HRXCT and SRμCT imaging, as available, could provide valuable new data on middle and inner ear anatomy. The mosasaur ear would benefit from a focused study that links ear anatomy (quadrate, stapes, braincase foramina, vestibular apparatus) to postcranial specializations in a more rigorous way.

Institutional Abbreviations

BAD	Badlands National Park collection, South Dakota, USA
BGS GSb	collections of the British Geological Survey, Keyworth, UK

FFHM	Fick Fossil and History Museum, Kansas
IVPP	Institute of Vertebrate Paleontology and Paleoanthropology, Beijing, China
KUVP	University of Kansas, Museum of Natural History
MACN	Museo Argentino de Ciencias Naturales 'Bernardino Rivadavia', Buenos Aires, Argentina
NHMG	Natural History Museum of Guangxi, Nanning, Guangxi, China, Zoology collections

Acknowledgements Thanks to Jenny Clack for the invitation to participate in this project; to Hilary Maddin for the advanced copy of Cuthbertson et al. on the ear and braincase of the mosasaur *Plioplatecarpus*; to the libraries of UCL and the Zoological Society of London for access to literature; and to James Turbett and Liping Dong for many hours of interesting discussions on squamate inner ears and for the HRXCT scan images of braincases and inner ears. Thanks also to colleagues who kindly provided original photographs/scan images of *Sineoamphisbaena hexatabularis* (Yuan Wang and Liping Dong, IVPP, Beijing, China); *Rhineura hatcherii* (Jessica Maisano, University of Texas, Austin, USA); *Tylosaurus* and *Platecarpus* (Mike Everhart, Sternberg Museum, Kansas, USA); and *Dinilysia patagonica* (Hussam Zaher, Museu de Zoologia, Universidade de São Paulo, Brazil). Jenny Clack and Art Popper provided helpful feedback on an earlier draft.

Compliance with Ethics Requirements Susan E. Evans had no conflicts of interest.

References

Apesteguia, S., & Novas, F. E. (2003). Large Cretaceous sphenodontian from Patagonia provides insight into lepidosaur evolution in Gondwana. *Nature*, 425, 609–612.

Apesteguía S., & Zaher, H. (2006). A Cretaceous terrestrial snake with robust hindlimbs and a sacrum. *Nature*, 440, 1037–1040.

Apesteguia, S., Gomez, R. O., & Rougier, G. W. (2014). The youngest South American rhynchocephalian, a survivor of the K/Pg extinction. *Proceedings of the Royal Society of London B: Biological Sciences*, 281, doi:10.1098/rspb.2014.0811.

Baird, I. L. (1970). The anatomy of the reptilian ear. In C. Gans & T. S. Parsons (Eds.), *Biology of the Reptilia, vol. 2. Morphology B* (pp. 193–275). New York: Academic Press.

Bell, G. L. (1997). A phylogenetic revision of North American and Adriatic Mosasauroidea. In J. M. Callaway & E. L. Nicholls (Eds.), *Ancient marine reptiles* (pp. 293–332). San Diego: Academic Press.

Berman, D. S. (1973). *Spathorhynchus fossorium*, a Middle Eocene amphisbaenian (Reptilia) from Wyoming. *Copeia*, 1973(4), 704–721.

Berman, D. S. (1976). A new amphisbaenian (Reptilia: Amphisbaenia) from the Oligocene-Miocene John Day Formation, Oregon. *Journal of Paleontology*, 50, 165–174.

Berman, D. S., & Regal, P. J. (1967). The loss of the ophidian middle ear. *Evolution*, 21, 641–643.

Boistel, R., Herrel, A., Lebrun, R., Daghfous, G., et al. (2011). Shake rattle and roll: The bony labyrinth and aerial descent in squamates. *Integrative and Comparative Biology*, 51, 957–968.

Borsuk-Białynicka, M. (1984). Anguimorphans and related lizards from the Late Cretaceous of the Gobi Desert, Mongolia. *Palaeontologica Polonica*, 46, 5–105.

Borsuk-Białynicka, M. (1988). *Globaura venusta* gen.et. sp. n. and *Eoxanta lacertifrons* gen et sp. n., non-teiid lacertoids from the Late Cretaceous of Mongolia. *Acta Palaeontologica Polonica*, 33, 211–248.
Borsuk-Białynicka, M. (1990). *Gobekko cretacicus* gen.et. sp. n., a new gekkonid lizard from the Cretaceous of the Gobi Desert. *Acta Palaeontologica Polonica*, 35, 67–76.
Borsuk-Białynicka, M. & Moody, S. M. (1984). Priscagaminae, a new subfamily of the Agamidae (Sauria) from the Late Cretaceous of the Gobi Desert. *Acta Palaeontologica Polonica*, 29, 51–81.
Borsuk-Białynicka, M., Lubka, M., & Bohme, W. (1999). A lizard from the Baltic amber (Eocene) and the ancestry of crown-group lacertids. *Acta Palaeontologica Polonica*, 44, 349–382.
Broschinski, A. (2001). The lizards from the Guimarota mine. In T. Martin & B. Krebs (Eds.), *Guimarota: A Jurassic ecosystem* (pp. 59–68). Munich: Dr Friedrich Pfeil.
Caldwell, M. W. (2001). On the aquatic squamate *Dolichosaurus longicollis* Owen, 1850 (Cenomanian, Upper Cretaceous), and the evolution of elongate necks in squamates. *Journal of Vertebrate Paleontology*, 20, 720–735.
Caldwell, M. W. (2006). A new species of *Pontosaurus* (Squamata, Pythonomorpha) from the Upper Cretaceous of Lebanon and a phylogenetic analysis of Pythonomorpha. *Memorie della Societa Italiana di Scienze Natrali e del Museo Civico di Storia Naturale di Milano*, 34(3), 1–42.
Caldwell, M. W., & Lee, M. S. Y. (1997). A snake with legs from the marine Cretaceous of the Middle East. *Nature*, 386, 705–709.
Caldwell, M. W., & Dal Sasso, C. (2004). Soft-tissue preservation in a 95 million year old marine lizard: Form, function, and aquatic adaptation. *Journal of Vertebrate Paleontology*, 24(4), 980–985.
Caldwell, M. W., & Palci, A. (2007). A new basal mosasauroid from the Cenomanian (U. Cretaceous) of Slovenia with a review of mosasauroid phylogeny and evolution. *Journal of Vertebrate Paleontology*, 27, 863–880.
Caldwell, M. W., & Calvo, J. (2008). Details of a new skull and articulated cervical column of *Dinilysia patagonica* Woodward 1901. *Journal of Vertebrate Paleontology*, 28, 349–362.
Callison, G. (1967). Intracranial mobility in Kansas mosasaurs. *The University of Kansas Paleontological Contributions*, 1967, Paper 26.
Camp, C. L. (1923). Classification of lizards. *Bulletin of the American Museum of Natural History*, 48, 289–481.
Camp, C. L. (1942). California mosasaurs. *University* of *California, Memoirs*, 13, 1–68.
Carroll, R. L. (1985). A pleurosaur from the Lower Jurassic and the taxonomic position of the Sphenodontida. *Palaeontographica A*, 189, 1–28.
Carroll, R. L., & DeBraga, M. (1992). Aigialosaurs: Mid-Cretaceous varanoid lizards. *Journal of Vertebrate Paleontology*, 12, 66–86.
Carroll, R. L., & Wild, R. (1994). Marine members of the Sphenodontia. In N. C. Fraser & H.-D. Sues (Eds.). *In the shadow of the dinosaurs: Early Mesozoic tetrapods* (pp. 70–83). Cambridge: Cambridge University Press.
Charig, A. J., & Gans, C. (1990). Two new amphisbaenians from the Lower Miocene of Kenya. *Bulletin of the British Museum of Natural History* (*Geology*), 46, 19–36.
Christensen-Dalsgaard, J., Brandt, C., Willis, K. L., Bech Christensen, C., et al. (2012). Specialization for underwater hearing by the tympanic middle ear of the turtle, *Trachemys scripta elegans*. *Proceedings of the Royal Society of London B: Biological Sciences*, 279, 2816–2824.
Clark, J. T., & Hernandez, R. (1994). A new burrowing diapsid from the Jurassic La Boca Formation of Tamaulipas, Mexico. *Journal of Vertebrate Paleontology*, 14, 180–195.
Conrad, J. (2008). Phylogeny and systematics of Squamata (Reptilia) based on morphology. *Bulletin of the American Museum of Natural History*, 310, 1–182.
Conrad, J. L., & Daza, J. D. (2015). Naming and rediagnosing the Cretaceous gekkonomorph (Reptilia, Squamata) from Öösh (Övörkhangai, Mongolia). *Journal of Vertebrate Paleontology*, 35:5, e980891. doi:10.1080/02724634.2015.980891

Conrad, J. L., & Norell, M. A. (2006). High-resolution X-ray computed tomography of an Early Cretaceous gekkonomorph (Squamata) from Öösh (Övörkhangai; Mongolia). *Historical Biology: An International Journal of Paleobiology*, 18(4), 405–431.
Conrad, J. L., & Norell M. A. (2008). The braincases of two glyptosaurines (Anguidae, Squamata) and anguid phylogeny. *American Museum Novitates*, 3613, 1–24.
Conrad, J. L., Rieppel, O., Gauthier, J. A., & Norell, M. A. (2011). Osteology of *Gobiderma pulchrum* (Monstersauria, Lepidosauria, Reptilia). *Bulletin of the American Museum of Natural History*, 362, 1–88.
Cope, E. D. (1869). On the reptilian orders Pythonomorpha and Streptosauria. *Proceedings of the Boston Society of Natural History*, 12, 250–266.
Cuthbertson, R., Maddin, H. C., Holmes, R., & Anderson, J. S. (2015). The braincase and osseous labyrinth of *Plioplatecarpus peckensis* (Mosasauridae, Plioplatecarpinae), with functional implications for locomotor behaviour. *The Anatomical Record*, 298, 1597–1611.
Daza, J. D., Bauer, A. M., & Snively, E. (2013). *Gobekko cretacicus* (Reptilia: Squamata) and its bearing on the interpretation of gekkotan affinities. *Zoological Journal of the Linnean Society*, 167, 430–448.
Daza, J. D., Stanley, E. L., Wagner, P., Bauer, A. M., & Grimaldi, D. A. (2016). Mid-Cretaceous amber fossils illuminate the past diversity of tropical lizards. *Science Advances*, 2(3), e1501080.
Dutchak, A. R., & Caldwell, M. W. (2006). Redescription of *Aigialosaurus dalmaticus* Kramberger, 1892, a Cenomanian mosasauroid lizard from Hvar Island, Croatia. *Canadian Journal of Earth Sciences*, 43, 1821–1834.
Estes, R., Frazzetta, T. H., & Williams, E. E. (1970). Studies on the fossil snake *Dinilysia patagonica* Woodward. Part 1. Cranial morphology. *Bulletin of the Museum of Comparative Zoology*, 140, 25–74.
Estes, R., de Queiroz, K., & Gauthier, J. (1988). Phylogenetic relationships within Squamata. In R. Estes & G. Pregill (Eds.), *Phylogenetic relationships of the lizard families—essays commemorating Charles L. Camp* (pp. 119–281). Stanford: Stanford University Press.
Evans, S. E. (1980). The skull of a new eosuchian reptile from the Lower Jurassic of South Wales. *Zoological Journal of the Linnean Society*, 70, 203–264.
Evans, S. E. (1991). A new lizard-like reptile (Diapsida: Lepidosauromorpha) from the Middle Jurassic of England. *Zoological Journal of the Linnean Society*, 103, 391–412.
Evans, S. E. (1994). A new anguimorph lizard from the Jurassic and Lower Cretaceous of England. *Palaeontology*, 37, 33–49.
Evans, S. E. (1998). Crown-group lizards (Reptilia, Squamata) from the Middle Jurassic of the British Isles. *Palaeontographica A*, 250, 123–154.
Evans, S. E. (2009). An early kuehneosaurid reptile (Reptilia: Diapsida) from the Early Triassic of Poland. *Palaeontologica Polonica*, 65, 145–178.
Evans, S. E., & Borsuk-Białynicka, M. M. (2009). A small lepidosauromorph reptile from the Early Triassic of Poland. *Palaeontologica Polonica*, 65, 179–202.
Evans, S. E., & Jones, M. E. H. (2010). The origins, early history and diversification of lepidosauromorph reptiles. In S. Bandyopadhyay (Ed.), *New aspects of Mesozoic biodiversity. Lecture notes in earth sciences 132* (pp. 27–44). Berlin, Heidelberg: Springer.
Evans, S. E., & Wang, Y. (2010). A new lizard (Reptilia: Squamata) with exquisite preservation of soft tissue from the Lower Cretaceous of Inner Mongolia, China. *Journal of Systematic Palaeontology*, 8(1), 81–95.
Evans, S. E., Raia, P., & Barbera, C. (2004). New lizards and rhynchocephalians from the Early Cretaceous of southern Italy. *Acta Palaeontologica Polonica*, 49(3), 393–408.
Fernandez, M. S., & Talevi, M. (2015). An halisaurine (Squamata: Mosasauridae) from the Late Cretaceous of Patagonia, with a preserved tympanic disc: Insights into the mosasaur middle ear. *Comptes Rendus Palevol*, 14, 483–493.
Fernandez, V., Buffetaut, E., Suteethorn, V., Rage, J.-C., Tafforeau, P., & Kundrat, M. (2015). Evidence of egg diversity in squamate evolution from Cretaceous anguimorph embryos. *PLoS ONE*, doi:10.1371/journal.pone.028610.

Folie, A, Smith, R., & Smith, T. (2013). New amphisbaenian lizards from the Early Paleogene of Europe and their implications for the early evolution of modern amphisbaenians. *Geologica Belgica*, 16(4), 227–235.
Fraser, N. C. (1982). A new rhynchocephalian from the British Upper Trias. *Palaeontology*, 25, 709–725.
Fraser, N. C. (1988). The osteology and relationships of *Clevosaurus* (Reptilia: Sphenodontida). *Philosophical Transactions of the Royal Society of London B: Biological Sciences*, 321, 125–178.
Friedel, P., Young, B. A., & van Hemmen, J. L. (2008). Auditory localization of ground-borne vibrations in snakes. *Physical Review Letters*, 100. doi:10.1103/PhysRevLett.100.048701.
Gans, C. (1978). The characteristics and affinities of the Amphisbaenia. *Transactions of the Zoological Society of London*, 34, 347–416.
Gans, C., & Wever, E. G. (1972). The ear and hearing in *Amphisbaena* (Reptilia). *Journal of Experimental Biology*, 179, 17–34.
Gans, C., & Wever, E. (1976). The ear and hearing in *Sphenodon punctatus*. *Proceedings of the National Academy of Sciences of London B: Biological Sciences*, 73, 4244–4246.
Gauthier, J., Kearney, M., Maisano, J. A., Rieppel, O., & Behlke, A. D. B. (2012). Assembling the squamate Tree of Life: Perspectives from the phenotype and the fossil record. *Bulletin of the Peabody Museum of Natural History*, 53(1), 3–308.
Georgi, J. A., & Sipla, J. S. (2008). Comparative and functional anatomy of balance in aquatic reptiles and birds. In J. G. M. Thewissen & S. Nummela (Eds.), *Sensory evolution on the threshold: Adaptations in secondarily aquatic vertebrates* (pp. 233–256). Berkeley: University of California Press.
Hetherington, T. (2008). Comparative anatomy and function of hearing in aquatic amphibians, reptiles and birds. In J. G. M. Thewissen & S. Nummela (Eds.), *Sensory evolution on the threshold: Adaptations in secondarily aquatic vertebrates* (pp. 183–224). Berkeley: University of California Press.
Hsiang, A. Y., Field, D. J., Webster, T. H., Behlke, A. D. B., et al. (2015). The origin of snakes: Revealing the ecology, behaviour and evolutionary history of early snakes using genomics, phenomics, and the fossil record. *BMC Biology*, 15(87), 1–22.
Hoffstetter, R. (1967). Coup d'oeil sur les Sauriens (= lacertliens) des couches de Purbeck (Jurassic supérieur d'Angleterre. Résumé d'un Memoire). *Colloques Internationaux du Centre National de la recherche Scientifique*, 163, 349–371.
Houssaye, A., Lindgren, J., Pellegrini, R., Lee, A. H., et al. (2013) Microanatomical and histological features in the long bones of mosasaurine mosasaurs (Reptilia, Squamata) – implications for aquatic adaptation and growth rates. *PLoS ONE* 8(10), doi:10.1371/journal.pone.0076741.
Jiang, D-Y., Motani, R., Tintori, A., Rieppel, O., et al. (2014). The Early Triassic eosauropterygian *Majiashanosaurus discocoracoidis*, gen. et sp. nov. (Reptilia, Sauropterygia) from Chaohu, Anhui Province, People's Republic of China. *Journal of Vertebrate Paleontology*, 34, 1044–1052.
Jones, M. E. H. (2006). The Early Cretaceous clevosaurs from China (Diapsida: Lepidosauria). *New Mexico Museum of Natural History and Science Bulletin*, 37, 548–561.
Jones, M. E. H. (2008). Skull shape and feeding strategy in *Sphenodon* and other Rhynchocephalia (Diapsida: Lepidosauria). *Journal of Morphology*, 269, 945–966.
Jones, M. E. H., & Cree, A. (2012). Tuatara: Quick guide. *Current Biology*, 22(23), 986–987.
Jones, M. E. H., Anderson, C. L., Hipsley, C. A., Müller, J., et al. (2013) Integration of molecules and new fossils supports a Triassic origin for Lepidosauria (lizards, snakes, and tuatara). *BMC Evolutionary Biology, Phylogenetics and Phylogeography*, doi:10.1186/1471-2148-13-208.
Jones, M. E. H., O'Higgins, P., Fagan, M. J., Evans, S. E., & Curtis, N. (2012). Shearing mechanics and the influence of a flexible symphysis during oral food processing in *Sphenodon* (Lepidosauria: Rhynchocephalia). *The Anatomical Record*, 295, 1075–1091.

Jones, M. E. H., Tennyson, A. J. D., Evans, S. E., & Worthy, T. H. (2009). A sphenodontine (Rhynchocephalia) from the Miocene of New Zealand and palaeobiogeography of the tuatara (*Sphenodon*). *Proceedings of the Royal Society of London B: Biological Sciences*, 276, 1385–1390.

Kearney, M., Maisano, J. A., & Rowe, T. (2005). Cranial anatomy of the extinct amphisbaenian *Rhineura hatcherii* (Squamata, Amphisbaenia) based on high-resolution computed tomography. *Journal of Morphology*, 264, 1–33.

Konishi, T., & Caldwell, M. W. (2009). New material of the mosasaur *Plioplatecarpus nichollsae* Cuthbertson et al., 2007, clarifies problematic features of the holotype specimen. *Journal of Vertebrate Paleontology*, 29, 417–436.

Lee, M. S. Y. (1997). The phylogeny of varanoid lizards and the affinities of snakes. *Philosophical Transactions of the Royal Society of London B: Biological Sciences*, 352, 53–91.

Lee, M. S. Y., & Caldwell, M. W. (1998). Anatomy and relationships of *Pachyrhachis problematicus*, a primitive snake with limbs. *Philosophical Transactions of the Royal Society of London B: Biological Sciences*, 353, 1521–1552.

Lindgren, J., Jagt, J. W. M., & Caldwell, M. W. (2007). A fishy mosasaur: The axial skeleton of *Plotosaurus* (Reptilia, Squamata) reassessed. *Lethaia*, 40, 153–160.

Lindgren, J., Polcyn, M. J., & Young, B. A. (2011) Landlubbers to leviathans: Evolution of swimming in mosasaurine mosasaurs. *Paleobiology*, 37(3), 445–469.

Lombard, R. E., & Hetherington, T. E. (1993). Structural basis of hearing and sound transmission. In J. Hanken & B. K. Hall (Eds.), *The skull, vol. 3* (pp. 241–302). Chicago: Chicago University Press.

Longrich, N. A., Bhullar, B. A., & Gauthier, J. A. (2012). A transitional snake from the Late Cretaceous period of North America. *Nature*, 488, 205–208.

Longrich, N. A., Vinther, J., Pyron, R. A., Pisani, D., & Gauthier, J. A. (2015). Biogeography of worm lizards (Amphisbaenia) driven by end-Cretaceous mass extinction. *Proceedings of the Royal Society of London B: Biological Sciences*, 282, doi:10.1098/rspb.2014.3034.

McDowell, S. B. (1967). The extracolumella and tympanic cavity of the 'Earless' Monitor lizard, *Lanthanotus borneensis*. *Copeia*, 1967(1), 154–159.

McDowell, S. B. (2008). The skull of Serpentes. In C. Gans, A. S. Gaunt, & K. Adler (Eds.), *Biology of the reptilia, vol. 21. Morphology I. The skull and appendicular locomotor apparatus of Lepidosauria* (pp. 467–620). Ithaca (NY): Society for the Study of Amphibians and Reptiles, Contributions to Herpetology, 24.

Manley, G. A. (2000). Cochlear mechanisms from a phylogenetic viewpoint. *Proceedings of the National Academy of Sciences of the USA*, 97, 11736–11743.

Marsh, O. C. (1880). New characters of mosasauroid reptiles. *American Journal of Science*, 109, 83–87.

Martill, D. M., Tischlinger, H., & Longrich, N. R. (2015). A four-legged snake from the Early Cretaceous of Gondwana. *Science*, 349, 416–419.

Motani, R., Jiang, D-Y., Chen G-B., Tintori, A., et al.(2015). A basal ichthyosauriform with a short snout from the Lower Triassic of China. *Nature*, 517, 485–488.

Müller, J., Hipsley, C. A., Head, J.J., Kardjilov, N., et al. (2011). Eocene lizard from Germany reveals amphisbaenian origins. *Nature*, 473, 364–367.

Norell, M. A., & Gao, K. (1997). Braincase and phylogenetic relationships of *Estesia mongoliensis* from the Late Cretaceous of the Gobi Desert and the recognition of a new clade of lizards. *American Museum Novitates*, 3211, 1–25.

Oelrich, T. M. (1956). The anatomy of the head of *Ctenosaura pectinata* (Iguanidae). *University of Michigan Museum of Zoology, Miscellaneous Publications*, 94, 1–122.

Palci, A., & Caldwell, M. W. (2014). The Upper Cretaceous snake *Dinilysia patagonica* Smith-Woodward, 1901, and the crista circumfenestralis of snakes. *Journal of Morphology*, 275, 1187–1200.

Palci, A., Caldwell M. W., & Albino, A. M. (2013a). Emended diagnosis and phylogenetic relationships of the Upper Cretaceous fossil snake *Najash rionegrina* Apesteguia and Zaher, 2006. *Journal of Vertebrate Paleontology*, 33, 131–140.

Palci, A., Caldwell, M. W., & Nydam, R. L. (2013b) Reevaluation of the anatomy of the Cenomanian (Upper Cretaceous) hind-limbed marine fossil snakes *Pachyrhachis, Haasiophis*, and *Eupodophis*. *Journal of Vertebrate Paleontology*, 33(6), 1328–1342.
Pierce, S. E., & Caldwell, M. W. (2004). Redescription and phylogenetic position of the Adriatic (Upper Cretaceous; Cenomanian) dolichosaur *Pontosaurus lesinensis* (Kornhuber, 1873). *Journal of Vertebrate Paleontology*, 24, 373–386.
Polcyn, M. (2008). Braincase evolution in plioplatecarpine mosasaurs. *Journal of Vertebrate Paleontology*, 28(3), 128A.
Polcyn, M. (2010). Sensory adaptations in mosasaurs. *Journal of Vertebrate Paleontology*, 30(3), 146A.
Polcyn, M. J., Jacobs, L. L., & Haber, A. (2005). A morphological model and CT assessment of the skull of *Pachyrhachis problematicus* (Squamata, Serpentes), a 98 million year old snake with legs from the Middle East. *Palaeontologia Electronica*, no. 8.1.26.
Prothero, D. R., & Estes, R. (1980). Late Jurassic lizards from Como Bluff, Wyoming, and their palaeobiogeographic significance. *Nature*, 286, 484–486.
Pyron, R. A., Burbrink, F. T., & Wiens, J. J. (2013). A phylogeny and revised classification of Squamata including 4161 species of lizards and snakes. *BMC Evolutionary Biology*, doi:10. 1186/1471.2148.13.93.
Rage, J. C., Métais, G., Bartolini, A., Brohi, I. A., et al. (2014). First report of the giant snake *Gigantophis* (Madtsoiidae) from the Paleocene of Pakistan: Paleobiogeographic implications. *Geobios*, 47, 147–153.
Reeder, T. W., Townsend, T. M., Mulcahy, D. G., Noonan, B. P., et al. (2015) Integrated analyses resolve conflicts over squamate reptile phylogeny and reveal unexpected placements for fossil taxa. *PLoS ONE*, 10(3), doi:10.1371/journal.pone.0118199.
Rieppel, O. (1980). The sound-transmitting apparatus in primitive snakes and its phylogenetic significance. *Zoomorphology*, 96, 45–62.
Rieppel O. (1985). The recessus scalae tympani and its bearings on the classification of reptiles. *Journal of Herpetology*, 19, 373–384.
Rieppel, O., & Zaher, H. (2000). The braincases of mosasaurs and *Varanus*, and the relationships of snakes. *Zoological Journal of the Linnean Society*, 129, 489–514.
Rieppel, O., Walker, A., & Odhiambo, I. (1992). A preliminary report on a fossil chamaeleonine (Reptilia: Chamaeleoninae) skull from the Miocene of Kenya. *Journal of Herpetology*, 26, 77–80.
Rieppel, O., Kluge, A. G., & Zaher, H. (2003a) Testing the phylogenetic relationships of the Pleistocene snake *Wonambi naracoortensis* Smith. *Journal of Vertebrate Paleontology*, 22(4), 812–829.
Rieppel, O., Zaher, H., Tchernov, E., & Polcyn, M. J. (2003b). The anatomy and relationships of *Haasiophis terrasanctus*, a fossil snake with well-developed hind limbs from the mid-Cretaceous of the Middle East. *Journal of Paleontology*, 77, 536–558.
Robinson, P. L. (1962). Gliding lizards from the Upper Keuper of Great Britain. *Proceedings of the Geological Society London*, 1601, 137–146.
Robinson, P. L. (1973). A problematic reptile from the British Upper Trias. *Journal of the Geological Society of London*, 129, 457–479.
Ross, C. F., Sues, H-D., & De Klerk, W. J. (1999). Lepidosaurian remains from the Lower Cretaceous Kirkwood Formation of South Africa. *Journal of Vertebrate Paleontology*, 19, 21–27.
Russell, D. A. (1967) Systematics and morphology of American mosasaurs. *Bulletin of the Peabody Museum of Natural History*, 23, 1–240.
Scanlon, J. D. (2005). Cranial morphology of the Plio-Pleistocene giant madtsoiid snake *Wonambi naracoortensis*. *Acta Palaeontologica Polonica*, 50, 139–180.
Scanlon, J. D. (2006). Skull of the large non-macrostomatan snake *Yurlunggur* from the Australian Oligo-Miocene. *Nature*, 439, 839–842.
Scanlon, J. D., & Lee, M. S. Y. (2000). The Pleistocene serpent *Wonambi* and the early evolution of snakes. *Nature*, 403, 416–420.

Simões, B. F., Sampaio, F. L., Jared, C., Antoniazzi, M. M., et al. (2015). Visual system evolution and the nature of the ancestral snake. *Journal of Evolutionary Biology*, doi:10.1111/eb.12663.
Sipla, J. S., & Spoor, F. (2008). The physics and physiology of balance. In J. G. M. Thewissen & S. Nummela (Eds.), *Sensory evolution on the threshold: Adaptations in secondarily aquatic vertebrates* (pp. 227–232). Berkeley (CA): University of California Press.
Tałanda, M. (2016). Cretaceous roots of the amphisbaenian lizards. *Zoologica Scripta*, 45, 1–8.
Tchernov, E., Rieppel, O., Zaher, H., Polcyn, M. J., & Jacobs, L. J. (2000). A new fossil snake with limbs. *Science*, 287, 2010–2012.
Townsend, T. M., Larson, A., Louis, E., & Macey, J. R. (2005). Molecular phylogenetics of Squamata: The position of snakes, amphisbaenians, and dibamids, and the root of the squamate tree. *Systematic Biology*, 53, 735–757.
Vidal, N., & Hedges, B. (2005). The phylogeny of squamate reptiles (lizards, snakes, and amphisbaenians) inferred from nine nuclear protein-coding genes. *Comptes rendus Biologies*, 328, 1000–1008.
Walls, G. L. (1940). Ophthalmological implications for the early history of the snakes. *Copeia*, 1940(1), 1–8.
Walsh, S. A., Luo, Z.-X., & Barrett, P. M. (2013). Modern imaging techniques as a window to prehistoric auditory worlds. In C. Koppl, G. A. Manley, A. N. Popper, & R. R. Fay. (Eds.), *Insights from comparative hearing research* (pp. 227–261). New York: Springer Science +Business Media.
Wever, E. G. (1973). The function of the middle ear in lizards: Divergent types. *Journal of Experimental Biology*, 184, 97–126.
Wever, E. G. (1978). *The reptile ear*. Princeton: Princeton University Press.
Wever, E. G., & Vernon, J. A. (1956) Sound transmission in the turtle's ear. *Proceedings of the National Academy of Sciences of the USA*, 42, 292–299.
Wever, E. G., & Werner, Y. L. (1970). The function of the middle ear in lizards: *Crotaphytus collaris* Iguanidae). *Journal of Experimental Zoology*, 175, 327–342.
Wever, E. G., & Gans, C. (1973) The ear in Amphisbaenia (Reptilia); further anatomical observations. *Journal of Zoology*, 171, 189–206.
Whiteside, D. I. (1986). The head skeleton of the Rhaetian sphenodontid *Diphydontosaurus avonis* gen.et.sp.nov. and the modernising of a living fossil. *Philosophical Transactions of the Royal Society of London B: Biological Sciences*, 312, 379–430.
Wiens, J. J., & Slingluff, J. L. (2001). How lizards turn into snakes: A phylogenetic analysis of body-form evolution in anguid lizards. *Evolution*, 55, 2303–2318.
Wiens, J. J., Brandley, M. C., & Reeder, T. W. (2006). Why does a trait evolve multiple times within a clade? Repeated evolution of snake-like body form in squamate reptiles. *Evolution*, 60, 123–141.
Wiens, J. J., Kuczynski, C. A., Townsend, T. M., Reeder, T. W., et al. (2010). Combining phylogenomics and fossils in higher-level squamate reptile phylogeny: Molecular data change the placement of fossil taxa. *Systematic Biology*, 59, 674–688.
Wiens, J. J., Hutter, C. R., Mulcahy, D. G., Noonan, B. P., et al. (2012). Resolving the phylogeny of lizards and snakes (Squamata) with extensive sampling of genes and species. *Biology Letters*, 8, 1043–1046.
Williston, S. W. (1914). *Water reptiles of the past and present*. Chicago: University of Chicago Press.
Wilson, J. A., Mohabey, D. M., Peters, S. E., & Head, J. J. (2010). Predation upon hatchling dinosaurs by a new snake from the Late Cretaceous of India. *PLoS Biology*, doi:10.1371/journal.pbio.1000322.
Wu, X.-C. (1994). Late Triassic-Early Jurassic sphenodontians (Clevosaurs) from China and the phylogeny of the Sphenodontida. In N. C. Fraser and H.-D. Sues (Eds.), *In the shadow of the dinosaurs: Early Mesozoic tetrapods* (pp. 38–69). Cambridge: Cambridge University Press.

Wu, X.-C., Brinkman, D. B., & Russell, A. P. (1996). *Sineoamphisbaena hexatabularis*, an amphisbaenian (Diapsida: Squamata) from the Upper Cretaceous redbeds at Bayan Mandahu (Inner Mongolia, People's Republic of China), and comments in the phylogenetic relationships of the Amphisbaenia. *Canadian Journal of Earth Sciences*, 33, 541–577.

Yi, H. & Norell, M. A. (2015). The burrowing origin of modern snakes. *Science Advances*, 1(10), e1500743.

Yi, H., Sampath, D., Schoenfeld, S., & Norell, M. A. (2012). Reconstruction of inner ear shape and size in mosasaurs (Reptilia: Squamata) reveals complex adaptation strategies in secondary aquatic reptiles. *Journal of Vertebrate Paleontology*, 32, 198A.

Zaher, H., & Rieppel, O. (2002). On the phylogenetic relationships of the Cretaceous snakes with legs, with special reference to *Pachyrhachis problematicus* (Squamata, Serpentes). *Journal of Vertebrate Paleontology*, 22, 104–109.

Zaher, H., & Scanferla, C. A. (2012). The skull of the Upper Cretaceous snake *Dinilysia patagonica* Smith-Woodward, 1901, and its phylogenetic position revisited. *Zoological Journal of the Linnean Society*, 164, 194–238.

Zaher, H., Apesteguia, S., & Scanferla, C. A. (2009). The anatomy of the Upper Cretaceous snake *Najash rionegrina* Apesteguia & Zaher, 2006, and the evolution of limblessness in snakes. *Zoological Journal of the Linnean Society*, 156, 801–826.

Chapter 10
Archosaurs and Their Kin: The Ruling Reptiles

Gabriela Sobral and Johannes Müller

Abstract Archosauria includes birds, crocodylians, and a number of fossil groups such as dinosaurs and pterosaurs. They first appeared in the Early Triassic and since then have dominated terrestrial ecosystems. This chapter is a compilation of the available information on the inner ear morphology of archosaur crown groups and stem groups and an exploration of different aspects of the evolution of their otic anatomy. It is still not clear whether tympanic hearing was present in the basalmost members of stem archosaur clades. However, more derived taxa show a number of modifications that certainly improved their hearing sense, such as a larger metotic foramen and a more elongate cochlea. Impedance-matching hearing appeared many times independently in archosaurs, although it is currently problematic to know at which point this happened. In theropods, impedance-matching hearing appeared before the origin of birds and was retained in the crown-group. Pneumatization must play an important role in directional hearing and is likely to have influenced skull pneumatization in crocodylians. Exquisite sound production capacities were present not only in hadrosaurids but also in ankylosaurids. Elongation of the semicircular canals seems to be linked to the acquisition of a more upright posture and a more active lifestyle in archosaurs. Many crown groups show further elongation of the canals, with birds representing an extreme condition.

Keywords Archosauria · Birds · Crocodylian · Hearing · Impedance matching · Inner ear · Middle ear · Tympanic hearing

Figure Label Abbreviations

aa anterior ampulla
amcv anterior middle cerebral vein

G. Sobral (✉)
Departamento de Ecologia e Zoologia, Universidade Federal de Santa Catarina, Campus Universitário s/n, Florianópolis 88040-970, Brazil
e-mail: gabisobral@gmail.com

G. Sobral · J. Müller
Museum für Naturkunde Berlin, Invalidenstraße 43, 10115 Berlin, Germany
e-mail: Johannes.Mueller@mfn-berlin.de

J.A. Clack et al. (eds.), *Evolution of the Vertebrate Ear—Evidence from the Fossil Record*, Springer Handbook of Auditory Research 59,
DOI 10.1007/978-3-319-46661-3_10

asc	anterior semicircular canal
atr.pc	anterior tympanic recess pneumatic cavity
bo	basioccipital
bt	basal tuber
car	carotid artery
cbl	cerebellum
cc	common crus
cer	cerebrum
ci	crista interfenestralis
c (= cl)	cochlea
clp	clinoid process
CN	cranial nerve (+ Roman numeral)
cp	cultriform process
de	dural expansion
dp.ps	dorsal process of the parabasisphenoid
ed	endolymphatic duct
eo	exoccipital
fa	fossa auditiva
fc.pa	facet for articulation with parietal
ff	floccular fossa
fio	foramen interorbitalis
fm	metotic foramen
fo/FO	fenestra ovalis (= fv fenestra vestibuli)
fp/fpd	fenestra pseudorotunda
fpr	fenestra pseudorotund
fv	fenestra vestibuli
iam	internal auditory meatus
ic/ica	internal carotid artery
jug	posterior cerebral vein
lab	endosseous labyrinth
ld	lateral depression
leu	lateral opening of the Eustachian tube
lg.cr	lagenar crest
lr	lateral ridge
lsc	lateral semicircular canal
meu	medial opening of the Eustachian tube
mf	metotic foramen
mfis	metotic fissure
mo	medulla oblongata
ob	olfactory bulb
oc	occipital condyle
ocv	orbitocerebral vein
op	opisthotic
pa	posterior ampulla

pd	perilymphatic duct
pf	perilymphatic foramen
pfo	pituitary fossa
pmcv	posterior middle cerebral vein
pnf	pneumatic foramen
pp	paroccipital process
pr	prootic
psc	PSC
psph	parasphenoid
ptf	post-temporal fenestra
qc	quadrate condyle
qu	quadrate
re	recess
rst	recessus scala tympan
rvcm	rostral middle cerebral vein
sd	semilunar depression
sin	blind dural venous sinus of hindbrain
sl.dp	semilunar depression
sp	pituitary sinus
spha	foramen of sphenoidal artery
st	stapes
tm	fossa tecti mesencephali
un.gp	unossified gap
vb	vestibule
vcm	vena cerebralis media
vr.op	ventral ramus of the opisthotic

10.1 Introduction

Archosaurs include not only the well-known modern groups of crocodylians and birds, but also dinosaurs, which gave rise to birds. Formally, Archosauria is "a monophyletic taxon composed of the living crocodiles and birds, and of fossil taxa that share their most recent common ancestor" (Gauthier and Padian 1985). The group first appeared during the Early Triassic, but it was not until the Late Cretaceous that major diversification events of crown birds (Neornithes) and crocodiles (Crocodylia) took place. Understanding of the anatomy, paleobiology, and phylogenetic relationships of basal archosaurs, as well as of the stem lineages, has increased substantially in the last two decades (Nesbitt et al. 2013). Archosaurs today show only a small part of the phylogenetic, ecological, and morphological diversity that can be seen from their long fossil record (Fig. 10.1).

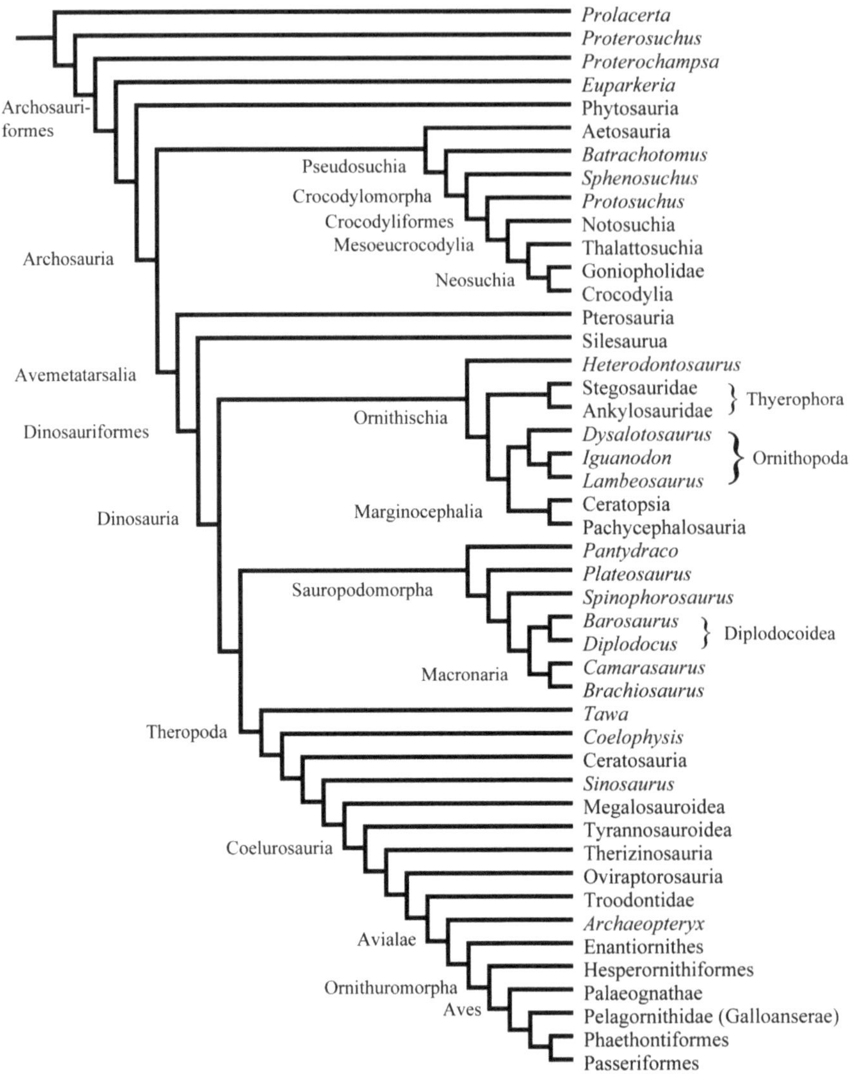

Fig. 10.1 Phylogenetic relationships of Archosauromorpha (Original by authors)

The end-Triassic mass extinction was an important event for archosaurs because of a major turnover that allowed them to become dominant among Mezosoic terrestrial faunas (Brusatte et al. 2010a). In their early evolutionary history during the Middle Triassic, archosaurs were restricted in taxonomic diversity. However, during the Late Triassic, the lineage leading to modern crocodiles, the Pseudosuchia, became more taxonomically and morphologically diverse and abundant than the group containing birds (i.e., the Avemetatarsalia). Later on, during the Early Jurassic, the avemetatarsalians became dominant. Although the Cretaceous–

Paleogene mass extinction event caused the disappearance of much of the clade, especially the dinosaurs, their avian descendants still represent the most diverse living tetrapod group.

This chapter reviews the morphological diversity of the braincase region related to hearing in archosaurs and related clades. A great amount of braincase material exists for the groups covered here, but although computed-tomography (CT) scanning techniques are being increasingly used by paleontologists, the general braincase morphology of archosaurs is still poorly explored, even in extant groups. Relatively little anatomical work has been done on the construction of the otic capsule of living birds and crocodiles. Current understanding of bird and crocodylian phylogeny is helping to build an evolutionary scenario in which their anatomy can be better interpreted, and the gaps in our anatomical knowledge are gradually being filled.

In the following pages, the general anatomy of the otic region of stem-group archosaurs is described first in order to provide the reader with a morphological background that is necessary to understand the discussion of the evolution of the hearing system at the end of the chapter. In order to deal with the diversity of archosauromorph otic morphology, the respective groups are treated separately. The most basal groups are treated first, followed by Pseudosuchia, Pterosauria, Dinosauria, and Avialae. Dinosaurs are divided into the classical ornithischians, sauropodomorphs, and (non-avialan) theropods (Fig. 10.1).

10.2 Stem Archosaurs: *Prolacerta, Euparkeria*, Choristodera, and Phytosauria

The evolutionary history of basal archosauromorphs began in the Permian, and there is strong evidence from fossil trackways that they had become substantially diverse by the end-Permian, although their body fossils are still quite rare from that time (Bernardi et al. 2015). They greatly diversified taxonomically and morphologically in the Triassic, becoming an important and dominant part of Mesozoic terrestrial ecosystems (Brusatte et al. 2010a).

The development and ossification of the bones of the otic capsule and occipital region are conserved among vertebrates (Evans 2008). Figure 10.2A shows the general relationship between the skull bones, the braincase, and otic capsule in an archosauromorph, including one of their uniting characters: an extra hole in front of the orbit, the antorbital fenestra. The metotic foramen is a foramen left between the embryonic occipital arch and the opisthotic (see Clack, Chap. 1) through which pass the cranial nerves (CN) IX–XI (glossopharyngeal, vagus, accessory) and, in some cases, also the posterior head vein. The borders of the metotic foramen are formed by the exoccipital posteriorly and by the opisthotic anteriorly. Similarly, those of the fenestra ovalis (FO) are formed by the opisthotic posteriorly and prootic anteriorly (Fig. 10.2A, B). The part of the opisthotic that separates these openings has different names throughout the literature, but in basal

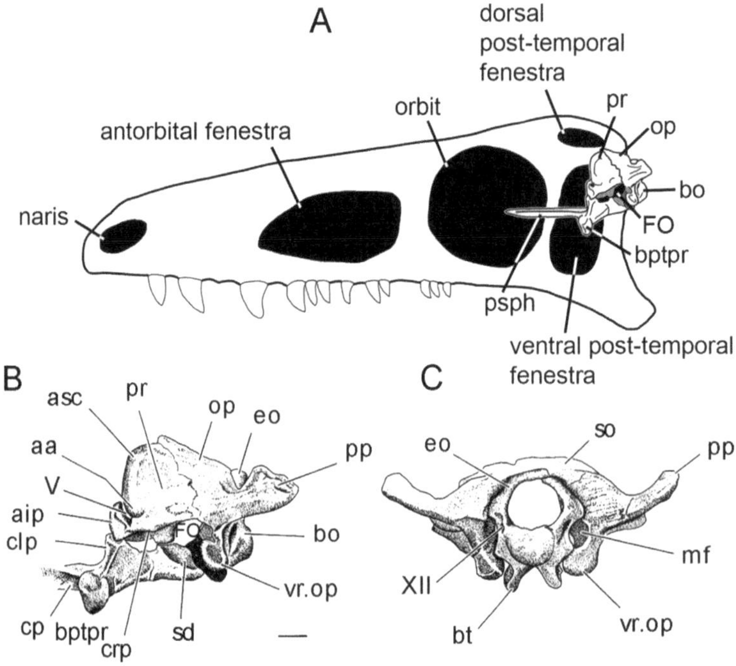

Fig. 10.2 (**A**) Generalized archosauromorph skull in lateral view, with left side skull bones removed to show the relationship between the braincase and the skull bones. (**B, C**) *Prolacerta* braincase: (**B**) left-lateral view (anterior to the left), and (**C**) posterior view. (**A** J. A. Clack original; **B**, **C** reproduced with permission by E. Schweizerbart'sche Verlagsbuchhandlung from Evans 1986)

archosauromorphs it is generally termed the ventral ramus. The basioccipital often participates in the formation of the floor of the metotic foramen and of the posteroventral part of the FO. Likewise, the parabasisphenoid may take part in the anteroventral rim of the FO. The anteroventral portion of the anterior semicircular canal (ASC) is associated with the prootic and its posterodorsal region with the supraoccipital. Likewise, the anterodorsal part of the posterior semicircular canal (PSC) lies in the supraoccipital, and the posteroventral part lies in the opisthotic. The anterior portion of the lateral semicircular canal is housed in the prootic, and its posterior part lies in the opisthotic. Usually, swellings on the lateral surface of the prootic mark the positions of the anterior and lateral semicircular canals.

There is little information on the otic regions of the most basal groups of archosauromorphs, the protorosaurs and the rhynchosaurs, but the braincase anatomy of basal archosauromorphs can be seen in *Prolacerta broomi* (Evans 1986). *Prolacerta* (Figs. 10.1 and 10.2) has been a key fossil since its discovery as it was first interpreted as a "missing link" between basal diapsids and modern lizards. With *Prolacerta* now confirmed as an archosauromorph, it became crucial to understanding early archosauromorph evolution, although the exact phylogenetic placement of *Prolacerta* is still disputed (e.g., Chen et al. 2014; Ezcurra et al. 2014).

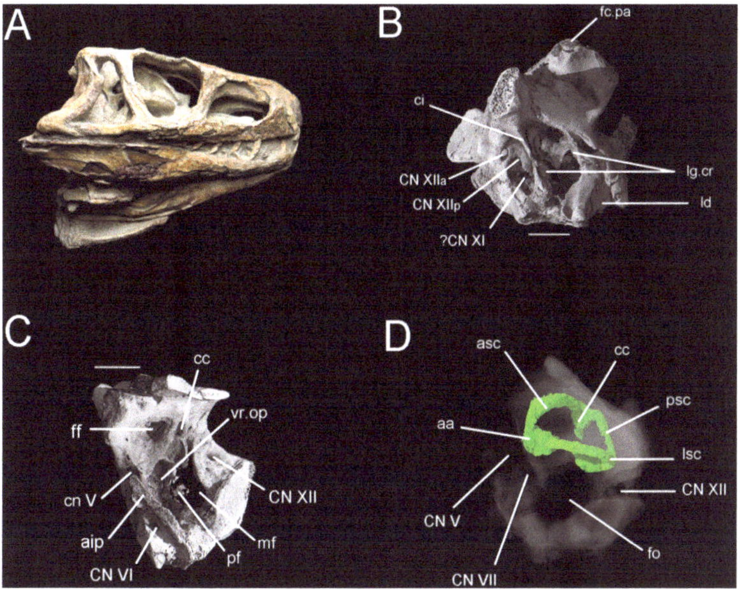

Fig. 10.3 *Euparkeria*. **(A)** South African Museum specimen SAM 5867, skull in right-lateral view. Details of the braincase: right-lateral **(B)** and right medial **(C)** views. **(D)** inner ear, with semicircular canals shown in green. (Modified from Sobral et al. 2016)

Most of the border of the FO of *Prolacerta* is better defined than in the early diapsid *Youngina* (Sobral, Reisz, Neenan, Müller, and Scheyer, Chap. 8), but not as much as more derived clades such as *Euparkeria* (compare Figs. 10.2, 10.3). There is a stout ventral ramus of the opisthotic, and the prootic is more dorsoventrally developed than in *Youngina*. Its medial surface bears indications of some brain and inner ear structures. The imperforate stapes is a slender rod with a small footplate. The quadrate bears a weakly developed embayment or conch to support the tympanic membrane.

Euparkeria capensis (Figs. 10.1, 10.3) is an important taxon for understanding the rise and early evolution of archosaurs, as it is either the sister-taxon to, or a very close relative of, Archosauria. The cursorial, upright morphology of *Euparkeria* has served as a model for the evolution of more active and specialized habits of archosaurs, in contrast to the more generalist ecology of earlier and smaller groups. The braincase of *Euparkeria* has been described in light of the phylogenetic relationships of birds and crocodylians (Gower and Weber 1998), and a more recent account includes finer details of its morphology with information on the inner ear structures from μCT scans of all available braincase material (Sobral et al. 2016). Medially, the basioccipital and parabasisphenoid contribute to the posteroventral, ventral, and anteroventral borders of the FO respectively, restricting the opisthotic and prootic to the posterior and anterodorsal borders. Compared to more basal taxa, the FO of *Euparkeria* is much enlarged and elongate (Fig. 10.3B). The open

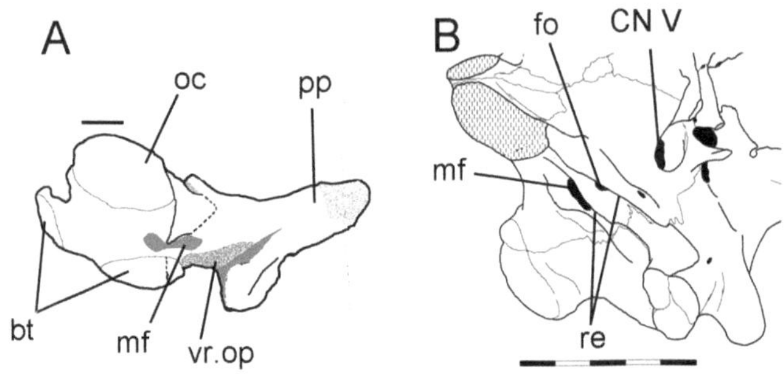

Fig. 10.4 (**A**) Right side of braincase of *Proterochampsa* in posterolateral view. (**B**) Braincase of *Mystriosuchus* in right-lateral view (anterior to the right). *Scale bar*, length = 50 mm. (**A** reproduced with permission of Springer from Trotteyn and Haro 2011; **B** reproduced with permission of The Royal Society of Edinburgh from Hungerbüghler 2002)

ventralmost tip of the lagenar recess lies within the "unossified gap" of Gower (2002) (also see Clack 1997). Medially, the perilymphatic duct passed through a rounded notch on the opisthotic ventral ramus. The semicircular canals are of equal length and much larger than those of *Youngina* (Fig. 10.3D). There is a well-developed floccular fossa (Fig. 10.3C). The tall but irregularly shaped metotic foramen is wider ventrally than dorsally (Fig. 10.3C). This shape indicates the positions of the associated internal soft tissue structure, and the ventral portion housed the perilymphatic sac that bulged into it. This area was occupied by the recessus scalae tympani. The narrower dorsal portion of the metotic foramen housed CNs IX–XI and possibly the posterior head vein.

In the early evolutionary history of archosauromorphs, some clades became secondarily aquatic. Fossil braincases still in articulation with the rest of the skull allow good occipital and ventral views. Choristoderes comprise an enigmatic and putatively archosauromorph group sometimes retrieved among stem diapsids or even lepidosauromorphs (Sobral et al. 2015). Their distinct skull morphology is marked by a widening of the temporal regions, especially around the parietal and squamosal bones. In the derived member *Proterochampsa* (Fig. 10.1), the ventral part of the metotic foramen is rather unusual in that it is anteroposteriorly long and a midpoint constriction gives it a figure-8 shape (Fig. 10.4A) (Trotteyn and Haro 2011). This anatomy superficially resembles that of some parareptiles (Sobral, Reisz, Neenan, Müller, and Scheyer, Chap. 8). The poorly defined borders of the FO had an open ventral rim, with the ventral tip of the cochlear recess closely approaching the basal tuber.

Phytosaurs (Fig. 10.1) form an aquatic archosauromoph group that, until recently, was classified as the basalmost pseudosuchian clade (see Sect. 10.3). However, recent analyses have placed it just outside the archosaur crown as the sister-clade of Archosauria (Nesbitt 2011). The cranial anatomy of phytosaurs is interesting because of its superficial resemblance to that of crocodylians, but the overall braincase anatomy and unique otic morphology are poorly known. The FO is small and recessed (Fig. 10.4B) (Hungerbühler 2002; Hungerbühler et al. 2012).

Its ventral border is open and confluent with the ventral part of the recess, which extends a long way anteroventrally on the parabasisphenoid. The metotic foramen is also recessed. The hypoglossal nerve exited the braincase through the posterior rim of the same recess. The ventral ramus of the opisthotic is also a thin and sharp, laterally projecting lamella. The roughly equal, rounded semicircular canals of a phytosaur are seen in the inner ear of *Pseudopalatus mccauleyi* (Holloway et al. 2013).

10.3 Pseudosuchia: The Crocodile Family Tree

The Triassic–Jurassic mass extinction affected the taxonomic diversity of pseudosuchians in that all but one clade, Crocodylomorpha (Fig. 10.1), became extinct. In the Early Jurassic, however, significant diversification events took place in most crocodyliform lineages, in particular "protosuchians" and thalattosuchians, the latter of which may have been facilitated by the invasion of new aquatic niches not explored by dinosaurs or pterosaurs (Bronzati et al. 2015). The Middle Jurassic seems to have been especially important for aquatic clades (Stubbs et al. 2013), but pseudosuchians also experienced great diversification during the Cretaceous represented by the notosuchians. This is a group of extremely morphologically diverse taxa that included large-sized, top predator species as well as herbivorous, cursorial, dog-like, and rodent-like forms.

Aetosaurs were an herbivorous group of crocodile-like pseudosuchians. Braincase material of derived aetosaurs (Fig. 10.1), such as *Stagonolepis*, shows that the metotic foramen is undivided (Fig. 10.5A) (Gower and Walker 2002). The ventral ramus of the opisthotic loops around the perilymphatic duct, suturing with itself and completely enclosing the perilymphatic foramen and resulting in a posterolaterally directed foramen. A ridge on the exoccipital, anterior to the foramina of CN XII (hypoglossal) forms the posterior border of the metotic foramen. In the braincase of later pseudosuchians ("rauisuchians"), the "unossified gap" is also present (Gower 2002). As in *Euparkeria*, there is a marked division between the vestibular part of the inner ear and the cochlear recess, in *Batrachotomus* (Figs. 10.1, 10.5B) marked by a low ridge. It is likely that the perilymphatic foramen was not entirely encircled by bone in this taxon (Gower and Nesbitt 2006).

The quadrate of early "sphenosuchid" crocodylomorphs has a line of attachment that indicates the presence of a tympanic membrane (Walker 1990). The enlarged middle-ear cavity and the extensive pneumaticity of the skull indicate the development of the paratympanic pneumatic system, including the bony enclosure of the Eustachian tubes. This arrangement in early crocodylomorphs is similar to that of extant clades, the main difference being the degree of development of the anterior tympanic recess (ATR), which is reduced in living taxa. In *Sphenosuchus* (Figs. 10.1, 10.5), the opisthotic loop is closed, completely encircling the perilymphatic foramen (Fig. 10.5C). The metotic foramen is laterally placed and not subdivided (Gower and Weber 1998). Medially, partial enclosure of the otic capsule forms an otic bulla that is more developed than in non-crocodylomorph clades. An

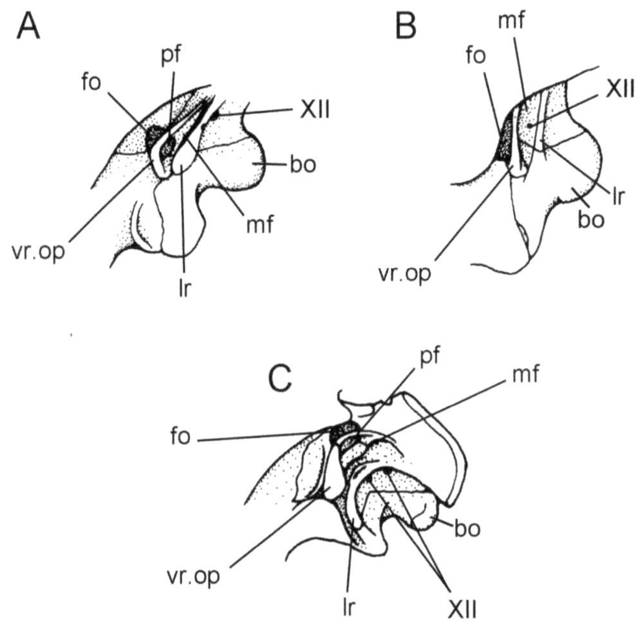

Fig. 10.5 Details of the otic regions of **(A)** *Stagonolepis*, **(B)** *Batrachotomus*, and **(C)** *Sphenosuchus*. (Reproduced with permission of Wiley from Gower and Walker 2002)

exoccipital-parabasisphenoid contact already subdivides the metotic foramen of later crocodylomorphs such as *Protosuchus* (Figs. 10.1, 10.6A, B).

Little information exists on the braincase morphology of early crocodyliforms between *Protosuchus* and the notosuchian *Simosuchus* or the thalattosuchian *Metriorhynchus*. This means that most of the evolutionary history of the crocodyliform otic region is unknown. The medial wall of the otic capsule of *Simosuchus* is completely ossified except for a small slit forming the "metotic fissure"—the internal opening for CN IX–XI (Fig. 10.6C) (Kley et al. 2010). Because of damage in the area, the presence of the fenestra pseudorotunda can only be inferred. Likewise, details of the morphology of the FO and other regions of the vestibule remain unknown. The semicircular canals of *Simosuchus* are taller and longer than those of extant crocodylians, with the lateral semicircular canal having the greatest radius of curvature. The middle ear of the thalattosuchian *Metriorhynchus* (Fernandéz et al. 2011) is enclosed by the craniopalate complex and isolated from the occipital region by a much-developed exoccipital-basioccipital contact, delimiting the braincase into lateral and occipital areas. Subdivision of the metotic foramen cannot be confirmed because the middle and inner ear regions are not well preserved.

Information about the braincase is almost equally scarce for the rest of the eusuchian tree (Neosuchia plus a few less derived taxa). Scattered neosuchian material includes some basal taxa like *Paluxysuchus* (Adams 2013) and some dyrosaurid materials such as *Rhabdognathus* (Brochu et al. 2002). Their braincases,

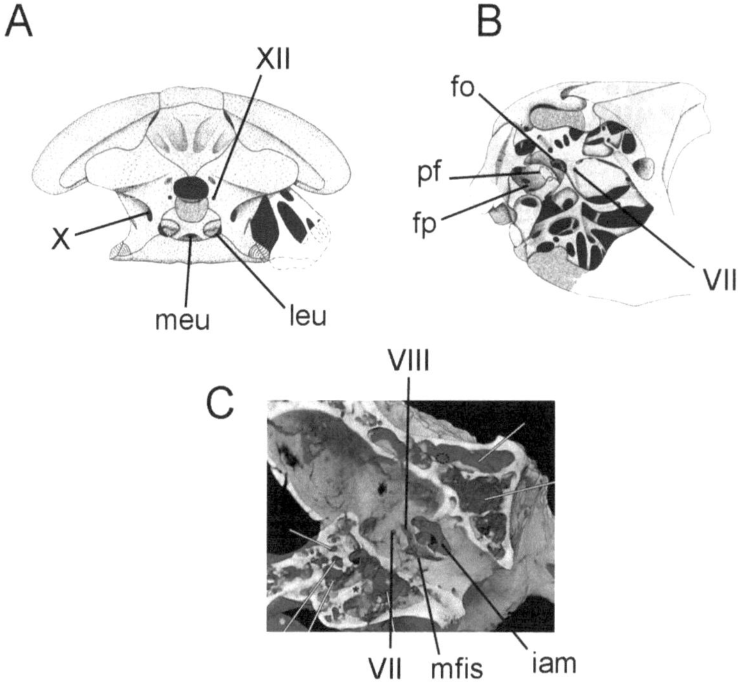

Fig. 10.6 Details of the braincase of *Protosuchus* in posterior view **(A)** and right-lateral view **(B)**. **(C)** Detail of the braincase of *Simosuchus* in medial view, anterior to the left. (**A**, **B** reproduced with permission of the Evolutionary Studies Institute at the University of the Witwatersrand from Busbey and Gow 1984; **C** reproduced with permission of Taylor and Francis from Kley et al. 2010)

however, are poorly explored because of poor preservation, but the morphology generally conforms to that of more basal forms.

The first records of the crown group Crocodylia are from the Upper Cretaceous (Brochu 2001), but its main diversification took place during the Paleocene. Much braincase material exists for the group; unfortunately, detailed anatomical descriptions are scarce. The unique anatomy of the cranio-quadrate articulation in all crocodyliforms hinders access to the internal anatomy of the skull because of the contact between quadrate and prootic.

The tympanic bulla of extant crocodylians lies medioventrally in the region of the "metotic fissure," as well as laterally in the region of the cochlear recess (Iordansky 1973; Gower and Weber 1998). Living taxa have an opisthotic loop enclosing the perilymphatic foramen. The vestibule is roughly triangular in shape, and the cochlea is elongate and strongly twisted medially at its midheight (Witmer et al. 2008). The stapes is extremely elongate. The braincase is highly pneumatized, although pneumatization can be relatively reduced in some taxa.

10.4 Avemetatarsalia: Pterosauria, Dinosauria, and Avialae

Avemetatarsalia (Fig. 10.1) is the sister-group of Pseudosuchia, and as such includes all clades that are more related to living birds than to crocodylians such as pterosaurs, dinosaurs, and a few other less known stem members.

10.4.1 Pterosauria: The Flying Reptiles

Pterosaurs were the first archosaurs to enter and dominate the air and to develop powered flight. Although their origins have been disputed (Bennett 1996; Renesto and Binelli 2006), recent analyses consider them as the sister-clade of dinosaurs (Fig. 10.1) (Nesbitt 2011). As such, they must have originated at latest during the late Middle Triassic, but the lack of obvious "transitional" forms makes their early evolutionary history obscure. The oldest pterosaur records date from the Late Triassic (Kellner 2015). General trends of diversity and disparity among pterosaurs increase up to the Late Jurassic–Early Cretaceous, followed by a Late Cretaceous decline (Butler et al. 2013). Early pterosaurs, such as *Rhamphorhynchus*, had long tails with a membrane called the cruropatagium that attached the hind limbs to the tail. Derived forms with short tails and without a cruropatagium formed the Pterodactyloidea, which originated in the Late Jurassic and produced most of pterosaur diversity. Pterosaur diversity spans from minute sparrow-sized forms to the giant azhdarchids with a wingspan up to 10 meters. Much information on pterosaur soft-part anatomy has been collected in the last 20 years (Hone 2012). Once seen as fragile gliders, it is now widely accepted that pterosaurs developed powered flight around 70 million years before birds, and that they had excellent mechanisms of flight control (Witmer et al. 2003; Middleton & English 2015).

All known pterosaur braincases are much modified in comparison to less derived archosauriforms. The pterosaur skull is generally highly pneumatized. While the straight quadrate bears no conch, the squamosal has an otic process that supported a tympanic membrane (Bennett 2001). The exit of CNs IX–XI lies on the occipital side of the skull, but there seems to have been no subdivision of the metotic foramen (Fig. 10.7). The only opening identified on the lateral braincase wall is the FO ("columellar recess" of Kellner 1996; Bennett 2001). In this case, the foramen for CNs IX–XI would be a posteriorly directed metotic foramen. No further information on the anatomy of the FO is available. The otic capsule is extensively ossified medially, the wall pierced by the foramina of VIII (vestibulo-acoustic nerve = vestibulocochlear nerve) ventral to the much-enlarged floccular fossa. In all three known pterosaur inner ears (*Rhamphorhynchus* and *Anhanguera*, Witmer et al. 2003; *Allkaruen*, Codorniú et al. 2016), all three semicircular canals are very elongate, especially the ASC (Fig. 10.7D, E). The cochlea is unknown in pterosaurs. The stapes is known only for *Pteranodon* (Bennett 2001), in which it is very

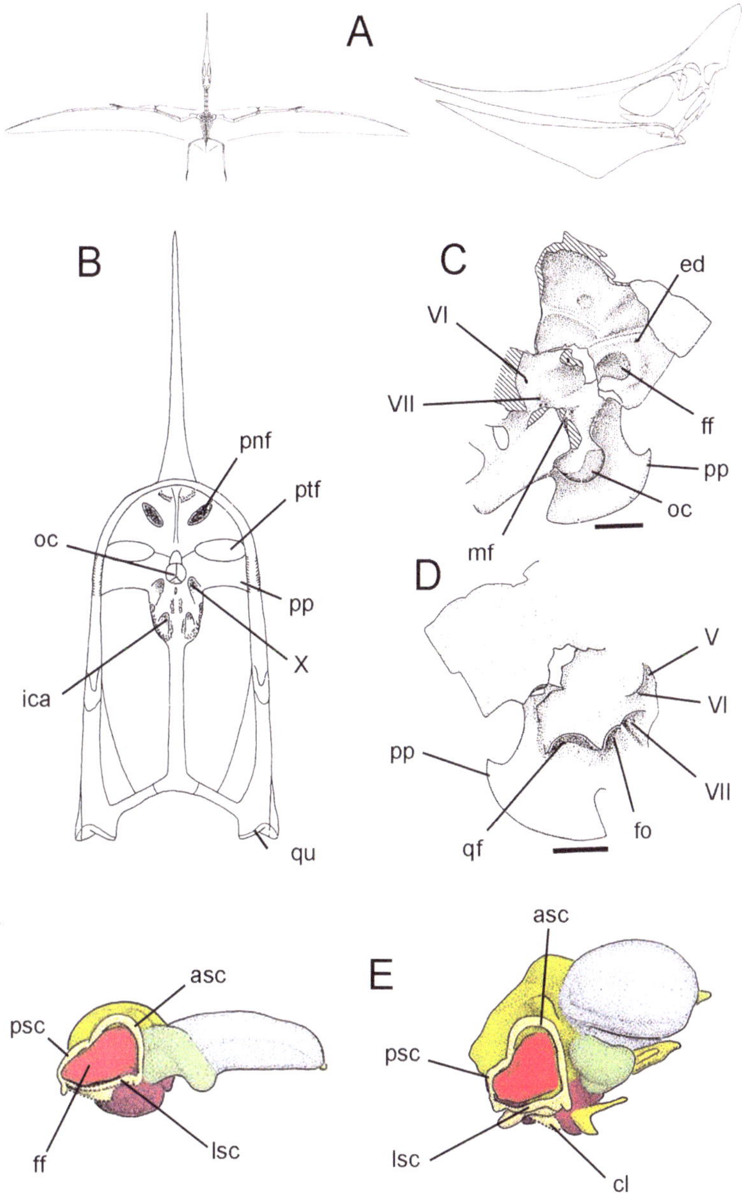

Fig. 10.7 Pterosaurs. **(A)** *Pteranodon* reconstructions (not to scale) of the body in dorsal view (*left image*) and head in lateral view (*right image*). **(B)** Skull of *Pteranodon* in posterior view. Note the relatively small contribution of the braincase to the skull. **(C)** Braincase of *Pteranodon* in right medial view. **(D)** Braincase of *Pteranodon* in right-lateral view. **(E)** Cranial endocasts of *Rhamphorhynchus* (*left image*) and *Anhanguera* (*right image*). (**A–C** reproduced with permission of E. Schweizerbart'sche Verlagsbuchhandlung from Bennett 2001; **D**, **E** reprinted by permission from Macmillan Publishers Ltd: Nature, from Witmer et al. 2003)

long and slender. Bennett's study interpreted the function of the inner region of *Pteranodon* in posture and flight control, and Witmer et al. (2003) did so for *Rhamphorhynchus* and *Anhanguera* (see Sect. 10.5.3 for functional morphology).

10.4.2 Dinosauria: Masters of the Mesozoic

Dinosauria is the most famous and well-studied fossil vertebrate group. The evolutionary history of dinosaurs begins with modest taxonomic and morphological diversity that seem to have remained unchanged during its early stages (Brusatte et al. 2010b). Diversity and disparity were not only lower than in pseudosuchians but also seem to have been little affected by the Triassic–Jurassic mass extinction. In the Jurassic and Cretaceous, however, dinosaurs diversified substantially, making archosaurs the dominant fauna of terrestrial ecosystems at the time. Upchurch et al. (2011) give a history of dinosaur rises, declines, and extinctions.

Braincases of early dinosauromorphs are almost unknown. The best preserved material is that of *Silesaurus* (Dzik 2003) from the Late Triassic of Poland (Fig. 10.8). An unossified gap seems present ventral to the FO (Fig. 10.8A). The metotic foramen was large and elongate. Medially, the most anterior portion of the vestibule is better ossified than in *Euparkeria* (Fig. 10.8B). Both a separate foramen for CN XI and a fenestra pseudorotunda have been reported for another basal taxon, *Lewisuchus*, which also preserves the proximal, rod-like portion of the stapes (Bitterncourt et al. 2014).

10.4.2.1 Ornithischia: The Bird-hipped Dinosaurs

Ornithischians were large-bodied herbivorous dinosaurs that dominated several terrestrial ecosystems in the Jurassic and Cretaceous. They have superficially similar pelvis construction to that of birds, in which the pubis points backwards. Ornithischia is roughly divided into two main clades: Thyreophora, which includes stegosaurids and ankylosaurids, and Cerapoda. Within the latter, there is Ornithopoda, the group of the now paraphyletic "iguanodontians" and hadrosaurids, and Marginocephalia, which comprises the frilled ceratopsians and the

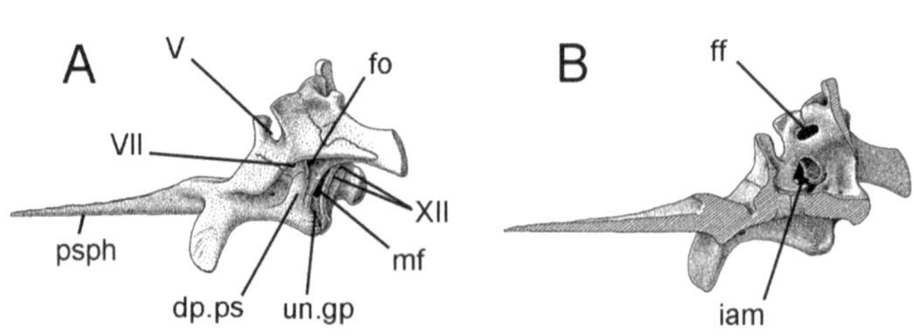

Fig. 10.8 Braincase of *Silesaurus* in left-lateral (**A**) and right medial views (**B**). Anterior to the left in both views. (Redrawn from Dzik 2003)

thick-domed pachycephalosaurs (Fig. 10.1). Notable among ornithischians is the appearance of elaborate postcranial armor and cranial ornamentation, such as the dorsal plates of stegosaurids or the hollow "horns" for sound production of hadrosaurids. They also developed elaborate social behavior and complex feeding mechanisms, such as the unique six-tissue teeth of hadrosaurids (Sereno 1997; Erickson et al. 2012). Ornithischia first appeared during the Late Triassic, but it was not until the Jurassic that the major clades originated and diversified. From the Early to the Middle Jurassic, thyreophorans evolved from small-bodied, bipedal generalists to bulky, obligate quadrupeds. Following their split with the ornithopods in the Early Jurassic, the marginocephalians next appear in the fossil record in the Early Cretaceous. The Early Jurassic was a period of global radiation for basal ornithischians, but their diversity declined in the Late Jurassic.

The basal ornithischian *Heterodontosaurus* (Fig. 10.1) shows a fenestra pseudorotunda located posteroventrally to the FO (Norman et al. 2011). There is a well-developed crista tuberalis (= ventral ramus of the opisthotic) (Fig. 10.9A), posterior of which is a foramen for the vagus nerve (CN X) and, more posteriorly still, the foramen for the accessory nerve (CN XI).

Braincases of basal thyreophorans are unknown so far but, in contrast, many specimens and endocasts are available for *Stegosaurus* (Fig. 10.9B, C) (Hopson 1979; Galton 2001) and *Kentrosaurus* (Galton 1988). A fenestra pseudorotunda is absent and the small FO might have opened ventrally. The crista interfenestralis (= ventral ramus of the opisthotic) is thin. The medial wall of the otic capsule was mostly ossified with a small, shallow floccular fossa. Reconstructions of the inner ear of *Stegosaurus* show an extremely short lateral semicircular canal (Hopson 1979; Galton 2001). The braincase of ankylosaurs is similar to that of stegosaurids (Fig. 10.9D, E) with an unsubdivided metotic foramen present posterior to a well-developed crista interfenestralis. However, a fenestra pseudorotunda might have been present in *Silvisaurus* (Eaton 1960) and *Pawpawsaurus* (Lee 1996). The ASC and the cochlea are more elongate than in stegosaurids (Miyashita et al. 2011).

Braincase anatomy of basal neornithischians seems generally conserved, as exemplified by *Dysalotosaurus* (Fig. 10.10A–C). The FO lies anterior to the fenestra pseudorotunda, separated from it by a thin crista interfenestralis. In comparison to the remainder of the inner ear, the cochlea was not particularly elongate and ran straight ventrally from the vestibule. Faint grooves mark the surface of the posterior margin of the fenestra pseudorotunda that may indicate the course of the glossopharyngeal nerve (CN IX). Posterior to the fenestra pseudorotunda is the foramen for CN X. The anteromedial wall of the otic capsule is ossified except for the region of the cochlear recess, and it was pierced by the foramina for CN VIII. The floccular fossa is very shallow. The stapes of basal ornithopods is known from *Thescelosaurus*, in which it is thin and slender (Boyd 2014). The braincases of derived ornithopod clades, the hadrosaurids, have general otic morphology similar to basal groups but with an enlargement of the ASC and of the cochlea (Fig. 10.10D) (Evans et al. 2009).

Fossils of the ear regions of basal Marginocephalia are unknown. No recent studies have been carried out on the pachycephalosaurs, although endocranial casts

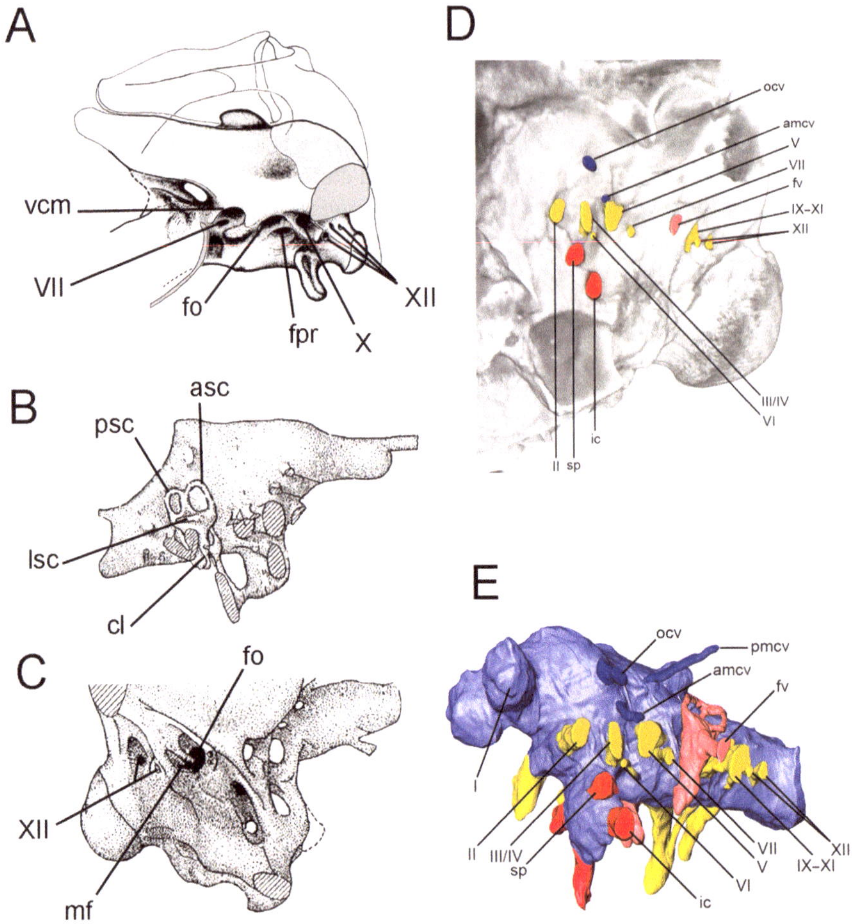

Fig. 10.9 Ornithischians 1. **(A)** *Heterodontosaurus* braincase in right-lateral view. **(B)** Right-lateral view of the cranial endocast and **(C)** braincase of *Stegosaurus* (*anterior to the right*). **(D)** Left anteroventral views of the braincase and **(E)** cranial endocast (*anterior to the left*) of *Euoplocephalus*. (**A** reproduced with permission of Wiley from Norman et al. 2011; **B**, **C** redrawn from Galton 2001; **D**, **E** reproduced with permission of Taylor and Francis from Miyashita et al. 2011)

were described and discussed by Hopson (1979). These show an extremely modified inner ear with strongly reduced ASCs and PSCs (Fig. 10.10E). On the other hand, ceratopsians show considerable variation in the width of the semicircular canals and in the cochlear length. The inner ears of some ceratopsians show a remarkably developed labyrinth with very elongate ASCs and PSCs, likely related to an unusually long common crus (Fig. 10.10F) (Witmer and Ridgely 2008a). In contrast, *Anchiceratops* shows a short common crus and a very short lateral semicircular canal (Hopson 1979).

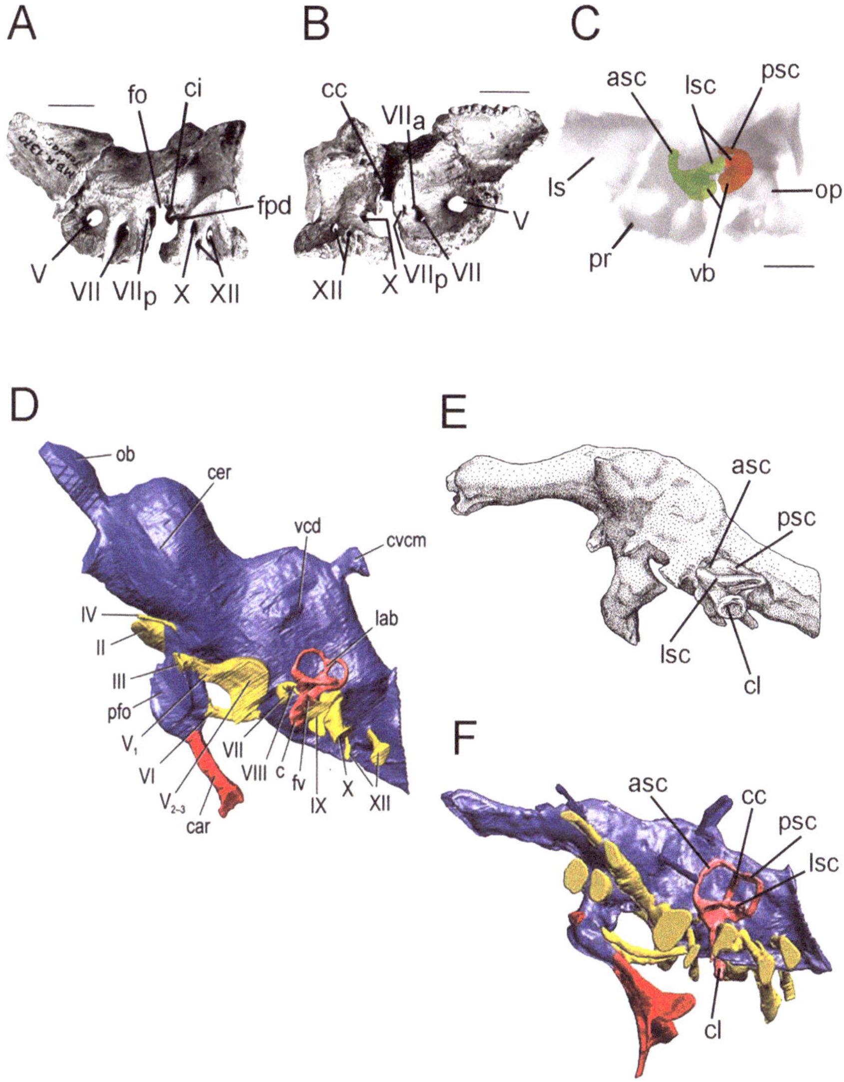

Fig. 10.10 Ornithischians 2. Braincase of *Dysalotosaurus* in left-lateral view (**A**) and medial view (**B**). (**C**) Inner ear in right-lateral view. (**D–F**) Left-lateral view of the cranial endocast of (**D**) *Hypacosaurus*, (**E**) *Pachycephalosaurus*, and (**F**) *Pachyrhinosaurus*. Anterior to the left in all three cases; *VIIa, VIIp* for anterior, posterior ramus of facial cranial nerve. (**A–C** reproduced with permission of Taylor and Francis from Sobral et al. 2012; **D** reproduced with permission of Wiley from Evans et al. 2009; **E** reproduced with permission of The Gans Collections and Charitable Fund from Hopson 1979; **F** reproduced with permission of Wiley from Witmer and Ridgely 2008a)

10.4.2.2 Saurischia: The Lizard-Hipped Dinosaurs

Pelvic construction in this group is basically a typical—though very large—version of the pelvic girdle common to most amniotes with a pubis that points forward. Although the ornithischian pelvis resembles that of birds, hence the name, birds belong to the saurischian clade. The resemblance to that of ornithischians is a result of convergent evolution (Fig. 10.1). Saurischians, as well as being a more diverse clade than ornithischians, are also the only group of dinosaurs that survived the end-Cretaceous extinction and whose descendants, the birds, persist to this day.

Sauropodomorpha

Sauropods are among the largest terrestrial vertebrates that have ever existed. Their extreme size and their elongated necks make them one of the most famous dinosaur groups for the general public. Nonetheless, the first members of the Sauropodomorpha (Fig. 10.1) were small and slender facultative bipeds. Their dentition was not specialized for herbivory, and the presence of gastroliths indicates that their digestive systems were equally simple. Sauropods, in contrast, were obligatorily quadrupedal with columnar limbs and partially pronated forearms (Sereno 1997). Neck and trunk vertebrae pneumatization occurs early in the evolutionary history of sauropods as well as a more specialized herbivorous dentition.

The basal sauropodomorph *Pantydraco* has a braincase that is very similar to that of basal dinosauriforms (Fig. 10.11A) (Galton and Kermack 2010). The FO is large, rounded, and ventrally open, with participation of the parabasisphenoid at its anteroventral border, and the metotic foramen is slit-like. On the other hand, other basal sauropodomorphs, such as *Plateosaurus* (Fig. 10.11B), have braincases similar to those of basal ornithischians (Galton 1984). The thin crista interfenestralis separates the large FO from a recess containing two foramina (Fig. 10.11B) whose identity is disputed (Galton 1984; Prieto-Márquez and Norell 2011). In *Plateosaurus*, the parabasisphenoid forms part of the margin of the metotic foramen and the FO (Prieto-Márquez and Norell 2011). The stapes of *Plateosaurus* is preserved in articulation with the FO, representing a rather robust element (Fig. 10.11B). The stapedial shaft is columnar and slightly flattened, and the footplate is subsquare with a flat surface. Medially, the wall of the otic capsule is well ossified and the floccular fossa rather deep. In *Masosspondylus*, the ASC and PSC are elongate (Knoll et al. 2012).

The metotic foramen of the more derived sauropod *Spinophorosaurus* (Fig. 10.11C) is elongate and not subdivided laterally, but CT scans suggest there

Fig. 10.11 Sauropodomorphs. (**A**) Braincase of *Pantydraco* in right-lateral view. (**B**) Braincase of *Plateosaurus* in right-lateral view. (**C**) Cranial endocast of *Spinophorosaurus* in left-lateral view. (**D**) Braincase of *Brachiosaurus* in left-lateral view. (**E**) Cladogram of saurischians with volumized labyrinth of corresponding taxa. (**A** reproduced with permission of Muséum d'Histoire Naturelle from Galton and Kermack 2010; **B** with permission of Schweizerbart'sche Verlagsbuchhandlung from Galton 1988; **C**, **E** from Knoll et al. 2012; **D** from Janensch 1935) ▶

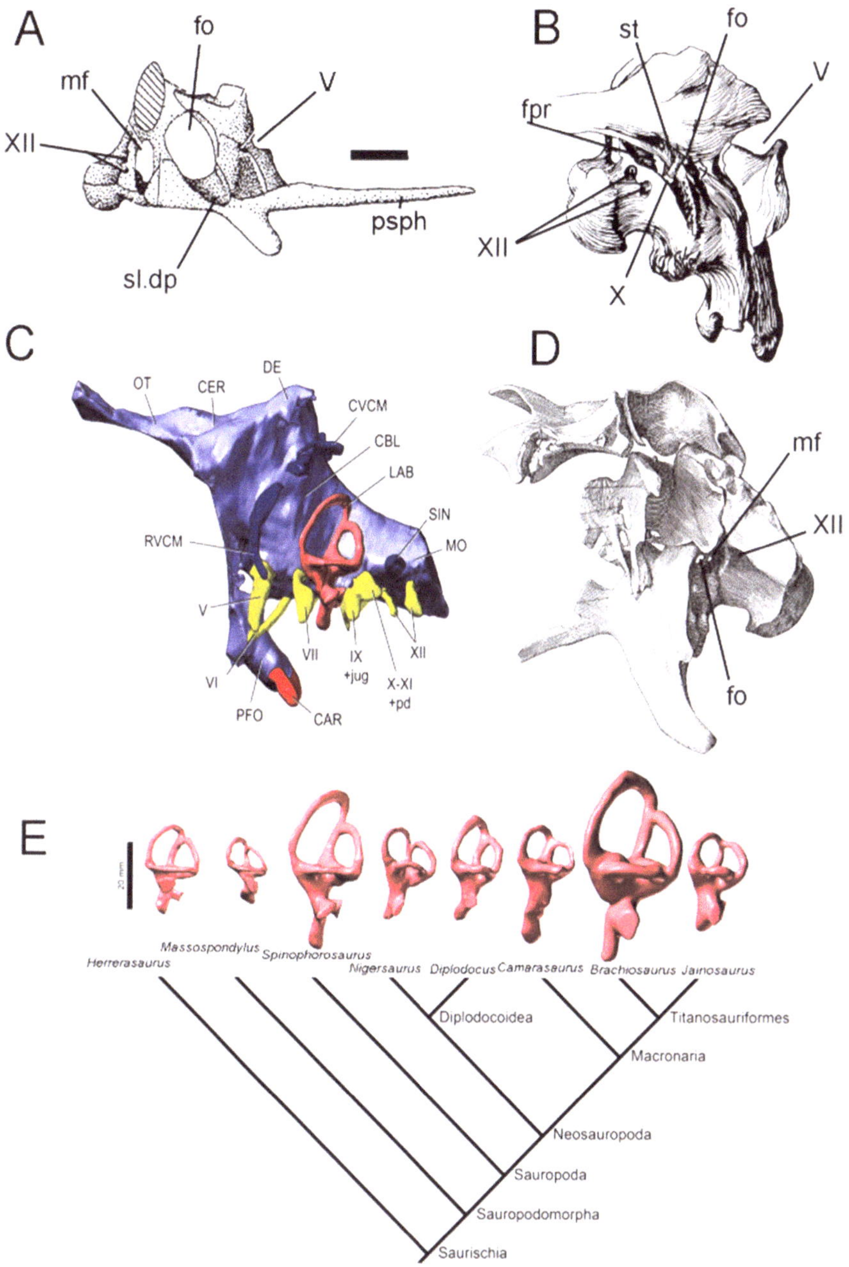
A
fo
mf
V
XII
psph
sl.dp
B
st
fo
V
fpr
XII
X
C
OT
CER
DE
CVCM
CBL
LAB
SIN
RVCM
MO
V
VII
IX
+jug
XII
X-XI
+pd
VI
PFO
CAR
D
mf
XII
fo
E
Herrerasaurus
Massospondylus
Spinophorosaurus
Nigersaurus
Diplodocus
Camarasaurus
Brachiosaurus
Jainosaurus
Diplodocoidea
Titanosauriformes
Macronaria
Neosauropoda
Sauropoda
Sauropodomorpha
Saurischia

was a bony subdivision medially (Knoll et al. 2012). The stapes is a relatively slender, straight element, oval in cross section. The distal end is missing, but the small footplate is preserved. The ASC is very elongate and significantly more elevated than the PSC (Fig. 10.11C). This general inner ear morphology is present in a number of other more derived clades, but a lot of variation exists. Although a fenestra pseudorotunda appears to have been absent in a number of sauropods, such as *Brachiosaurus* (Fig. 10.11D), it could have been present in *Barosaurus* (Janensch 1935, 1936). In Diplodocoidea, the ASCs and PSCs are more asymmetrical in derived taxa, such as *Diplodocus*, in which the vestibule is enlarged, and the cochlea short (Fig. 10.11E). In basal macronarians (Fig. 10.1), such as *Camarasaurus*, the semicircular canals are also short, but they are not short in "intermediate" clades like *Brachiosaurus* (Knoll et al. 2012). In *Camarasaurus*, the stapes is similar to that of basal sauropods (Madsen et al. 1995).

Theropoda

Theropod dinosaurs retain some of the plesiomorphic features of early dinosauromorphs such as bipedalism and carnivorous habits. They, however, developed specialized adaptations for predation such as a flexible bite through a mid-mandibular joint, a refined balance through a distally stiffened tail, and light bones through pneumatic systems of diverticula (Sereno 1997). The evolutionary history of theropods has received considerable attention in the past two decades since the discovery of feathered forms from China. Studies have focused on the evolution of traits that were *sine qua non* for the origin and evolution of powered flight (Benson and Choiniere 2013; Dececchi and Larsson 2013). Theropod diversity began with a peak in the Late Triassic, after the Carnian–Norian extinction event (Brusatte et al. 2010b). The earliest theropods were small, slender forms, but large-bodied forms evolved independently in several lineages such as Megalosauroidea and Tyrannosauroidea (Fig. 10.1). Successive sister-clades of birds include many small-sized arboreal taxa.

Information on the braincases of basal theropods is limited and contradictory. A very small fenestra pseudorotunda was described by Tykoski (1998) in *Coelophysis* (= *Syntarsus*), but this may be a misidentification.

A fenestra pseudorotunda cannot be demonstrated in basal ceratosaurs. The ASC extends more dorsally than the PSC in *Ceratosaurus*, and the short cochlea is directed ventrally (Sander and Smith 2005). The floccular fossa is also small. Braincase and inner ear anatomy have been thoroughly covered in the derived ceratosaur *Majungasaurus* (Fig. 10.12B, C) (Sampson and Witmer 2007). In this taxon there is a subdivided metotic foramen with the robust crista tuberalis separating the vagus foramen from the fenestra pseudorotunda. Its anteroventral portion is overlapped by the parabasisphenoid, forming the ventral borders of both fenestrae. The FO was located in a deep recess. The cochlea is not particularly elongate. The braincase of *Majungasaurus* is well pneumatized, but only the anterior tympanic, basisphenoid, and subsellar recesses are present.

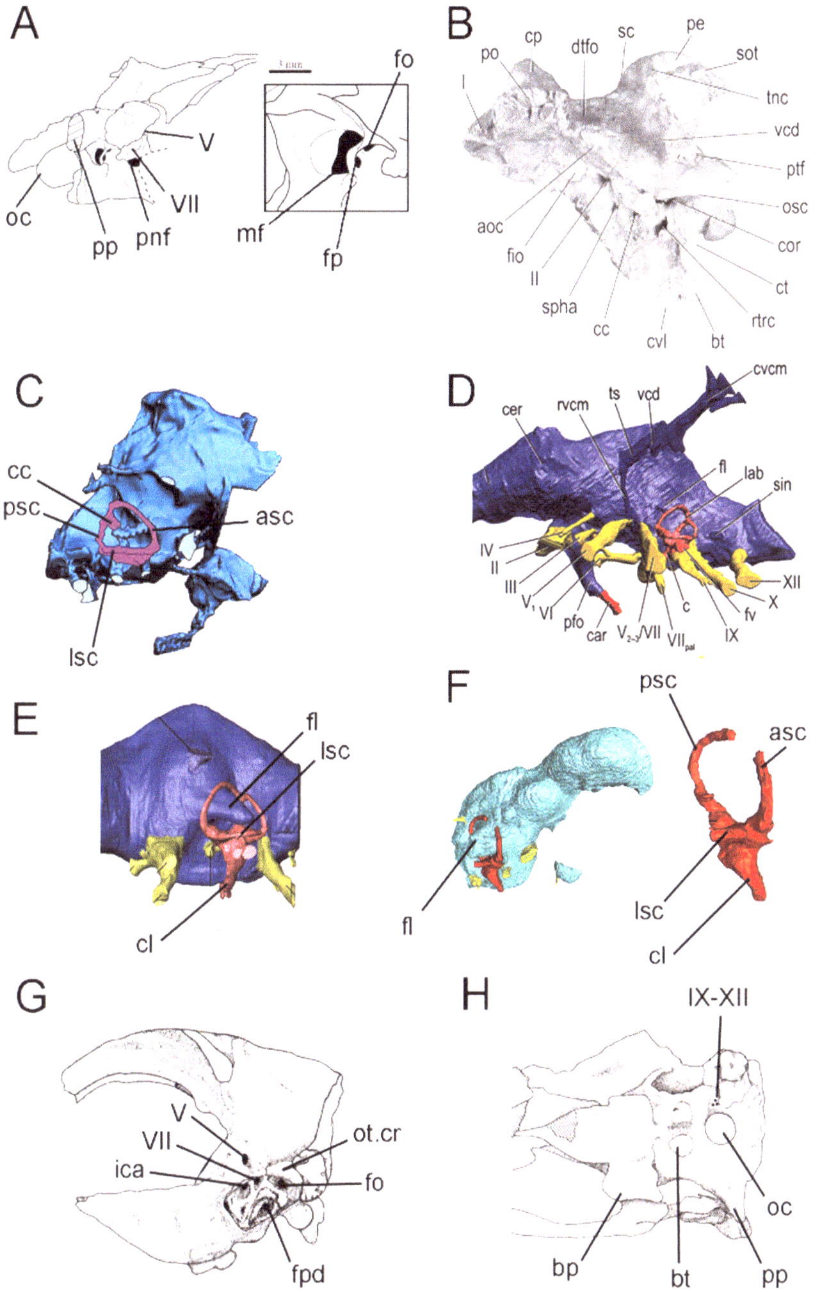
A
V
VII
oc
pp
pnf
fo
mf
fp
B
po
cp
dtfo
sc
pe
sot
tnc
vcd
ptf
osc
cor
ct
rtrc
bt
cvl
cc
spha
II
fio
aoc
l
C
cc
psc
asc
lsc
D
cer
rvcm
ts
vcd
cvcm
fl
lab
sin
IV
II
III
V1
VI
pfo
car
V2-3/VII
VIIpal
c
IX
fv
X
XII
E
fl
lsc
cl
F
psc
asc
fl
lsc
cl
G
V
VII
ica
ot.cr
fo
fpd
H
IX-XII
oc
pp
bt
bp

◂**Fig. 10.12** Theropods. **(A)** Braincase of *Coelophysis* in right-lateral view (*right*) and detail of otic region (*left*). **(B)** Braincase of *Majungasaurus* in left-lateral view. Cranial enocasts: **(C)** *Sinosaurus* in right latertal view. **(D)** *Tyrannosaurus* in left-lateral view. **(E)** Detail of *Erlikosaurus* in left-lateral view. **(F)** *Oviraptorosaurus* (details of the labyrinth on the *left*). **(G, H)** Braincase of *Saurornithoides* in left-lateral view **(G)** and ventral view **(H)**. (**A** redrawn from Tykoski 1998; **B** redrawn from Sampson and Witmer 2007; **C** reproduced with permission from Xing et al. 2014; **D** reprinted with permission of Indiana University Press from Witmer and Ridgely 2009; **E** from Lautenschlager et al. 2012; **F** from Balanoff et al. 2014; **G, H** reproduced with permission of the Institute of Palaeobiology of the Polish Academy of Sciences from Barsbold 1974)

Continuing on the avialan lineage, a foramen located posterior to the FO in *Sinosaurus* (Fig. 10.12C) has been identified as an independent exit for the glossopharyngeal nerve, CN IX (Xing et al. 2014), but it should probably be more precisely identified as the fenestra pseudorotunda. The foramina for CN X and CN XII were located in a common recess posterior to a well-developed crista tuberalis. The lateral tympanic recess is present laterally on the parabasisphenoid, albeit poorly developed. The ASC has a rather small radius of curvature and the floccular recess is very small, but it is very deep and well-marked in the endocast (Fig. 10.12C). The braincase of the basal megalosauroid *Piatnitzkysaurus* also conforms to this pattern (Rauhut 2004). Braincase morphology of more derived megalosauroids is not well known, but the braincases of the *Allosaurus* (Rogers 1998) and *Tyrannosaurus* (Fig. 10.12D) (Witmer and Ridgely, 2009) also do not deviate much from that pattern. The most striking difference in the braincases of both groups is the degree of pneumaticity present in tyrannosaurs, especially with regard to the posterior tympanic recess. It is the most developed of the tympanic recesses and occupies the prootic, the supraoccipital, the otoccipital, and, in particular, the paroccipital processes (see Sect. 10.5.4).

Braincase and otic morphology exhibit much variation in coelurosaurs that are more derived than tyrannosaurids, but these pertain to more subtle details such as the degree of elongation of the semicircular canals and their relative position to each other, length and direction of the cochlea, and degree of development of the tympanic pneumatic recesses. *Falcarius* (Smith et al. 2011) and *Erlikosaurus* (Lautenschlager et al. 2012) are examples of this kind of variation.

The inner ear of the oviraptorosaur *Conchoraptor* resembles that of some birds in that the inner ear is tilted anterodorsally relative to the braincase, and the ASC is much extended posteriorly (Fig. 10.12F) (Balanoff et al. 2014). Derived paravians, such as the troodontid *Saurornithoides*, exhibit an overall braincase morphology that is similar to those of avialans but different from that of *Archaeopteryx* (Fig. 10.12G, H) (see Sect. 10.4.3). The otic region of *Saurornithoides* is recessed in an area protected by an otosphenoidal crest formed by the opisthotic, prootic, and basisphenoid (Barsbold 1974). Both the FO and the fenestra pseudortunda are small and seem to be recessed. The exit of the vagus nerve has a unique morphology in that it leaves the braincase in a posteroventral position, so it is not visible in occipital view. The paroccipital processes are well developed dorsoventrally, delimiting a small post-temporal fenestra. This extension, however, is limited to the dorsal region of the skull, thus not delimiting the middle ear cavity posteriorly.

10.4.3 Avialae: Birds and Their Closest Relatives

Avialans include living birds and a number of extinct forms like the aquatic hesperornithiforms and the giant terror birds, Phorusrhacidae. Cladistic analysis recovers *Archaeopteryx* as the basalmost avialan in many phylogenetic trees (Xu and Pol 2014), although other studies suggest closer affinities to dromaeosaurids (Godefroit et al. 2013). The relationships of birds to non-avian theropods had been suggested as early as 1868 by Thomas Huxley shortly after the publication of "On the Origin of Species," and further evidence came with the discovery and description of *Deinonychus* by Ostrom in 1969. Ostrom's work was followed by an intense debate on the origin of birds until the mid-1990s when a number of discoveries of feathered fossils from China settled the debate (e.g., Currie et al. 1998; Ji and Ji 1996; Wu et al. 1999).

The origin of birds and of powered flight has driven many recent studies on theropod evolution. The main achievements in this field refute the earlier belief that powered flight was a single and dramatic point in the evolutionary transition into air, or that bird anatomy was constrained by flight. It is now known that not only transformations in the anatomy of theropods were underway long before the origin of birds or flight but also that morphological novelties associated with flight do not represent key innovations for the group (Benson and Choiniere 2013; Dececchi and Larsson 2013).

Archaeopteryx is the best known early fossil bird and its braincase and otic region are known in some detail from μCT scans of a three-dimensional preserved specimen (Alonso et al. 2004). It comes from the late Jurassic lagerstätte of Solnhofen, from which many excellent fossil specimens of vertebrates and invertebrates have come. There, soft tissue is represented in the fossils, including the feather impressions of *Archaeopteryx*.

The otic region of *Archaeopteryx* (Fig. 10.1) is not recessed, unlike that of more derived clades, and thus it is confluent with the anterior region of the lateral surface of the braincase (Fig. 10.13A) (Walker 1985). Posteriorly, the isolation of the middle ear from the occipital region is not made by a ventral expansion of the paroccipital process but by a lateral expansion of the exoccipital. Micro-CT scanning of the available three-dimensional braincase of *Archaeopteryx* revealed that the lateral semicircular canal enters the anterior ampulla in its posterior region, and it does not share the posterior ampulla with the PSC—the latter extends further ventrally than the former (Alonso et al. 2004). The ASC is much elongated, extending further posteriorly than the common crus, flexing it.

Diversification events in Avialae occurred twice during the Mesozoic, with peaks occurring both in enantiornithine (a group of extinct birds that still showed many "dinosaur-like" features such as toothed beaks) and ornithurine (includes hesperornithiforms and living birds) lineages in the Early Cretaceous. Although molecular estimates place the origin of Neornithes in the mid-Cretaceous, only a few Mezosoic neornithine fossils have been found (Stein et al. 2015). A better record is found in the Paleogene, indicating that the radiation of modern birds

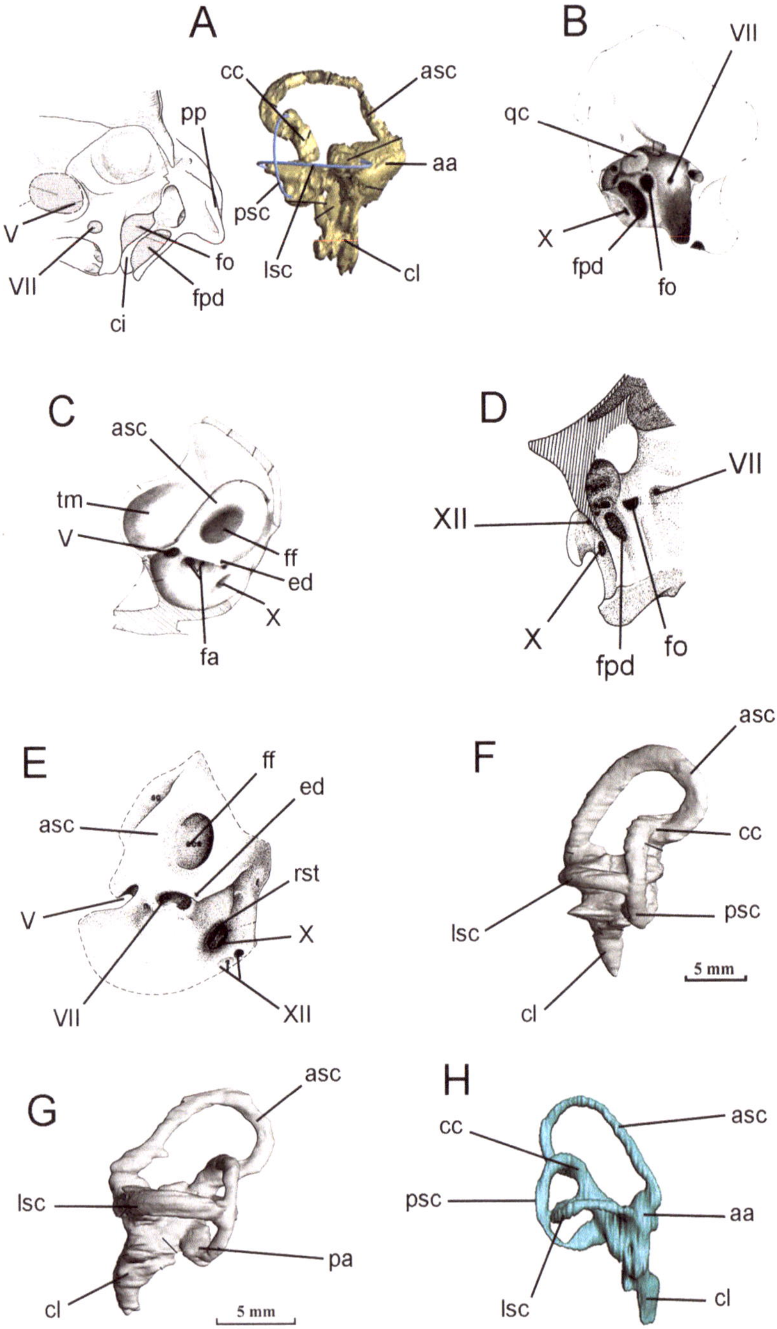
A
cc
asc
pp
aa
psc
V
fo
lsc
cl
VII
fpd
ci
B
VII
qc
X
fpd
fo
C
asc
tm
V
ff
ed
X
fa
D
VII
XII
X
fpd
fo
E
ff
ed
asc
rst
V
X
VII
XII
F
asc
cc
psc
lsc
5 mm
cl
G
asc
lsc
pa
cl
5 mm
H
cc
asc
psc
aa
lsc
cl

◄**Fig. 10.13** Avialans. **(A)** *Archaeopteryx* braincase, left-lateral view (*left*), and inner ear, right-lateral view (*right*). **(B)** *Enaliornis* braincase, left-lateral view and **(C)** left medial view. **(D)** *Hesperornis* braincase, right-lateral view and **(E)** right medial view. **(F–H)** Inner ear of the crown birds: **(F)** *Odontopteryx,* **(G)** *Prophaethon,* and **(H)** *Paraptenodytes.* (**A** *left* reproduced with permission of the Jura-Museum Eichstätt from Walker 1985 and *right* by permission of Macmillan Publishers Ltd: Nature, from Alonso et al. 2004; **B**, **C** reprinted with permission from Taylor and Francis from Elzanowski and Galton 1991; **D**, **E** from Elzanowski 1991; **F**, **G** reproduced with permission of Wiley from Milner and Walsh 2009; **H** from Ksepka et al. 2012)

probably took place after the extinction of non-avian theropods. Two events are thought to have been responsible for peaks in diversification rates of neornithines: the Paleocene–Eocene thermal maximum and the Eocene–Oligocene mass extinction—although geological events linked to the splitting of parts of Gondwana might also have played an important role (Selvatti et al. 2015; Stein et al. 2015). Fossils of Paleocene birds are also scarce, but giant ground birds, such as *Phorusrhacos*, dominated the ecosystems of the early Tertiary. The ecology of these birds as top predators have populated the imagination of the public, and phorusrhacids are often informally called "terror birds". However, while their beak morphology is indeed similar to that of extant birds of prey, recent studies indicate that members of Gastornithidae and Dromornithidae were actually herbivorous (Angst et al. 2014).

Hesperornithiforms (Fig. 10.1) are aquatically adapted, flightless birds from the Cretaceous from the Northern Hemisphere. *Enaliornis*, a basal member of the group, shows a laterally developed otoccipital dividing the braincase into occipital and lateral regions (Elzanowski and Galton 1991). The middle ear is also different from that of non-avialan theropods in possessing an intricate system of ridges and expansions (Fig. 10.13B). Both the FO and fenestra pseudorotunda are located in a recess together with one large pneumatic foramen. The floccular fossa is rounded and large and is deeply excavated on the medial side of the braincase (Fig. 10.13C). Instead of being located dorsal to the foramen of CN V (trigeminal nerve), the fossa is shifted more posteriorly. The paroccipital processes are short and narrow. The braincase of *Hesperornis* shows a similar morphology but, in medial view, the long axis of the elongate floccular fossa is vertical (Fig. 10.13D, E) (Elzanowski 1991). Medially, the fenestra pseudorotunda and the vagus foramen have a common wide opening. In *Enaliornis*, in contrast, the foramen for the vagus nerve is small and separated from the internal opening of the fenestra pseudodotunda.

A major gap exists not only in the fossil record, but also in the literature on avialan braincase anatomy. The descriptions above are of stem-group members, but the only other available material for crown-group members show essentially a modern anatomy of the otic region. Pelagornithids, or pseudotoothed birds, were albatross-like seabirds with outgrowths of the premaxilla and dentary. Fossils are found worldwide in deposits dating from the Late Paleocene to the Pliocene-Pleistocene boundary (Bourdon et al. 2010). Their phylogenetic position is controversial, but they are part of the fowl birds (Galloanserae, Fig. 10.1). The braincase of *Dasornis* shows a deeply recessed middle ear and a laterally directed vagus foramen (Bourdon et al. 2010). The inner ear of *Odontopteryx*, another pelagornithid, is quite similar to that of *Archaeopteryx* with a possibly longer PSC

(Fig. 10.13F) (Milner and Walsh 2009). The cochlear recess is straight and extends strongly anteroventrally.

Prophaetodontids are closely related to tropicbirds, which, in turn, are related to penguins and pelicans. In the *Lithoptila*, the middle ear also is deeply recessed due to the development of the lateral cotylar articulations with the quadrate and has a more ventral position in relation to the rest of the braincase (Bourdon et al. 2005). The inner ear of *Prophaethon* shows semicircular canals that are narrower than those of pelagornithids, but that are otherwise similar to them and to *Archaeopteryx* (Figs. 10.9, 10.13G) (Milner and Walsh 2009). Another type of morphology in the crown group is seen in the inner ear of penguins (sphenisciforms) (Ksepka et al. 2012). Peguins are flightless but developed "underwater flying." In the inner ear of the early Miocene *Paraptenodytes* of Argentina, the ASCs and PSCs are more elongate than in previously described taxa, and the ampullae are well-developed (Fig. 10.13H). The lateral semicircular canal is almost circular in outline and, in dorsal view, describes an arch that reaches further laterally than the floccular lobe. Because of the extreme flexure of the common crus, the posterior portion of the lateral semicircular canal joins the vestibule close to the posterior portion of the crus. The cochlear recess is directed anteromedially and is elongate. The anterior and lateral semicircular canals of the dodo (*Raphus cucullatus*) show bends at about their midpoint that make them even more elongate than in penguins (Gold et al. 2016). This is a rather common feature of some birds. The cochlea of the dodo is also long and strongly anteriorly directed.

10.5 Morphological Evolution of the Otic Region in Archosauromorpha

The hearing systems of Archosauromorpha show considerable changes in the course of their evolutionary history with abundant and independent changes in the two major archosaur lineages. It is therefore very difficult to talk about general archosaur trends if most similarities were acquired independently in birds and crocodylians. Generalizations ignore the great diversity in their morphologies and evolutionary history. Thus, the topics listed below will cover both archosaur branches separately.

10.5.1 Ossification of the Otic Capsule and Its Significance

There is a general process of isolation of the ear from the surrounding structures in archosaurs. Energy can be lost due to excessive flexion of soft structures during sound transmission. Therefore, anatomical features of the walls of the ear (muscles, connective tissue, cartilage, bone) will influence frequency response. Stiffening of these structures produces acoustic isolation, hindering sound transmission along routes where sound detecting tissues are not located (Lombard and Hetherington 1993). Isolation of the middle ear cavity in crocodylians happened with the abutting of the medial wing of the quadrate against the posterior portion of the prootic anteriorly and by the contact between exoccipital and basioccipital posteriorly (Iordansky 1973). The middle ear of birds, on the other hand, is enclosed by the basitemporal plate of the parabasisphenoid anteroventrally. Posteriorly, the contact of the braincase with the suspensorium happens not by means of the paroccipital process but through the development of secondary condyles of the prootic (Livezey and Zusi 2006). How these contacts were formed in the bird skull has been explored little. In contrast, the stepwise evolution of the braincase contacts of crocodyliforms can be tracked in the fossil record (Pol et al. 2013). The most important feature of the crocodylian skull is the tightening of the cranio-palate complex because the increased roles of the lower jaw and its muscles are the main forces that shape their skull anatomy. Pterosaurs also show isolation of the middle ear cavity from the occipital region of the skull, but this seems to have been led by a ventral extension of the paraoccipital process.

One otic character explored in vertebrates is the relative contribution of each bone forming the FO. This issue is important among early tetrapodomorphs (Clack 1997; Clack and Allin 2004), but the configuration in archosaurs has become more stable. The FO is formed by the prootic and opisthotic with a contribution from the parabasisphenoid. For archosaurs and their stem groups, an "unossified gap" in the ventrolateral border of the cochlear recess has been explored more. It seems to correspond to a similar structure in *Sphenodon* (Gower and Weber 1998) and also is found in many stem archosaurs and in pseudosuchians, including extant crocodylians. The exact function of this structure is unknown and deserves a more careful survey. Fossil avialans, however, seem to be fully ossified in this region. The parabasisphenoid, prootic, and opisthotic of bird-line dinosaurs had extensive lateral contact with each other, although the cochlear recess of *Archaeopteryx* may have been "notched."

In addition to the unossified gap, the FO is incomplete ventrally in stem groups, although not in extant birds and crocodylians. The timing of closure of the FO is unknown in both archosaur clades, but it may have happened independently in major dinosaur clades. In some basal sauropodomorphs like *Thecodontosaurus* (Benton et al. 2000), the ventral ramus of the opisthotic contacts the parabasisphenoid, closing the FO ventrally, with early ornithischians and theropods having a similar morphology. In *Pantydraco*, in contrast, the FO remains open, which may be either a reversal or a character unique to that taxon.

The otic capsule also becomes progressively more ossified medially. In stem groups, the semicircular canals and vestibular part of the inner ear are enclosed by the prootic, opisthotic, and supraoccipital, but ventrally, the medial braincase wall remains essentially unossified. It seems to ossify gradually without significant reversals in either archosaur lineage, presumably convergently, because this does not happen before the origin of the crown group. In pseudosuchians, the medial wall of the otic capsule is only open posteroventrally, as a slit termed the "metotic fissure." This is also true in dinosaurs. The otic capsules of pterosaurs were extensively ossified medially, resembling those of avialans.

Ossification of the otic capsule and of the middle ear is connected to mechanical tuning. Soft tissues do not fossilize, but the increase in the degree of ossification of the otic capsule indicates the importance of mechanical tuning. Extant mammals use mechanical tuning via an elongate cochlea (Manley and Clack 2004), whereas birds use electrical tuning in which hair cells actively set up a voltage gradient (Mann and Kelley 2011). It would thus be interesting to compare the stages of otic capsule ossification of diapsids and synapsids to find out at which point these mechanisms diverged in the inner ears of birds and mammals.

10.5.2 The Development of Impedance-Matching Hearing

Details of tympanic hearing are provided in Chap. 8 (Sobral, Reisz, Neenan, Müller, and Scheyer). As in stem diapsids, a tympanic membrane may have been absent in the basalmost clades of stem archosaurs, although it was present in *Prolacerta*. Either the membrane has been lost in early taxa, or it was acquired only later in the history of the group. Other characteristics associated with the development of tympanic and impedance-matching hearing also appear to have arisen convergently among several groups within Archosauromorpha (Clack 1997). While the origin of tympanic hearing within diapsids, including archosaurs, remains obscure, that of impedance-matching hearing seems now clearer. In impedance matching a pressure-relief window, termed the fenestra pseudorotunda, developed in addition to a tympanic mechanism (Rieppel 1985). An unossified otic capsule allows movements of the inner ear structures, decreasing impedance and facilitating sound transmission (Evans 2008). However, ossification directs the sound along routes where sound detecting tissues are located (Lombard and Hetherington 1993). Thus, to avoid limiting hearing capacity, acoustic isolation requires a compensatory mechanism, which was achieved by the development of the fenestra pseudorotunda. Cranial nerves IX, X, and XI leave the brain cavity through the metotic foramen. This may become subdivided by the metotic strut, creating a vagus foramen posteriorly and a fenestra pseudorotunda anteriorly (Rieppel 1985). The fenestra pseudorotunda is the lateral opening of the space formed by the subdivision of the metotic foramen, the recessus scala tympani. Medially, the recess communicates with the brain cavity through the apertura medialis (or perilymphatic aqueduct), and dorsally (or anteriorly) with the cochlear recess through the perilymphatic foramen. The perilymphatic duct leaves the otic capsule through this foramen and bulges as the perilymphatic sac into the recessus. The perilymphatic sac covers the interior of the recessus and, at the opening of the fenestra pseudorotunda, it forms the secondary tympanic membrane together with the membrane lining the middle ear cavity. During sound transmission, the fluid displaced runs through the cochlea from the FO to the fenestra pseudorotunda, and the fluid pressure so created is relieved by bulging of the secondary tympanic membrane.

In the basalmost pseudosuchians there is no subdivision of the metotic foramen, and it only appears later in crocodylomorphs. It cannot yet be confirmed in pterosaurs. The subdivision of the foramen is much more plastic in dinosaurs. Unknown in basal taxa, it is likely that all three classic dinosaur groups acquired a fenestra pseudorotunda independently. Probably present in basal ornithischians, it was retained in many subsequent groups but lost in pachycephalosaurs, ankylosaurs, and stegosaurs. Basal sauropodomorphs also may have possessed one, but it was then lost in derived sauropods. In theropods, a fenestra pseudorotunda may not have appeared until the origin of coelurosaurs, and it was retained in avialans. It may also have developed independently in derived abelisauroids. The anatomy of the otic region in each of the three main groups differs substantially, resulting in a differently positioned vagus foramen among various ornithischians and theropods, including birds.

In groups without a fenestra pseudorotunda, its pressure-relief function is, to some extent, carried out by the metotic foramen. In these, although a recessus scalae tympani is absent, there is an area into which the perilymphatic sac bulges, which is separate from where the CN X leaves the braincase. This is seen, for example, in non-diapsid reptilians such as *Captorhinus* (Sobral, Reisz, Neenan, Müller, and Scheyer, Chap. 8). When the metotic foramina of stem archosaurs are compared, it is seen that the foramen becomes increasingly enlarged, especially in the area of the perilymphatic sac. The elongate foramina of proterochampsids and phytosaurs may have had similar functions. The perilymphatic sac also becomes structurally separated from the inner ear with the thickening of the ventral ramus of the opisthotic. All this suggests more specialized hearing abilities through the enhanced pressure relief function of the metotic foramen.

One last issue on the morphology of the metotic foramen is its location in relation to the ventral ramus of the opisthotic. In clades with a metotic foramen, it is found posterior to the ventral ramus. Its subdivision should then result in a fenestra pseudorotunda that is also located posterior to it. This is the case in crocodyliforms but, in dinosaurs, this is only true in sauropodomorphs. In both ornithischians and theropods, the fenestra pseudorotunda lies anterior to the ventral ramus, which in these groups is called crista tuberalis. This implies an unusual anterior displacement of the perilymphatic sac and corresponding structures.

10.5.3 Evolution of the Inner Ear and Its Implications

Although not related to hearing, the anatomy of the vestibular and canalicular portions of the inner ear can be informative for the palaeobiology of fossil groups. Unfortunately, in the stem archosaur clades, including pseudosuchians and pterosaurs, the anatomy of the inner ear is basically unknown. For dinosaurs, however, it is better known, although a comprehensive study, rather than just raw data, is lacking. In agile animals, like birds and pterosaurs, living in complex three-dimensional environments, the semicircular canals are elongate (Witmer et al. 2003). Long semicircular canals are more sensitive to motion in their plane of action because the inertia of the fluid inside them is lower (Sampson and Witmer 2007). The presence of elongate canals in highly mobile taxa indicates both features are connected, although it remains obscure as to how this relationship works and which aspects of the canalicular morphology better express this mobility. There is also a positive relationship between the length of the ASC with the size of the floccular fossa. It houses the floccular lobe, a structure that plays a role in coordinating the movements of the head, neck, and eyes (Cox and Jeffery 2010).

In comparison to the non-saurian diapsid *Youngina* (Gardner et al. 2010; also see Sobral, Reisz, Neenan, Müller, and Scheyer, Chap. 8), stem archosaurs show a general trend of elongation in the ASCs and PSCs but a shortening of the lateral canal. In most archosaur groups, the inner ear starts out with a morphology similar to that of *Euparkeria*. In *Euparkeria*, a terrestrial animal, the ASC is taller than the PSC. *Euparkeria* has an upright body posture, in contrast to the sprawling posture of *Youngina*, and elongation of its canals may thus be related to an ecological change in habitat permitted by the new body posture of stem archosaurs. The canals are subequal in length in the aquatic phytosaur *Pseudopalatus*, which may be connected to its aquatic habits and limited neck movement.

In Pseudosuchia, the anatomy of the inner ear is only available for the notosuchian *Simosuchus*, and it is very similar to that of *Euparkeria*. This may indicate that *Simosuchus* retained the plesiomorphic state; although, since information for pseudosuchians is so scarce, it could equally indicate a reversal to that state. Extant crocodylians have shorter semicircular canals, in particular the PSC. Notosuchians, with their notably gracile anatomy, upright posture, and scansorial habits, probably reflected these features in the anatomy of their semicircular canals. Extant crocodylians, by contrast, have very restricted head mobility because of their unique braincase morphology (see Sect. 10.5.9).

The basalmost pterosaur for which the inner ear is known, *Rhamphorhynchus*, has longer canals than *Euparkeria*. In more derived clades, the canals become even longer—also longer than many bird canals. Pterosaurs, like birds, were also very agile animals capable of active flight, and thus it is no surprise that their inner ears are equally developed (see Sect. 10.5.6).

In ornithischian dinosaurs, the basalmost taxon for which the inner ear is known is *Dysalotosaurus*. Its inner ear is very similar to that of *Euparkeria*, whereas derived hadrosaurids have longer canals. Basal ceratopsians were agile animals with elongate canals, but in more derived clades, generally larger animals, the canals are shorter. Derived ceratopsian clades also have characteristic well-developed cranial ornamentation, such as frills and horns, which must have hindered head mobility. Pachycephalosaurs have extremely modified canals, with such short anterior and posterior ones that they are barely distinguishable from the rest of the vestibule, resembling those of some burrowing snakes (Olori 2010). It is unknown how this relates to the ecology of pachycephalosaurs, but it may be connected to their possible head-butting behavior. As short semicircular canals imply higher inertia of the fluids of the inner ear, it may be important for dealing with the post-impact stress. Stegosaurids and ankylosaurids also have short canals, with the lateral one being particularly short. Ankylosaurids had extensive armor, which must also have restricted head movements.

Although basal sauropods have inner ears similar to those of *Euparkeria*, derived clades show such different morphologies that there is no obvious general trend. Some taxa, like *Spinophorosaurus* or *Diplodocus*, have very elongate semicircular canals, while *Camarasaurus* and *Brachiosaurus* have very short ones. The only thing in common to these taxa seems to be the shortening of the lateral semicircular canal. These different morphologies may be related to neck length and mobility of the head.

A general trend of elongation is also seen in the semicircular canals of theropods —although in therizinosaurs the canals, in fact, become shorter in the more derived taxa. In some theropod groups, elongation of the ASC is so extreme that it leads to a twisting of the common crus. This may have happened as early as in the origin of tetanurans like *Sinosaurus* and was retained in avialans and crown birds. Birds show extremely elongate canals. The ASC twists the common crus so strongly that it lies closer to the parallel plane of the lateral canal. Birds also show the development of independent ampullae for the lateral semicircular canal. In *Deinonychus*, the lateral canal becomes detached from the anterior ampulla, while in *Archaeopteryx* there is detachment of the lateral canal and the posterior canals. Therizinosaurs may have convergently developed an independent lateral ampulla.

A great elongation of the ASCs and PSCs is characteristic among archosaur groups such as notosuchians, pterosaurs, hadrosaurids, and maniraptorans, including avialans. These are generally agile animals capable of refined locomotor abilities, with pterosaurs and birds inhabiting very complex three-dimensional environments. Correspondingly, they have the most complex canalicular anatomy of all archosaurs. On the other hand, groups of large and heavy animals with limited neck mobility, such as ankylosaurids and derived ceratopsians and sauropods, have shorter and wider semicircular canals.

The lateral semicircular canal is usually used to deduce the head posture of some fossil groups. Witmer et al. (2003) used the lateral canal to interpret the angle at which the head was held in pterosaurs. They showed that in the more primitive taxon, *Rhamphorhynchus*, the head was held horizontally, whereas that of the more derived *Anhanguera* was held at a 45° angle downwards. The greater angle was thought to have given better lateral vision and a less obstructed view, but it was also probably related to ground-based quadrupedal locomotion.

A downward head position is also reported for sauropods in general (Sereno et al. 2007) and for the theropod *Majungasaurus* (Sampson and Witmer 2007). In contrast, an upward head posture has been calculated for the ornithischian *Dysalotosaurus* (Sobral et al., 2012) and for the sauropod *Massospondylus* (Sereno et al. 2007). However, although it is tempting to use the anatomy of the inner ear to infer aspects of behavior of these extinct animals, some authors suggest the lateral semicircular canal is not always reliable for these purposes (Marugán-Lobón et al. 2013).

Another feature of the inner ear observed in archosaurs in general is the elongation of the cochlea, connected to the extension of the hearing range to lower frequencies. In ceratopsians and sauropods, the cochlea becomes relatively shorter; however, in ankylosaurs, hadrosaurids, therizinosaurs, but in crown birds, the cochlea becomes longer. Since these groups are thought to have relied heavily on sound communication, it is clear why their hearing organ is also more elaborate. In crown archosaurs, the cochlea is very elongate. In crocodylians, it is twisted at its midlength and this is thought to extend the sound detecting tissue. Since elongation of the cochlea is related to mechanical tuning, this suggests that this mechanism was important in the hearing of these fossil groups.

10.5.4 The Importance of Bone Pneumatization

Not only birds but also crocodylians exhibit complex behaviors between individuals that are based on sound production (Thewissen and Nummela 2008). Pneumatization of the bones has usually been seen as a mechanism of weight reduction. In the case of birds, the connection to flight is obvious. In crocodylians, it has been harder to explain, but buoyancy also has been cited. Recently, however, anatomical and ontogenetic studies of the pneumatic sinuses in crocodylians suggested that the sinuses had auditory functions (Dufeau and Witmer 2015), with their appearance and further development proposed to increase the volume of the middle ear. That would be an efficient way to enhance hearing capacities without giving up on the protection offered by a bony otic capsule.

Directional hearing, a consequence of pneumatization, is a mechanism present early in the evolutionary history of the archosaurs. The middle ear cavities of tetrapods are coupled internally to each other and the external environment. Internally coupled ears provide two different signals to the brain: sound arriving in the inner ear from the external environment, and sound reaching the inner ear by an internal pathway through the Eustachian tubes or the pharynx. The different times of the waves arriving in the inner ear produces a delay with which the brain calculates the direction of the source (Wever 1978; Vedurmudi et al. 2016).

Developmentally, the middle ear cavity is formed by expansion of the first pharyngeal pouch. The Eustachian tubes connect the cavity with the pharynx. Living crocodylians, however, possess not only these connections but also a median pharyngeal system that is already present in stem archosaur groups, such as *Euparkeria*. In these early taxa, the median tube forms a recess on the ventral surface of the basisphenoid and basioccipital bones. In the evolution of pseudosuchians, both lateral and median systems become completely enclosed in these bones—but at exactly which point this happened is unknown because the intermediate stages are so far unknown. The evolution of pneumatic sinuses should be clarified by further research (Witmer and Ridgely 2009; Dufeau and Witmer 2015).

10.5.5 Formation of the Opisthotic Loop

The perilymphatic foramen of pseudosuchians is completely enclosed by bone. However, the bony framework is formed neither by a median extension of the ventral ramus of the opisthotic nor by a lateral displacement of the perilymphatic duct, but by the looping of the opisthotic around the duct. The opisthotic sutures with itself, forming an "opisthotic loop." In stem archosaur clades, the perilymphatic foramen is either not visible on the medial side of the ventral ramus of the opisthotic or it forms only a notch in the ramus. In pseudosuchians, however, the foramen results from a contact of the opisthotic with itself dorsal to the duct. The foramen becomes visible through the lateral opening of the metotic foramen. The development of the opisthotic loop was noticed by Gower and Walker (2002), who envisaged a closer relationship between aetosaurs and crocodylomorphs. However, aetosaurs are now considered a basal pseudosuchian clade. Thus, the loop could be interpreted as a convergence between them. In birds, the opisthotic also forms the perilymphatic foramen, but no sign of looping has been found. No information is available as to when the complete enclosure of the perilymphatic duct happened in birds.

10.5.6 Flight and the Interpretation of the Floccular Lobe

Birds and pterosaurs are the only flying archosaur groups. Flight is a complex behavior that requires coordination of several neurological systems, and thus many aspects of the inner ear and brain morphology of both groups are probably strongly connected to flying. These include the elongation of the ASC and the size of the floccular fossa. In all amniotes, the anterior canal runs around the floccular lobe, and it has been suggested that an increase in the size of the lobe will result in an enlargement of the canal (Witmer et al. 2003). However, many large-bodied dinosaurs, including some ornithischians and saurischians, seem to have an ASC longer than the size of their floccular fossae.

The size of the floccular lobe has been suggested to be connected to motor coordination because it acts in maintaining a stable gaze during movements of the head and in stabilizing the head during movements of the body through the neck muscles (Walsh et al. 2013). However, when the volume of the floccular fossa and the type of flight behavior were contrasted, no significant correlation was found. Furthermore, the morphology of the floccular lobe showed no close connection to the floccular fossa due to the amount of vascular elements within it. Thus, a large fossa does not necessarily mean a large lobe. An alternative hypothesis is that the floccular lobe is related to proprioception: the perception of the relative position of parts of the body. The lobe of both birds and mammals is responsible for processing a large amount of cutaneous sensory information (Cox and Jeffery 2010). In the case of birds, this could be related to the flight feathers, while Witmer et al. (2003) have suggested that, in pterosaurs, it is related to the wing membrane. The wing membranes of pterosaurs contain muscles and tendons and, since these span over large areas, would send a large amount of somatosensory information to the brain. This could perhaps explain why some dinosaurs show discrepant ASC and floccular fossa size: their heads would be relatively mobile but not their big, heavy bodies.

10.5.7 Sound Production and Communication

It has long been suggested by Hopson (1975) that hadrosaurids possessed sound-producing abilities because of their unique cranial crests. Sound-producing capacities have been proven for the group (Weishampel 1981), although it also has been shown that mechanisms and anatomies were likely decoupled (Evans et al. 2009). The internal morphology of the crest was thought to be the most important feature for sound production, but it seems that the form of the nasal cavities and air passageways are actually more significant. The sound produced was likely of low-frequency, and the long cochleae of hadrosaurids were adapted to detect such sounds (Evans et al. 2009).

The nodosaurid *Panoplosaurus* and the ankylosaurid *Euoplocephalus* were recently scanned and showed complex nasal cavities with extensive looping of the airways (Witmer and Ridgely 2008b; Miyashita et al. 2011). The loops of *Euoplocephalus* are extensive and complex and, as in hadrosaurids, they were not connected to the olfactory region of the nasal cavity. This suggests functions other than olfaction, such as thermoregulation. However, this anatomy is also suitable for sound production, and this could have been a secondary use of the loops. The long cochlea of *Euoplocephalus* seems to support that.

10.5.8 Acquisition of Aquatic Hearing

Very little research has been done on the aquatic adaptations of hearing in extant archosaurs (Thewissen and Nummela 2008). For mammals, these adaptations seem to include the anatomy of the external ear and middle ear cavity, such as size decrease of the outer ear, smaller tympanic membrane to FO ratio, and modifications of the ossicle chain (e.g., enlargement of some of its components). The related soft parts do not fossilize and are difficult to follow in the fossil record; moreover, such features may not have an equivalent in archosaurs.

Extant birds and crocodylians that live in aquatic environments still rely mostly on aerial hearing, as suggested by the anatomy of their middle ear cavity and organization of the sound transmission system (Thewissen and Nummela 2008). However, some convergences can be seen, such as protection of the canal of the external ear and connection between left and right middle ears (Thewissen and Nummela 2008). To protect the entrance of the ear canal from water, birds usually have thick feathers covering the outer ear that collapse underwater, whereas crocodylians, uniquely among archosaurs, close their external ears by ear flaps controlled by muscles (Thewissen and Nummela 2008). Feathers are found only rarely in the fossil record, and thus their presence is difficult to prove, but muscles usually leave scars and other markings on bones that could reveal their first appearance in pseudosuchians. Regarding the connection between the middle ears, the evolution of the Eustachian tube system could be of much help, as well as the results of skull pneumatization for sound transmission in aquatic animals. Unfortunately, none of these aspects of the anatomy of archosaurs has been approached as yet. Many aquatic mammals rely on the transmission of sound through the bones enclosing the middle ear and through the ossicular chain, enhanced with greater bone density (Thewissen and Nummela 2008). Although this pathway is not usually used in extant birds and crocodylians, it is currently unknown whether such a mechanism could have been developed by any archosaur or stem group, such as phytosaurs.

10.5.9 The Crocodylian Braincase

The most striking feature of the crocodylian skull is its akinetic nature. This lack of mobility is represented in modern crocodylians by the suture of the cranio-quadrate-palate complex. Recent fossil discoveries are helping to build a more complete picture of the evolution of these characters. The pterygoid of basal crocodylomorphs has a well-developed flange, but no contact between it and the quadrate. There is also limited contact between the quadrate and the prootic dorsally. In more derived crocodylomorphs, an exoccipital-quadrate suture and a subsequent basisphenoid-quadrate contact appear (Pol et al. 2013). In crocodyliforms, there is a strengthening of the exoccipital-basisphenoid suture and the further development of the quadrate-braincase interaction through the contact of the quadrate with the laterosphenoid. Later, in mesoeucrocodylians, two other features appear: the secondary palate and the palate-braincase suture. In the latter, the basisphenoid loses its basipterygoid processes and becomes sutured with the pterygoid.

The hearing system was unlikely to have been the force driving these changes in the crocodylian skull, but it was deeply affected by it. The intimate relationships of the quadrate, the braincase, and the palate result in an akinetic skull and played a much more important role in the evolution of the unique feeding behavior of crocodylians. However, it also delimited the middle ear cavity. In modern crocodylians, the cavity lies deep within the skull, communicating with the external ear through a channel formed by the quadrate and the squamosal bones (Iordansky 1973). These anatomical changes are quite similar to the ones seen in the evolution of mammals. The convergences between mammalian and crocodylian anatomy and ecology have been pointed out many times in the literature (Pol et al. 2013) but have not been quantitatively compared.

10.6 Summary

Archosaurs have dominated terrestrial ecosystems since the Mesozoic, but the evolution of the hearing sense in the group still remains poorly known. This chapter compiles the available information on the middle and inner ear morphologies of archosaurs and related groups and explores different evolutionary scenarios. It is still unknown whether a tympanic membrane was present in the basalmost members of the stem clades. If not, it appeared early in their evolutionary history. Many derived taxa show several inner ear modifications that helped to improve their hearing sense, such as an enlarged metotic foramen, partial ossification of the medial wall of the otic capsule, and elongate cochlea. The subdivision of the metotic foramen led to further modifications of the ear: the appearance of the fenestra pseudorotunda and the secondary tympanic membrane. These characters were acquired in pseudosuchians and likely in all three major dinosaur groups independently. In theropods, modifications of the ear appeared before the origin of birds, and is retained in the crown group. Pneumatization of the braincase is important for directional hearing, and it seemed to have played a very important role in crocodylian evolution. Ankylosaurids, like hadrosaurs, were likely capable of producing low-frequency sounds though lacking elaborated cranial crests. The inner ear also has a balance function. Elongation of the semicircular canals is connected to the acquisition of an upright posture in archosaurs. In pseudosuchians, further elongation of the canals is seen only in notosuchians, but it is found in many avemetatarsalians. Crown birds show even further elongation of the canals, highlighting their superb flight capacities.

Roman numerals indicate corresponding cranial nerves (e.g., CN VIII)
(For other abbreviations, please consult referred papers)

Acknowledgements G. S. was partially supported by the DAAD and CAPES program with funding provided by CAPES (BEX 3474/09-7). J. M. was funded by the Deutsche Forschungsgemeinschaft (MU 1760/2-3). We also thank M. Bronzatti and F. C. Montefeltro for discussions.

Compliance with Ethics Requirements G. Sobral and J. Müller have declared that they had no conflicts of interest.

References

Adams, T. L. (2013). A new neosuchian crocodyliform from the lower cretaceous (Late Aptian) Twin Mountains Formation of North-Central Texas. *Journal of Vertebrate Paleontology*, 33, 85–101.

Alonso, P. D., Milner, A. C., Ketcham, R. A., Cookson, M. J., & Rowe, T. B. (2004). The avian nature of the brain and inner ear of *Archaeopteryx*. *Nature*, 430, 666–669.

Angst, D., Lécuyer, C., Amiot, R., Buffetaut, E., et al. (2014). Isotopic and anatomical evidence of an herbivorous diet in the Early Tertiary giant bird *Gastornis*. Implications for the structure of Paleocene terrestrial ecosystems. *Naturwissenschaften*, 101, 313–322.

Barsbold, R. (1974). Saurornithoididae, a new family of small theropod dinosaurs from Central Asia and North America. In Z. Kielan-Jaworowska (Ed.), *Results of the Polish-Mongolian palaeontological expeditions, Part V. Palaeontologia Polonica*, 30, 5–22.

Balanoff, A. M., Bever, G. S., & Norell, M. A. (2014). Reconsidering the avian nature of the oviraptorosaur brain (Dinosauria: Theropoda). *PLoS ONE*, 9(12), e113559. doi:10.1371/journal.pone.0113559

Bennett, S. C. (1996). The phylogenetic position of the Pterosauria within the Archosauromorpha. *Zoological Journal of the Linnean Society*, 118, 261–308.

Bennett, S. C. (2001). The osteology and functional morphology of the Late Cretaceous pterosaur *Pteranodon*. *Palaeontographica Abteilung A*, 260, 1–112.

Benson, R. B. J., & Choiniere, J. N. (2013). Rates of dinosaur limb evolution provide evidence for exceptional radiation in Mesozoic avialans. *Proceedings of the Royal Society of London B: Biological Sciences*, 280, 20131780. http://dx.doi.org/10.1098/rspb.2013.1780

Benton, M. J., Juul, L., Storrs, G. W., & Galton, P. M. (2000). Anatomy and systematics of the prosauropod dinosaur *Thecodontosaurus antiquus* from the Late Triassic of Southwest England. *Journal of Vertebrate Paleontology*, 20, 77–108.

Bernardi, M., Klein, H., Petti, F. M., & Ezcurra, M. D. (2015). The origin and early radiation of archosauriforms: Integrating the skeletal and footprint record. *PLoS ONE*, 10(6), e0128449. doi:10.1371/journal.pone.0128449

Bittencourt, J. S., Arcucci, A. B., Marsicano, C. A., & Langer, M. C. (2014). Osteology of the Middle Triassic archosaur *Lewisuchus admixtus* Romer (Chañares Formation, Argentina), its inclusivity, and relationships amongst early dinosauromorphs. *Journal of Systematic Palaeontology*. doi:10.1080/14772019.2013.878758.

Bourdon, E., Bouya, B., & Iarochene, M. (2005) Earliest African neornithine bird: A new species of prophaethontidae (Aves) from the Paleocene of Morocco. *Journal of Vertebrate Paleontology*, 25, 157–170.

Bourdon, E., Amaghzaz, M., & Bouya, B. (2010). Pseudotoothed Birds (Aves, Odontopterygiformes) from the Early Tertiary of Morocco. *American Museum Novitates*, 3704, 1–71.

Boyd, C. A. (2014). The cranial anatomy of the neornithischian dinosaur *Thescelosaurus neglectus*. *PeerJ*, 2, e669. doi:10.7717/peerj.669

Brochu, C. A. (2001). Crocodilian snouts in space and time: Phylogenetic approaches toward adaptive radiation. *American Zoologist*, 41, 564–585. doi:10.1093/icb/41.3.564

Brochu, C. A., Bouaré, M. L., Sissoko, F., Roberts, E. M., & O'Leary M. A. (2002). A dyrosaurid crocodyliform braincase from Mali. *Journal of Paleontology*, 76, 1060–1071.

Bronzati, M., Montefeltro, F. C., & Langer, M. C. (2015). Diversification events and the effects of mass extinctions on Crocodyliformes' evolutionary history. *Royal Society Open Science*, 2, 140385. http://dx.doi.org/10.1098/rsos.140385 (accessed 21 July 2016)

Brusatte, S. L., Benton, M. J., Lloyd, G. T., Ruta M., & Wang, S. C. (2010a). Macroevolutionary patterns in the evolutionary radiation of archosaurs (Tetrapoda: Diapsida). *Earth and Environmental Science Transactions of the Royal Society of Edinburgh*, 101, 367–382. doi:10.1017/S1755691011020056

Brusatte, S.L., Nesbitt, S. J., Irmis, R. B., Butler, R. J., et al. (2010b). The origin and early radiation of dinosaurs. *Earth-Science Reviews*, 101, 68–100.

Busbey, A. B., & Gow C. (1984). A new protosuchian crocodilian from the Late Triassic Elliot Formation of South Africa. *Palaeontologia Africana*, 25, 127–149.

Butler, R. J., Benson, R. B. J., & Barrett, P. M. (2013). Pterosaur diversity: Untangling the influence of sampling biases, Lagerstätten, and genuine biodiversity signals. *Palaeogeography, Palaeoclimatology, Palaeoecology*, 372, 78–87.

Chen, X.-H., Motani, R., Cheng, L., Jiang, D.-Y., & Rieppel, O. (2014). The enigmatic marine reptile *Nanchangosaurus* from the Lower Triassic of Hubei, China and the phylogenetic affinities of Hupehsuchia. *PLoS ONE* 9(7), e102361. doi:10.1371/journal.pone.0102361

Clack, J. A. (1997). The evolution of tetrapod ears and the fossil record. *Brain, Behavior and Evolution*, 50, 198-212.

Clack, J. A., & Allin, E. (2004). The evolution of single- and multiple-ossicle ears in fishes and tetrapods. In G. A. Manley, A. N. Popper, & R. R. Fay (Eds.), *Evolution of the vertebrate auditory system* (pp. 128–163). New York: Springer Science+Business Media.

Codorniú, L. S., Carabajal, A. P., Pol, D., Unwin, D., & Rauhut, O. W. M. (2016). A Jurassic pterosaur from Patagonia and the origin of the pterodactyloid neurocranium. *PeerJ*, 4, e2311. https://doi.org/10.7717/peerj.2311.

Cox, P. G., & Jeffery, N. (2010). Semicircular canals and agility: The influence of size and shape measures. *Journal of Anatomy*, 216, 37–47.

Currie, P. J., Qiang, J., Norell, M. A., & Shu-An, J. (1998). Two feathered dinosaurs from northeastern China. *Nature*, 393, 753–761.

Deccechi, A. T., & Larsson, H. C. E. (2013). Body and limb size dissociation at the origin of avialans: Uncoupling allometric constraints across a macroevolutionary transition. *Evolution*, 67, 2741–2752.

Dufeau, D. L., & Witmer, L. M. (2015) Ontogeny of the middle-ear air-sinus system in *Alligator mississippiensis* (Archosauria: Crocodylia). *PLoS ONE*, 10(9), e0137060. doi:10.1371/journal.pone.0137060

Dzik, J. (2003). A beaked herbivorous archosaur with dinosaur affinities from the early Late Triassic of Poland. *Journal of Vertebrate Paleontology*, 23, 556–574.

Eaton, T. H. (1960). A new armored dinosaur from the Cretaceous of Kansas. *University of Kansas Paleontological Contributions*, 25, 1–24.

Erickson, G. M., Krick, B. A., Hamilton, M., Bourne,G. R., et al. (2012). Complex dental structure and wear biomechanics in hadrosaurid dinosaurs. *Science*, 338, 98–101.

Evans, S. E. (1986). The braincase of *Prolacerta broomi* (Reptilia; Triassic). *Neues Jahrbuch für Geologie und Paläontologie Abhandlungen*, 173, 181–200.

Evans, S. E. (2008). The skull of lizards and tuatara. In C. Gans, A. S. Gaunt, & K. Adler (Eds.), *Biology of the Reptilia: The skull of Lepidosauria, vol. 20. Morphology H* (pp. 1–347). Salt Lake City: Society for the Study of Amphibians and Reptiles.

Evans, D. C., Ridgely, R., & Witmer, L. M. (2009). Endocranial anatomy of lambeosaurine hadrosaurids (Dinosauria: Ornithischia): A sensorineural perspective on cranial crest function. *The Anatomical Record*, 292, 1315–1337.

Elzanowski, A. (1991). New observations on the skull of *Hesperornis* with reconstructions of the bony palate and otic region. *Postilla*, 207, 1–20.

Elzanowski, A., & Galton, P. M. (1991). Braincase of *Enaliornis*, an Early Cretaceous bird from England. *Journal of Vertebrate Paleontology*, 11, 90–107.

Ezcurra, M. D., Scheyer, T. M., & Butler, R. J. (2014). The origin and early evolution of Sauria: Reassessing the Permian saurian fossil record and the timing of the crocodile-lizard divergence. *PLoS ONE* 9(2), e89165. doi:10.1371/journal.pone.0089165

Fernández, M. S., Carabajal, A. P., Gasparini Z., & Díaz, G. C. (2011). A metriorhynchid crocodyliform braincase from northern Chile. *Journal of Vertebrate Paleontology*, 31, 369–377.

Galton, P. M. (1984). Cranial anatomy of the prosauropod dinosaur *Plateosaurus* from the Knollenmergel (Middle Keuper, Upper Triassic) of Germany. *Geologica et Palaeontologica*, 18, 139–171.

Galton, P. M. (1988). Skull bones and endocranial casts of stegosaurian dinosaur *Kentrosaurus* Hennig, 1915 from Upper Jurassic of Tanzania, East Africa. *Geologica et Palaeontologica*, 22, 123–143.

Galton, P. M. (2001). Endocranial casts of the plated dinosaur *Stegosaurus* (Upper Jurassic, Western USA): A complete undistorted cast and the original specimens of Othniel Charles Marsh. In K. Carpenter & P. M. Galton (Eds.), *The armoured dinosaurs* (pp. 103–129). Bloomington (IN): Indiana University Press.

Galton, P. M., & Kermack, D. (2010). The anatomy of *Pantydraco caducus*, a very basal sauropodomorph dinosaur from the Rhaetian (Upper Triassic) of South Wales, UK. *Revue de Paléobiologie*, 29, 341–404.

Gardner, N. M., Holliday, C. M., & O'Keefe, F. R. (2010). The braincase of *Youngina capensis* (Reptilia: Diapsida): New insights from high-resolution CT scanning of the holotype. *Palaeontologia Electronica* 13. http://palaeoelectronica.org/2010_3/217/index.html

Gauthier, J., & K. Padian. (1985). Phylogenetic, functional, and aerodynamic analyses of the origin of birds and their flight. In J. H. Ostrom, M. K. Hecht, G. Viohl, & P. Wellnhofer (Eds.), *The beginning of birds* (pp. 185–197). Eichstatt: Freunde des Jura Museums.

Godefroit, P., Cau, A., Dong-Yu, H., Escuillié, F., Wenhao, W., & Dyke, G. (2013). A Jurassic avialan dinosaur from China resolves the early phylogenetic history of birds. *Nature*, 498, 359–362. doi:10.1038/nature12168

Gold, M. E. L., Bourdon, E., & Norell, M. A. (2016). The first endocast of the extinct dodo (*Raphus cucullatus*) and an anatomical comparison amongst close relatives (Aves, Columbiformes). *Zoological Journal of the Linnean Society*, 177: 950–963. doi:10.1111/zoj.12388

Gower, D. J. (2002). Braincase evolution in suchian archosaurs (Reptilia: Diapsida): Evidence from the rauisuchian *Batrachotomus kupferzellensis*. *Zoological Journal of the Linnean Society*, 136, 49–76.

Gower, D. J., & Weber, E. (1998). The braincase of *Euparkeria*, and the evolutionary relationships of avialans and crocodilians. *Biological Reviews*, 73, 367–411.

Gower, D. J., & Walker, A. D. (2002). New data on the braincase of the aetosaurian archosaur (Reptilia: Diapsida) *Stagonolepis robertsoni* Agassisz. *Zoological Journal of the Linnean Society*, 136, 7–23.

Gower, D. J., & Nesbitt, S. J. (2006). The braincase of *Arizonasaurus babbitti*—further evidence for the non-monophyly of 'rauisuchian' archosaurs. *Journal of Vertebrate Paleontology*, 26, 79–87.

Holloway, W. L., Claeson, K. M., & O'Keefe, F. R. (2013). A virtual phytosaur endocast and its implications for sensory system evolution in archosaurs. *Journal of Vertebrate Paleontology*, 33, 848–857.

Hone, D. W. E. (2012). Pterosaur research: Recent advances and a future revolution. *Acta Geologica Sinica English Edition*, 86, 1366–1376.

Hopson, J. A. (1975). The evolution of cranial display structures in hadrosaurian dinosaurs. *Paleobiology*, 1, 21–43.

Hopson, J. A. (1979). Paleoneurology. In C. Gans, R. G. Northcutt, & P. Ulinski (Eds.), *Biology of the reptilia, vol. 9* (pp. 39–146). New York: Academic Press.

Hungerbühler, A. (2002). The Late Triassic phytosaur *Mystriosuchus westphali*, with a revision of the genus. *Palaeontology*, 45, 377–418.

Hungerbühler, A., Chatterjee, S., & Cunningham, D. P. (2012). Cranial anatomy of the Late Triassic phytosaur *Machaeroprosopus*, with the description of a new species from West Texas. *Earth and Environmental Science Transactions of the Royal Society of Edinburgh*, 103, 269–312.

Iordansky, N. N. (1973). The skull of the Crocodilia. In C. Gans & T. Parsons (Eds.), *Biology of the reptilia, vol. 4. Morphology D* (pp. 201–262). London: Academic Press.

Janensch, W. (1935). Die Schädel der Sauropoden *Brachiosaurus*, *Barosaurus* und *Dicraeosaurus* aus den Tendaguru-Schichten Deutsch-Ostafrikas. *Palaeontographica*, 2 (Supplement to 7), 147–298.

Janensch, W. (1936). Über Bahnen von Hirnvenen bei Saurischiern und Ornithischiern, sowie einigen anderen fossilien und rezenten Reptilien. *Palaeontologische Zeitschrift*, 18, 181–198.

Ji Q., & Ji S. (1996). On the discovery of the earliest bird fossil in China and the origin of birds. *Chinese Geology*, 10, 30–33.

Kellner, A. W. A. (1996). Description of the braincase of two early cretaceous pterosaurs (Pterodactyloidea) from Brazil. *American Museum Novitates*, 3175, 1–34.

Kellner, A. W. A. (2015). Comments on Triassic pterosaurs with discussion about ontogeny and description of new taxa. *Anais da Academia Brasileira de Ciências*, 87, 669–689.

Kley, N. J., Sertich, J. J. W., Turner, A. H., Krause, D. W., et al. (2010). Craniofacial morphology of *Simosuchus clarki* (Crocodyliformes: Notosuchia) from the Late Cretaceous of Madagascar. *Society of Vertebrate Paleontology Memoir*, 10, 13–98.

Knoll, F., Witmer, L. M., Ortega, F., Ridgely, R. C., & Schwarz-Wings, D. (2012). The braincase of the basal sauropod dinosaur *Spinophorosaurus* and 3D reconstructions of the cranial endocast and inner ear. *PLoS ONE*, 7, e30060. doi:10.1371/journal.pone.0030060

Ksepka, D. T., Balanoff, A. M., Walsh, S., Revan, A., & Ho, A. (2012). Evolution of the brain and sensory organs in Sphenisciformes: New data from the stem penguin *Paraptenodytes antarcticus*. *Zoological Journal of the Linnean Society*, 166, 202–219.

Lautenschlager, S., Witmer, L. M., Altangerel, P., Zanno, L. E., & Rayfield, E. J. (2012). The endocranial anatomy of Therizinosauria and its implications for sensory and cognitive function. *PLoS ONE*, 7, e52289. doi:10.1371/journal.pone.0052289

Lee, Y. N. (1996). A new nodosaurid ankylosaur (Dinosauria: Ornithischia) from the Paw Paw Formation (Late Albian) of Texas. *Journal of Vertebrate Paleontology*, 16, 232–245.

Livezey, B. C., & Zusi, R. L. (2006). Higher-order phylogeny of modern avialans (Theropoda, Aves: Neornithes) based on comparative anatomy. I. Methods and characters. *Bulletin of the Carnegie Museum of Natural History*, 37, 501–544.

Lombard, R. E., & Hetherington, T. E. (1993). Structural basis of hearing and sound transmission. In J. Hanken & B. K. Hall (Eds.), *The skull, vol. 3. Functional and evolutionary mechanisms* (pp. 241–302). Chicago: The University of Chicago Press.

Madsen, J. H., McIntosh, J. S., & Berman, D. S. (1995). Skull and atlas-axis complex of the Late Jurassic sauropod *Camarasaurus* Cope (Reptilia: Saurischia). *Bulletin of the Carnegie Museum of Natural History*, 31, 1–115.

Manley, G. A., & Clack, J. A. (2004). An outline of the evolution of vertebrate hearing. In G. A. Manley, A. N. Popper, & R. R Fay (Eds.), *Evolution of the vertebrate auditory system* (pp. 1–26). New York, Springer Science+Business Media.

Mann, Z. F., & Kelley, M. W. (2011). Development of tonotopy in the auditory periphery. *Hearing Research*, 276, 2–15.

Marugán-Lobón, J., Chiappe, L. M., & Farke, A. A. (2013), The variability of inner ear orientation in saurischian dinosaurs: Testing the use of semicircular canals as a reference system for comparative anatomy. *PeerJ*, 1, e124. doi:10.7717/peerj.124

Middleton, K. M., & English, L. T. (2015). Challenges and advances in the study of pterosaur flight. *Canadian Journal of Zoology*, 93, 945–959. doi:10.1139/cjz-2013-0219

Milner, A. C., & Walsh, S. A. (2009). Avian brain evolution: New data from Palaeogene birds (Lower Eocene) from England. *Zoological Journal of the Linnean Society*, 155, 198–219.

Miyashita, T., Arbour, V. M., Witmer, L. M., & Currie, P. J. (2011). The internal cranial morphology of an armoured dinosaur *Euoplocephalus* corroborated by X-ray computed tomographic reconstruction. *Journal of Anatomy*, 219, 661–675.

Nesbitt, S. J. (2011). The early evolution of archosaurs: Relationships and the origin of the major clades. *Bulletin of the American Museum of Natural History*, 352, 1–292.

Nesbitt, S. J., Desojo, J. B., & Irmis, R. B. (2013). Anatomy, phylogeny and palaeobiology of early archosaurs and their kin. In S. J. Nesbitt, J. B. Desojo, & R. B. Irmis (Eds.), *Anatomy, phylogeny and palaeobiology of early archosaurs and their kin* (pp. 1–7). Special Publication 379. London: Geological Society of London.

Norman, D. B., Crompton, A. W., Butler, R. J., Porro, L. B., & Charig, A. J. (2011). The Early Jurassic ornithischian dinosaur *Heterodontosaurus tucki* Crompton & Charig, 1962: Cranial anatomy, functional morphology, taxonomy, and relationships. *Zoological Journal of the Linnean Society*, 163, 182–276.

Olori, J. C. (2010). Digital endocasts of the cranial cavity and osseous labyrinth of the burrowing snake *Uropeltis woodmasoni* (Alethinophidia: Uropeltidae). *Copeia*, 2010, 14–26.

Pol, D., Rauhut, O. W. M., Lecuona, A., Leardi, J. M., et al. (2013). A new fossil from the Jurassic of Patagonia reveals the early basicranial evolution and the origins of Crocodyliformes. *Biological Reviews*, 88, 862–872.

Prieto-Márquez, A., & Norell, M. A. (2011). Redescription of a nearly complete skull of *Plateosaurus* (Dinosauria: Sauropodomorpha) from the Late Triassic of Trossingen (Germany). *American Museum Novitates*, 3727, 1–58.
Rauhut, O. W. M. (2004). Braincase structure of the Middle Jurassic theropod dinosaur *Piatnitzkysaurus*. *Canadian Journal of Earth Sciences*, 41, 1109–1122.
Renesto, S., & Binelli, G. (2006). *Vallesaurus cenensis* Wild, 1991, a drepanosaurid (Reptilia, Diapsida) from the Late Triassic of Northern Italy. *Rivista Italiana di Paleontologia e Stratigrafia*, 112(1), 77–94.
Rieppel, O. (1985). The recessus scalae tympani and its bearing on the classification of reptiles. *Journal of Herpetology*, 19(3), 373–384.
Rogers, S. W. (1998). Exploring dinosaur neuropaleobiology: Computed tomography scanning and analysis of an *Allosaurus fragilis* endocast. *Neuron*, 21, 673–679.
Sampson, S. D., & Witmer, L. M. (2007). Craniofacial anatomy of *Majungasaurus crenatissimus* (Theropoda: Abelisauridae) from the Late Cretaceous of Madagascar. *Journal of Vertebrate Paleontology*, 27, 32–102.
Sanders, R. K., & Smith, D. K. (2005). The endocranium of the theropod dinosaur *Ceratosaurus* studied with computed tomography. *Acta Palaeontologica Polonica*, 50, 601–616.
Selvatti, A. P., Gonzaga, L. P., & Russo, C. A. M. (2015). A Paleogene origin for crown passerines and the diversification of the Oscines in the New World. *Molecular Phylogenetics and Evolution*, 88, 1–15.
Sereno, P. C. (1997). The origin and evolution of dinosaurs. *Annual Review of Earth and Planetary Sciences*, 25, 435–489.
Sereno, P. C., Wilson, J. A., Witmer, L. M., Whitlock, J. A., et al. (2007). Structural extremes in a Cretaceous dinosaur. *PLoS ONE*, 2(11), e1230. doi:10.1371/journal.pone.0001230
Smith, D. K., Zanno, L. E., Sander, P. K., Deblieux, D. D., & Kirkland, J. I. (2011). New information on the braincase of the North American therizinosaurian (Theropoda, Maniraptora) *Falcarius utahensis*. *Journal of Vertebrate Paleontology*, 31, 387–404.
Sobral, G., Hipsley, C. A., & Müller, J. (2012). Braincase redescription of *Dysalotosaurus lettowvorbecki* (Dinosauria, Ornithopoda) based on computed tomography. *Journal of Vertebrate Paleontology*, 32, 1090–1102.
Sobral, G., Sues, H.-D., & Müller, J. (2015). Anatomy of the enigmatic reptile *Elachistosuchus huenei* Janensch, 1949 (Reptilia: Diapsida) from the Upper Triassic of Germany and its relevance for the origin of Sauria. *PLoS ONE*, 10(9), e0135114. doi:10.1371/journal.pone.0135114
Sobral, G., Sookias, R. B., Butler, R. J., Bhullar, B. A. S., et al. (2016). New information on the braincase and inner ear of *Euparkeria capensis* Broom: Implications for diapsid and archosaur evolution. *Royal Society Open Science*, 3: 160072. http://dx.doi.org/10.1098/rsos.160072
Stein, R. W., Brown, J. W., & Mooers, A. Ø. (2015). A molecular genetic time scale demonstrates Cretaceous origins and multiple diversification rate shifts within the order Galliformes (Aves). *Molecular Phylogenetics and Evolution*, 92, 155–164.
Stubbs, T. L., Pierce, S. E., Rayfield, E. J., & Anderson, P. S. L. (2013). Morphological and biomechanical disparity of crocodile-line archosaurs following the end-Triassic extinction. *Proceedings of the Royal Society of London B: Biological Sciences*, 280, 20131940. http://dx.doi.org/10.1098/rspb.2013.1940
Thewissen, J. G. M. H., & Nummela, S. (2008). *Sensory evolution on the threshold – adaptations in secondarily aquatic vertebrates*. Oakland (CA): University of California Press.
Trotteyn, M. J., & Haro, J. A. (2011). The braincase of a specimen of *Proterochampsa* Reig (Archosauriformes: Proterochampsidae) from the Late Triassic of Argentina. *Paläontologische Zeitschrift*, 85, 1–17.
Tykoski, R. S. (1998). The osteology of *Coelophysis kayentakatae* and its implications for ceratosaurid phylogeny. *Master of Science Thesis*. Austin, TX: The University of Texas.
Upchurch, P., Mannion, P. D., Benson, R. J. B., Butler, R. J., & Carrano, M. T. (2011). Geological and anthropogenic controls on the sampling of the terrestrial fossil record: A case study from the Dinosauria. In A. J. McGowan & A. B. Smith (Eds.). *Comparing the geological and fossil*

records: Implications for biodiversity studies (pp. 209–240). Special Publication 358. London: Geological Society of London.

Vedurmudi, A. P., Goulet, J., Christensen-Dalsgaard, J., Young, B. A., Williams, R., & van Hemmen, J. L. (2016). How internally coupled ears generate temporal and amplitude cues for sound localization. *Physical Review Letters*, 116, 028101.

Walker, A. D. (1985). The braincase of *Archaeopteryx*. In M. K. Hecht, J. H. Ostrom, G. Viohl, & P. Wellnhofer (Eds.), *The beginnings of birds* (pp. 123–134). Eichstätt, Germany: Freunde des Jura-Museums.

Walker, A. D. (1990). A revision of *Sphenosuchus acutus* Haughton, a crocodylomorph reptile from the Elliott Formation (Late Triassic or Early Jurassic) of South Africa. *Philosophical Transactions of the Royal Society of London B: Biological Sciences*, 330, 1–120.

Walsh, S. A., Iwaniuk, A. N., Knoll, M. A., Bourdon, E., Barrett, P. M., Milner, A. C., Nudds, R. L., Abel, R. L., & Sterpaio, P. D. (2013). Avian cerebellar floccular fossa size is not a proxy for flying ability in birds. *PLoS ONE*, 8(6), e67176. doi:10.1371/journal.pone.0067176

Weishampel, D. B. (1981). Acoustic analyses of potential vocalization in lambeosaurine dinosaurs (Reptilia: Ornithischia). *Paleobiology*, 7, 252–261.

Wever, E. G. (1978). *The reptile ear: Its structure and function*. Princeton: Princeton University Press.

Witmer, L. M., & Ridgely, R. C. (2008a). Structure of the brain cavity and inner ear of the centrosaurine ceratopsid dinosaur *Pachyrhinosaurus* based on CT scanning and 3D visualization. In P. J. Currie (Ed.), *A new horned dinosaur from an Late Cretaceous bone bed in Alberta* (pp. 117–144). Ottawa, Canada: National Research Council Research Press.

Witmer, L. M., & Ridgely, R. C. (2008b). The paranasal air sinuses of predatory and armored dinosaurs (Archosauria: Theropoda and Ankylosauria) and their contribution to cephalic structure. *The Anatomical Record*, 291, 1362–1388.

Witmer, L. M., & Ridgely, R. C. (2009). New insights into the brain, braincase, and ear region of tyrannosaurs (Dinosauria, Theropoda), with implications for sensory organization and behaviour. *The Anatomical Record*, 292, 1266–1296.

Witmer, L. M., Chatterjee, S., Franzosa, J., & Rowe, T. (2003). Neuroanatomy of flying reptiles and implications for flight, posture and behavior. *Nature*, 425, 950–953.

Witmer, L. M., Ridgely, R. C., Dufeau, D. L., & Semones, M. C. (2008). Using CT to peer into the past: 3D visualization of the brain and ear regions of avialans, crocodilians, and nonavian dinosaurs. In H. Endo & R. Frey (Eds.). *Anatomical imaging – Towards a new morphology* (pp. 67–87). New York: Springer Science+Business Media.

Wu, X.-C., Xu, X., & Wang, X.-L. (1999). A dromaeosaurid dinosaur with a filamentous integument from the Yixian Formation of China. *Nature*, 401, 262–266.

Xing, L., Paulina-Carabajal, A., Currie, P. J., Xing, X., et al. (2014). Braincase anatomy of the basal theropod *Sinosaurus* from the Early Jurassic of China. *Acta Geologica Sinica English Edition*, 88, 1653–1664.

Xu, X., & Pol, D. (2014). *Archaeopteryx*, paravian phylogenetic analyses, and the use of probability-based methods for palaeontological datasets. *Journal of Systematic Palaeontology*, 12, 323–334. doi:10.1080/14772019.2013.764357

Chapter 11
Amphibia: A Case of Diversity and Convergence in the Auditory Region

Rainer R. Schoch and Jason S. Anderson

Abstract The ears of extant amphibians are remarkably diverse and when fossil taxa are considered, the picture becomes even more complicated. Anurans have a differentiated stapes inside a middle ear cavity associated with a eustachian tube and tympanum. Instead, salamanders and caecilians have rudimentary stapes connected to the cheek or jaw articulation, and they lack the tympanum and middle ear cavity. At the same time, batrachians (salamanders and frogs) share a second ear ossicle, the batrachian operculum, whereas all lissamphibians have a second receptor in the inner ear, the amphibian papilla. The largest fossil clade and probable stem group of Lissamphibia, the temnospondyls, had a stapes similar to that of anurans, consistent in the possession of a ventral process and an elongate and slender distal shaft that probably attached to a tympanum. The evolutionary sequence of ear types forms a puzzle with several of the major groups each sharing features that others lack. The primitive condition is exemplified by the temnospondyl ear, especially that of dissorophoids. We argue that the loss of the tympanic system was an evolutionary option only available after the batrachian operculum had evolved.

Keywords Amphibamid · Amphibian papilla · Anuran · Basilar papilla · Batrachian operculum · Caecilian · Caudate · Dissorophoid · Perilymphatic canal · Squamosal embayment · Stapes · Stereospondyl · Temnospondyl · Tympanic ring · Urodele

R.R. Schoch (✉)
Staatliches Museum für Naturkunde, Rosenstein 1, 70191 Stuttgart, Germany
e-mail: rainer.schoch@smns-bw.de

J.S. Anderson
Department of Comparative Biology and Experimental Medicine, University of Calgary, 3330 Hospital Drive, Calgary, AB T2N 4N1, Canada
e-mail: janders@ucalgary.ca

J.A. Clack et al. (eds.), *Evolution of the Vertebrate Ear—Evidence from the Fossil Record*, Springer Handbook of Auditory Research 59,
DOI 10.1007/978-3-319-46661-3_11

11.1 Introduction

Because amphibians are ecologically and morphologically diverse, their ears encompass a remarkable variety of structures and functions (Duellman and Trueb 1994). In the three extant taxa—frogs, salamanders, and caecilians—the ear has been studied in variable detail, but in frogs it is relatively well understood (Wever 1985). In contrast, the fossilized parts of the ear in the amphibian stem group (temnospondyls; but for a different view, see Marjanovic and Laurin 2013) have revealed numerous interesting structures; however, these are often difficult to understand.

In extant amphibians (Lissamphibia) and their stem taxa, the anatomical diversity of ears reflects the structural diversity of the animals themselves. There are five major groups comprising lissamphibians and their stem group (Fig. 11.1): temnospondyls (itself a very diverse group), the extinct albanerpetontids, the extant frogs (anurans: "toads" are a subgroup of frogs), salamanders (caudates: "newts" are a subgroup of salamanders), and caecilians (gymnophionans).

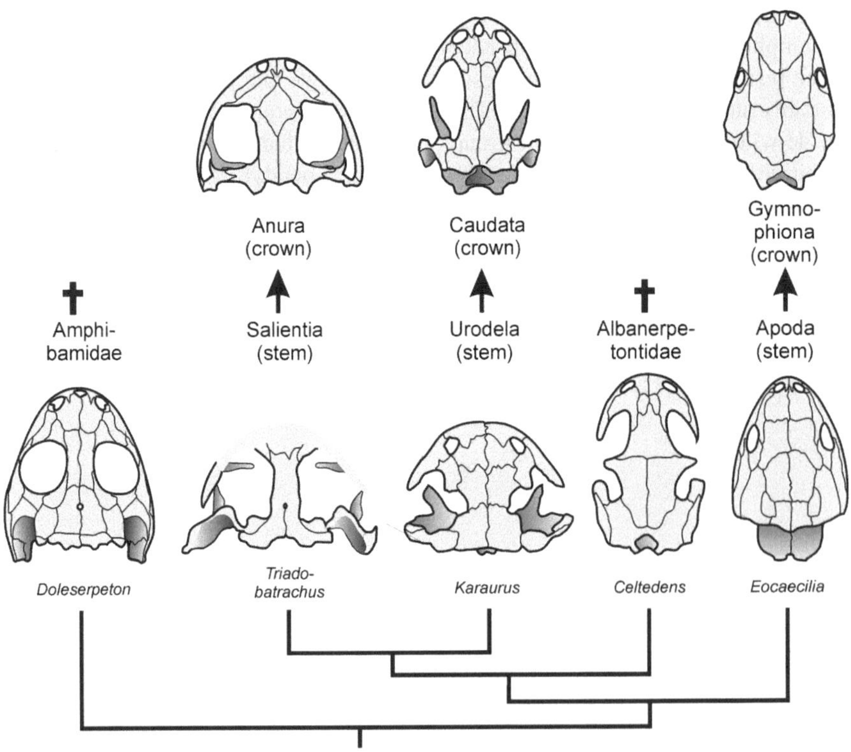

Fig. 11.1 Relationships of lissamphibians according to the temnospondyl hypothesis. The extinct Albanerpetontidae form a separate branch within the lissamphibians. (Based on Milner 1988; Ruta and Coates 2007)

Although frogs have probably the most highly derived locomotor mechanism among amphibians, they retain an auditory apparatus that closely resembles the plesiomorphic condition of temnospondyls in many aspects (Bolt and Lombard 1985). The otic capsule is formed from anterior prootics and posterior opisthotics, frequently fused, and the fenestra vestibuli (FV) is floored by the parasphenoid. The FV is partially filled by the footplate of the stapes, which articulates ventrally with the parasphenoid. In frogs (and some temnospondyls) this articulation forms a hinge with a fibrous connective-tissue reinforcement (Lombard and Bolt 1979; Wever 1985).

Frogs and salamanders (= batrachians) differ from temnospondyls and caecilians in having a second element in the FV posterior to the stapedial footplate, the batrachian operculum, which is variously ossified or cartilaginous (Duellman and Trueb 1994). The stapedial columella (the main shaft as opposed to the flat footplate located within the FV) in frogs is a slender, elongate structure that projects dorsolaterally into a space defined by a notch in the posterior margin of the squamosal that suspends a tympanic membrane. Contact is made between the stapes and tympanum via a cartilaginous extracolumellar cap.

In contrast, salamanders, caecilians, and albanerpetontids have modified ears. The stapes has a broad footplate but short columella that is directed dorsolaterally toward the underside of the skull roof (a squamosal embayment being absent) or laterally toward the quadrate. Since lissamphibians derive from amphibamid temnospondyls (Bolt 1977; Ruta et al. 2003; Maddin and Anderson 2012), the shared features in the ears of temnospondyls and anurans are plesiomorphies, whereas the ears of salamanders, caecilians, and albanerpetontids are secondarily reduced, structurally simplified versions of the temnospondyl ear (see Sect. 11.5).

Lissamphibians also are interesting because their inner ears include a second receptor organ in addition to the standard basilar papilla of amniotes (Wever 1985): the amphibian papilla. This receptor reacts to low-frequency environmental sound stimuli (with the lagenar macula), whereas the basilar papilla covers high-frequency tympanic hearing (Mason, 2007). This is in contrast with the amniote condition, which has a greatly elongated basilar papilla in order to receive both high and low frequencies. It is thought that this difference is due to the mechanical tuning mechanism used by amniotes (Manley 2000), whereas amphibians also tune neurologically (Lewis 1988). In addition the receptor frequency ranges are split between different epithelia and the lissamphibian basilar papilla is relatively small. The amphibian papilla forms an important synapomorphy of lissamphibian clades, but it is difficult to establish the existence of such a papilla in fossil groups. Even in taxa that have been studied using μCT (Maddin et al. 2012a; Szostakiwskyj et al. 2015) impressions indicative of these sensory organs are absent because of a lack of ossification of the medial wall of the otic capsule. Frogs, salamanders, and caecilians also share a unique pressure relief mechanism by which the periotic canals pass posteriorly and the periotic sacs extend into the braincase, ventral to the brain (Maddin and Anderson 2012).

Furthermore, in amphibians these two sound reception pathways are correlated with different structures of the middle ear. The stapes articulates with the tympanum in frogs and is linked with stimulation of the basilar papilla for high-frequency aerial sound reception (and impedance matching) (Wever 1985; Mason 2007). The batrachian operculum in frogs and salamanders, on the other hand, has been linked with low-frequency sound reception of the lagenar system (Hetherington et al. 1986). This link is further bolstered by the muscular connection of this middle ear structure with the pectoral apparatus.

Understanding the diversification of amphibian ears, therefore, involves a full range of problems and offers various insights into principles and patterns of evolution. Modularity, heterochrony, and miniaturization play important roles in this domain.

11.2 Temnospondyls

Temnospondyls form a disparate group of some 300 species spanning an enormous time range in the fossil record (Schoch and Milner 2000, 2014). They first appear in the early Carboniferous (Viséan, 337 Ma) and reach at least up to the Early Cretaceous (Aptian, 115 Ma). Fossils of these animals were among the first Paleozoic tetrapods to be found. This was in the mid-nineteenth century. As such, they provided the model for all subsequent discoveries of fossil tetrapods for many decades. They were collectively known as "labyrinthodonts" because of the structure of their teeth, but more importantly for this chapter, their ear regions, superficially at least, were considered to resemble not only those of frogs but also those of lizards and many diapsid reptiles as described later in this section and in Sect. 11.3.1 (also see Sobral, Reisz, Neenan, Müller and Scheyer, Chap. 8; Evans, Chap. 9). Therefore, they provided the model for ear structure that was then considered primitive for all tetrapods and was known as the "labyrinthodont ear." As a result, this type of ear was thought to have been acquired early in the evolution of tetrapods in the Paleozoic (e.g., Parrington 1949). Subsequent discoveries have overturned this view, beginning in the late 1970s (Lombard and Bolt 1979; Clack 1983), which is discussed further in Chap. 4 by Clack and Anderson.

Considering their species numbers and long existence through geologic time, temnospondyls retained a relatively conservative morphology: flattened and massive skull, heavy pectoral girdle, rather small and feeble limbs, and an elongate trunk and tail. The most common morphotype was an aquatic predator, usually in the 1–3 m size range but in the Middle Triassic well up to 5 m in some clades. There were also highly terrestrial forms, such as the amphibamids from which lissamphibians probably arose (Anderson et al. 2008).

From these amphibious or fully water-dwelling animals, which often retained lateral lines and sometimes gills as adults, a large number of taxa are known with well-preserved ear regions. In temnospondyls, the braincase and otic capsules were usually poorly ossified, and their internal structure is little known in most taxa. The auditory capsules ossified slowly during ontogeny, and they were fully mineralized in only a few large genera (Fig. 11.2). The prootic and opisthotic usually fused into a single bone in such forms, and they sometimes merged with the braincase (sphenethmoid) to form a single composite ossification (e.g., *Edops*, *Eryops*, *Mastodonsaurus*, *Gerrothorax*) (see Schoch and Milner 2014). In some taxa, semicircular canals were traced in the roof of the auditory capsule (Sawin 1941; Schoch 2002).

The single ear ossicle of temnospondyls is the stapes, an element whose homology to that of other early tetrapods is undisputed. The characteristic features of temnospondyl stapes (Fig. 11.2) are as follows: (1) an elongated shaft that is not rounded but is anteroposteriorly compressed in cross section; (2) a large expanded footplate that has a flat and oval medial side that pointed into the FV; and (3) a distinctive ventral process that evidently articulated with a thorn-like projection on the dorsal side of the parasphenoid (preserved in several taxa). This process was formed of dermal bone, an outgrowth of the basal plate, slightly posteromedial to the basipterygoid articulation. The nature of this joint appears to have varied across temnospondyls: whereas in some it formed a ball-and-socket joint, it was usually more fixed by a sutural connection or even fused in adults of some taxa (*Mastodonsaurus*, *Benthosuchus*) (see Schoch and Milner 2000). The other end of the stapes was usually unexpanded, sometimes even slightly constricted, and capped by cartilage. In specimens preserved in three dimensions and full articulation, the distal end of the stapes pointed into the squamosal embayment (SE)—not in its center but rather close to the tabular bone. This is known from genera as distant as *Cacops*, *Eryops*, *Benthosuchus*, and *Mastodonsaurus* (Sawin 1941; Schoch 2000; Reisz et al. 2009), which is why it may be considered the primitive condition for the clade. Despite the consistent set of features shared by most temnospondyls, the stapes varied considerably across taxa, as is outlined in the following sections and exemplified by well-preserved taxa. In a few genera, which all lack the SE entirely, the stapes attained a highly derived morphology (*Dvinosaurus*, Shishkin 1973; *Batrachosuchus*, Watson 1956; *Gerrothorax*, Witzmann et al. 2012).

The SE of temnospondyls is more variable in morphology than their stapes, and in them and in other early tetrapods, it has a number of different names as well as SE, including “otic notch,” “tympanic notch,” or “spiracular notch.” By inference from anurans, the SE is traditionally interpreted as having housed a tympanic membrane. This was probably the case in dissorophoids, the small-bodied temnospondyls from which lissamphibians probably evolved (Milner 1993; Anderson 2008; Schoch 2014). However, in taxa as distant as the late Carboniferous edopoids and basal eryopiforms, the SE looks quite different from that of dissorophoids and appears to have evolved in different directions. It also needs to be stressed that the SE is bordered in large part by the tabular but also the supratemporal in more basal temnospondyls. These bones are always absent in anurans.

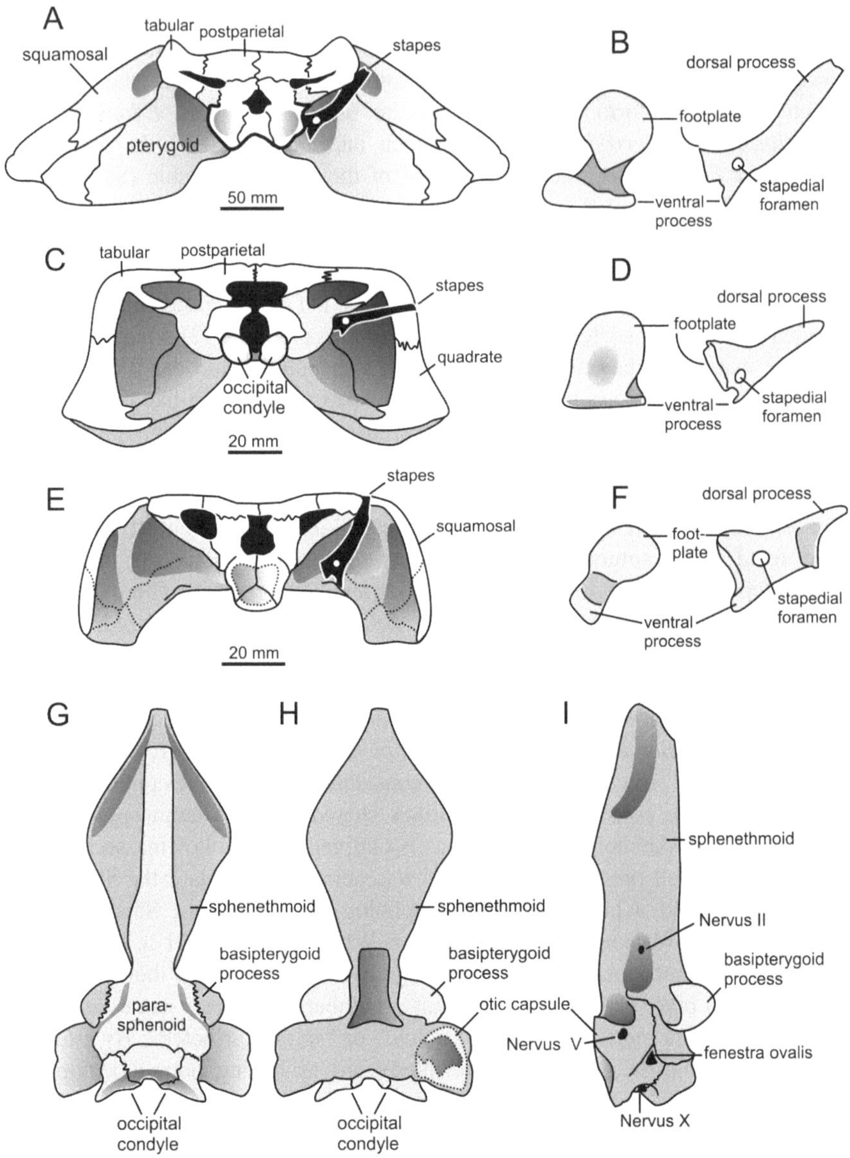

Fig. 11.2 Braincase and ear region of temnospondyls. **(A, C, E)** occipital view; **(B, D, F)** stapes in proximal (*left*) and posterior view (*right*); **(G–I)** braincase: **(G)** ventral, **(H)** dorsal, **(I)** lateral view. (Species: **A**, **B** *Eryops megacephalus*, modified from Sawin 1941; **C** *Acheloma dunni,* modified from Polley and Reisz 2011; **D** *Doleserpeton annectens*, modified from Bolt and Lombard 1985; **E** *Trimerorhachis* insigni, modified from Schoch and Milner 2014; **F** *Dvinosaurus primus*, modified from Shishkin 1973; **G–I** *Eryops megacephalus*, modified from Sawin 1941)

11.2.1 Edopoids

The basalmost temnospondyls were the edopoids, 1–3 m long crocodiliform predators in lakes and rivers with heavy skulls and long parabolic snouts (Fig. 11.3). They dwelled in lakes in late Carboniferous rainforest settings and some survived into the Early Permian. Their ear is best known from the giant genus *Edops* (Early Permian of Texas), which possessed the most massive ear ossicle in any amphibian, reaching 10 cm in length (Romer and Witter 1942). This stapes already had the three main features of temnospondyl ear ossicles: a long and anteroposteriorly flattened shaft, a massive oval footplate, and a prominent ventral process. Corresponding with this giant ossicle was the SE, which formed a huge, broad parabolic opening. If seen from the anuran perspective, it might have housed a gigantic tympanum. However, even in the largest frog (*Conraua goliath*), which has an enormous SE, the tympanum is actually tiny and confined to the dorsalmost end of the embayment (RRS, personal observation). Although there is no doubt that the fossilized components of the auditory apparatus were homologous to those of frogs, the hearing capacity of the edopoid ear therefore remains unknown.

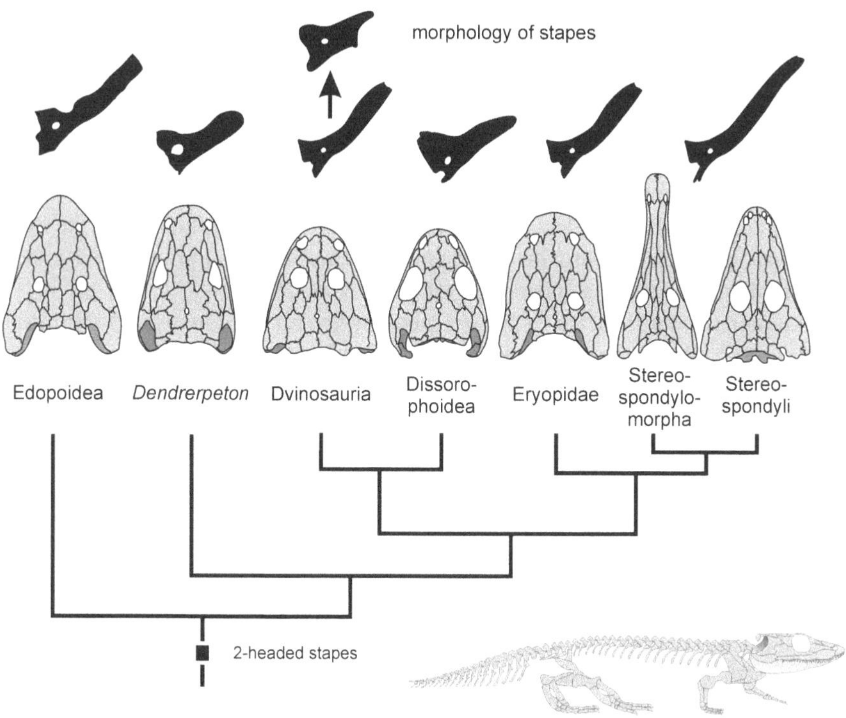

Fig. 11.3 Phylogeny of temnospondyls. The most common type of stapes is shown above the group.

11.2.2 Dendrerpetidae and Balanerpeton

Slightly more advanced temnospondyls were the small genera *Dendrerpeton* (Holmes et al. 1998), and *Balanerpeton* (Milner and Sequeira 1994). *Balanerpeton* is the oldest known temnospondyl from the Viséan aged deposits of East Kirkton, Scotland (Milner and Sequeira 1994). It is significant because it first shows an external ear morphology approaching that seen in anurans, and the base of a typically temnospondyl-style stapes is present. The squamosal embayment is expanded in both *Balanerpeton* and *Dendrerpeton* compared with the condition seen in more stemward temnospondyls and early tetrapods. It is the first squamosal embayment likely to be an otic notch rather than a plesiomorphic notch related to the spiracle of sarcopterygian "fishes." The embayment extends anterodorsally toward the orbit. The superficial surface is covered by dermal ornamentation, indicative of a tight integration between the dermis and periosteum, which is continuous with that of other bones on the same plane. The c-shaped squamosal also has dorsal and anteroventral surfaces that extend medially within the notch itself. These surfaces are smooth and presumably were placed deep in soft tissue. Between these two internal rami is an open slit. It is through this gap, usually dorsally, that the stapes extends in other temnospondyls. The tympanum was presumably attached to a raised rim at the interface between the superficial and internal surfaces and possibly was supported posteroventrally by the dorsal process of the quadrate.

Balanerpeton is known mostly from the superficial dermal skull. Because of the nature of preservation at East Kirkton, it is not a good candidate for study by CT. The younger fossils of *Dendrerpeton* from the Langsettian (Westphalian A) of Joggins, Nova Scotia and other localities (Milner 1996; Godfrey et al. 1987), however, are preserved in three-dimensions without the thermal diagenesis seen at East Kirkton and are excellent CT subjects. The external morphology of the SE in *Dendrerpeton* is consistent with that seen in *Balanerpeton* and previously has been suggested as supporting a tympanum (Holmes et al. 1998).

The stapes and middle ear of *Dendrerpeton* were described in detail by Robinson et al. (2005). As in many temnospondyls the prootic and opisthotic cannot be separated into distinct ossifications in *Dendrerpeton*. The otic capsules suture to the occipital complex but are dorsally separated from each other by a space, unlike edopids. Ventrally, the articulation with the parasphenoid presumably was completed by cartilage, as evidenced by another space found there. The horizontal semicircular canal is represented externally by a ridge that has a complicated shape dorsally to accommodate the curved columella and below which is the FV. The μCT scans also revealed a perfectly preserved stapes within the braincase. The footplate is subdivided between a larger dorsal portion and smaller, triangular, ventral surface by the stapedial canal. The columella is elongate, anteroposteriorly compressed, and slightly arched anteriorly. A concavity at the distal end of the stapes was presumably finished in an extracolumellar cartilage. On the posterior surface a scar is present, similar to that seen in other temnospondyls, including *Doleserpeton* (Robinson et al. 2005), which pertains to either a columellaris or slip from the levator scapulae (as in anurans), and a groove extends distally from that. The extent to which the footplate would have filled the FV is unknown because the margins of the FV clearly were finished in cartilage. Similarly, it is impossible to determine whether a batrachian operculum was present.

11.2.3 Dvinosaurians

An early clade of fully aquatic temnospondyls were the dvinosaurians, which include late Carboniferous to Early Triassic taxa (Fig. 11.3). These had elongate bodies, particularly small limbs, and short-faced skulls with extensive gill skeletons. A stereotyped basal taxon is *Trimerorhachis* from the Texas Red Beds (Early Permian), which is known from hundreds of specimens. The stapes is consistent in all features with the temnospondyl stapes as described in Sects. 11.2, 11.2.1, and 11.2.2 (Bolt and Lombard 1985). However, its proximal region was somewhat more massive than in later temnospondyls (eryopiforms, dissorophoids), and the SE was accordingly very wide, both features more similar to the situation in edopoids. In later dvinosaurians, such as *Dvinosaurus*, the stapes was an abbreviated and flattened element that retained a ventral process but had lost the slender shaft. In place of this shaft, there was a broad dorsal and a robust lateral process that pointed toward the quadrate (Shishkin 1973). This feature is especially interesting because it also occurs in many salamanders. Rather than indicating any closer relationship with caudates, this observation adds to the hypothesis that the salamander stapes may be a modified or simplified version of the temnospondyl ear ossicle (Schoch 2014). The SE of *Dvinosaurus* was completely absent, suggesting the lack of a tympanum that is consistent with its aquatic adaptations (Bystrow 1938).

11.2.4 Eryopiforms

The best-known temnospondyl is *Eryops*, a large (more than 2 m long), heavily built amphibious species from the Early Permian of the Western United States. Like *Edops*, this fish-eating predator had a flattened crocodile-like skull with a fully ossified braincase and ear region. The stapes was much smaller and more lightly built than that of *Edops*, albeit with the same set of features as described for temnospondyls in Sects. 11.2–11.2.2 (Fig. 11.3) (Sawin 1941). Correlating with this was the small, slit-like SE, in stark contrast to that of *Edops*. The whole middle ear apparatus of edopoids appears to have been miniaturized in eryopiforms.

The ontogeny of the stapes has been studied in detail in the genus *Sclerocephalus*, a distant relative of the large Triassic taxon Stereospondyli (Boy 1988; Schoch 2002). *Sclerocephalus* was similar to *Eryops* in general anatomy, but its mode of life appears to have been more aquatic. The stapes in this taxon underwent a remarkable transformation. Starting as a short spool-shaped bone, it became a four-headed element in larvae. In addition to the footplate, ventral process, and shaft, there was also a lateral process that probably contacted the quadrate region by some ligamentous attachment. This condition resembles the one described from adult *Dvinosaurus* (Shishkin 1973). Finally, the adult stapes of *Sclerocephalus* closely resembles that of *Eryops*, having an elongated compressed shaft and a reduced, rudimentary quadrate process. This forms an interesting observation because the larval stapes of *Sclerocephalus* resembles the cheek-anchored stapes of salamanders, but the adult stapes of the same taxon is closer to that of anurans.

11.2.5 Stereospondyls

The stereospondyls formed a speciose clade of temnospondyls that lived in shallow aquatic habitats throughout the Mesozoic supercontinent Pangea, sometimes in large numbers (Schoch and Milner 2000). Most of them reached a body length greater than 2 m and were piscivores or apex predators in lakes, streams, and even brackish lagoons. Their huge skulls assumed many different shapes, ranging from forms with extremely long and slender snouts (trematosaurs) that resembled modern gharials, through broader snouted crocodilian and giant salamander types (capitosaurs, metoposaurs, brachyopoids) to almost D-shaped heads with very short snouts (plagiosaurs).

The stereospondyl ear region was usually well ossified, often forming a single compound ossification of the ear capsule, and a slender and delicate stapes with a very distinct offset ventral process and a much-elongated shaft (Fig. 11.3). In all taxa with a well-established SE, the stapes was of this type. In some taxa lacking the SE, the ear ossicle was stout and apparently articulated with the underside of the tabular, but not the quadrate (*Batrachosuchus*: Watson 1956; *Gerrothorax*: Witzmann et al. 2012).

More detailed anatomical information has been gathered from the giant capitosaur *Mastodonsaurus* (Schoch 2002). In this heavily ossified genus, the Eustachian or auditory tube is preserved, running between the exoccipital and pterygoid to the floor of the auditory capsule. The articulation of the ventral process of the stapes also is well preserved, involving a moveable joint in juveniles and apparently a tighter suturing in very large (probably old) individuals. The middle ear cavity also is reconstructed with more detail in various stereospondyls because the pterygoid formed a vertical lamella that supported the floor of the middle ear tube, whose course may be traced with more confidence than in other temnospondyl taxa.

11.2.6 Dissorophoids

Dissorophoids form a diverse clade of small temnospondyls that were present, with several families, by the late Carboniferous. They apparently gave rise to lissamphibians some time in the Late Permian (Milner 1993; Anderson et al. 2008). In all but the most primitive taxa (Micromelerpetidae, see Boy 1995), the ear region was highly modified, involving a larger SE and an abbreviated and laterally directed stapes (Figs. 11.2E, 11.3, 11.4). Dissorophids have long served in discussions of the origins of modern amphibian ears (Bolt 1977; Bolt and Lombard 1985), so only highlights and recent findings will be presented here.

Three groups are particularly significant to discussions of ear morphology: the dissorophids, trematopids, and amphibamids. Trematopids and dissorophids are medium to large temnospondyls with heavily ornamented dermal skull bones (Figs. 11.4, 11.5). Amphibamids are small to miniaturized species with little to no ornamentation that progressively acquire more lisamphibian characteristics closer to the likely root of the modern groups. All bear enlarged ears with respect to more basal temnospondyls, with some amphibamids reducing the postorbital region to the point at which the otic notches very closely approach the relatively large orbits.

Trematopids have distinctive nasal regions but are otherwise the most generalized dissorophoids (Olson 1941). Their ear morphology is consistent with that described for *Dendrerpeton* (Sect. 11.2.2). One recently described species, *Rotaryus*, shows the distal end of a slender potential stapes (the internal anatomy being inaccessible) emerging through the dorsal portion of the slit in the otic notch (Berman et al. 2011). Maddin et al. (2010) described a partial growth series of

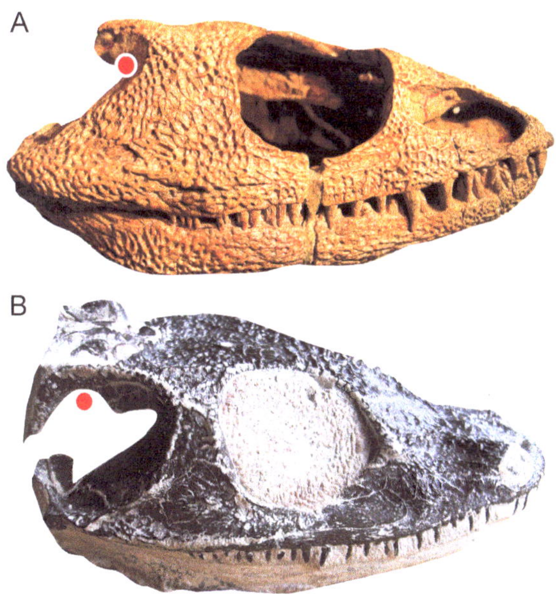

Fig. 11.4 Ear morphology and position of stapes (*red circle*) in two dissorophoid temnospondyls: **(A)** Trematopid *Acheloma cumminsi*; **(B)** Dissorophid *Cacops morrisi*. (Original from Schoch)

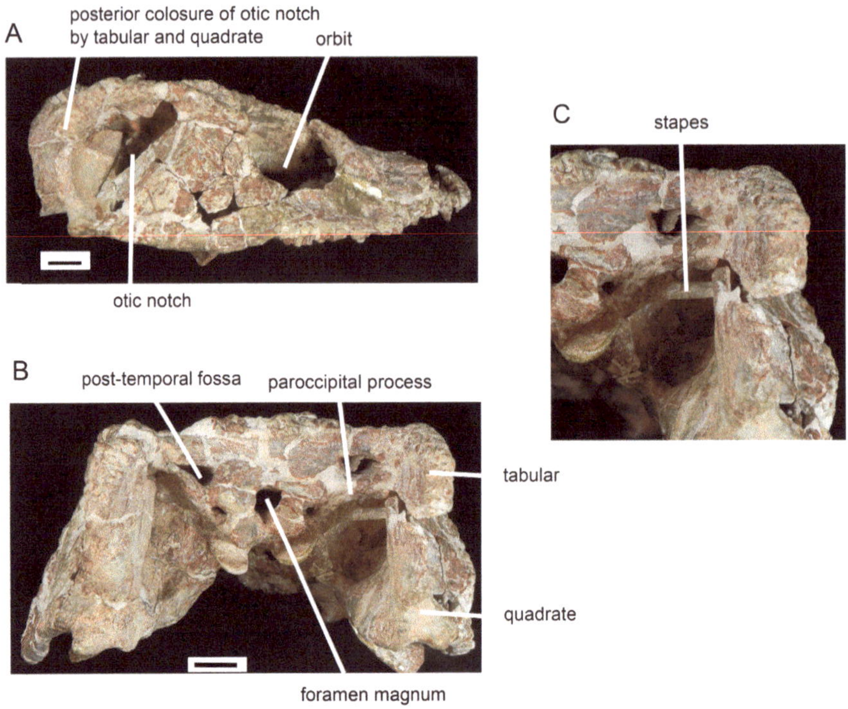

Fig. 11.5 Closed squamosal embayment (SE) and articulated stapes in the heavily ossified dissorophid *Cacops aspidephorus*: **(A)** lateral view; **(B)** posterior view; and **(C)** close-up of **B**. (Original from Schoch)

Acheloma from Richards Spur (Polley and Reisz 2011) that demonstrated separate prootics and opisthotics in juvenile specimens. This is surprising because they are usually fused when ossified in other temnospondyls. Although they are paedomorphic in many features, including in the otic capsule, the fusion at early growth stages suggests that the miniaturized amphibamids are peramorphic (Maddin et al. 2010).

Dissorophids are distinctive in their armored covering of osteoderms located dorsal to the vertebral column. As characterized by the recently described, exquisitely preserved *Cacops* specimens from Richards Spur (Reisz et al. 2009; Fröbisch and Reisz 2012), what had been thought to be a posteriorly closed otic notch has been shown to have a gap that persists between the elongate dorsal process of the quadrate and the ventral occipital process of the tabular. Other species, such as *Broiliellus* (Holmes et al. 2013), have a more typical "open" notch. The otic notch in lateral view is well rounded with a prominent margin and a more or less prominent subtemporal flange on the interior of the notch, the construction of which has systematic importance for dissorophids (Bolt 1974). In both *Cacops* and *Broiliellus*, the stapes is preserved in the FV. It extends toward the slit between the internal rami of the notch. The so-called "desert varnish" is an encrustation of the surface by a red, iron-rich layer and is typical for specimens from the Texas Permian. It is difficult to remove, and it obscures details of the fossil, which means that more specific details of the nature of the footplate's relationship with the FV are indeterminable.

As described previously in this section, amphibamids have proportionately enormous orbits and otic notches, no doubt partially the result of allometry since their skulls are so small (typically 10 cm or less). However, within amphibamids there is a derived group that has much larger ears than would be predicted from scaling alone. This clade includes taxa such as *Amphibamus*, *Doleserpeton*, and *Gerobatrachus* (Bolt 1979; Sigurdsen and Bolt 2010), which form a progressive grade into a stem lissamphibian (Anderson et al. 2008; Maddin et al. 2012a) or stem batrachian (Maddin and Anderson 2012) position, and the otic notch becomes proportionately more frog-like as a result. The stapes, where known, is a slender rod that extends into the massive otic notch from the otic capsules. *Doleserpeton*, in particular, has been intensively studied (Bolt 1969; Sigurdsen 2008). It shows features of the otic capsule, such as a ventrolateral ridge on the opisthotic (Sigurdsen 2008), that are possibly reflective of a frog-like tympanic annulus attachment not seen on other amphibamids, although it should be noted that other specimens are not as exquisitely preserved as is typical for Richards Spur fossils. *Doleserpeton* also shows evidence that the passage of the perilymphatic duct had a lissamphiban position and a close association with the jugular foramen as in frogs. Despite its extremely small size, the prootics and opisthotics are frequently fused together dorsal to the FV. Some of the endosseous labyrinth was described by Sigurdsen (2008), including possible recesses in the auditory portion that mark the position of the basilar papilla, but as yet no CT study of the endocast has been published. As seen from visual inspection of an isolated opisthotic, an ossified wall sets the saccule apart from the utricle.

11.3 Lissamphibia: The Modern Amphibians

Molecular studies have consistently shown that modern amphibians comprise a natural or monophyletic group to the exclusion of amniotes (e.g., San Mauro 2010). Despite this, the internal relationships of the extant groups, Anura, Caudata, and Gymnophiona are still contested, although the sister-group relationship between Anura and Caudata is accepted here (Pyron and Wiens 2011) (see also Sect. 11.3.1). Most of the morphological characters that unite Lissamphibia are characters of soft tissues that are not preserved in fossils and, in fact, are absent or reduced and thus difficult to determine in caecilians. A few characters uniting anurans and caudatans are related to the ear region, at least one of which may be determinable by modern techniques in some fossil taxa (see Sect. 11.2.5). One of these is the course of the periotic canal which, in lissamphibians, passes anterior to the vertical axis of the otic labyrinth. This character has recently been found in caecilians as well

(Sect. 11.3.4). In amniotes it passes posteriorly (Lombard and Bolt 1979). Anurans and caudatans have an additional ossification, the operculum, as part of the ossicular chain in the ear (Duellman and Trueb 1994).

The earliest fossil representatives of all three groups of modern Lissamphibia are dated to around the Early Jurassic, about 190 Ma. Preservational or collector bias may account for the lack of earlier representatives, but given the modern appearance of these fossil taxa, it is likely that the group as a whole originated earlier, perhaps much earlier. As it stands, there is a very large temporal gap between the potential stem group, the dissorophoids, and the origin of the modern forms.

11.3.1 Anura: Frogs

The anurans, or frogs and "toads" (the term "toad" has no real biological meaning), appear as the most derived lissamphibians in having fundamentally modified both larval and adult stages. Most fossil anurans can be placed in modern families and sometimes in extant genera, apart from one extinct family, the Palaeobatrachia (Estes and Reig 1973). Their ear regions, therefore, are consistent with those of modern forms, whose ear regions are described here. The earliest known frog fossils date from the Early Jurassic (Evans and Borsuk-Bialynicka 1998) and a jumping frog called *Prosalirus* is a key example (Shubin and Jenkins 1995). Unfortunately, its ear region was not described in detail.

Whereas tadpoles rely on the lateral line system for underwater orientation, the auditory system is highly important for adult frogs in large part because it is used to detect vocalizations during courtship and mating (Duellman and Trueb 1994). Obviously, this activity has played a major role in their evolution and diversification. All available data on extinct frogs and their stem group are consistent with the morphology of the extant anuran auditory system, indicating that this condition evolved very early (Sanchiz 1998).

Salamanders are the closest relatives of anurans, according to the favored hypothesis of relationships (Ruta and Coates 2007; Anderson 2008; Pyron and Wiens 2011), but share only the stapes and a second ear ossicle, the batrachian operculum, with them (Schoch 2014). This element is located posterior to the oval window and forms the origin of a muscle that attaches to the suprascapula (musculus opercularis). Effectively, this opercular system has been hypothesized to permit ground vibrations to be perceived by the ear independent of the tympanic system that, in frogs, transmits vibrations produced by airborne sounds (Kingsbury and Reed 1908; Duellman and Trueb 1994). However, experimental studies have also revealed that the opercularis muscle serves as an antagonist of another muscle (musculus collumellaris) and reduces transmission of sounds inward to the auditory papillae, guarding the inner ear against overstimulation (Wever 1985). This appears to be a highly useful device, especially in mating anurans. The two hypotheses need not contradict one another, and there are observations on the frequencies conducted by the opercular system (see Sect. 11.3.3).

Anurans share with amniotes a full set of middle ear components, both skeletal and soft anatomy, which are only partially present in other amphibians (Figs. 11.6, 11.7). These include the stapes, an air-filled middle ear cavity, a Eustachian tube that connects the middle ear with the buccal cavity, and the ear drum (tympanum), which is located between the cheek and skull table, usually in an emargination. The stapes is slender and resembles that of temnospondyls (Fig. 11.6) in the following features: (1) it has an anteroposteriorly compressed shaft; (2) the distal end points into the squamosal embayment very close to the margin of the squamosal and attaches to the tympanum, not in the center but near its medial margin; and (3) the proximal region is divided into a footplate and a ventral process. Unlike in temnospondyls, the ventral process does not contact (or even come close to) the parasphenoid, but frames the batrachian operculum. The orientation of the footplate also differs from all temnospondyls in pointing ventrally, not medially (Duellman and Trueb 1994).

Unique to anurans is the attachment of the tympanum to a cartilaginous ring, the annulus tympanicus. This forms from an outgrowth of the quadrate cartilage and is only fully established in adults (Bolt and Lombard 1985). The dorsal quadrate process, a structure unique to dissorophoid temnospondyls, has been suggested as an early homologue of the annulus.

Functionally, the tympanal and opercular sound-conducting systems perceive sounds with different ranges of frequencies (Lombard and Straughan 1974): the opercular system below 1 kHz and the tympanal system largely above that value. The opercular system thus enhances perception of environmental sounds (vibrations of the ground), whereas the tympanal system is associated with reproductive communication. This is also reflected by the developmental sequence in which the two systems are formed: the opercular before metamorphosis and the tympanal long after that phase (Smirnov and Vorobyeva 1988).

Smirnov and Vorobyeva (1988) showed that the two sound-conducting systems in anurans evolved in divergent ways in different anuran clades, which may have contributed to their enormous diversity. They found that in species in which one system was not established, the other system adopted the function of the lost one. For instance, the tympanic system was lost in *Ascaphus*, *Pelobates*, *Bombina*, and species of *Microhyla*. A general pattern is that the opercular system is very conservative across anuran clades, whereas the tympanic system appears more flexible. Smirnov and Vorobyeva (1988) argued that the accelerated development of the opercular system in the short phase before the climax of metamorphosis leaves no room for variation, restricting the options for morphological diversity in this system.

11.3.2 Triadobatrachus

The oldest definitive stem anuran is *Triadobatrachus* from the Triassic of Madagascar (but the precise geologic provenance is unknown) (Piveteau 1937;

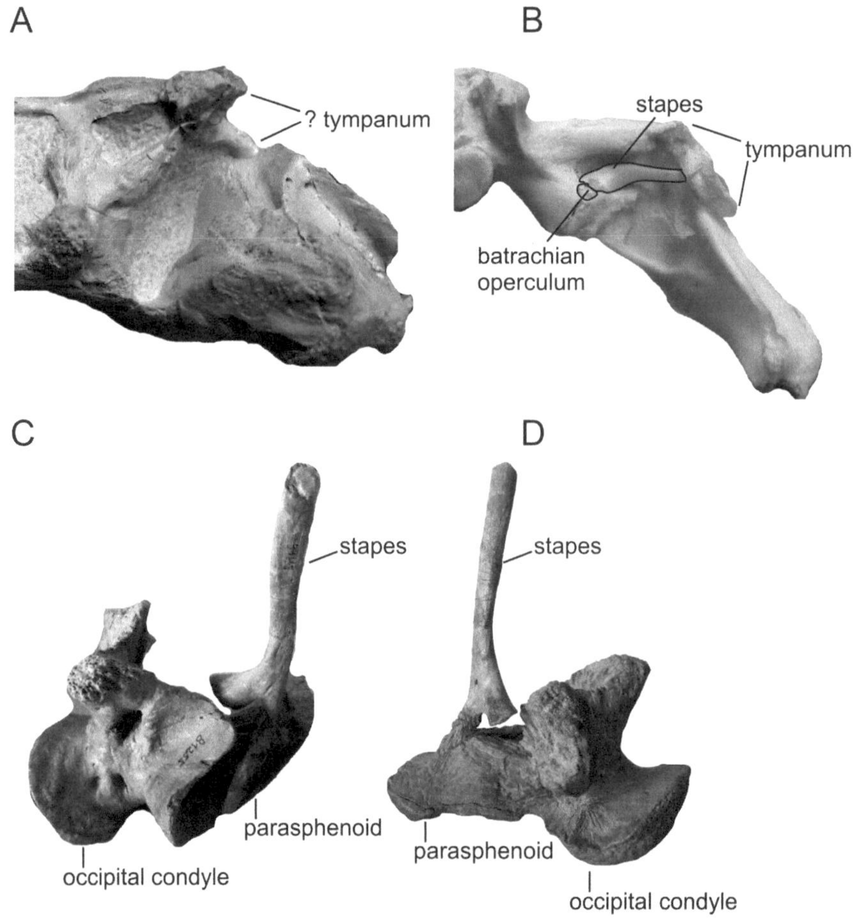

Fig. 11.6 Ear region and articulation of the stapes in two stereospondyl temnospondyls in comparison to a large anuran. **(A)** Stereospondyl *Angusaurus dentatus* (posterior view, stapes not preserved). **(B)** Goliath frog *Conraua goliath* (posterior view, stapes and batrachian operculum attached). **(C, D)** Parasphenoid, the floor of the braincase, and articulating stapes in the stereospondyl *Mastodonsaurus giganteus* (**C** lateral view; **D** medial view). (Original from Schoch)

Rage and Roček 1989). This fossil is a natural mold inside a nodule. Only superficial data are available for study, most commonly via latex peels. Nevertheless, details of the ear are observable. The specimen shows a continuation of reduction of the extent of the posterior dermal skull (seen progressively in dissorophoid temnospondyls and within amphibamids toward *Doleserpeton* and *Gerobatrachus* on the lissamphibian stem), so the otic capsules are only partially roofed dorsally. The otic capsules are well ossified, except the medial walls appear unfinished (JSA, personal observation in 2007) and are not attached solidly to the lateral walls of the braincase. The squamosals bear large embayments but are

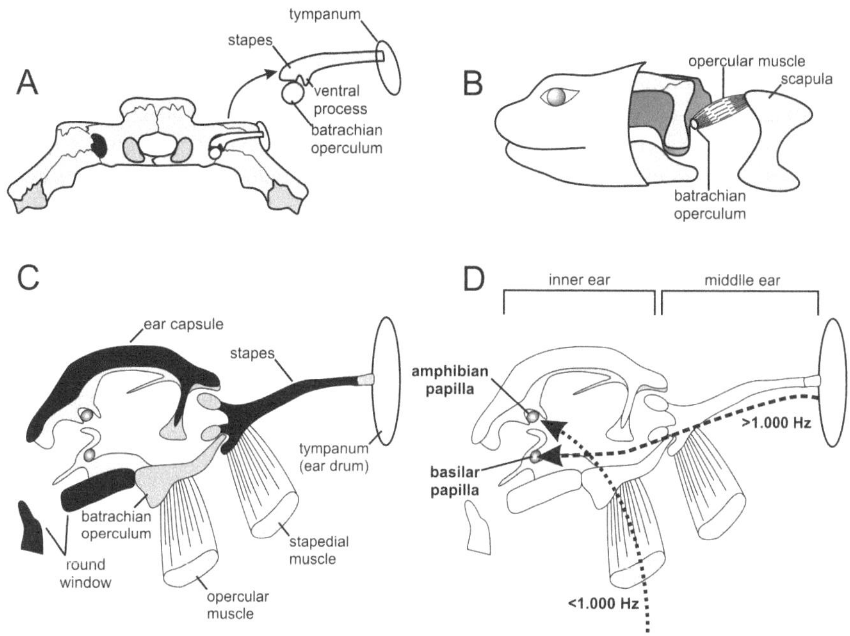

Fig. 11.7 Anatomy of the amphibian ear. (**A**) Stapes and batrachian operculum in the extant goliath frog *Conraua*. (**B**) Batrachian operculum and opercularis muscle in a salamander. (**C, D**) Inner ear and middle ear of a frog in cross section. *Pap am*, papilla amphibiorum; *pap bas*, papilla basilaris. (**A** unpublished data; **B** adapted from Duellman and Trueb 1994; C, **D** adapted from Wever 1985)

preserved only loosely articulated to the dorsal dermal skull roof. Two stapes are preserved slightly displaced from their articulations with the FV. These are oriented somewhat dorsolaterally and, on the right, the tip of the columella can be seen emerging near the dorsalmost portion of the otic notch, as seen in temnospondyls and frogs. No ossified batrachian operculum was present.

11.3.3 Caudata: Salamanders and Newts

One of the earliest fossil salamanders is *Karaurus* from Russia, dated to the Late Jurassic (Ivachnenko 1978). Salamanders that are basal members of the modern family Cryptobranchidae, and thus the earliest known members of the crown group, are also known from the Middle Jurassic of China (Gao and Shubin 2003). Isolated bones are known from earlier in the Middle Jurassic in the United Kingdom (Evans et al. 1988), suggesting that the group evolved prior to this. Again, as with early fossil anurans, their ear regions have not been described in detail but were probably consistent with those of their modern counterparts.

In salamanders, the stapes is more robust and rudimentary, attaching either to the quadrate or the squamosal distally. The shaft is short and round in cross section (Fig. 11.8). Proximally, there is only a large footplate but no ventral process. The ear drum, middle ear cavity, and Eustachian tube are consistently absent. The stapes is already present in tiny larvae of salamanders and, despite morphological changes in ontogeny, it does not change its connections to neighboring elements during metamorphosis. In transformed specimens, the stapes is larger because the distance between cheek and auditory capsule has increased in proportion (Larsen 1963). As in anurans, the stapedial footplate only fills the dorsal part of the FV. Ventrally, it is capped by the batrachian operculum, which may remain cartilaginous in some taxa.

Salamanders perceive only sounds from the ground, for which they use two different auditory systems. The opercular system, as in frogs, makes use of the connection between the forelimb, shoulder blade, opercular muscle, and batrachian

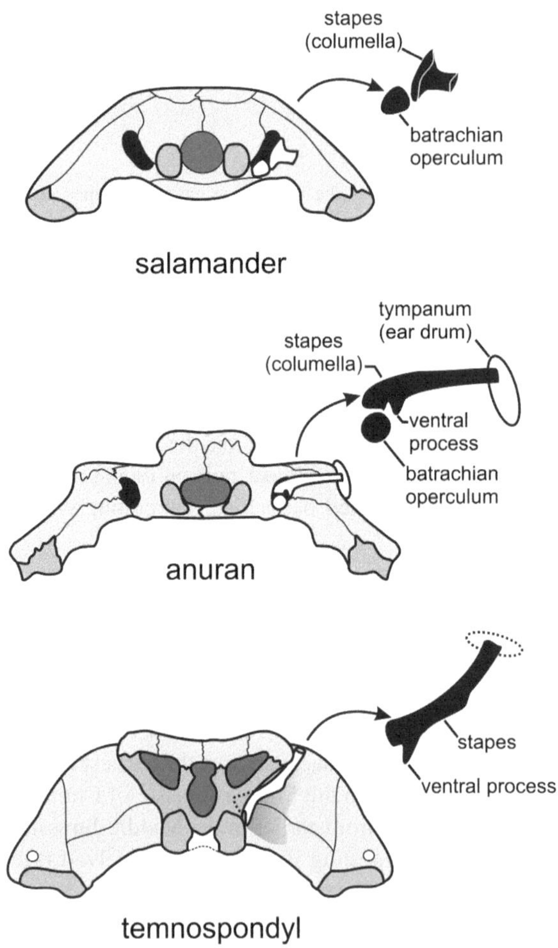

Fig. 11.8 Three hypothesized steps in the evolution of amphibian ear systems: (**1**) Temnospondyl ear with stapes pointing into squamosal embayment (housing a tympanum?). (**2**) Anuran ear with stapes attached to tympanum and batrachian operculum. (**3**) Salamander ear with stapes articulating with cheek and operculum retained. (Modified from Schoch 2014)

operculum. This unique system is so similar in the two groups that it forms a robust synapomorphy. As outlined before, the operculum and its muscular connections may not serve only a sound-conducting function but also may protect the ear from excessive stimulation. Although salamanders rarely vocalize and do not use sounds for reproductive communication, the same protective function has been proposed (Wever 1985). Conversely to anurans, the opercular system develops much later in salamanders than the stapedial system, as it is used only in terrestrial animals. This is consistent with the fact that the opercular muscle is absent in all the neotenic taxa (cryptobranchids, amphiumids, sirenids, and proteids) that never leave the water (Monath 1965). Thus, the same pattern is found in salamanders as in anurans in that the opercular and stapedial systems may mutually replace each other and take over the function of a lost system. For instance, in hynobiids and ambystomatids, the batrachian operculum tends to fuse with the otic capsule, which is a reversal to the primitive condition of tetrapods. Instead, salamandrids have lost the stapes, and in these animals the batrachian operculum fills the entire FV (Duellman and Trueb 1994).

The stapedial system of salamanders is not connected to a tympanum but to the mandible by means of articulation with the quadrate or squamosal, which in turn is connected to the lower jaw. Therefore, in salamanders the stapes plays an entirely different role from that which it has in frogs and probably also different from the role it had in many temnospondyls. As outlined in Sect. 11.4.3, the salamander condition is likely to be derived rather than plesiomorphic. This is also the case in the basalmost caudates, the hynobiids and cryptobranchids (giant salamanders), in which the stapes is abbreviated and attached by means of ligaments to the squamosal (Larsen 1963; Lebedkina 1979). In the stem group (urodeles), this condition was already established (Ivachnenko 1978).

11.3.4 Gymnophiona: Caecilians or Limbless Amphibians

Eocaecilia, whose ear region is described in Sect. 11.3.5, is the earliest fossil caecilian and is a transitional form in that it still retained limbs (Jenkins and Walsh 1993; Jenkins et al. 2007). It comes from the Early Jurassic of the United States. There are some differences also in the construction of the otic region between it and those of modern caecilians, which are described next.

In caecilians, the exoccipitals, prootics, and parasphenoid form a composite structure, the os basale (Marcus et al. 1935; Müller 2007). The stapes is a large element filling the FV completely. On the proximal end, it forms an articulation with the os basale by means of an anteroventral process. The nature of this process is unknown. A possible homology with the ventral process in anurans and temnospondyls has not been discussed by previous authors. Distally, it articulates with the quadrate, paralleling the situation in salamanders and various Palaeozoic tetrapod clades. Specific details of morphologic variation of the stapes in extant caecilians were described by Maddin et al. (2012b).

The existence of the batrachian operculum in caecilians remains unclear. There is no such element present at any stage of development in gymnophionans. The massive footplate of the caecilian stapes has been suggested to be a compound structure that includes both stapes and operculum. This is because some authors have reported components of the ear capsule as contributing to the stapedial footplate (Marcus et al. 1933, 1935). This was not corroborated in some later studies (Lawson 1963; Müller 2007); however, Ahlberg and Clack (Chap. 4) discuss the embryogenesis of the region as studied by O'Gorman (2005) and Thompson et al. (2012) in which the compound nature of the stapes is shown.

A remnant of an operculum may be present in the oldest stem caecilian *Eocaecilia* (see Sect. 11.3.5) (Jenkins et al. 2007). It seems evident that the opercular system is not established in any living caecilian because these limbless animals lack the shoulder girdle and forelimb. With the loss of the batrachian operculum and the tympanic stapes, the sensory epithelia associated with both auditory pathways show a phylogenetically correlated loss of innervation from basalmost families to highly derived families (Fritzsch and Wake 1988; Maddin and Anderson 2012).

The inner ear of caecilians has been reviewed recently (Maddin and Anderson 2012; Maddin and Sherratt 2014). Detailed three-dimensional reconstructions of histological slides, coupled with μCT based investigations, have revealed that the inner ear of caecilians is largely similar to that of other lissamphibians. Caecilians have a posteriomedially located periotic foramen for the periotic canal, and the periotic sacs extend into the braincase, ventral to the brain, providing pressure relief (Maddin and Anderson 2012). The semicircular canals are strongly rounded, similar to other animals that live in multiple dimensions (Maddin and Sherratt 2014). However, the short canals suggested lower sensitivity since caecilians have degenerated eyes and are less dependent on vestibulo-ocular reflexes. Specific variations in shapes may be related to different modes of head-first burrowing or different substrate preferences, but we currently lack the basic natural history data to see how well the vestibular morphology correlates with ecology. Maddin and Sherratt (2014) also found that caecilians have a uniquely expanded saccule, possibly as an adaptation to the perception of ground-based vibrations through the skull.

11.3.5 Eocaecilia

Like batrachians, caecilians have a long evolutionary history but an extraordinarily poor fossil record. The oldest stem group taxon and perhaps best preserved fossil species is *Eocaecilia micropodia*, a tiny form with feeble limbs and girdle elements and a skull that retains several dermal bones that are lost in all crown caecilians (Jenkins et al. 2007). *Eocaecilia* has an otic-occipital complex of co-ossified bones that is very similar to that termed the os basale in extant caecilians.

The ear of *Eocaecilia* is unusual for tetrapods, so determining the precise homology of elements is difficult. There is a large bone with a foramen at its base that fills the FV and articulates with the lower jaw. Jenkins et al. (2007) identified this bone as a composite stapes-quadrate, which are known to fuse in extant species. A second small, oval ossicle was found associated with the stapes-quadrate and near the FV, which Jenkins et al. (2007) interpreted as an operculum. However, working from μCT derived data, Maddin et al. (2012b) suggested an alternative homology, wherein the entire quadrate-stapes was in fact the quadrate, and the small ossicle a remnant of the stapes (also see Maddin and Anderson 2012). Other fossil caecilians are known from much more recent deposits, but none has been studied using μCT to date.

11.3.6 Albanerpetontids

Albanerpetontids have a long fossil record, going back 160 million years, and only narrowly missed surviving until present times. The youngest fossil comes from Pliocene-aged deposits (Venczel and Gardner 2005). Many fossils are known, both disarticulated (from screen washing sediments) and nearly complete articulated skeletons, but most fossils are crushed and so the ear region was relatively unknown. What was clear was that albanerpetontids lack a squamosal embayment, similar to salamanders and caecilians. A single three-dimensional preserved braincase was known from the Miocene of France (Estes and Hoffstetter 1976). It showed co-ossification of elements of the otic capsule and occiput and had a posteroventrally located FV. Unfortunately, this specimen is now missing from the collections of le Museum national d'Histoire naturelle in Paris (Maddin et al. 2013).

Recently, a new braincase was described from Pliocene deposits from Hungary (Maddin et al. 2013). It is in the same genus, *Albanerpeton,* as the braincase described by Estes and Hoffstetter (1976) and appears consistent in external morphology. It presents a co-ossification of the antotic wall, prootics, opisthotics, exoccipitals, and ossification of the synotic tectum, although it is unknown whether it represents a dorsal continuation from the exoccipitals, as described by Maddin et al. (2011), or a separate supraoccipital bone. It also cannot be said whether there were co-ossified basioccipital and basisphenoid ossifications, but they are usually considered absent in lissamphibians (Duellman and Trueb 1994). The FV is located in a urodele-like position and has a urodele-like shape, but no stapes is known in any albanerpetontids, and so the presence or absence of an operculum cannot be determined.

This specimen was studied using μCT, which permitted the creation of an endocast of the endosseous labyrinth. This shows features of all lissamphibian groups. Medial to the FV *Albanerpeton* shows an anuran-like lateral chamber; however, because the surface is smooth the presence of the various sensory epithelia located in the amphibian recess, basilar recess, and lagenar recess cannot be known. There is evidence for an anuran and urodele-like close association of the facial and vestibulo-acoustic cranial nerves (VII, VIII, respectively) in the pattern of the medial foramina and endosseous canals. The endolymphatic foramen is in a position consistent with salamanders and frogs, whereas the perilymphatic foramen is in a position in the range of all extant lissamphibians.

The arcing and length of the semicircular canals were interpreted to strongly indicate a terrestrial lifestyle, especially when using the criteria of Lombard (1977). *Albanerpeton* has highly arced canals, similar to those seen in anurans, which approach the form seen in arboreal mammals, presumably because of their use of the third dimension in leaping. However, *Albanerpeton* has shorter canals, especially the horizontal canal, which is more similar to caecilians. Maddin et al. (2013) speculate that this may be due to the use of three dimensions in burrows, but they note that *Albanerpeton* lacks the derived expansion of the saccular region seen in caecilians (Sect. 11.3.4), which they attributed to the retention of reasonably well-developed limbs that kept the head off the substrate.

11.4 The Evolutionary Sequence of Amphibian Ear Types: Making Sense of a Puzzle

The study of the evolution of amphibian ears offers a range of difficulties, including unresolved questions of homology, unknown pathways of evolutionary transformation, simplification, parallelism and convergence, and complete loss of components. The presence or absence of components in the auditory apparatus form a large puzzle when viewed from a distance, with several of the five major groups each sharing features that others lack. The common theme is the question of whether this complicated pattern is the result of convergent reduction and loss, or did shared components evolve independently? More specifically, the major problems in our understanding of amphibian ears include the following questions:

1. How did the diversity of components and functions evolve?
2. At which node did the anuran hearing system originate?
3. How did the apparent structural simplification and reduction of the ear in caecilians, albanerpetontids, and salamanders come about and what was its functional background?
4. What was the original function of the hearing system in temnospondyls, and what impact did the frequent change between aquatic and terrestrial lifestyles have on the auditory system?

In the following sections, attempts are made to address some of these points, referring to the current state of knowledge, especially from a phylogenetic perspective.

11.4.1 The Tetrapod Ear and the Primitive Condition for Lissamphibians

Lombard and Bolt (1979) interpreted the similarities between the anuran and amniote ears as convergences. They argued that the different course and position of nerves and blood vessels relative to the middle ear and tympanum indicate an independent origin of the middle ear cavity, tympanum, and Eustachian tube. This conclusion was drawn at a time when the large-scale phylogeny of tetrapods, the diversification of early crown tetrapods, and the origin of lissamphibians were still largely white areas on the map. After more than three decades of cladistic analyses and new fossil discoveries (e.g., *Eocaecilia*, *Gerobatrachus*) (see review in Anderson 2008), the situation has changed, and we can justifiably study the ear region of dissorophoids and amphibamids in order to find out more about the primitive condition of lissamphibians (for an alternative view see Marjanović and Laurin 2013). These taxa all share large tympana, lightly built and short stapes that were directed laterally, but as far as is known, they still lacked a batrachian operculum (Fig. 11.8). The presence of a middle ear cavity, consistent with the orientation of the stapedial shaft and the tympanum, is further indicated by the morphology of the squamosal and pterygoid, which encircled that cavity. Thus an air-filled middle ear tube is likely to have existed. Therefore, the available evidence indicates that the immediate ancestors of lissamphibians possessed an anuran-like auditory apparatus (Bolt and Lombard 1985; Clack 1992). The earliest known temnospondyl *Balanerpeton* is also the first taxon that had this type of apparatus (Milner and Sequeira 1994). It remains unclear how this apparatus was used in early temnospondyls, which frequencies it was able to perceive, and whether it also was used for the perception of mating calls. The sequence of fossil taxa also indicates that the batrachian operculum must have evolved at a later stage (Schoch 2014).

11.4.2 The Batrachian Ear

Despite so many differences in their ears, anurans and salamanders share a second ear ossicle, the batrachian operculum, which is not found (or identified) in other groups (Fig. 11.8). This tiny plate attaches to the opercularis muscle, connecting the inner ear with the shoulder girdle and forelimb. It forms an independent hearing apparatus from that of the stapes. The opercular apparatus transmits low-frequency vibrations from the ground to the inner ear, which are perceived by the papilla amphibiorum (Wever 1985). The fact that both the papilla amphibiorum and opercular apparatus are functionally coupled suggests that the ancestors of caecilians probably possessed an operculum, although the extant taxa lack it. At any rate, the lissamphibian invention was the addition of a new receptor, which permitted airborne vibrations and ground vibrations to be perceived by different mechanisms. Whether the invention of the opercular system evolved by some correlated progression (shared by lissamphibians, but lost in caecilians), or was a later event (evolved in batrachians), requires more careful analysis of stem caecilians such as *Eocaecilia*. The element described as a possible operculum in *Eocaecilia* is not located ideally for a definitive homology to be established. In extant frogs and salamanders the operculum occupies the posteroventral FV, but in *Eocaecilia* it is medial to the footplate of the quadrate-stapes. This led Maddin and Anderson (2012) to posit that the element was itself a remnant stapes, and the compound bone simply the quadrate that had massively grown, perhaps as a reinforcement against stresses from burrowing. This interpretation would mean *Eocaecilia* lacked an operculum, which suggests it is a batrachian synapomorphy. The discovery of more stem caecilians and well preserved stem lissamphibians would permit these alternatives to be tested.

11.4.3 How and Why Was the Tympanic System Lost?

Salamanders and caecilians share a short and massive stapes, which has a rudimentary distal shaft and attaches to the cheek rather than the skull table or within a squamosal embayment (cheek-anchored stapes). In salamanders and caecilians, the stapes is not enclosed in a middle ear tube but is surrounded by various layers of soft tissue (unknown in the extinct albanerpetontids). Interestingly, within both salamanders and caecilians there is a phylogenetically correlated loss of innervation to the basilar papilla, which is associated with the perception of high-frequency tympanic sound (Lombard 1977; Fritzsch and Wake 1988), and suggests that tympanic hearing was present at one point in their evolutionary history but subsequently lost (Maddin and Anderson 2012). Conceivably, the loss of the tympanic system was made possible only after the opercular system had been established. It could readily take over an essential part of auditory function, namely, the transmission of vibrations of the ground, which are vital for protection in all lissamphibians. Probably, the ancestors of salamanders and caecilians had lost the necessity to hear airborne sounds, although it is possible this segregation is due to chemosensory, rather than auditory, mate attractants (Duellman and Trueb 1994). In caecilians, the burrowing mode of life is consistent with that hypothesis. However, caecilians also lack an operculum, and they show a phylogenetically correlated loss of innervation to the lagenar macula (Fritzsch and Wake 1988; Maddin and Anderson 2012), which is most sensitive to the low-frequency, vibration-based sound that the opercularis system is thought to receive. At the same time, caecilians show an expansion in the saccular region of the labyrinth (Maddin 2011; Maddin and Sherratt 2014), which might be compensatory for the loss of the opercularis system. As for salamanders, burrowing occurs occasionally but does not appear to be the primitive condition (Schoch 2014). Hence, the adaptive background for the reduction of the tympanic system in salamander ancestors remains unknown. However, neoteny (= paedomorphosis, sexual reproduction in the larval phase) is a frequent and recurrent adaptive strategy in salamanders. One possibility is that the ancestral salamanders might have lost the tympanic system while undergoing an evolutionary phase of neoteny, comparable to that of extant sirenids or proteids, in which anatomy associated with metamorphosis was lost.

11.5 Summary

The amphibian ear falls into several types, each represented by a major group (clade). The three extant taxa, frogs (anurans), salamanders (caudates), and caecilians (gymnophionans), have very diverse ears. However, they share a second receptor, the amphibian papilla, which perceives sounds with a different range of frequencies than the tetrapod basilar papilla. Anurans have a delicate stapes located inside a middle ear cavity that is associated with a eustachian tube and a tympanum located between the skull table and cheek. A generally consistent apparatus is found in temnospondyls, which are considered the closest relatives of lissamphibians. Frog and temnospondyl stapes share a ventral process, clearly offset from the footplate, which articulates with the floor of the braincase. Temnospondyls probably possessed a tympanum, for which there is rich evidence especially in dissorophoids, the immediate stem group of extant amphibians. The shape and size of the stapes and tympanum varied enormously within temnospondyls, indicating some still unknown functional variation in that vast clade. At any rate, anurans and temnospondyls share an ear type referred to as the tympanal system.

The ears of salamanders and caecilians are very different from those of frogs in having a rudimentary stapes connected to the squamosal or quadrate, and they always lack a tympanum and middle ear cavity. This ear is referred to as a cheek-anchored type, which also has been reported from some juvenile temnospondyls and thus may represent the result of paedomorphosis. The cheek-anchored ear of salamanders and caecilians is probably apomorphic because the tympanal system forms the most plausible primitive condition of Lissamphibia, as indicated by temnospondyls. The ear of albanerpetontids, an extinct clade of salamander-like lissamphibians, remains unknown.

Batrachians (salamanders and frogs) share a second ear ossicle, the batrachian operculum, which connects via the opercularis muscle to the scapula and permits the forelimb and pectoral girdle to transmit vibrations from the ground. It is possible that the loss of the tympanic system was an evolutionary option that had become available only after the batrachian operculum had evolved, and hearing could be taken over by the opercular system when the tympanal system had to be given up. This is likely at least for the early evolution of caecilians when they adapted to a burrowing mode of life.

Compliance with Ethics Requirements Rainer R. Schoch and Jason S. Anderson declared that they had no conflicts of interest.

References

Anderson, J. S. (2008). Focal review. The origin(s) of modern amphibians. *Evolutionary Biology*, 35, 231–247.

Anderson, J. S., Reisz, R. R., Scott, D., Fröbisch, N. B., & Sumida, S. S. (2008). A stem batrachian from the Early Permian of Texas and the origin of frogs and salamanders. *Nature*, 453, 515–518.

Berman, D. S., Henrici, A. C., Martens, T., Sumida, S. S., & Anderson, J. S. (2011). *Rotaryus gothae*, a new trematopid (Temnospondyli: Dissorophoidea) from the Lower Permian of Central Germany. *Annals of the Carnegie Museum*, 80(1), 49–65.

Bolt, J. R. (1969). Lissamphibian origins: Possible protolissamphibian from the Lower Permian of Oklahoma. *Science*, 166, 888–891.

Bolt, J. R. (1974). Evolution and functional interpretation of some suture patterns in Paleozoic labyrinthodont amphibians and other lower tetrapods. *Journal of Paleontology*, 48(3), 434–458.

Bolt, J. R. (1977). Dissorophoid relationships and ontogeny, and the origin of the Lissamphibia. *Journal of Paleontology*, 51(2), 235–249.

Bolt, J. R. (1979). *Amphibamus grandiceps* as a juvenile dissorophid: Evidence and implications. In M. H. Nitecki (Ed.), *Mazon creek fossils* (pp. 529–563). New York: Academic Press.

Bolt, J. R., & Lombard, R. E. (1985). Evolution of the amphibian tympanic ear and the origin of frogs. *Biological Journal of the Linnean Society*, 24, 83–99 .

Boy, J.A. (1988). Über einige Vertreter der Eryopoidea (Amphibia: Temnospondyli) aus dem europäischen Rotliegend (? höchstes Karbon–Perm). 1 . *Sclerocephalus. Paläontologische Zeitschrift*, 62, 107–132.

Boy, J. A. (1995). Über die Micromelerpetontidae (Amphibia: Temnospondyli). 1. Morphologie und Paläoökologie des *Micromelerpeton credneri* (Unter-Perm; SW-Deutschland). *Paläontologische Zeitschrift*, 69, 429–457.

Bystrow, A. P. (1938). *Dvinosaurus* als neotenische Form der Stegocephalen. *Acta Zoologica*, 19, 209–295.

Clack, J. A. (1983). The stapes of the Coal Measures embolomere *Pholiderpeton scutigerum* Huxley and otic evolution in early tetrapods. *Zoological Journal of the Linnean Society*, 79, 121–148.

Clack, J. A. (1992). The stapes of *Acanthostega gunnari* and the role of the stapes in early tetrapods. In D. B. Webster, R. R. Fay, and A. N. Popper (Eds.), *The evolutionary biology of hearing* (pp. 405–420). New York: Springer-Verlag.

Duellman, W. W., & Trueb, L. (1994). *Biology of amphibians, 2nd ed.* Baltimore: Johns Hopkins University Press.

Estes, R., & O. A. Reig. (1973). The early fossil record of frogs: a review of the evidence. In J. L. Vial (Ed.), *Evolutionary biology of the anurans: Contemporary research on major problems* (pp. 11–63). Columbia (MO): University of Missouri Press.

Estes, R., & Hoffstetter, R. (1976). Les urodèles du Miocène de La Grive-Saint-Alban (Isère, France). *Bulletin du Muséum National d'Histoire Naturelle, Série 3, Sciences de la Terre*, 57, 297–343.

Evans, S. E., Milner A. R., & Mussett, F. (1988). The earliest known salamanders (Amphibia, Caudata): A record from the Middle Jurassic of England. *Geobios*, 21, 539–552.

Evans, S. E., & Borsuk-Bialynicka, M. (1998). A stem-group frog from the Early Triassic of Poland. *Acta Palaeontologica Polonica*, 43, 573–580.

Fritzsch, B., & Wake, M. H. (1988). The inner ear of gymnophione amphibians and its nerve supply: A comparative study of regressive events in a complex sensory system (Amphibia, Gymnophiona). *Zoomorphology*, 108(4), 201–217.

Fröbisch, N. B., & Reisz, R. R. (2012). A new species of dissorophid (*Cacops woehri*) from the Lower Permian Dolese Quarry, near Richards Spur, Oklahoma. *Journal of Vertebrate Paleontology*, 32(1), 35–44.

Gao, K.–Q., & Shubin, N. H. (2003). Earliest known crown-group salamanders. *Nature*, 422, 424–428.
Godfrey, S. J., Fiorillo, A. R., & Carroll, R. L. (1987). A newly discovered skull of the temnospondyl amphibian *Dendrerpeton acadianum* Owen. *Canadian Journal of Earth Science*, 24, 796–805.
Hetherington, T. E., Jaslow, A. P., & Lombard, R. E. (1986). Comparative anatomy of the amphibian opercularis system: I. General design features and functional interpretation. *Journal of Morphology*, 190, 43–61.
Holmes, R., Berman, D. S., & Anderson, J. S. (2013). A new dissorophid (Temnospondyli, Dissorophoidea) from the Early Permian of New Mexico (United States). *Comptes Rendus Palevol*, 12, 419–435.
Holmes, R. B., Carroll, R. L., & Reisz, R. R. (1998). The first articulated skeleton of *Dendrerpeton acadianum* (Temnospondyli, Dendrerpetontidae) from the Lower Pennsylvanian locality of Joggins, Nova Scotia, and a review of its relationships. *Journal of Vertebrate Paleontology*, 18, 64–79.
Ivachnenko, M. F. (1978). Urodelians from the Triassic and Jurassic of Soviet Central Asia. *Paleontological Journal*, 12, 362–368. [in Russian]
Jenkins, F. A., & Walsh, D. M. (1993). An Early Jurassic caecilian with legs. *Nature* 365, 246–250.
Jenkins, F. A., Walsh, D. M., & Carroll, R. L. (2007). Anatomy of *Eocaecilia micropodia*, a limbed caecilian of the Early Jurassic. *Bulletin of the Museum of Comparative Zoology*, 158, 285–366.
Kingsbury, B. F., & Reed, H. D. (1908). The columella auris in Amphibia. *Journal of Morphology*, 22, 459.
Larsen, J. H., Jr. (1963). *The cranial osteology of neotenic and transformed salamanders and its bearing on interfamilial relationships*. Dissertation, University of Washington, Seattle, WA.
Lawson, R. (1963). The anatomy of *Hypogeophis rostratus* Cuvier. Part I. Skin and skeleton. *Proceedings of the University of Durham Philosophical Society*, 13, 254–273.
Lebedkina, N. S. (1979). *Evolution of the amphibian skull*. Moscow, Russia: Nauka.
Lewis, E. R. (1988). Tuning in the bullfrog ear. *Biophysical Journal*, 53, 441–447.
Lombard, R. E. (1977). Comparative morphology of the inner ear in salamanders (Caudata: Amphibia). *Contributions to Vertebrate Evolution*, 2, 1–143.
Lombard, R. E., & Straughan, I. R. (1974). Functional aspects of anuran middle ear structures. *Journal of Experimental Biology*, 61(1), 71–93.
Lombard, R. E., & Bolt, J. R. (1979). Evolution of the tetrapod ear: An analysis and reinterpretation. *Biological Journal of the Linnean Society*, 11, 19–76.
Maddin, H. C. (2011). Deciphering morphological variation in the braincase of caecilian amphibians (gymnophiona). *Journal of Morphology*, 272(7), 850–871.
Maddin, H. C., & Anderson, J. S. (2012). The evolution of the amphibian ear with implications for lissamphibian phylogeny: Insight gained from the caecilian inner ear. *Fieldiana*, 2012(5), 59–76.
Maddin, H. C., & Sherratt, E. (2014). Influence of fossoriality on inner ear morphology: Insights from caecilian amphibians. *Journal of Anatomy*, 225(1), 83–93.
Maddin, H. C., Reisz, R. R., & Anderson, J. S. (2010). Evolutionary development of the neurocranium in Dissorophoidea (Tetrapoda: Temnospondyli), an integrative approach. *Evolution and Development*, 12(4), 393–403.
Maddin, H. C., Jenkins, F. A., Jr., & Anderson, J. S. (2012a). The braincase of *Eocaecilia micropodia* (Lissamphibia, Gymnophiona) and the origin of caecilians. *PLoS One*, 7, e50743.
Maddin, H. C., Russell, A. P., & Anderson, J. S. (2012b). Phylogenetic implications of the morphology of the braincase of caecilian amphibians (Gymnophiona). *Zoological Journal of the Linnean Society*, 166, 160–201.
Maddin, H. C., Venczel, M., Gardner, J. D., & Rage, J. C. (2013). Micro-computed tomography study of a three-dimensionally preserved neurocranium of *Albanerpeton* (Lissamphibia,

Albanerpetontidae) from the Pliocene of Hungary. *Journal of Vertebrate Paleontology*, 33(3), 568–587.
Marcus, H., Stimmelmayr, E., & Porsch, G. (1935). Die Ossifikation des Hypogeophisschädels. Beitrag zur Kenntnis der Gymnophionen XXV. *Gegenbaurs Morphologisches Jahrbuch*, 76, 375–420.
Manley, G. A. (2000). Cochlear mechanisms from a phylogenetic viewpoint. *Proceedings of the National Academy of Sciences of the USA*, 97, 11736–11743.
Marcus, H., Winsauer, O., & Hueber, A. (1933). Der kinetische Schädel von *Hypogeophis* und die Gehörknöchelchen. Beitrag zur Kenntnis der Gymnophionen XVIII. *Gegenbaurs Morphologisches Jahrbuch*, 100, 149–193.
Marjanović, D., & Laurin, M. (2013). The origin(s) of extant amphibians: A review with emphasis on the "lepospondyl hypothesis." *Geodiversitas*, 35(1), 207–272.
Mason, M. J. 2007. Pathways for sound transmission to the inner ear in amphibians. In P. M. Narins, A. S. Feng, R. R. Fay, & A. N. Popper (Eds.), *Hearing and sound communication in amphibians* (pp. 147–183). New York: Springer Science+Business Media
Monath, T. (1965). The opercular apparatus of salamanders. *Journal of Morphology*, 116, 149–170.
Milner, A. R. (1988). The relationships of temnospondyl amphibians. In M. J. Benton (Ed.), *The phylogeny and classification of the tetrapods* (pp. 59–102). Oxford: Clarendon.
Milner, A. R. (1993). The Paleozoic relatives of lissamphibians. In: D. Cannatella & D. Hillis (Eds.), *Amphibian relationships. Phylogenetic analysis of morphology and molecules. Herpetological Monograph*, 7, 8–27.
Milner, A. R. (1996). A revision of the temnospondyl amphibians from the Upper Carboniferous of Joggins, Nova Scotia. *Special Papers in Palaeontology*, 52, 81–103.
Milner, A. R., & Sequeira, S. E. K. (1994). The temnospondyl amphibians from the Visean of East Kirkton, West Lothian, Scotland. *Transactions of the Royal Society of Edinburgh: Earth Sciences*, 84, 331–361.
Müller, H. (2007). *Developmental morphological diversity in caecilian amphibians*. Leiden: Leiden University Press.
O'Gorman, S. (2005). Second branchial arch lineages of the middle ear of wild-type and *Hoxa2* mutant mice. *Developmental Dynamics*, 234, 124–131.
Olson, E. C. (1941). The Family Trematopsidae. *Journal of Geology*, 49, 149–176.
Parrington, F. R. (1949) Remarks on a theory of the evolution of the tetrapod middle ear. *Journal of Laryngology and Otology*, 63, 580–595.
Piveteau, J. (1937). Paléontologie de Madagascar. XXIII. Un Amphibien du Trias inferieur. *Annales de Paléontologie*, 26, 135–177.
Polley, B. and Reisz, R. R. (2011). A new Lower Permian trematopid (Temnospondyli: Dissorophoidea) from Richards Spur, Oklahoma. *Zoological Journal of the Linnean Society*, 161, 789–815.
Pyron, R. A., & Wiens, J. J. (2011). A large-scale phylogeny of Amphibia including over 2800 species and a revised classification of extant frogs, salamanders and caecilians. *Molecular Phylogenetics and Evolution*, 61, 543-583.
Rage, J.-C., & Roček, Z. (1989). Redescription of *Triadobatrachus massinoti* (Piveteau, 1936), an anuran amphibian from the Early Triassic. *Palaeontographica*, A, 206, 1–16.
Reisz, R., Schoch, R., & Anderson, J. S. (2009). The armoured dissorophid *Cacops* from the Early Permian of Oklahoma and the exploitation of the terrestrial realm by amphibians. *Naturwissenschaften*, 96(7), 789–796.
Robinson, J., Ahlberg, P. E., & Koentges, G. (2005). The braincase and middle ear region of *Dendrerpeton acadianum* (Tetrapoda: Temnospondyli). *Zoological Journal of the Linnean Society*, 143, 577–597.
Romer, A. S., & Witter, R. V. (1942). *Edops*, a primitive rhachitomous amphibian from the Texas red beds. *Journal of Geology*, 50, 925–960.
Ruta, M., & Coates, M. I. (2007). Dates, nodes and character conflict: Addressing the lissamphibian origin problem. *Journal of Systematic Palaeontology*, 5, 69–122.

Ruta, M., Coates, M. I., & Quicke, D. L. J. (2003). Early tetrapod relationships revisited. *Biological Reviews*, 78, 251–345.
San Mauro, D. (2010). A multilocus timescale for the origin of extant amphibians. *Molecular Phylogenetics and Evolution*, 56, 554–561.
Sanchiz, B. (1998). Salientia. In P. Wellnhofer (Ed.), *Encyclopedia of Paleoherpetology, vol. 4* (pp. 1–275). Munich: Pfeil.
Sawin, H. J. (1941). The cranial anatomy of *Eryops megacephalus*. *Bulletin of the Museum of Comparative Zoology*, 89, 407–463.
Schoch, R. R. (2000). The stapes of *Mastodonsaurus giganteus* (Jaeger, 1828) – structure, articulation, and functional implications. *Neues Jahrbuch Geologie Paläontologie Abhandlungen*, 214, 201–231.
Schoch, R. R. (2002). The neurocranium of the stereospondyl *Mastodonsaurus giganteus*. *Palaeontology*, 45, 627–645.
Schoch, R. R. (2014). *Amphibian evolution. The life of early land vertebrates*. Hoboken (NJ): Wiley.
Schoch, R. R., & Milner, A. R. (2000). Stereospondyli. In P. Wellnhofer (Ed.), *Handbuch der Paläoherpetologie, vol. 3B* (pp.1–203 + 16 plates). Munich: Pfeil.
Schoch, R. R., & Milner, A. R. (2014). Temnospondyli. In H. D. Sues (Ed.), *Handbuch der Paläoherpetologie, vol. 3A2* (pp. 1–150). Munich: Pfeil.
Shishkin, M. A. (1973). *The morphology of the early amphibia and some problems of lower tetrapod evolution*. Moscow: Nauka.
Shubin, N. H., & Jenkins F. A. (1995). An Early Jurassic jumping frog. *Nature*, 377, 49–52.
Sigurdsen, T. (2008). The otic region of *Doleserpeton* (Temnospondyli) and its implications for the evolutionary origin of frogs. *Zoological Journal of the Linnean Society*, 154(4), 738–751.
Sigurdsen, T., & Bolt, J. R. (2010). The Lower Permian amphibamid *Doleserpeton* (Temnospondyli: Dissorophoidea), the interrelationships of amphibamids, and the origin of modern amphibians. *Journal of Vertebrate Paleontology*, 30(5), 1360–1377.
Smirnov, S. V., & Vorobyeva, E .I. (1988). Morphological grounds for diversification and evolutionary change in the amphibian sound-conducting apparatus. *Anatomischer Anzeiger Jena*, 166, 317–322.
Szostakiwskyj, M., Pardo, J. D., & Anderson, J. S. (2015). Micro-CT study of *Rhynchonkos stovalli* (Lepospondyli, Recumbirostra), with description of two new genera. *PLoS ONE*, 10 (6): e0127307.
Thompson, J., Ohazama, A., Sharpe, P. T., & Tucker, A. S. (2012). The origin of the stapes and relationship to the otic capsule and oval window. *Developmental Dynamics*, 241, 1396-1404. doi:10.1002/dvdy.23831
Venczel, M., &. Gardner J. D. (2005). The geologically youngest albanerpetontid amphibian from the lower Pliocene of Hungary. *Palaeontology*, 48, 1273–1300.
Watson, D. M. S. (1956). The brachyopid labyrinthodonts. *Bulletin of the British Museum (Natural History): Geology*, 2, 315–392.
Wever, E. G. (1985) *The amphibian ear*. Princeton: Princeton University Press.
Witzmann, F., Schoch, R. R., Hilger, R., & Kardjilov, N. (2012). The braincase, palatoquadrate and ear region of the plagiosaurid *Gerrothorax pulcherrimus* from the Middle Triassic of Germany. *Palaeontology*, 55, 31–50.

MIX
Papier aus verantwortungsvollen Quellen
Paper from responsible sources
FSC® C105338

If you have any concerns about our products,
you can contact us on
ProductSafety@springernature.com

In case Publisher is established outside the EU,
the EU authorized representative is:
Springer Nature Customer Service Center GmbH
Europaplatz 3, 69115 Heidelberg, Germany

Printed by Libri Plureos GmbH
in Hamburg, Germany